6

Microbiology Monographs

Series Editor: Alexander Steinbüchel

Microbiology Monographs

Volumes published in the series

Inclusions in Prokaryotes
Jessup M. Shively (Editor)
Volume 1 (2006)

Complex Intracellular Structures in Prokaryotes
Jessup M. Shively (Editor)
Volume 2 (2006)

Magnetoreception and Magnetosomes in Bacteria
Dirk Schüler (Editor)
Volume 3 (2007)

Predatory Prokaryotes – Biology, Ecology and Evolution
Edouard Jurkevitch (Editor)
Volume 4 (2007)

Amino Acid Biosynthesis – Pathways, Regulation and Metabolic Engineering
Volker F. Wendisch (Editor)
Volume 5 (2007)

Molecular Microbiology of Heavy Metals

Volume Editors: Dietrich H. Nies, Simon Silver

With 60 Figures, 10 in Color

Springer

Volume Editor:

Prof. Dr. Dietrich H. Nies
Institut für Biologie/Mikrobiologie
Universität Halle
Kurt-Mothes-Str. 3
06099 Halle/Saale
Germany
e-mail: dietrich.nies@mikrobiologie.uni-halle.de

Professor Dr. Simon Silver
Department of Microbiology and Immunology
University of Illinois at Chicago
M/C 790, Room E-704
835 S. Wolcott Avenue
Chicago, IL 60612, USA
e-mail: simon@uic.edu

Series Editor:

Professor Dr. Alexander Steinbüchel
Institut für Molekulare Mikrobiologie und Biotechnologie
Westfälische Wilhelms-Universität
Corrensstraße 3
48149 Münster
Germany
e-mail: steinbu@uni-muenster.de

Library of Congress Control Number: 2006940357

ISSN 1862-5576
ISBN 978-3-540-69770-1 Springer Berlin Heidelberg New York
DOI 10.1007/978-3-540-69771-8

Springer is a part of Springer Science+Business Media

springer.com

Editor: Dr. Christina Eckey, Heidelberg, Germany
Desk Editor: Dr. Jutta Lindenborn, Heidelberg, Germany
Cover Design: WMXDesign GmbH, Heidelberg, Germany
Typesetting and Production: LE-TEX Jelonek, Schmidt & Vöckler GbR, Leipzig, Germany

Printed on acid-free paper SPIN 11741800 149/3100 YL – 5 4 3 2 1 0

Preface

The understanding of bacterial inorganic metabolism has advanced so rapidly that one can no longer attempt to be comprehensive. Nevertheless, some authors have tried, with emphasis on the chemistry but relatively little attention given to molecular genetics or microbiology (Faústo da Silva and Williams, 1991; Lippard, and J.M. Berg, 1994). Even with an emphasis on gene regulation only, a single volume that brings together microbial studies with plant and animal studies seems no longer to be practical (Silver and Walden, 1998). The closest precursor of this book is a dedicated special journal issue (Brown et al., 2003). Here we have done what can be done within reasonable space confines for 17 logically related articles divided into two parts.

The first half of this monograph mainly contains summaries of global aspects of metal cation homeostasis. Discussed are, on a sound physico-chemical basis, the properties of the regulatory proteins that activate or turn off genes involved in metal transport or allocation. The contrast between the two points of view by Tottey et al. and Nies allows readers to make their own decision, or do their own *synthesis* (derived from *thesis* and *antithesis*). Some chapters are more technical (Tottey et al., Soosanga et al.) and others take an end view of the eventual result of mineral ion homeostasis (Hobman, Harms, Wesenberg et al.). More work is needed in the future, for example, on the post-transcriptional translational control and effects of phosphate kinase control of regulatory proteins and gene silencing.

The second half of the volume consists mostly of an in-depth presentation of the current understanding of the molecular biology of key specific elements from a microbial view, element by element. This includes chapters on essential nutrient elements, such as Fe, Mn, Cu, Ni, Zn and Mo; and Ag, As and Cr as toxic elements for which microbes have specific resistance and detoxification genes.

We hope readers will find this book useful and of interest. Starting with the initial microbiology of heavy metal ions, which focused on growth requirements and tolerance levels, we have now moved far beyond to protein structures from X-ray and NMR analysis. Together with the results from genomics, in depth molecular genetic analysis from gene deletions to residue-specific mutagenesis, and chemical reaction mechanisms on the basis of protein structures, we are now rapidly approaching the level of real mechanistic understanding.

Moreover, treating metal–microbe interaction by a molecular systems theoretical approach may make this process even more perceptible and integrate the enormous amount of data. The rapid increase in sophistication over the last decade makes it easy to expect that the next years will continue to add more in-depth insights. Furthermore, comprehending microbial metal homeostasis may be a small step towards understanding of the fascinating process we call life.

Halle and Chicago
November 2006

Dietrich H. Nies
Simon Silver

References

Brown NL, Morby AP, Robinson NJ (eds) (2003) Thematic issue. Interactions of bacteria with metals. FEMS Microbiol Rev 27:129–447

Faústo da Silva JRR, Williams RJP (1991) The Biological Chemistry of the Elements. The Inorganic Chemistry of Life. Clarendon, Oxford UK

Lippard SJ, Berg JM (1994) Principles of bioinorganic chemistry. University Science Books, Mill Valley, CA

Silver S, Walden W (eds) (1998) Metal Ions in Gene Regulation. Chapman and Hall, New York

Contents

Part I
Molecular Physiology of Metal-Microbe Interactions According to Mechanisms

Microbiol Monogr (6)
D. H. Nies, S. Silver: Molecular Microbiology of Heavy Metals
DOI 10.1007/7171_2006_072/Published online: 13 February 2007

Understanding How Cells Allocate Metals

Stephen Tottey · Duncan R. Harvie · Nigel J. Robinson (✉)

Institute for Cell and Molecular Biosciences, The Medical School,
University of Newcastle, Newcastle NE2 4HH, UK
n.j.robinson@newcastle.ac.uk

Abstract Life depends upon multiple metals. It is estimated that approximately one-third of all gene products require a metal for folding and/or catalysis. How does the correct metal locate to the correct protein? Provision of sufficient atoms of each of the metals required by protein metal-binding sites is a challenge for cell biology. This is often especially true for iron, which is poorly soluble under aerobic conditions. Protein metal-binding sites follow universal affinity series. Under such a regime, exclusion of the wrong metals from metalloproteins is arguably an even greater challenge. High-fidelity homeostasis must match the number of some metal cations to the number of bonafide metal-binding sites. Selective protein–protein interactions also limit access of some atoms to the required subsets of proteins. Here we provide an overview of the contributions of metal sensors, metallochaperones, metal transporters and metal-storage proteins to the allocation of metals in cells. In this chapter an emphasis is placed on studies of the cell biology of metals in cyanobacteria.

1 The Irving–Williams Series and the Challenge for Homeostasis

The principles that govern the chemical speciation of metals in cells are clearly laid out in "*The Biological Chemistry of the Elements*" by Fraústo da Silva and Williams (2001) to which readers are directed. The underlying issue is that the metal sites of individual proteins are not sufficiently selective to solely bind the correct metal and exclude all others. Multiple partitioning events are required. Viewed as systems, cells monitor and modulate the number of atoms of each metal ion and the respective number of ligands. The homeostasis of different metal ions is also somehow integrated to avoid a surplus or deficiency of one element interfering with the speciation of another.

The Irving–Williams series was initially based upon empirical observations of the divalent transition metal-binding properties of model complexes and has been extended to include monovalent copper (Fraústo da Silva and Williams 2001). It specifies that the binding constants of proteins for essential metals will approximately follow a standard global order:

$$Mg^{2+} < Mn^{2+} < Fe^{2+} < Co^{2+} < Ni^{2+} < Cu^{2+}(Cu^{+}) > Zn^{2+} .$$

In a simple cell model, where all metal ions are equally available and in surplus to proteins, all metallo-proteins would become copper proteins. From this binding affinity series it follows that concepts of all nascent metalloproteins plucking the correct metal ions from free solution as they emerge from the ribosome, and of all transporters releasing metal ions to a cytosol that contains some concentration of the free ions, are grossly naive. The full suite of protein ligand chemistries, coordination geometries, and other physico-chemical properties, is inadequate for such a simplistic model, and therefore cells must have evolved to manage metal-protein speciation (Tottey et al. 2005).

From the perspective of bioinorganic chemistry, a series of pseudo-equilibria between the number of atoms of each metal ion and the number of ligands for each metal ion in any compartment can substantially resolve metal ion selectivity by becoming a function of relative affinity between different proteins in vivo, as opposed to absolute affinity. However, for molecular cell biology, this poses many more questions. What are the sensors? What are the regulatory circuits and how do they achieve the correct balance between metal ion number and ligand number? To date, some progress has been made towards understanding the control of the number of metal ions in cells. In bacterial systems, metal-responsive transcriptional regulators detect surplus or deficiency of specific metal ions and modulate production of metal-selective transporters and storage proteins and this is discussed in (Sect. 3). Several metal-responsive transcriptional regulators are also known in yeast (Rutherford and Bird 2004) but in higher eukaryotes, MTF-1 is the only well-described metal ion responsive transcriptional regulator (Lichtlen and Schaffner 2001). Much of the metalloregulation in eukaryotes acts at a post-transcriptional level, modulation of RNA stability or modulation of translation as occurs with IRP and iron-sulphur clusters (Pantopoulos 2004), or direct modulation of protein activity, localization or stability as occurs with several metal-transport proteins (Eisenstein 2000).

Control of the ligand number can sometimes be predominantly determined by modulation of a protein that functions to buffer surplus metal ion such as iron in ferritin (Hintze and Theil 2006), or zinc in bacterial metallothionein (Sect. 4.1). The ligand number can also to be adjusted by switching metabolic demand for a particular metal ion. For example, cyanobacteria, but not higher plants (Molina-Heredia et al. 2003; Weigel et al. 2003), undergo a metal-dependent switch between iron in cytochrome c_6 and copper in plastocyanin for photosynthetic electron transport but the control mechanism remains to be described (Raven et al. 1999) (discussed in Sect. 4.2). In the eukaryotic algae *Chlamydomonas reinhardtii*, an analogous response (Merchant 1998) involves the action of the Crr1 transcriptional regulator (Kropat et al. 2005). Crr1 also regulates tetrapyrrole biosynthesis in a copper dependent manner.

Kinetic factors influence the distribution of metal ions on proteins as reviewed by others (Finney and O'Halloran 2003; Fraústo da Silva and Williams 2001). Many metal-sites are buried within proteins and therefore exchange rates can be extremely slow. Once a metal ion is inserted into such a site it becomes trapped and unlikely to be replaced even with a metal ion that has greater affinity. Macromolecular assemblies and molecular crowding create chemically distinct microenvironments within the cytoplasm. Here we consider how the specificity of protein–protein interactions can influence the metallochaperone-mediated distribution of copper (Sect. 5), and where considerations of the relative affinities and allosteric mechanisms of metal-ion sensors are indicative of kinetic components to the control of selective metal-detection (Sect. 3).

It is worth noting that some proteins may only be occupied by the "correct" metal ion for some fraction of the time without becoming rate-limiting. In addition, while many enzymes show optimal activities with a specific metal ion in vitro, it is probable that some metal ions that are suboptimal for catalysis are nonetheless adequate to sustain reaction rates required in vivo.

2
Cyanobacteria in the Study of Metals in Cells

"To begin with, for you to be here now trillions of drifting atoms had somehow to assemble in an intricate and curiously obliging manner to create you", "why atoms take this trouble is a bit of a puzzle" (Bryson 2003). Among these atoms are essential metal ions and much of the work to assemble them into larger molecules is, of course, done for you by photoautotrophs that synthesize organic molecules from inorganic atoms and energy derived from photosynthesis.

2.1
The Assimilation of Inorganic Atoms Into Organic Molecules

The chloroplasts of higher plants and algae arose via a single endosymbiotic event involving the ancestors of modern cyanobacteria approximately 1.26 billion years ago (Yoon et al. 2002). A typical sequenced cyanobacterial genome encodes in the region of 3000 proteins (3168 in *Synechocystis* PCC 6803, 2653 in *Synechococcus* PCC 7942) while *Anabaena* PCC 7942 has 5368 genes, which is consistent with the additional complexity associated with the differentiation of nitrogen fixing heterocysts under nitrogen-limiting conditions. Chloroplast genomes typically contain 110–120 predicted open reading frames. However, a substantial number of endosymbiont-derived genes have been retained within the plant genome; with up to 4500 open reading frames (18% of the total) in the *Arabidopsis thaliana* genome predicted to be derived from the cyanobacterial ancestor of plastids (Martin et al. 2002). One of the crucial activities of the products of these retained genes is the coupling of photosynthesis to the assimilation of inorganic elements. The inorganic cell biology of cyanobacteria therefore assumes a special significance.

By taking advantage of entire genome sequences it is possible to construct a most likely origin for the ancestral eukaryotic genome itself. It is widely accepted that the transcriptional machinery of eukaryotes is more akin to that of archaea than eubacteria. Genome-based phylogenies support the idea that the progenitor of eukaryotes was derived from a fusion of two diverse prokaryotic genomes, one of which was indeed related to archaeal prokaryotes (Rivera and Lake 2004). However, the other lies deep within an ancient photosynthetic clade that includes the cyanobacteria and α-proteobacteria, which are the groups that gave rise to chloroplasts and mitochondria.

The photosynthetic machinery has substantial demand for metal ions. It has been estimated that 80% of the iron content of plant leaves is contained within chloroplasts (Terry 1983) and the cyanobacterium *Synechocystis* PCC 6803 requires ten times more iron than *Escherichia coli* (Keren et al. 2004). Photosystem I alone contains three separate iron-sulphur clusters of the Fe_4S_4-type and heme-iron is required for the associated cytochromes b_6f and *f*. The two photosystems and associated cytochromes require a total of 22 atoms of iron (Fig. 1). The oxygen evolving, water-splitting, complex of photosystem II requires four atoms of manganese. The manganese requirements of *Synechocystis* PCC 6803 are consequently 100-fold greater than a photosynthetic bacterium such as *Rhodobacter capsulatus*, which lacks the water-splitting enzyme (Keren et al. 2002).

Cyanobacteria have significance in relation to studies of metal-selection via insertion of ions into pre-formed chelating rings. These include tetrapyrroles of which cyanobacteria produce an atypically large repertoire that includes iron containing heme and siroheme, cobalt containing corrin for vitamin B_{12} plus magnesium containing chlorophyll. Higher plant chloroplasts do not produce vitamin B_{12} and contain no vitamin B_{12}-requiring enzymes. Many algae do have a requirement for vitamin B_{12} but possess no cobalamin biosynthetic machinery. It is proposed that these algae acquire vitamin B_{12} through a symbiotic relationship with bacteria (Croft et al. 2005). Association of the metal with the ring chelate is irreversible in vivo, other than via degradation of the organic molecule. The different rings are differently decorated and so metal-ion selection now becomes based upon proteins recognizing the

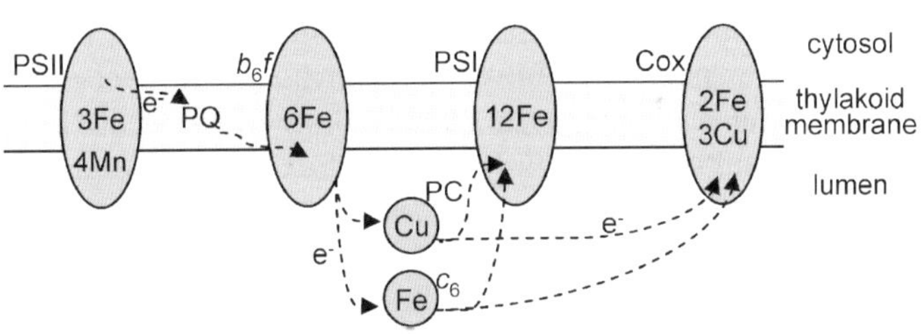

Fig. 1 *Synechocystis* PCC 6803 electron transfer in photosynthesis and respiration. During photosynthesis, electrons (e^-) are transferred (*dashed line*) from photosystem II (PSII) via the plastoquinone pool (PQ) to the b_6f complex, before being shuttled to photosystem I (PSI) via the soluble electron carriers plastocyanin (PC) or cytochrome c_6 (c_6). In some cyanobacteria, including *Synechocystis* PCC 6803, the soluble electron transporters can also donate their electrons to cytochrome oxidase (Cox). The location of the Mn, Fe, and Cu atoms are shown. PSII contains four Mn atoms in the water splitting complex, plus three Fe atoms consisting of two cytochromes and one haem Fe (Zouni et al. 2001). Cytochrome b_6f contains six Fe atoms comprising one Fe_2-S_2 cluster plus four haems (Kurisu et al. 2003). The soluble electron transporters PC and c_6 contain one Cu and Fe atom, respectively, while PSI contains 12 Fe atoms in the form of three Fe_4-S_4 clusters (Jordan et al. 2001). Cytochrome oxidase contains three Cu atoms and two Fe atoms made from its haem a and a_3 sites (Ferguson-Miller and Babcock 1996)

correct tetrapyrrole cofactor, which is much less challenging than discerning the individual elements alone.

2.2 The Liberators of Dioxygen Altered Metal Speciation in the Environment

During early evolution, prior to the development of photosynthesis, life was restricted to habitats where energy could be derived from inorganic chemical sources. Molecular phylogenies imply that the early ancestors of cyanobacteria arose approximately 2.7 billion years ago (Brocks et al. 1999) and underwent a rapid adaptive radiation, presumably due to the greater autonomy afforded by photosynthesis and hence the opportunity to colonize previously vacant niches. This coincided with a global rise in di-oxygen as a by-product of the activity of the water splitting enzyme, changing many habitats from anaerobic to aerobic. In turn, this altered the chemical forms and solubility of metals with, for example, ferric-iron becoming largely insoluble. Only small amounts of ferric-iron bound to organic molecules would remain in aerobic solution. Thus, cyanobacteria were evolving swiftly at a time when the availabilities of essential metals were changing.

2.3 Thylakoids and Cytoplasmic Compartmentalization

Cyanobacteria contain internal membrane-bound compartments, thylakoids, which house the photosynthetic machinery. Some slow-growing strains, such as *Gleobacter*, are devoid of thylakoids and photosynthetic electron transport occurs at the plasma membrane. The thylakoid lumen, at least in the model organism *Synechocystis* PCC 6803, is known to be distinct from the periplasm since a GFP-fusion to the TAT specific targeting sequence of *E. coli* TorA gives detectable fluorescence in the periplasm but not in the thylakoid lumen (Spence et al. 2003). Cryo-EM studies have observed internal lipid bodies from which thylakoid membranes emanate (van de Meene et al. 2006), suggesting independent processes of biogenesis. Such cytosolic compartmentalization provides additional opportunities for metal-partitioning in cyanobacteria, especially in relation to copper, iron, and manganese where the predominant demand lies within the thylakoids.

3 Metal Sensors

Cells somehow detect when critical metal thresholds, either excess or deficiency, have been exceeded. In bacteria, this commonly involves metal-responsive DNA-binding transcriptional regulators.

3.1 Control of Metal Ion and Ligand Concentrations

By controlling the number of atoms of each metal ion [M] and the number of ligands [L] within a compartment, it becomes possible to restrict metal ions to bonafide sites, even under the constraints of the Irving–Williams series. How are the set points monitored and adjusted? Regulated production of metal-transporters is one option for controlling [M], and control of the production of proteins involved in metal-sequestration is one approach for controlling [L]. In this respect, metal-responsive transcriptional regulators are presumed to perform central roles in bacteria and several families of sensors are known (Fig. 2).

Auto-regulated activity, stability, or localization of the transport proteins is another approach for controlling [M]. The soluble amino-terminal regions of some P_1-type ATPases are, for example, thought to perform sensory, regulatory-roles. In theory, the selective advantage of metal sensors could relate to energetic savings associated with accumulating proteins required for metal-ion homeostasis solely when required, rather than directly determining metal-ion sufficiency.

3.2 Inferences Made from Metal-Ion Thresholds of Metal Sensors

By determining the properties (DNA affinities, metal affinities) of metal-sensor proteins in vitro, it is possible to make inferences about the disposition of metals in cells.

3.2.1 The Tight Metal Affinities of Copper-Sensing CueR and Multiple Zinc Sensors

Cytosolic copper is detected in *E. coli* by CueR, (Stoyanov et al. 2001) as reviewed in other chapters. In common with other members of the MerR-like family of transcriptional regulators, CueR binds to a promoter with abnormally long spacing between the – 10 and – 35 promoter elements. Upon binding copper, CueR distorts the promoter region, under-winding the DNA such that both elements are more optimally aligned for RNA polymerase. The concentration of copper at which CueR responds in vitro is peculiarly low, 10^{-21} M, or zeptomolar (Changela et al. 2003). Hypothetically, CueR responds when the concentration of copper exceeds $\sim$ 600 atoms per liter. This, of course, equates to substantially less than one atom per *E. coli* cell volume (Changela et al. 2003). CueR activation will lead to production of proteins (CopA and CueO) associated with copper efflux from the cytoplasm, and therefore if CueR is the dictator of copper sufficiency, these data imply that a single free copper atom in the *E. coli* cytoplasm represents an excess. This is

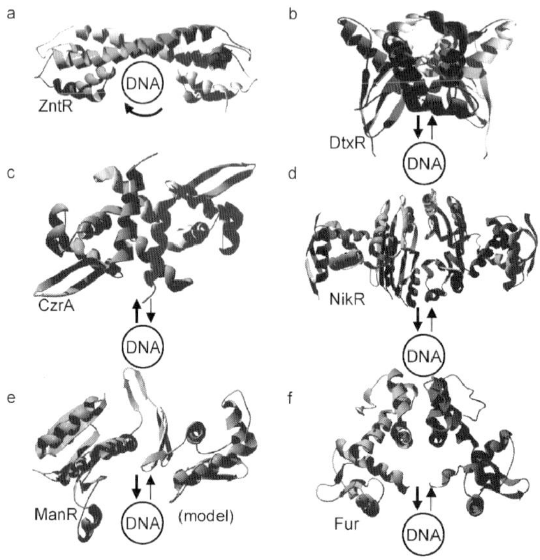

Fig. 2 Diversity of action of bacterial metal sensors. **a** Structure of ZntR (Changela et al. 2003) a Zn-sensing member of the MerR family of regulators. The addition of Zn to ZntR from *E. coli* converts the protein to a transcriptional activator, which works by distorting the bound DNA molecule. This makes the promoter a better substrate for the RNA polymerase (Outten et al. 1999). **b** Structure of DtxR (Qui et al. 1995). DtxR is a Fe-binding dimeric holo-repressor. DtxR from *Corynebacterium diphtheriae* regulates the expression of genes required for Fe uptake and the diphtheria toxin, a major virulence factor. Some DtxRs also contain an additional C-terminal SH3 domain, which is believed to be rare in prokaryotes (White et al. 1998). **c** Structure of CzrA (Eicken et al. 2003). CzrA from *Staphylococcus aureus* is a member of the SmtB family of DNA repressors which dissociates from DNA upon binding of Zn or Co. The metal-ion binding site is derived from two ligands from the $\alpha5$ helix of one subunit and two from the $\alpha5$ helix of the other subunit. A hydrogen-bonding network has been described which forms an intersubunit quaternary structure. Other members of the SmtB family have been characterized that require different ligands from different parts of the protein for derepression. **d** Structure of NikR (Chivers and Tahirov 2005). NikR from *Pyrococcus horikoshii* is a member of the NikR family of metal-responsive transcriptional regulators which represses the expression of genes in response to Ni. It is a homotetramer in the presence and absence of Ni. **e** Homology model of ManR, made using SWISS-MODEL (Schwede et al. 2003) based on the structure of DrrD (Buckler et al. 2002). ManR is a member of the two-component family of regulatory proteins, which involves phosphorylation of an intracellular response regulator protein by a sensory kinase that is usually integral to the inner cell membrane. The regulatory proteins are composed of a conserved phosphorylation-activated regulatory domain, and a DNA-binding effector domain. **f** The structure of Fe-binding homodimeric repressor Fur (Pohl et al. 2003). Fur from *Pseudomonas aeruginosa* and other prokaryotic homologues regulates the transcription of genes required for Fe uptake. Two metal ions are bound per subunit, with four metal ions bound per dimer at two sites. Only the metal-ion binding site at the dimer interface is required for regulation, while the second remote site is structural. All proteins are at equilibrium with their DNA binding sites. *Heavy arrow* indicates protein movement/action upon metal-ion binding

consistent with the position of copper at the apex of the Irving–Williams series, and the linked proposition that the *E. coli* cytoplasm is devoid of soluble copper proteins (Sect. 4.2).

Cytosolic zinc is detected by ZntR (Brocklehurst et al. 1999) and Zur (Patzer and Hantke 1998) in *E. coli*, as reviewed in other chapters. Analysis of the in vitro metal-ion concentrations at which these sensors respond places the detection thresholds at femtomolar concentrations (Outten and O'Halloran 2001). At this zinc concentration, ZntR would therefore activate the *zntA* promoter, leading to efflux of surplus zinc while Zur would repress the *znu* promoter, leading to inhibition of zinc influx. The *E. coli* cell volume is in the region of one femtoliter (Akerlund et al. 1995), such that one free atom per *E. coli* cell volume formally equates to ~ 1.7 nM. In common with the calculations for CueR, this seems to imply that the set-point for the cellular metal sensor is so sensitive that the cytosol will be devoid of any atoms of free zinc. Once the number of atoms of zinc exceeds the number of ligands (L_{Zn}), the surplus will begin to populate sites that should be occupied by other, less competitive metal ions, rather than forming a pool of free zinc. One simple model is that the affinity of the zinc sensors is set to compete with these adventitious sites, not to modulate the concentration of some hypothetical free zinc pool. Similar arguments can be applied to other types of sensors such as the first characterized zinc sensor SmtB (Huckle et al. 1993), which is also estimated to have an affinity for zinc tighter than picomolar (VanZile et al. 2000).

3.2.2 The Weak Affinity of Cobalt Sensor NmtR

Mycobacterium tuberculosis NmtR is a member of the ArsR-SmtB family of metal-responsive transcriptional repressors that regulates expression from the *nmt* operator-promoter of a deduced metal-exporter similar to cobalt-exporting CoaT from *Synechocystis* PCC 6803 (Cavet et al. 2002; Rutherford et al. 1999). Repression from the *nmt* operator-promoter is solely alleviated by nickel and cobalt in vivo in mycobacterial cells. Purified NmtR has an estimated affinity for cobalt in the region of 7.1×10^{-5} M (Cavet et al. 2002). Using similar arguments to those applied to CueR, ZntR and Zur, this implies that a cell would have to accumulate more than $\sim 43\,000$ free atoms of cobalt in order to gain 50% cobalt saturation of NmtR. Under such a regime, cobalt will form associations with many adventitious sites before NmtR responds and surplus cobalt atoms are expelled from the cytoplasm. This seems implausible, and such disparity between in vitro thermodynamic measurements and metalloregulation observed in vivo is indicative of kinetic factors acting in vivo. These could include targeted cobalt delivery to the sensor either by a small molecule, such as an intermediate in vitamin B_{12} metabolism, a metallochaperone such as the cobalt chelatase CbiX (Leech et al. 2003) or direct ligand exchange of cobalt from the cobalt importer to the sensor.

3.2.3
Intra- and Inter-Family Comparisons in Cellular Systems

The multiple families of bacterial metal-sensing transcriptional regulators include the MerR-like activators and the ArsR-SmtB like DNA-binding metal-alleviated repressors (Fig. 2). In addition, there are several types of DNA-binding metal-requiring repressors such as the Fur-family, DtxR, NikR, which will be described in detail in the following chapters. Different representatives of each family show different metal specificities and the reasons why these are especially valuable as probes of metal-ion selectivity have been discussed elsewhere (Tottey et al. 2005). By contrasting the properties of related sensors that detect different metal ions, it has been possible to observe how cells discern different metal ions and more generally to gain insight into the contributions of different processes to the distribution of metal ions in cells. In the next section, this is considered in some detail in relation to the ArsR-SmtB family of sensors. Pair-wise comparisons between manganese-sensing MntR versus the related iron-sensing DtxR (Guedon and Helmann 2003; Glasfield et al. 2003) (see next chapter), and between copper-sensing CueR versus zinc-sensing ZntR (Changela et al. 2003) have identified differences in the metal-ion coordination environments that modify binding preferences for these elements. These include changes in the ligand types, ligand number or coordination geometries. However, second coordination spheres can also exert a subtle influence on metal-preference. For example, the binding site for monovalent copper in CueR is located at the end of an α-helix (Changela et al. 2003). The helix dipole creates an environment that disfavors divalent metal ions such as zinc, and in this respect the metal-site of CueR is unlike ZntR. The Irving–Williams series implies that, conversely, features of the metal-binding site of ZntR will not be adequate to exclude copper.

Analyses of selectivity have tended to focus on comparisons of the properties of different members within single families. Of course, the cellular system functions without these distinctions, each sensor responding to the correct metal ion regardless of its mode of action or evolutionary history. There is now an opportunity to make inter-family comparisons. For example, it should become possible to better understand metal-homeostasis if the relative affinities for every metal ion of all sensors within a cell are inter-compared.

3.3
Metal Selectivity in ArsR-SmtB Metal Sensors

Expression of the bacterial metallothionein gene, *smtA*, in response to surplus zinc, or cadmium, is mediated by a divergently transcribed gene encoding a DNA-binding repressor, SmtB (Huckle et al. 1993). Binding of SmtB to the *smt* operator-promoter was enhanced by the addition of the zinc chelator 1,10-phenanthroline to binding reactions (Morby et al. 1993). SmtB is similar in sequence to ArsR, which is required for arsenite responsive regulation of the *ars* operator-promoter (Wu and Rosen 1991), and to CadC, which is required for cadmium-responsive expression of *cadA*, encoding a cadmium-efflux ATPase (Endo and Silver 1995). All of these proteins (SmtB, ArsR and CadC) have conspicuous cysteine residues associated with deduced helix-turn-helix DNA-binding regions and these were assumed to provide ligands of the metal-sensing sites (Bairoch 1993). However, subsequent studies have identified diversity in the metal-binding sites of these proteins, with the sites often located elsewhere in the protein structure (Tottey et al. 2005). Intriguingly, the initial alignments of SmtB, ArsR, and CadC also included two other proteins; HlyU associated with the regulation of haemolysin production in *Vibrio cholerae* (Williams et al. 1993) and NolR associated with the control of nodulation in *Rhizobium meliloti* (Kondorosi et al. 1991). These were initially dismissed as metalloregulators partly due to the absence of cysteine residues adjacent to the deduced helix-turn-helix. The possibility that HlyU, NolR, and/or closely similar proteins from other organisms respond to metal ions should now be reconsidered.

Structures are known for several representatives of the ArsR-SmtB family of sensors, SmtB (Cook et al. 1998), CzrA (Busenlehner et al. 2003), and CadC (Ye et al. 2005). These are all dimeric winged-helix proteins (Fig. 3). SmtB has two pairs of metal-ion binding sites, both formed at dimer interfaces. Helices $\alpha3$ and $\alpha4$ form the DNA-binding helix-turn-helix region but the α3N site involving cysteine$_{61}$ is not required for inducer recognition in SmtB (Turner

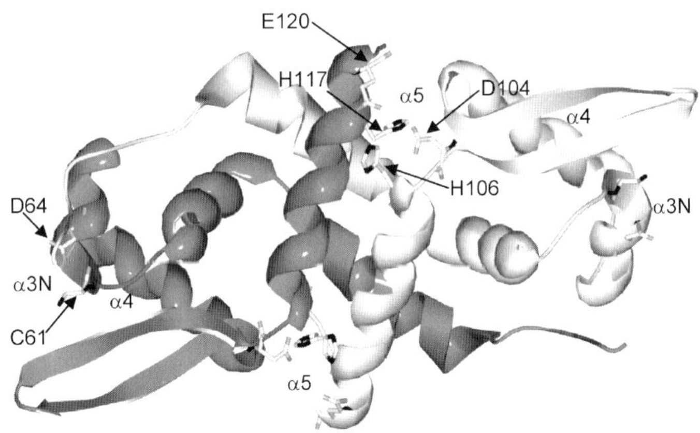

Fig. 3 The cytoplasmic zinc sensor SmtB from *Synechococcus* PCC 7942. SmtB is a homodimeric repressor that regulates Zn-responsive production of metallothionein SmtA. The regulator contains the classical helix-turn-helix motif seen in many DNA-binding proteins. Side chains of the amino acids forming the Zn-binding sites are shown. The $\alpha5$ site, at the dimer interface, is essential for function, while the α3N site is not. There are additional residues at the N-terminus that contribute to the α3N site, which were not observed in the original crystal structure (Cook et al. 1998)

et al. 1996). Both in vivo (Turner et al. 1996) and in vitro (VanZile et al. 2002) experiments confirm that ligands from anti-parallel α5-helices are required to impair SmtB DNA-binding in response to zinc. In contrast, ligands associated with the helix-turn-helix region are required for inducer recognition by ArsR (Shi et al. 1994). Six alternative sensory sites have been described to date in ArsR-SmtB family members using a combination of site-directed mutagenesis coupled with in vivo studies of reporter gene expression and in vitro studies of DNA binding or biophysical assays of metal binding. The sites have been defined according to the known or predicted structural elements where the ligands are located, and there is therefore some variability in nomenclature (Ye et al. 2005), but the most common designations are; α3 (ArsR), α5 (SmtB), α5C (NmtR), α4 (CmtR), α3 and α5 (ZiaR), α3N (CadC) (Tottey et al. 2005). Our ongoing structural studies of CmtR suggest that a most important feature of many of these sensory sites might simply be that they straddle the dimer interface.

3.3.1 DNA Association Versus Disassociation and the Abundance of Sensor Molecules

The DNA affinities of some transcriptional repressors are relatively weak even in the forms that confer repression. For example, K_{app} for apo-NmtR is $\sim 5 \times 10^{-7}$ M (Pennella et al. 2003). This means that 50% occupancy of the *nmt* operator promoter requires a concentration of NmtR, which equates to $\sim$ 300 molecules of apo-repressor per cell, and a greater number of molecules will be required in order to confer full repression. A crucial implication is that the cells contain a substantial excess of molecules of DNA-free NmtR. This indicates that nickel and cobalt do not form complexes with NmtR-DNA adducts to promote disassociation from DNA in vivo, but rather they predominantly form NmtR-nickel or NmtR-cobalt complexes to impair association with DNA.

Metal-ion affinities of metal-responsive transcriptional regulators have mostly been determined in free solution, rather than for the DNA adducts. Such affinity values will be valid for sensors that act in a manner analogous to that proposed for NmtR and have similar weak affinities for DNA. Different allosteric mechanisms can also apply when the starting state is free repressor rather than repressor-DNA adduct. Metal-binding need not switch the protein from a single DNA-associated conformation to a single metal-bound form, but rather metal binding can merely act to preclude the subset of conformers that are competent to associate with DNA.

There is a need to directly determine the number of molecules of NmtR and other transcriptional regulators in vivo, for example using quantitative Western blots. It has been estimated that *E. coli* contains 5000 molecules of the Fur repressor during exponential growth, 10 000 at stationary phase (Zheng et al. 1999). Unlike NmtR, Fur does have multiple operator-promoter targets, but the number of copies of Fur per cell is still vastly in excess of the number of DNA-binding sites. These observations also hint that there may be some secondary role(s) for the regulatory proteins that do not involve promoter binding, perhaps in metal-ion storage or metal-ion trafficking.

3.3.2 Affinity, Allostery, and Access

The coordination of oxyanions, arsenite, and antimonite to ArsR is trigonal, which is a preferred ligand number for these ions (Shi et al. 1996). The coordination of zinc to SmtB is tetrahedral (VanZile et al. 2000; Eicken et al. 2003), while that of nickel and cobalt to NmtR is octahedral (Cavet et al. 2002; Pennella et al. 2003), again ideal for these metal ions. Cysteinyl-thiol ligands predominate in sensors such as CadC (Busenlehner et al. 2002) and CmtR (Cavet et al. 2003a) that detect "soft" metal ions such as cadmium, while histidine-nitrogen and aspartate-carboxylate ligands are more prevalent in sensors for metal ions with more "hard" characteristics such as zinc in SmtB (VanZile et al. 2000).

When the metal-ion affinities of the nickel and cobalt sensor NmtR were compared with those of the zinc sensor SmtB, it emerged that SmtB has the tighter affinity for cobalt, which it does not detect in vivo (Cavet et al. 2002). NmtR also has a tighter affinity for zinc, which it does not detect in vivo, than it does for cobalt by at least three orders of magnitude (Cavet et al. 2002). Furthermore, in vitro competition experiments indicate that the cobalt affinities of two ArsR-SmtB sensors that are both from *M. tuberculosis* are closely matched even though one of them (NmtR) detects these ions and the other (CmtR) does not (Cavet et al. 2003a). The implication is that selectivity does not solely operate at the level of metal binding and partitioning mediated by absolute, or indeed relative, affinities. In view of predictions about the limitations of metal selectivity in proteins made by the Irving–Williams series, these observations should not be surprising. In some proteins, metal ions bind to rigid pre-formed sites. However, in metal sensors, a function of metal binding is to induce some form of conformational switch, which leads to altered gene expression, and therefore the metal sites of metal sensors must have flexibility. This allows even greater potential for the "wrong" metal ions to bind by selecting a preferred subset of the available ligands, plus preferred coordination geometry. For example, when zinc binds tightly to NmtR, it does so via a tetrahedral geometry rather than the octahedral geometry preferred by nickel and cobalt (Cavet et al. 2002; Pennella et al. 2003). However, DNA binding by zinc-NmtR is not inhibited in the manner of cobalt-NmtR (Cavet et al. 2002) and thus selectivity operates at the level of the allosteric switch.

In mycobacterial cells, NmtR responds to nickel and cobalt with nickel being the more potent effector. However, when NmtR is introduced into a cyanobacterium, it only responds to cobalt, and not to nickel (Cavet et al. 2002). The nickel content of mycobacterial cells rises from 2×10^4 to 7.4×10^5 atoms per cell at maximum permissive concentrations while cyanobacterial cells rise from 0.2×10^5 to only 0.7×10^5 atoms per cell at maximum permissive nickel concentrations (Cavet et al. 2002). In part, this relates to cyanobacteria being more sensitive to elevated nickel than mycobacteria, but nonetheless it formally demonstrates that selectivities observed in vivo can be dictated by access to metal ions in the cytosol rather than inherent properties of the sensor proteins. Deletion of the *E. coli* nickel-importer, *nikA-E*, inactivates hydrogenase due to lack of inward nickel supply. However, the thresholds at which NikR responds to nickel are unaltered by this mutation (Rowe et al. 2005). NikR must have access to a different pool of nickel than that available to the hydrogenase. HypCDEF interact with immature hydrogenase and HypC facilitates the GTP-dependent insertion of nickel in association with other accessory factors (Atanassova and Zamble 2005). Presumably, the specificity of interactions of such accessory factors could influence which metal sensors gain access to which metal ions in vivo. To date, CopZ from *Enterococcus hirae* is the only precedent for a metallochaperone that interacts with a metal sensor, CopY (Odermatt and Solioz 1995; Solioz and Stoyanov 2003).

3.3.3 Allosteric Inhibition by Competitive Metal Ions

Where selectivity operates at the level of allostery, there is potential for a non-effector to bind tightly to sensory site ligands and thereby inactivate the sensor. This has been observed in vitro using CzrA from *Bacillus subtilis* (Harvie et al. 2006). Formation of CzrA-DNA complexes was detected using fluorescence anisotropy, consistent with specific binding of dimeric CzrA, with $K_{app} \sim 5 \times 10^{-7}$ M. In the presence of 2 μM zinc, no binding was detected up to the addition of 20 μM CzrA, establishing that zinc weakened DNA binding by greater than one order of magnitude. In the presence of 2 μM copper, K_{app} was similar to apo-CzrA, and crucially this remained unaltered after the subsequent addition of 2 μM zinc. Consistent with this observation, in the presence of equimolar copper and zinc, copper co-migrated with CzrA by size exclusion chromatography while zinc migrated as the free ion. Thus copper binds to CzrA more tightly than zinc and prevents the protein from responding to its effector. It will be interesting to determine whether copper inhibits the detection of zinc by CzrA in vivo. The lack of such an effect may be indicative of kinetic control of the distribution of copper in vivo, for example due to copper being solely trafficked to the correct destinations by the CopZ metallochaperone and therefore preventing the ligands of CzrA from gaining access to the copper (Radford et al. 2003).

3.4 Integration of Metal Sensing with Metabolism and CoaR

Under iron-replete conditions, Fe-Fur represses *sodA*, encoding Mn-SOD, but under iron-limiting conditions, Mn-SOD functionally replaces Fe-SOD (Niederhoffer et al. 1990). This exemplifies the adjustment of metabolism to accommodate changing metal-ion supply. Iron levels, via the action of Fur, also control the expression of fumarase isoenzymes. Fumarases A and B contain a $Fe_4 - S_4$ cluster but can be functionally replaced by non-iron containing fumarase C, allowing the Krebs cycle to continue under low-iron conditions (Tseng 1997). In the converse situation, a change in metabolism creates an altered metal-ion demand. For example, the principal demand for cobalt is for cobalamin and hence vitamin B_{12}. The cobalt sensor CoaR from *Synechocystis* PCC 6803 (Rutherford et al. 1999) has a domain with sequence similarity to precorrin isomerase, an enzyme of cobalamin biosynthesis (Shipman et al. 2001). Cobalt detection by CoaR involves carboxyl-terminal cysteine and histidine ligands and a model was proposed in which the precorrin isomerase-like domain of CoaR binds pathway intermediates to inhibit CoaR activation of cobalt export when there is a greater metabolic demand. An alternate suggestion is that this domain binds a small molecule cobalt donor.

4 Sequestration of Surplus Competitive Metal Ions

Metals can be sequestered in specialized metal-binding proteins or within compartments.

4.1 Zinc and Metallothionein in *Synechococcus* PCC 7942

In response to surplus zinc, a few species of bacteria are known to produce metallothioneins. These proteins tightly bind the excess zinc atoms and are capable of binding other metal ions including copper (Shi et al. 1992). Bacterial zinc metallothionein was first purified and sequenced from *Synechococcus* TX-20 (Olafson et al. 1988) and the *smtA* metallothionein gene was subsequently isolated from *Synechococcus* PCC 7942 (Robinson et al. 1990). Related genes are present in the genomes of several cyanobacteria (Cavet et al. 2003b). A polypeptide with features of metallothioneins, including Cd-NMR spectral features indicative of cysteine-thiol and imidazole-nitrogen ligands, was also identified in extracts from *Pseudomonas putida* strains recovered from a Cd-contaminated industrial site (Higham et al. 1984, 1986). With the completion of genome sequences, pseudomonads have emerged as a second bacterial group containing metallothionein-like genes (Blindauer et al. 2002). It remains to be established whether bacterial metallothioneins have a limited species distribution, or whether such small, poorly conserved genes have been overlooked during gene annotation. A Blast search of the genome of *Anabaena* PCC 7120, prior to annotation, identified an ORF encoding a metallothionein-like protein, which was subsequently overexpressed in *E. coli* and structurally characterized (Blindauer et al. 2002). However, a larger open reading frame on the complementary strand was identified in the genome annotation, not the metallothionein gene. Reverse transcriptase PCR has established that the metallothionein gene is transcribed, rather than the complement. Furthermore, a SmtB/ZiaR-like protein encoded by the same genome binds to the respective operator-promoter region plus the promoter of a gene encoding a putative zinc-exporting P_1-type ATPase (Bowness and Robinson unpublished) (Fig. 4). This illustrates the potential for cryptic metallothionein genes to be hidden in other bacterial genomes.

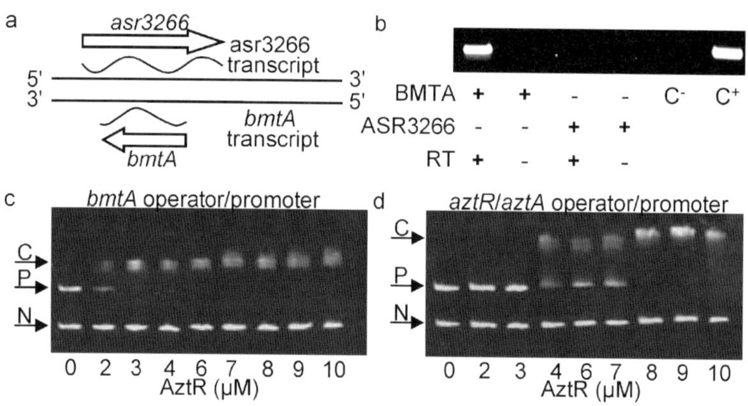

Fig. 4 Identification of *bmtA* in *Anabaena* PCC 7120. **a** The *bmtA* gene, identified via a Blast search of the pre-annotated *Anabaena* PCC 7120 genome sequences, is encoded in the inverse complement of the CyanoBase-designated ORF *asr3266*. **b** Reverse transcriptase PCR performed using transcript specific primers designed against the potential *asr3266* or *bmtA* mRNA transcripts. RNA was harvested from cultures exposed to maximum non-inhibitory concentrations of Zn for 24 h. +/− indicates presence/absence in the reaction mixture of *bmtA* specific primer, *asr3266* specific primer or reverse transcriptase. C^+/C^- indicates PCR control reaction with (C^+) or without (C^-) *Anabaena* PCC 7120 template genomic DNA. Bands indicate the presence of PCR-generated DNA products. **c** Electrophoretic gel mobility shift assay of AztR binding to *bmtA* operator/promoter 30-bp DNA fragment. The SmtB-like protein AztR (Liu et al. 2005) from *Anabaena* PCC 7120 was titrated at increasing concentrations to aliquots of (N) control (134 bp) DNA or (P) probe (164 bp, control DNA fragment plus extra 30 bp of *bmtA* operator/promoter) DNA. Purified AztR generates AztR-*bmtA* operator/promoter complexes (C). **d** Electrophoretic gel mobility shift assay of AztR binding to *aztR*/*aztA* operator/promoter 30-bp fragment. Increasing concentrations of AztR were mixed with (N) control or (P) probe (control DNA fragment plus 30 bp of *aztR*/*aztA* operator/promoter) DNA. Purified AztR generates AztR-*aztR*/*aztA* operator/promoter complexes (C)

Several previous reviews have described in some detail the characteristics of bacterial metallothionein genes (Cavet et al. 2003; Robinson et al. 1997, 2001; Turner and Robinson 1995) and their products (Blindauer and Sadler 2005). The solution structure of SmtA was determined by high-field NMR (Blindauer et al. 2001), which revealed a cluster containing four atoms of zinc, similar to one of the domains of eukaryotic metallothioneins but with the exception that two of the bacterial ligands are derived from histidine imidazolium groups (Daniels et al. 1998), rather than exclusively cysteinyl thiols. A second feature was the presence of small stretches of secondary structure, absent from eukaryotic metallothioneins. These formed a GATA-like Zn-finger around one of the four zinc atoms of the cluster (Fig. 5). This zinc atom is less labile than the other three (Blindauer et al. 2003). A reasonable expectation is that this zinc finger engages in macromolecular interactions but a partner remains to be found. A bioinformatic analysis of available genomes, including *B. subtilis* and *Streptomyces coelicolor* has identified a family of genes encoding zinc-finger-containing ribosome-binding proteins (Panina et al. 2003). These proteins are predicted to be regulated by the intracellular zinc sensor Zur. Paralogs of these proteins, devoid of zinc fin-

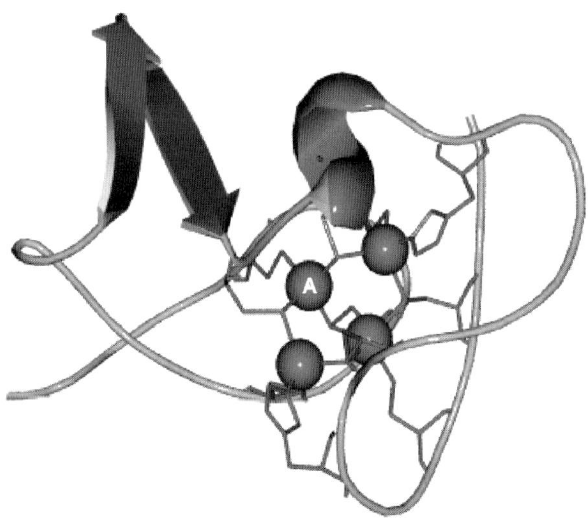

Fig. 5 Bacterial metallothionein, SmtA. SmtA is a 56-residue, cysteine-rich, Zn-binding metallothionein from *Synechococcus* PCC 7942. The metallothionein binds four Zn ions in a nine-cysteine and two-histidine cluster. The thiolate sulphurs of five of the cysteine residues form bridged ligands between the bound metal ions (Blindauer et al. 2001). SmtA is mostly unstructured, but contains three short regions of secondary structure. Areas of α-helix and β-strand are shown forming a Zn finger that resembles the Zn fingers of GATA and LIM-domain proteins. The Zn ion (A), which is buried within the Zn finger, is not readily exchanged with exogenous Zn, while the remaining Zn ions are. The solution structure of SmtA revealed the existence of GATA-like Zn fingers in prokaryotes

gers, are also present. It is suggested that under conditions of limiting zinc, the zinc-binding peptides are released from the ribosome and replaced by the non-zinc-binding paralogs. This has the effect of mobilizing an intracellular store of zinc ions. It is tempting to notice some analogy between this model, in which the zinc-regulated ribosome-binding proteins act as a zinc reservoir, and SmtA with its zinc finger.

4.1.1
The Contribution of Zinc Metallothionein to Metal Selectivity

According to the Irving–Williams series, there will be a propensity for zinc to mis-populate binding sites for metal ions lower down the series, such as iron. It is possible that zinc is restricted from these sites via the zinc affinity, k_{Zn}, of *bonafide* zinc ligands, L_{Zn}, in zinc-requiring enzymes, being greater than the zinc affinity of iron ligands, L_{Fe}. This requires that the total number of zinc atoms within the cell, ΣZn, be no greater than the total number of zinc-requiring ligands (Fig. 6). It is presumed that metal-ion buffering by SmtA is regulated by SmtB such that all surplus zinc atoms in excess of those required by L_{Zn} are bound to SmtA, thereby preventing excess zinc from occupying erroneous sites such as those of L_{Fe}. The k_{Zn} of SmtA must be tighter than the zinc affinities of L_{Fe}. By this mechanism, SmtA would withhold zinc from erroneous sites. A simplistic model would also infer that the zinc affinity of SmtB be weaker than that of L_{Zn}, but tighter than k_{Zn} of L_{Fe}. In this way, SmtB would only be able to out-compete erroneous ligands for zinc and

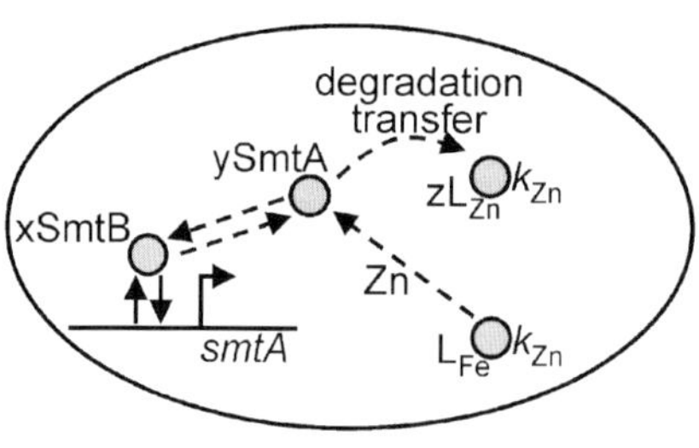

$$L_{Fe}\,k_{Fe} < L_{Fe}\,k_{Zn} \quad (1)$$

$$L_{Fe}\,k_{Zn} < L_{Zn}\,k_{Zn} \quad (2)$$

$$L_{Fe}\,k_{Zn} < \mathrm{SmtA}\,k_{Zn} \quad (3)$$

$$2x + 4y + z = \Sigma\,Zn \quad (4)$$

Fig. 6 Hypothetical relationships between zinc-binding sites of *Synechococcus* PCC 7942. SmtA protects cytosolic ligands that bind metal ions lower down the Irving–Williams series from being mis-populated by Zn. Fe is used as an example of a lower-series metal ion. Statement 1. The Irving–Williams series predicts that Fe requiring ligand, L_{Fe} will have a higher affinity for Zn (k_{Zn}) than for Fe (k_{Fe}). Statement 2. Zn requiring ligand, L_{Zn} will have a higher affinity for Zn than L_{Fe}. Statement 3. SmtA, which is produced in response to elevated Zn by the action of the regulator SmtB, has a k_{Zn} higher than that of L_{Fe}. Eq. (4). There will be no free Zn atoms to bind to L_{Fe}, provided statements 1 to 3 hold and provided the total number of Zn atoms, ΣZn, does not exceed the total number of Zn requiring ligands (two per SmtB, four per SmtA and total L_{Zn}). The potential direction of Zn transfer is shown by *dashed arrows*

the *smtA* gene would become transcribed when the number of zinc atoms in the cell exceeded the number required by L_{Zn} (Fig. 6). Unfortunately, in vitro measurements imply that SmtB has an affinity for zinc tighter than picomolar, which is tighter than some known zinc-requiring proteins. Presumably, the buried nature of many bonafide zinc sites is one kinetic factor that accounts for why this simplistic model does not apply.

4.2 Compartmentalization of Copper

In eukaryotes, copper is found in extra-cellular proteins (which acquire the metal in the trans-Golgi network) in mitochondria and in superoxide dismutase. The latter is a rare example of a cytosolic copper-requiring protein. In bacteria, opportunities for compartmentalization are, of course, more restricted.

4.2.1 Copper Proteins in the Periplasm and Plasma Membrane

The highly competitive nature of copper imposes a requirement that the cell somehow restricts protein access to this metal ion. Maintaining a cytosol free of readily exchangeable copper appears to be one solution. *E. coli* has no known copper-requiring enzymes in the cytosol and no known plasma-membrane uptake system for copper (Rensing and Grass 2003). There are, however, copper proteins within the *E. coli* periplasm plus a cytosolic copper detector (CueR) and efflux-protein (CopA). It is formally possible that these systems solely detect "stray" copper atoms, this element being non-essential within the *E. coli* cytosol. However, there is evidence that eukaryotic enzymes of molybdopterin biosynthesis require copper, and furthermore these plant enzymes are active when expressed in *E. coli*, suggesting that they are somehow able to acquire copper in this cellular environment (Kuper et al. 2004). It remains to be established whether bacterial enzymes of molybdopterin biosynthesis also have an absolute requirement for copper.

Copper is required for intramembraneous Cu_B-sites of quinol oxidase in *E. coli*. These atoms are proximal to the cytosolic face of the plasma membrane and an expectation is that the copper is inserted from this side. Whether these atoms are recruited from the cytosol or trafficked directly to the oxidase from the periplasm by ligand exchange via membraneous accessory proteins is unresolved.

Methanotrophs have a copper requirement for particulate methane monooxygenase (pMMO) that catalyzes the conversion of methane to methanol. In many strains, the copper-requiring pMMO replaces the iron containing soluble enzyme (sMMO) when copper is sufficient. This enzyme is located within internalized membranes and a siderophore-like compound, methanobactin, plays a role in scavenging copper from the environment to supply pMMO (Kim et al. 2004). The 3D crystal structure of pMMO has recently been solved but some uncertainty about copper coordination remains (Lieberman and Rosenzweig 2005). It is formally possible that copper-methanobactin forms part of the active site of the enzyme and this has implications for whether or not free copper is recruited by the enzyme. It remains to be determined whether the membranous invaginations are contiguous with the periplasm and hence whether copper must traverse the cytosol to reach pMMO.

4.2.2
Trafficking to Thylakoids in Cyanobacteria

Much of the evidence in support of the copper trafficking pathway to thylakoids in *Synechocytis* PCC 6803 (Fig. 7) has already been reviewed (Tottey et al. 2005; Cavet et al. 2003b). In summary, the pathway involves two P_1-type ATPases, CtaA, and PacS (Tottey at al. 2001), and a metallochaperone that interacts with them both (Tottey et al. 2002). The metallochaperone has been variously called Atx1 or scAtx1 (Banci et al. 2004). Mutants in any of the three components have impaired cytochrome oxidase activity in isolated membranes and impaired switching from cytochrome c_6 to plastocyanin even in the presence of copper concentrations normally sufficient to sustain use of plastocyanin. Evidence of additivity in double mutants implies that the metallochaperone is not solely dependent upon CtaA for copper supply, suggesting either a second plasma-membrane copper importer and/or a role for

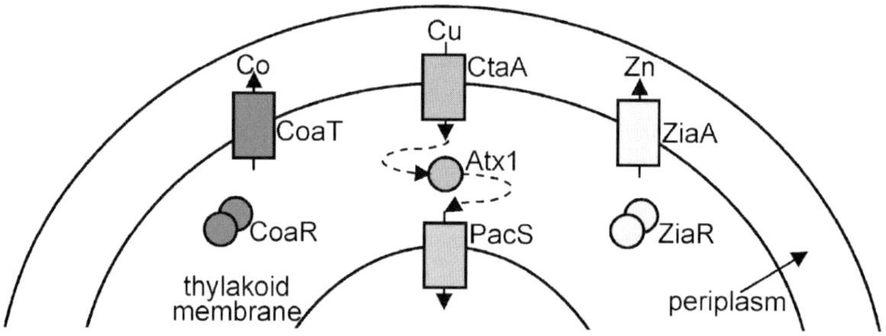

Fig. 7 Copper-trafficking pathway to thylakoids, multiple P-type ATPases and their regulation. Inner membrane located P-type ATPases CtaA and thylakoid membrane located PacS in conjunction with Cu chaperone Atx1, deliver Cu *dashed line* from the periplasm to the thylakoid compartment. Two further P-type ATPases, ZiaA (Zn) and CoaT (Co) export metal ions out of the cytosol into the periplasm. The elevation of CoaT production in response to increased Co is regulated by the MerR-like protein CoaR (Rutherford et al. 1999), while ZiaA production in response to elevated Zn is controlled by the SmtB-like protein ZiaR (Thelwell et al. 1998). Atx1 has been shown to interact with the cytosolic amino-terminal domain of CtaA and PacS, but does not interact with ZiaA or CoaT

the metallochaperone in recycling endogenous copper released, for example, at turnover of copper metalloenzymes.

CtaA from *Synechocystis* PCC 6803 and PAA1, the analogous transporter located at the chloroplast envelope (Abdel-Ghany et al. 2005), plus CopA from *Enterococcus hirae* (Magnani and Solioz 2005), are the only CPx-type ATPases that have been assigned roles in metal-ion import. This raises questions about their atypical mode of action. The crystal structures of calcium ATPases (Olesen et al. 2004) have been used to model the catalytic cycles of P_1-type ATPases. The manner of the predicted coupling between substrate-binding, changes in phosphorylation, switching between E_1 and E_2 states, and hence transport need not preclude import provided the effective K_{IN} is weaker than K_{OUT}. Where metallochaperones are present and interact with the ATPase, K_{IN} becomes some function of the interaction as considered previously (Cavet et al. 2003b).

5 Metallochaperones and the Specificity of Protein–Protein Interactions

The selection of partner proteins by metallochaperones can influence metal-specificity in vivo.

5.1 Metallochaperones as Copper Insertases

There are several reports describing the delivery of copper by metallochaperones (Huffman and O'Halloran 2001) (see chapter from Magnani and Solioz). In *E. hirae* CopZ assists copper acquisition by the transcriptional repressor CopY (Magnani and Solioz 2005) while in eukaryotes Cox17, CCS (Cobine et al. 2006; Culotta et al. 1997) and Atx1 (Pufahl et al. 1997) assist copper acquisition by mitochondrial cytochrome oxidase, cytosolic superoxide dismutase, and (potentially) P_1-type ATPases located at the *trans*-Golgi network, respectively, (Fig. 8a). In vitro studies have established that a soluble domain of the copper importer Ctr1 can interact with the Atx1 metallochaperone and the thermodynamics are favorable for copper donation from the importer to the metallochaperone (Xiao et al. 2004). It is known that yeast superoxide dismutase has femtomolar affinity for copper and yet in vivo fails to become active in the absence of CCS (Rae et al. 1999). This was taken to infer that the cytosol is devoid of freely exchangeable copper and that the chaperone was therefore essential to donate the metal ion to superoxide dismutase. More recently, it has been established that CCS is needed to form an essential, kinetically stable disulphide bond in superoxide dismutase and hence CCS activates superoxide dismutase via a mechanism other than solely copper donation (Fig. 8b) (Furukawa et al. 2004). Furthermore, in mammalian

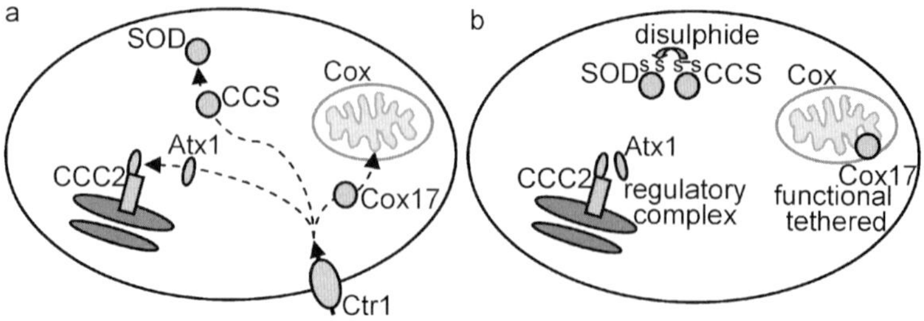

Fig. 8 Proposed modes of action of copper chaperones in a yeast cell. **a** Standard model. Cu is delivered from Cu importer Ctr1, to its cytosolic destination via soluble metal ion chaperones. Cox17 delivers Cu to cytochrome oxidase (Cox). CCS delivers Cu to Cu/Zn superoxide dismutase (SOD). Atx1 delivers Cu to the cytosolic domain of *trans*-Golgi network located P-type ATPase CCC2. Chaperone-Cu complex pathways represented as *dashed lines*. **b** Alternative model. CCS functions in the activation of SOD via the reallocation of a disulphide bond (Furukawa et al. 2004). The disulphide bond in CCS is transferred to form an intermolecular disulphide bond between CCS and SOD before being resolved to an intramolecular bond in SOD. This process is vital for the function and stability of SOD and allows for the sensing of local oxygen concentrations. The protein Cox17 is fully functional in the activation of Cox even when it is tethered to the mitochondrial inner membrane via a fusion to the transmembrane domain of the inner membrane protein, Sco2 (Maxfield et al. 2004). Atx1 has been shown to form a stable complex with the amino-terminal soluble domain of CCC2. The thermodynamic gradient of Cu transfer between these two proteins is shallow (Huffman and O'Halloran 2001). With the exception of cytosolic Cox17, the models are not mutually exclusive

species, CCS is not obligatory for superoxide dismutase activation, and the enzyme can obtain copper through an alternative pathway involving reduced glutathione (Carroll et al. 2004). Cox17 is capable of activating cytochrome oxidase even when tethered to the mitochondrial inner membrane, implying that copper trafficking across the cytoplasm is not its role (Maxfield et al. 2004). It remains to be established whether copper associated with Atx1 is donated to P_1-type ATPases for transport, or whether the metallochaperone solely forms regulatory complexes with the cognate soluble domains of the transporter. In summary, unequivocal data to confirm that copper metallochaperones act to traffic the metal ion across a copper-free cytosol (Fig. 8a) are currently lacking, although for both Cox17 and CCS there is compelling evidence that they do assist in the insertion of copper into the respective enzymes (Horng et al. 2004; Schmidt et al. 2000).

5.2
Atx1, Copper-, Zinc- and Cobalt-P_1-Type ATPases of *Synechocystis* PCC 6803

In common with copper metallochaperones such as CopZ (Banci et al. 2003) and eukaryotic Atx1 (Rosenzweig et al. 1999), the metallochaperone from *Synechocystis* PCC 6803 has a ferredoxin-like fold with two Cu(I) coordinating cysteine residues located on the first loop (Banci et al. 2004). A peculiar feature of the cyanobacterial protein is the presence of a histidine residue on loop five, which provides a third ligand to stabilize the copper ion (Borrelly et al. 2004a; Banci et al. 2004). The soluble domain of PacS also shows a ferredoxin-like global fold. The structure of each partner in the complex has been analyzed individually by mixing proteins in which only one of the pair is $^{1}H-^{15}N$ labelled and by monitoring NMR spectral changes coincident with hetero-dimer formation (Banci et al. 2006), to generate a structural model of the complex (Fig. 9). The complex is stabilized by a network of intermolecular hydrogen bonds and salt bridges plus, most importantly, intermolecular copper liganding. Removal of copper with the chelator bathocuproine disulphonate disrupts the hetero-dimer. The histidine residue is displaced from the copper coordination sphere in the complex providing a trigger for copper release from the metallochaperone to the ATPase. Additionally, in the docked complex, a glutamate residue from the metallochaperone forms a hydrogen bond to one of the copper-coordinating cysteine residues of the ATPase, consistent with this residue acting in proton abstraction to assist copper migration from the thiols of Atx1 to those of the transporter. These data imply a favored vector for copper transfer. The conformational changes in the cyanobacterial metallochaperone that occur upon contact with the thylakoid localized ATPase are consistent with copper release, providing support to the notion that the function of this metallochaperone is to supply copper to its metal-transporting partner.

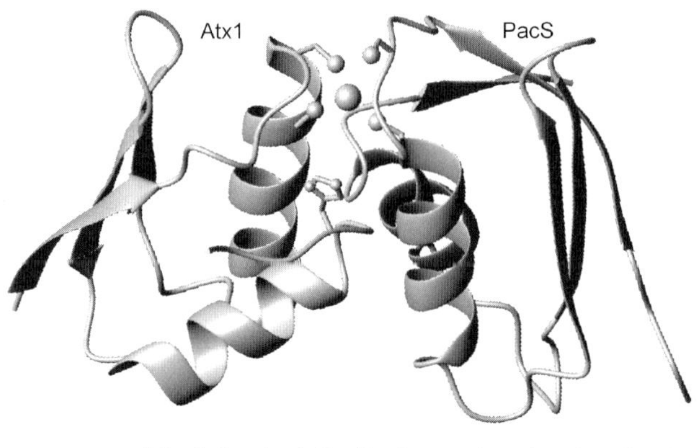

Fig. 9 Lowest-energy model of the Atx1-Cu-PacS complex showing the transfer of copper to thylakoids. The structure of the cytosolic amino-terminal domain of PacS, and of Atx1 has been solved. Both proteins form ferredoxin-like folds. In the Atx1-Cu complex, Cu is coordinated to Atx1 via two cysteine thiols and one histidine residue. Upon interaction with the amino terminal domain of PacS, the Cu ion is coordinated by both proteins. Atx1 is converted to an apo-like structure, with the histidine residue displaced from Cu in the complex. This structure provides evidence at the molecular level, of Cu-vectoring by a metallochaperone to its cognate target

There are two metal-transporting ATPases related to CtaA and PacS in *Synechocystis* PCC 6803, ZiaA, which exports zinc into the periplasm and CoaT, which exports cobalt (Thelwell et al. 1998; Rutherford et al. 1999). ZiaA has a soluble amino-terminal region that includes a deduced ferredoxin-like fold, analogous to the amino-terminal regions of PacS and CtaA. This domain binds zinc via the anticipated cysteine pair, but also binds Cu(I) under anaerobic conditions. Addition of zinc to copper-ZiaA$_N$ does not bleach ligand to metal charge transfer features that arise from copper-thiol bonding (Borrelly et al. 2004b). Furthermore, in the presence of equimolar Cu(I) and zinc, under anaerobic conditions, Cu(I) co-migrates with the protein by size exclusion chromatography while zinc migrates as a free ion. The metal-ion-binding site of the zinc transporter has a higher affinity for copper than for zinc, which is wholly consistent with predictions made by the Irving–Williams series. Atx1 does not interact with the amino-terminal region of ZiaA, nor its ferredoxin-fold subdomain in isolation, in bacterial two-hybrid assays (Tottey et al. 2002; Borrelly et al. 2004b). Thus, provided there is no freely available cytosolic copper in *Synechocystis* PCC 6803 the specificity of metallochaperone interaction can impose a kinetic barrier to prevent aberrant copper binding to ZiaA. An aspartate residue in ZiaA is in an equivalent location to an arginine in PacS. The arginine provides complementary charge, attractive to a glutamate residue of Atx1, while the aspartate will be repulsive. Once Atx1 docks with the ATPase, it must correctly orientate and undergo conformational changes to release copper to the transporter, and while PacS has evolved to facilitate these events, ZiaA has not. On this basis, metal specificity can be determined by the specificity of protein–protein contact surfaces, rather than the inherent metal affinities of the metal-binding sites. In theory, such a mechanism can also operate in the presence of freely exchangeable metal-ion pools, provided the metal-binding sites are not readily accessible to solvent. In this manner, the metallochaperone would operate to insert the correct metal ion into an otherwise protected binding site.

5.3 Substrate-Binding Proteins and Metal Partitioning in the Periplasm

In several sections of this review, selective protein interactions have been proposed as a mediator of metal specificity in vivo. However, this still requires that the metal ions are loaded onto the correct delivery routes in the first place. In Gram-negative bacteria this would involve metal-ion partitioning in the periplasm to the correct importers. ABC-type transporters are responsible for the import of several different metal ions and it has been noted that the residues of the metal-ion-binding domains of periplasmic substrate-binding proteins show remarkable similarity even when they contribute to the uptake of different metal ions (Banerjee et al. 2003). In *Synechocystis* PCC 6803, periplasmic substrate-binding proteins are known for manganese,

MntC (Bartsevich and Pakrasi 1995), zinc, ZnuA (Banerjee et al. 2003) and iron, FutA1 (Katoh et al. 2001b). All three are associated with the plasma-membrane (Fig. 10). In addition, there is a fourth related protein designated FutA2, which is abundant and soluble in the periplasm. As would be expected, MntC can bind zinc, raising a question about why this does not happen in vivo, and one proposal is that the lack of a highly charged mobile loop in MntC makes it less effective at scavenging zinc compared to ZnuA (Banerjee et al. 2003). We suggest that one advantage associated with a periplasm is that metal-ion specificity can now be dictated by relative affinities between proteins rather than absolute affinities. For example, provided ZnuA has a greater affinity for zinc than do MntC or FutA1, zinc will partition to ZnuA. In this respect, the proteins responsible for sequestering the most competitive metal ions for import in cyanobacteria are especially interesting.

In other bacteria, two-component sensors such as CusS (Munson et al. 2000) and PcoS (Brown et al. 1995) are known to detect periplasmic copper and regulate production of transporters that export copper from the

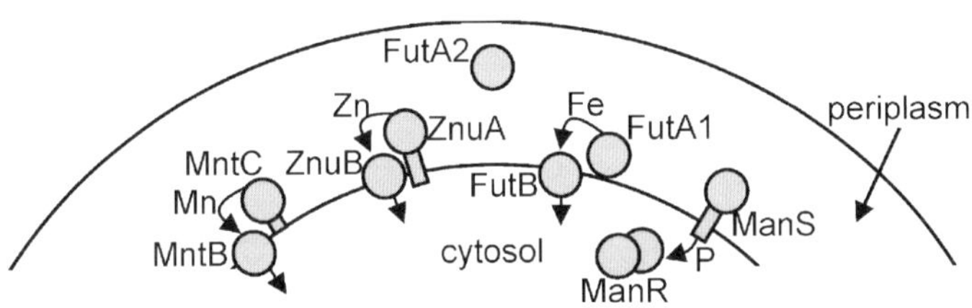

Fig. 10 Metal-binding proteins in the periplasm of *Synechocystis* PCC 6803. Within the periplasm are multiple proteins involved in the uptake and sensing of metal ions. In *Synechocystis* PCC 6803 Fe, Zn and Mn are take into the cytosol via representatives of the ABC transporter family of proteins. Periplasmic Mn is bound by MntC, which is tethered to the periplasmic membrane via a lipid anchor. The metal ion is transported into the cell via MntA and MntB (Bartsevich and Pakrasi 1995). ZnuA is the periplasmic substrate binding protein of the Zn transport system. The amino-terminus of ZnuA contains a hydrophobic membrane anchor. This protein has a highly charged and mobile loop, which has been proposed to act in the acquisition of Zn (Banerjee et al. 2003). Two proteins, FutA1 and FutA2, have been suggested to function as Fe periplasmic substrate binding proteins. FutA1 is associated with the periplasmic membrane, and is particularly abundant in salt stressed cells (Huang et al. 2006). FutA2 is one of the most abundant soluble proteins in the periplasm (Fulda et al. 2000). Deletion of FutA1 causes a 63% decrease of Fe uptake, while deletion of FutA2 caused only a 13% decrease when compared to the wild-type (Katoh et al. 2001a). The FutA associated Fe is transported into the cell via FutB and FutC (Katoh et al. 2001b). No Cu periplasmic substrate-binding protein has been identified to date. The periplasm is a site of metal ion partitioning. Two-component signal transduction pathways also operate in *Synechocystis* PCC 6803. ManS, which is attached to the plasma membrane via two transmembrane domains, is a Mn sensor (Ogawa et al. 2002). Upon metal-ion binding ManS phosphorylates (P) ManR, the cytosolic regulatory component. Metal-ion detection by related proteins may be important for sensing and minimizing the periplasmic free concentrations of competitive metal ions

periplasm. Surplus competitive metal-ion atoms may thus be avoided in the periplasm, allowing less competitive metal ions to bind to substrate-binding proteins such as MntC.

6
Prospective

Several opportunities for future studies have already been noted. One recurrent theme is the need for more enumeration: the number of atoms of each metal ion per cell, the number of molecules of sensor proteins per cell, the affinities of sensors for DNA, and the precise affinities of proteins of metal homeostasis for both the correct metal ions and for other metal ions. Comparisons of these values for a complement of metal-binding proteins will provide greater insight into how cells function as systems to encourage "correct" metal-protein speciation.

Acknowledgements We thank the BBSRC for funding and Peter Bowness for the data that is included in Fig. 4.

References

Abdel-Ghany SE, Muller-Moule P, Niyogi KK, Pilon M, Shikanai T (2005) Two P-type ATPases are required for copper delivery in *Arabidopsis thaliana* chloroplasts. Plant Cell 17:1233–1251

Akerlund T, Nordstrom K, Bernander R (1995) Analysis of cell size and DNA content in exponentially growing and stationary-phase batch cultures of *Escherichia coli*. J Bacteriol 177:6791–6797

Atanassova A, Zamble DB (2005) *Escherichia coli* HypA is a zinc metalloprotein with a weak affinity for nickel. J Bacteriol 187:4689–4697

Bairoch A (1993) A possible mechanism for metal-ion induced DNA-protein dissociation in a family of prokaryotic transcriptional regulators. Nucleic Acids Res 21:2515

Banci L, Bertini I, Del Conte R (2003) Solution structure of apo CopZ from *Bacillus subtilis*: further analysis of the changes associated with the presence of copper. Biochemistry 42:13422–13428

Banci L, Bertini I, Ciofi-Baffoni S, Su XC, Borrelly GP, Robinson NJ (2004) Solution structures of a cyanobacterial metallochaperone: insight into an atypical copper-binding motif. J Biol Chem 279:27502–27510

Banci L, Bertini I, Ciofi-Baffoni S, Kandias NG, Robinson NJ, Spyroulias GA, Su XC, Tottey S, Vanarotti M (2006) The delivery of copper for thylakoid import observed by NMR. Proc Natl Acad Sci USA 103:8320–8325

Banerjee S, Wei B, Bhattacharyya-Pakrasi M, Pakrasi HB, Smith TJ (2003) Structural determinants of metal specificity in the zinc transport protein ZnuA from *Synechocystis* 6803. J Mol Biol 333:1061–1069

Bartsevich VV, Pakrasi HB (1995) Molecular identification of an ABC transporter complex for manganese: analysis of a cyanobacterial mutant strain impaired in the photosynthetic oxygen evolution process. EMBO J 14:1845–1853

Blindauer CA, Harrison MD, Parkinson JA, Robinson AK, Cavet JS, Robinson NJ, Sadler PJ (2001) A metallothionein containing a zinc finger within a four-metal cluster protects a bacterium from zinc toxicity. Proc Natl Acad Sci USA 98:9593–9598

Blindauer CA, Harrison MD, Robinson AK, Parkinson JA, Bowness PW, Sadler PJ, Robinson NJ (2002) Multiple bacteria encode metallothioneins and SmtA-like zinc fingers. Mol Microbiol 45:1421–1432

Blindauer CA, Polfer NC, Keiper SE, Harrison MD, Robinson NJ, Langridge-Smith PR, Sadler PJ (2003) Inert site in a protein zinc cluster. J Am Chem Soc 125:3226–3227

Blindauer CA, Sadler PJ (2005) How to hide zinc in a small protein. Acc Chem Res 38:62–69

Borrelly GP, Blindauer CA, Scmid R, Butler CS, Cooper CE, Harvey I, Sadler PJ, Robinson NJ (2004a) A novel copper site in a cyanobacterial metallochaperone. Biochem J 378:293–297

Borrelly GP, Rondet SA, Tottey S, Robinson NJ (2004b) Chimeras of P-type ATPases and their transcriptional regulators: contributions of a cytosolic amino-terminal domain to metal specificity. Mol Microbiol 53:217–227

Brocklehurst KR, Hobman JL, Lawley B, Blank L, Marshall SJ, Brown NL, Morby AP (1999) ZntR is a Zn(II)-responsive MerR-like transcriptional regulator of *zntA* in *Escherichia coli*. Mol Microbiol 31:893–902

Brocks JJ, Logan GA, Buick R, Summons RE (1999) Archaen molecular fossils and the early rise of eukaryotes. Science 285:1033–1036

Brown NL, Barrett SR, Camakaris J, Lee BT Rouch DA (1995) Molecular genetics and transport analysis of the copper-resistance determinant (pco) from *Escherichia coli* plasmid pRJ1004. Mol Microbiol 17:1153–1166

Bryson B (2004) A short history of nearly everything. Black Swan, London

Buckler DR, Zhou Y, Stock AM (2002) Evidence of intradomain and interdomain flexibility in an OmpR/PhoB homolog from *Thermatoga maritime*. Structure 10:153–164

Busenlehner LS, Weng TC, Penner-Hahn JE, Giedroc DP (2002) Elucidation of primary (alpha(3)N) and vestigial (alpha(5)) heavy metal-binding sites in *Staphylococcus aureus* pI258 CadC: evolutionary implications for metal ion selectivity of ArsR/SmtB metal sensor proteins. J Mol Biol 319:685–701

Busenlehner LS, Pennella MA, Giedroc DP (2003) The SmtB/ArsR family of metalloregulatory transcriptional repressors: structural insights into prokaryotic metal resistance. FEMS Microbiol Rev 27:131–143

Carroll MC, Girouard JB, Ulloa JL, Subramaniam JR, Wong PC, Valentine JS, Culotta VC (2004) Mechanisms for activating Cu- and Zn-containing superoxide dismutase in the absence of the CCS Cu chaperone. Proc Natl Acad Sci USA 101:5964–5969

Cavet JS, Meng W, Pennella MA, Appelhoff RJ, Giedroc DP, Robinson NJ (2002) A nickel-cobalt-sensing ArsR-SmtB family repressor. Contributions of cytosol and effector binding sites to metal selectivity. J Biol Chem 277:38441–38448

Cavet JS, Graham AI, Meng W, Robinson NJ (2003a) A cadmium-lead-sensing ArsR-SmtB repressor with novel sensory sites. Complementary metal discrimination by NmtR and CmtR in a common cytosol. J Biol Chem 278:44560–44566

Cavet JS, Borrelly GP, Robinson NJ (2003b) Zn, Cu and Co in cyanobacteria: selective control of metal availability. FEMS Microbiol Rev 27:165–181

Changela A, Chen K, Xue Y, Holschen J, Outten CE, O'Halloran TV, Mondragon A (2003) Molecular basis of metal-ion selectivity and zeptomolar sensitivity by CueR. Science 301:1383–1387

Chivers PT, Tahirov TH (2005) Structure of *Pyrococcus horikoshii* NikR: nickel sensing and implications for the regulation of DNA recognition. J Mol Biol 348:597–607

Cobine PA, Pierrel F, Winge DR (2006) Copper trafficking to the mitochondrion and assembly of copper metalloenzymes. Biochim Biophys Acta 1763:759–772
Cook WJ, Kar SR, Taylor KB, Hall LM (1998) Crystal structure of the metallothionein repressor SmtB: a model for metalloregulatory proteins. J Mol Biol 275:337–346
Croft MT, Lawrence AD, Raux-Deery E, Warren MJ, Smith AG (2005) Algae acquire vitamin B_{12} through a symbiotic relationship with bacteria. Nature 438:90–93
Culotta VC, Klomp LW, Strain J, Casareno RL, Krems B, Gitlin JD (1997) The copper chaperone for superoxide dismutase. J Biol Chem 272:23469–23472
Daniels MJ, Turner-Cavet JS, Selkirk R, Sun H, Parkinson JA, Sadler PJ, Robinson NJ (1998) Coordination of Zn^{2+} (and Cd^{2+}) by prokaryotic metallothionein. Involvement of His-imidazole. J Biol Chem 273:22957–22961
Eicken C, Pennella MA, Chen X, Koshlap KM, VanZile ML, Sacchettini JC, Giedroc DP (2003) A metal-ligand-mediated intersubunit allosteric switch in related SmtB/ArsR zinc sensor proteins. J Mol Biol 333:683–695
Eisenstein RS (2000) Iron regulatory proteins and the molecular control of mammalian iron metabolism. Annu Rev Nutr 20:627–662
Endo G, Silver S (1995) CadC, the transcriptional regulatory protein of the cadmium resistance system of *Staphylococcus aureus* plasmid pI258. J Bacteriol 177:4437–4441
Ferguson-Miller S, Babcock GT (1996) Heme/copper terminal oxidases. Chem Rev 96:2889–2908
Finney LA, O'Halloran TV (2003) Transition metal speciation in the cell: insights from the chemistry of metal ion receptors. Science 300:931–936
Fraústo da Silva JJR, Williams RJP (2002) The biological chemistry of the elements: the inorganic chemistry of life. Clarendon Press, Oxford
Fulda S, Huang F, Hagemann M, Norling B (2000) Proteomics of *Synechocystis* sp. Strain PCC 6803. Identification of periplasmic proteins in cells grown at low and high salt concentrations. Eur J Biochem 267:5900–5907
Furukawa Y, Torres AS, O'Halloran TV (2004) Oxygen-induced maturation of SOD1: a key role for disulfide formation by the copper chaperone CCS. EMBO J 23:2872–2881
Glasfeld A, Guedon E, Helmann JD, Brennan RG (2003) Structure of the manganese-bound transport regulator of *Bacillus subtilis*. Nat Struct Biol 10:652–657
Guedon E, Helmann JD (2003) Origins of metal ion selectivity in the DtxR/MntR family of metalloregulators. Mol Microbiol 48:495–506
Harvie DR, Andreini C, Cavallaro G, Meng W, Connolly BA, Yoshida K, Fujita Y, Harwood CR, Radford DS, Tottey S, Cavet JS, Robinson NJ (2006) Predicting metals sensed by ArsR-SmtB repressors: allosteric interference by a non-effector metal. Mol Microbiol 59:1341–1356
Higham DP, Sadler PJ, Scawen MD (1984) Cadmium-resistant *Pseudomonas putida* synthesizes novel cadmium proteins. Science 225:1043–1046
Higham DP, Sadler PJ, Scawen MD (1986) Cadmium-binding proteins in *Pseudomonas*: pseudothioneins. Environ Health Perspect 65:5–11
Hintze KJ, Theil EC (2006) Cellular regulation and molecular interactions of the ferritins. Cell Mol Life Sci 63:591–600
Horng YC, Cobine PA, Maxfield AB, Carr HS, Winge DR (2004) Specific copper transfer from the Cox17 metallochaperone to both Sco1 and Cox11 in the assembly of yeast cytochrome C oxidase. J Biol Chem 279:35334–35340
Huang F, Fulda S, Hagemann M, Norling B (2006) Proteomic screening of salt-stress-induced changes in plasma membranes of *Synechocystis* sp. strain PCC 6803. Proteomics 6:910–920

Huckle JW, Morby AP, Turner JS, Robinson NJ (1993) Isolation of a prokaryotic metalothionein locus and analysis of transcriptional control by trace metal ions. Mol Microbiol 7:177–187

Huffman DL, O'Halloran TV (2001) Function, structure, and mechanism of intracellular copper-trafficking proteins. Annu Rev Biochem 70:677–701

Jordan P, Fromme P, Witt HT, Klukas O, Saenger W, Krauss N (2001) Three-dimensional structure of cyanobacterial photosystem I at 2.5 Å resolution. Nature 411:909–917

Katoh H, Hagino N, Grossman AR, Ogawa T (2001a) Genes essential to iron transport in the cyanobacterium *Synechocystis* sp. strain PCC 6803. J Bacteriol 183:2779–2784

Katoh H, Hagino N, Ogawa T (2001b) Iron-binding activity of FutA1 subunit of an ABC-type iron transporter in the cyanobacterium *Synechocystis* sp. strain PCC 6803. Plant Cell Physiol 42:823–827

Keren N, Kidd MJ, Penner-Hahn JE, Pakrasi HB (2002) A light-dependent mechanism for massive accumulation of manganese in the photosynthetic bacterium *Synechocystis* sp. PCC 6803. Biochemistry 41:15085–15092

Keren N, Aurora R, Pakrasi HB (2004) Critical roles of bacterioferritins in iron storage and proliferation of cyanobacteria. Plant Physiol 135:1666–1673

Kim HJ, Graham DW, DiSpirito AA, Alterman MA, Galeva N, Larive CK, Asunskis D, Sherwood PM (2004) Methanobactin, a copper-acquisition compound from methane-oxidizing bacteria. Science 305:1612–1615

Kondorosi E, Pierre M, Cren M, Haumann U, Buire M, Hoffmann B, Schell J, Kondorosi A (1991) Identification of NolR, a negative transacting factor controlling the nod regulon in *Rhizobium meliloti*. J Mol Biol 222:885–896

Kropat J, Tottey S, Birkenbihl RP, Depege N, Huijser P, Merchant S (2005) A regulator of nutritional copper signaling in *Chlamydomonas* is an SBP domain protein that recognizes the GTAC core of copper response element. Proc Natl Acad Sci USA 102:18730–18735

Kuper J, Llamas A, Hecht H-J, Mendel RR, Schwarz G (2004) Structure of the molybdopterin-bound Cnx1G domain links molybdenum and copper metabolism. Nature 430:803–806

Kurisu G, Zhang H, Smith JL, Cramer WA (2003) Structure of the cytochrome *b6f* complex of oxygenic photosynthesis: tuning the cavity. Science 302:1009–1014

Leech HK, Raux E, McLean KJ, Munro AW, Robinson NJ, Borrelly GP, Malten M, Jahn D, Rigby SE, Heathcote P, Warren MJ (2003) Characterization of the cobaltochelatase CbiXL: evidence for a 4Fe – 4S center housed within an MXCXXC motif. J Biol Chem 278:41900–41907

Lichtlen P, Schaffner W (2001) Putting its fingers on stressful situations: the heavy metal-regulatory transcription factor MTF-1. Bioessays 23:1010–1017

Lieberman RL, Rosenzweig AC (2005) Crystal structure of a membrane-bound metalloenzyme that catalyses the biological oxidation of methane. Nature 434:177–182

Liu T, Golden JW, Giedroc DP (2005) A zinc (II)/lead(II)/cadmium(II)-inducible operon from the cyanobacterium *Anabaena* is regulated by AztR, an alpha3N ArsR/SmtB metalloregulator. Biochemistry 44:8673–8683

Magnani D, Solioz M (2005) Copper chaperone cycling and degradation in the regulation of the *cop* operon of *Enterococcus hirae*. Biometals 18:407–412

Martin W, Rujan T, Richly E, Hansen A, Cornelsen S, Lins T, Leister D, Stoebe B, Hasegawa M, Penny D (2002) Evolutionary analysis of *Arabidopsis*, cyanobacterial, and chloroplast genomes reveals plastid phylogeny and thousands of cyanobacterial genes in the nucleus. Proc Natl Acad Sci USA 99:12246–12251

Maxfield AB, Heaton DN, Winge DR (2004) Cox17 is functional when tethered to the mitochondrial inner membrane. J Biol Chem 279:5072–5080

Merchant S (1998) Synthesis of metalloproteins involved in photosynthesis: plastocyanin and cytochromes. In: Rochaix J-D, Goldschmidt-Clermont M, Merchant S (eds) The Molecular Biology of Chloroplasts and Mitochondria in *Chlamydomonas*. Kluwer, Dordrecht, pp 597–611

Molina-Heredia FP, Wastl J, Navarro JA, Bendall DS, Hervas M, Howe CJ, De La Rosa MA (2003) Photosynthesis: a new function for an old cytochrome? Nature 424:33–34

Morby AP, Turner JS, Huckle JW, Robinson NJ (1993) SmtB is a metal-dependent repressor of the cyanobacterial metallothionein gene *smtA*: identification of a Zn inhibited DNA-protein complex. Nucleic Acids Res 21:921–925

Munson GP, Lam DL, Outten FW, O'Halloran TV (2000) Identification of a copper-responsive two-component system on the chromosome of *Escherichia coli* K-12. J Bacteriol 182:5864–5871

Niederhoffer EC, Naranjo CM, Bradley KL, Fee JA (1990) Control of *Escherichia coli* superoxide dismutase (*sodA* and *sodB*) genes by the ferric uptake regulation (*fur*) locus. J Bacteriol 172:1930–1938

Odermatt A, Solioz M (1995) Two trans-acting metalloregulatory proteins controlling expression of the copper-ATPases of *Enterococcus hirae*. J Biol Chem 270:4349–4354

Ogawa T, Bao DH, Katoh H, Shibata M, Pakrasi HB, Bhattacharyya-Pakrasi M (2002) A two-component signal transduction pathway regulates manganese homeostasis in *Synechocystis* 6803, a photosynthetic organism. J Biol Chem 277:28981–28986

Olafson RW, McCubbin WD, Kay CM (1988) Primary- and secondary-structural analysis of a unique prokaryotic metallothionein from a *Synechococcus* sp. cyanobacterium. Biochem J 251:691–699

Olesen C, Sørensen TLM, Nielsen RC, Møller JV, Nissen P (2004) Dephosphorylation of the calcium pump coupled to counterion occlusion. Science 306:2251–2255

Outten CE, Outten FW, O'Halloran TV (1999) DNA distortion mechanism for transcriptional activation by ZntR, a Zn (II)-responsive MerR homologue in *Escherichia coli*. J Biol Chem 274:37517–37524

Outten CE, O'Halloran TV (2001) Femtomolar sensitivity of metalloregulatory proteins controlling zinc homeostasis. Science 292:2488–2492

Panina EM, Mironov AA, Gelfand MS (2003) Comparative genomics of bacterial zinc regulons: enhanced ion transport, pathogenesis, and rearrangement of ribosomal proteins. Proc Natl Acad Sci USA 100:9912–9917

Pantopoulos K (2004) Iron metabolism and the IRE/IRP regulatory system: an update. Ann NY Acad Sci 1012:1–13

Patzer SI, Hantke K (1999) The ZnuABC high-affinity zinc uptake system and its regulator Zur in *Escherichia coli*. Mol Microbiol 32:887–889

Pennella MA, Shokes JE, Cosper NJ, Scott RA, Giedroc DP (2003) Structural elements of metal selectivity in metal sensor proteins. Proc Natl Acad Sci USA 100:3713–3718

Pohl E, Haller JC, Mijovilovich A, Meyer-Klauke W, Garman E, Vasil ML (2003) Architecture of a protein central to iron homeostasis: crystal structure and spectroscopic analysis of the ferric uptake regulator. Mol Microbiol 47:903–915

Pufahl RA, Singer CP, Peariso KL, Lin SJ, Schmidt PJ, Fahrni CJ, Culotta VC, Penner-Hahn JE, O'Halloran TV (1997) Metal ion chaperone function of the soluble Cu(I) receptor Atx1. Science 278:853–856

Qiu X, Verlinde CL, Zhang S, Schmitt MP, Holmes RK, Hol WG (1995) Three-dimensional structure of the diphtheria toxin repressor in complex wih divalent cation co-repressors. Structure 3:87–100

Radford DS, Kihlken MA, Borrelly GP, Harwood CR, Le Brun NE, Cavet JS (2003) CopZ from *Bacillus subtilis* interacts in vivo with a copper exporting CPx-type ATPase CopA. FEMS Microbiol Lett 220:105–112

Rae TD, Schmidt PJ, Pufahl RA, Culotta VC, O'Halloran TV (1999) Undetectable intracellular free copper: the requirement of a copper chaperone for superoxide dismutase. Science 284:805–808

Raven JA, Evans MCW, Korb RE (1999) The role of trace metals in photosynthetic electron transport in O_2-evolving organisms. Photosynth Res 60:111–149

Rensing C, Grass G (2003) *Escherichia coli* mechanisms of copper homeostasis in a changing environment. FEMS Microbiol Rev 27:197–213

Rivera MC, Lake JA (2004) The ring of life provides evidence for a genome fusion origin of eukaryotes. Nature 431:152–155

Robinson NJ, Gupta A, Fordham-Skelton AP, Croy RR, Whitton BA, Huckle JW (1990) Prokaryotic metallothionein gene characterisation and expression: chromosome crawling by ligation-mediated PCR. Proc Biol Sci 242:241–247

Robinson NJ, Bird AJ, Turner JS (1997) Metallothionein gene regulation in cyanobacteria. In: Silver S, Walden W (eds) Metal Ions in Gene Regulation. Chapman and Hall, New York, pp 372–397

Robinson NJ, Whitehall SK, Cavet JS (2001) Microbial metallothioneins. Adv Microb Physiol 44:183–213

Rosenzweig AC, Huffman DL, Hou MY, Wernimont AK, Pufahl RA, O'Halloran TV (1999) Crystal structure of the Atx1 metallochaperone protein at 1.02 Å resolution. Structure 7:605–617

Rowe JL, Starnes GL, Chivers PT (2005) Complex transcriptional control links NikABCDE-dependent nickel transport with hydrogenase expression in *Escherichia coli*. J Bacteriol 187:6317–6323

Rutherford JC, Cavet JS, Robinson NJ (1999) Cobalt-dependent transcriptional switching by a dual-effector MerR-like protein regulates a cobalt-exporting variant CPx-type ATPase. J Biol Chem 274:25827–25832

Rutherford JC, Bird AJ (2004) Metal-responsive transcription factors that regulate iron, zinc and copper homeostasis in eukaryotic cells. Eukaryotic Cell 3:1–13

Schmidt PJ, Kunst C, Culotta VC (2000) Copper activation of superoxide dismutase 1 (SOD1) in vivo. Role for protein–protein interactions with the copper chaperone for SOD1. J Biol Chem 275:33771–33776

Schwede T, Kopp J, Guex N, Peitsch MC (2003) SWISS-MODEL: an automated protein homology-modeling server. Nucleic Acids Res 31:3381–3385

Shi J, Lindsay WP, Huckle JW, Morby AP, Robinson NJ (1992) Cyanobacterial metallothionein gene expressed in *Escherichia coli*: metal-binding properties of the expressed protein. FEBS Lett 303:159–163

Shi W, Wu J, Rosen BP (1994) Identification of a putative metal binding site in a new family of metalloregulatory proteins. J Biol Chem 269:19826–19829

Shi W, Dong J, Scott RA, Ksenzenko MY, Rosen BP (1996) The role of arsenic-thiol interactions in metalloregulation of the *ars* operon. J Biol Chem 271:9291–9297

Shipman LW, Li D, Roessner CA, Scott AI, Sacchettini JC (2001) Crystal structure of precorrin-8x methyl mutase. Structure 9:587–596

Solioz M, Stoyanov JV (2003) Copper homeostasis in *Enterococcus hirae*. FEMS Microbiol Rev 27:183–195

Spence E, Sarcina M, Ray N, Moller SG, Mullineaux CW, Robinson C (2003) Membrane-specific targeting of green fluorescent protein by the Tat pathway in the cyanobacterium *Synechocystis* PCC6803. Mol Microbiol 48:1481–1489

Stoyanov JV, Hobman JL, Brown NL (2001) CueR (YbbI) of *Escherichia coli* is a MerR family regulator controlling expression of the copper exporter CopA. Mol Microbiol 39:502–511

Terry N (1983) Limiting factors in photosynthesis: IV. Iron stress-mediated changes in light-harvesting and electron transport capacity and its effects on photosynthesis in vivo. Plant Physiol 71:855–860

Thelwell C, Robinson NJ, Turner-Cavet JS (1998) An SmtB-like repressor from *Synechocystis* PCC 6803 regulates a zinc exporter. Proc Natl Acad Sci USA 95:10728–10733

Tottey S, Rich PR, Rondet SAM, Robinson NJ (2001) Two Menkes-type ATPases supply copper for photosynthesis in *Synechocystis* PCC 6803. 276:19999–20004

Tottey S, Rondet SA, Borrelly GP, Robinson PJ, Rich PR, Robinson NJ (2002) A copper metallochaperone for photosynthesis and respiration reveals metal-specific targets, interaction with an importer, and alternative sites for copper acquisition. J Biol Chem 277:5490–5497

Tottey S, Harvie DR, Robinson NJ (2005) Understanding how cells allocate metals using metal sensors and metallochaperones. Acc Chem Res 38:775–783

Turner JS, Robinson NJ (1995) Cyanobacterial metallothioneins: biochemistry and molecular genetics. J Ind Microbiol 14:119–125

Turner JS, Glands PD, Samson AC, Robinson NJ (1996) Zn^{2+}-sensing by the cyanobacterial metallothionein repressor SmtB: different motifs mediate metal-induced protein-DNA dissociation. Nucleic Acids Res 24:3714–3721

Tseng CP (1997) Regulation of fumarase (*fumB*) gene expression in *Escherichia coli* in response to oxygen, iron and heme availability: role of the *arcA, fur,* and *hemA* gene products. FEMS Microbiol Lett 157:67–72

Van de Meene AM, Hohmann-Marriott MF, Vermaas WF, Roberson RW (2006) The three-dimensional structure of the cyanobacterium *Synechocystis* sp. PCC 6803. Arch Microbiol 184:259–270

VanZile ML, Cosper NJ, Scott RA, Giedroc DP (2000) The zinc metalloregulatory protein *Synechococcus* PCC7942 SmtB binds a single zinc ion per monomer with high affinity in a tetrahedral coordination geometry. Biochemistry 39:11818–11829

VanZile ML, Chen X, Giedroc DP (2002) Allosteric negative regulation of *smt* O/P binding of the zinc sensor, SmtB, by metal ions: a coupled equilibrium analysis. Biochemistry 41:9776–9786

Weigel M, Varotto C, Pesaresi P, Finazzi G, Rappaport F, Salamini F, Leister D (2003) Plastocyanin is indespensable for photosynthetic electron flow in *Arabidopsis thaliana.* J Biol Chem 278:31286–31289

White A, Ding X, van der Spek JC, Murphy JR, Ringe D (1998) Structure of the metal-ion-activated diphtheria toxin repressor/tox operator complex. Nature 394:502–506

Williams SG, Attridge SR, Manning PA (1993) The transcriptional activator HlyU of *Vibrio cholerae*: nucleotide sequence and role in virulence gene expression. Mol Microbiol 9:751–760

Wu J, Rosen BP (1991) The ArsR protein is a trans-acting regulatory protein. Mol Microbiol 5:1331–1336

Xiao Z, Loughlin F, George GN, Howlett GJ, Wedd AG (2004) C-terminal domain of the membrane copper transporter Ctr1 from *Saccharomyces cerevisiae* binds four Cu(I) ions as a cuprous-thiolate polynuclear cluster: sub-femtomolar Cu(I) affinity of three proteins involved in copper trafficking. J Am Chem Soc 126:3081–3090

Ye J, Kandegedara A, Martin P, Rosen BP (2005) Crystal structure of the *Staphylococcus aureus* pI258 CadC Cd(II)/Pb(II)/Zn(II)-responsive repressor. J Bacteriol 187:4214–4221

Yoon HS, Hackett JD, Pinto G, Bhattacharya D (2002) The single, ancient origin of chromist plastids. Proc Natl Acad Sci USA 99:15507–15512

Zheng M, Doan B, Schneider TD, Storz G (1999) OxyR and SoxRS regulation of *fur*. J Bacteriol 181:4639–4643

Zouni A, Witt HT, Kern J, Fromme P, Krauss N, Saenger W, Orth P (2001) Crystal structure of photosystem II from *Synechococcus elongatus* at 3.8 Å resolution. Nature 409:739–743

Microbiol Monogr (6)
D. H. Nies, S. Silver: Molecular Microbiology of Heavy Metals
DOI 10.1007/7171_2006_073/Published online: 1 February 2007

Metalloregulators: Arbiters of Metal Sufficiency

John D. Helmann (✉) · Sumarin Soonsanga · Scott Gabriel

Department of Microbiology,
Cornell University, 372 Wing Hall, Ithaca, NY 14853-8101, USA
jdh9@cornell.edu

Abstract Metal homeostasis relies on the ability of metalloregulatory proteins to coordinate the expression of transport and storage functions. Metalloregulatory proteins can be divided into two major groups: those that regulate the uptake of essential metals (the Fur, DtxR/MntR, and NikR families) and those that regulate metal efflux and detoxification mechanisms (the ArsR/SmtB and MerR families). Within each metalloregulator protein family, there is a tremendous diversity in metal selectivity and the corresponding biological responses. The availability of at least one protein structure from each family is beginning to provide insights into the origins of metal selectivity. Biochemical measurements of metal ion selectivity and affinity provide a window into the ambient metal ion conditions within the cytosol: metalloregulators that sense nutrient metals must be poised to bind the metal ion once the essential functional sites are saturated, but before adventitious associations begin to interfere with cellular function. Similarly, sensors of metal ion excess, whether for non-essential toxic metals or nutrient metals, must respond to metals, at levels below those that will inhibit or prevent cell growth, to activate appropriate defensive measures. Recent insights highlight the global nature of stress responses elicited by metal ion deficiency. In addition to the expected derepression of high affinity uptake systems, metal ion starvation leads to a large-scale remodeling of the proteome that includes: (i) metal-sparing, (ii) metal-substitution, and (iii) metal-mobilization responses.

1
Introduction: Metalloregulation and its Guises

Metal ions are essential cofactors for perhaps one-third of all enzymes and are essential for life (Holm et al. 1996). To facilitate the acquisition of essential metal ions from the environment, bacteria have evolved a suite of high affinity transport systems. Typically, these uptake systems are induced when a particular metal ion is limiting for growth. Conversely, when metal ions are in excess, bacteria respond by the induction of efflux pumps or sequestration proteins to mitigate the toxic effects of metal ion overload. The regulation of metal transport and storage proteins is tightly controlled, most often at the transcriptional level. Here, we present an overview of the major families of metalloregulatory proteins operative in bacteria, with an emphasis on those regulating the uptake of nutrient metal ions. We then draw some general conclusions regarding the biological roles of metalloregulators and their remarkable ability to discriminate against chemically similar metal ions in the complex milieu of the cytoplasm.

Metalloregulation refers to the regulation of gene expression in direct response to metal ion availability (O'Halloran 1993). Metalloregulation, as defined here, does not include systems that sense metal ion complexes, changes in metal ion valence state or coordination chemistry, or indirect effects of perturbed metal homeostasis. In contrast with the complex regulatory cascades that abound in eukaryotic cells, metalloregulation is often remarkably simple. For example, many genes involved in metal ion uptake are repressed by a protein that directly binds the cognate metal ion as a co-repressor (Fur /DtxR/ NikR families), thereby providing a direct "read" on the level of available metal ion in the cell. In other cases, metal ion resistance, efflux, or detoxification genes are induced when the cognate metal ion triggers derepression of a repressor (ArsR/SmtB family) or activates an activator protein (MerR family). Naturally, there are numerous variations on these common themes and, in some cases, proteins can function as both transcriptional repressors and activators in response to metal-ion binding.

Not all metalloregulation involves transcription factors. Regulation may also be exerted at the level of translation or mRNA stability. One of the best characterized examples of such a post-transcriptional control mechanism is the iron-responsive element (IRE)-binding protein in eukaryotes (Hentze et al. 2004). IRE-BP is an enzyme (aconitase) containing a 4Fe-4S cluster whose integrity is sensitive to iron availability. When iron is limiting, the IRE-BP is inactive as an enzyme, but binds with high affinity to RNA stem loop structures known as IREs. This binding serves to stabilize the mRNA for the transferrin receptor (thereby increasing iron uptake) and blocks translation of the iron storage protein, ferritin. Recent results are consistent with the notion that aconitase may also play a role in the post-transcriptional control of gene expression in bacterial systems (Alen and Sonenshein 1999; Hantke 2001). Although operative at a post-transcriptional level, this clearly fits the definition of metalloregulation since it is responsive directly and specifically to iron availability. In select cases, it may also be possible for metal ions to directly control RNA structure by modulating RNA structure. The first example of such a metal-sensing riboswitch is the Mg^{2+}-responsive leader region of the *Salmonella enterica* serovar Typhimurium *mgtA* gene encoding a Mg^{2+} uptake transporter (Cromie et al. 2006).

In some cases, metalloregulators control gene expression through an intermediary regulator. For example, genes for the synthesis and uptake of the ornibactin siderophore in *Burkholderia cenocepacia* are expressed only in the presence of the OrbS σ factor, which is itself repressed by Fur (Agnoli et al. 2006). Other examples include genes regulated by the *trans*-acting RyhB small RNA under the control of Fur in *Escherichia coli* (Masse and Gottesman 2002). In this system, RyhB acts to destabilize mRNA molecules that encode iron-rich enzymes (e.g., succinate dehydrogenase) and storage proteins (ferritin) and, as a result, these proteins are (indirectly) positively regulated by Fur (Masse and Arguin 2005; Masse et al. 2005). Despite the fact that these regulatory systems have an intermediary, these genes are still regulated in a highly metal-selective manner. We refer to such genes as *indirect targets* of metalloregulation.

The dramatic effects of metal ions on cells have become clearer with the widespread use of transcriptome and proteome approaches to characterization of metal starvation and stress responses (see chapter by Hobman et al. in this volume). Often, the genes that are most strongly and rapidly regulated in response to changes in metal ion status are controlled by metalloregulatory systems. However, numerous other effects quickly become apparent. Depletion or toxic levels of a particular metal ion may interfere with the function of one or more metal-dependent enzymes and this, in turn, can lead to the activation of stress responses. Indeed, the inclusion of many genes in metal starvation and stress stimulons may result from such indirect effects. In *Bacillus subtilis* we have noted, for example, that elevated Mn^{2+} activates both the σ^B-dependent general stress response and the nitrogen starvation response regulated by TnrA (Guedon et al. 2003). These effects are postulated to reflect the role of Mn^{2+} as a cofactor for regulatory protein phosphatases that control σ^B and the altered regulatory properties of the Mn-cofactored form of glutamine synthase, respectively. Similarly, regulation of the Spo0F phosphotransfer protein by metal ions has recently been suggested to influence the efficiency of sporulation in *B. subtilis* (Kojetin et al. 2005). We view these as collateral effects of perturbed metal ion levels, rather than as part of a specific, adaptive response. Sometimes, these secondary effects can be distinguished by monitoring transcriptional changes as a function of time: genes directly controlled by metalloregulatory proteins often show rapid changes in mRNA levels, whereas those resulting from metabolic perturbations may not display altered regulation until later.

2 Metal Uptake Regulators

Most bacterial metalloregulatory proteins fall into one of five families of DNA-binding proteins. These families are each named for the first and prototypic family member(s): Fur, DtxR, NikR, MerR, and ArsR. Although proteins from each of these families are homologs (they are derived from a common ancestor), representatives may respond to a variety of different metals (Table 1). One of the challenges for biochemists is to explain how often subtle changes in protein sequence can contribute to alterations in metal selectivity. For a comprehensive overview of this aspect of metalloregulation the reader is referred to the recent review by Pennella and Giedroc (2005). In addition to these families, new regulators are being discovered all the time. Other proteins likely to be regulated by direct interactions with metal ions include AdcR (Loo et al. 2003) and RirA (Todd et al. 2005).

In general, there is a good correspondence between the metal-selectivity of metalloregulators and the functional roles of the regulated proteins. For example, both DtxR and Fur respond with high selectivity to iron in vivo, as befitting their roles as repressors of iron uptake functions. Similarly, MerR is highly specific for the mercuric ion, to which the MerR-regulated gene products confer resistance. There are numerous instances, however, where there is apparently inappropriate regulation. The ability of Mn^{2+} to function as corepressor for *E. coli* Fur is apparently not adaptive (Hantke 2001), and sev-

Table 1 Major families of bacterial metalloregulatory proteins

Family	Representative members	Metals (metalloids) sensed[a]
Fur	Fur, Zur, Mur, Nur, (PerR)	Fe^{2+}, Zn^{2+}, Mn^{2+}, Ni^{2+}
DtxR	DtxR, IdeR, MntR, ScaR, SirR	Fe^{2+}, Mn^{2+}, (Zn^{2+}), (Cd^{2+})
NikR	NikR	Ni^{2+}
MerR	MerR, ZntR, CueR, PbrR	Hg^{2+}, Zn^{2+}, Cu^{1+}, Pb^{2+}
ArsR	ArsR, SmtB, CadC, CzrA, NmtR	As^{3+}, Bi^{3+}, Zn^{2+}, Cd^{2+}, Pb^{2+}, Co^{2+}

[a] Metals where the physiological relevance of the induction is uncertain are in parentheses

eral ArsR family members respond to metal ions to which the regulated gene products are unable to confer resistance (Moore et al. 2005). In other cases, metal ions may occupy a regulatory site, but not elicit the required conformational changes to alter DNA-binding or protein activity (Cavet et al. 2002). In these cases, the "wrong" metal ion can act as an antagonist (Harvie et al. 2006).

2.1
The Fur Family

The Fur family of metalloregulators includes sensors of iron (Fur), zinc (Zur), manganese (Mur), nickel (Nur) and oxidative stress (PerR) (Table 2). In many cases, multiple Fur paralogs are present within a single genome, a situation first clarified by analyses of the three Fur paralogs encoded in the *B. subtilis* genome (Bsat et al. 1998; Gaballa and Helmann 1998). It is important to note that many genes (and therefore proteins) detected as a result of genome sequencing efforts may be annotated as *fur* when, in fact, they actually encode a different member of the Fur family.

2.1.1
Fur: Ferric Uptake Regulator

The ferric uptake regulator (Fur) protein was amongst the first metalloregulatory proteins to be characterized, both genetically (Hantke 1981) and biochemically (Bagg and Neilands 1987). *E. coli* null mutants in the *fur* locus constitutively express siderophores and siderophore uptake proteins, and are unable to grow on succinate as carbon source. In addition, null mutants are resistant to elevated levels of Mn^{2+}, perhaps because Mn^{2+} inappropriately activates the Fur repressor leading to an inability of the cell to obtain sufficient iron for growth (Hantke 1987). This observation led to the development of a convenient genetic selection that allowed the identification of *fur* genes from several different Gram-negative bacteria (Prince et al. 1993; Thomas and Sparling 1996; Murphy et al. 1999; Funahashi et al. 2000).

In *Corynebacterium diphtheriae*, a Gram-positive pathogenic bacterium, Fur is replaced by the functionally analogous DtxR (diphtheria toxin repressor) protein (see below) leading to the initial suggestion that Gram-positive bacteria may not contain Fur homologs (Tao et al. 1994). However, iron-sensing DtxR homologs are now known to be limited to the high GC subgroup of the Gram-positive bacteria (although a broad range of bacteria appear to contain Mn-sensing DtxR homologs). We initially suspected the presence of a Fur homolog in *Bacillus subtilis* based on the finding of a *cis*-acting element resembling a Fur box preceding the iron-repressible gene, *mrgC* (Chen et al. 1993). Similarly, the sequencing of the regulatory region for the ferric hydroxamate uptake genes in *B. subtilis* provided support for a Fur-

Table 2 Fur family members

Protein (organism)	ID #	Corepressor(s)[a]	Function	Noted features	Refs.[b]
Fur (*E. coli*)	CAA26429	Fe^{2+}, (Mn^{2+})	Iron homeostasis	Structural Zn^{2+} with C92 and C95 as ligands	Escolar et al. 1999; Andrews et al. 2003
Fur (*P. aeruginosa*)	AAC05679	Fe^{2+}	Iron homeostasis	Only structure of a Fur homolog	Vasil and Ochsner 1999; Pohl et al. 2003[b]
Fur (*B. japonicum*)	AAR12905	Fe^{2+}	Iron homeostasis	Metal binding sites apparently different than Fur_{PA}	Friedman and O'Brian 2004
Zur (*E. coli*)	P0AC51	Zn^{2+}	Zinc uptake	Zn^{2+} sensitivity in femtomolar range (10^{-15})	Patzer and Hantke 1998; Outten and O'Halloran 2001; Outten et al. 2001
Zur (*B. subtilis*)	P54479	Zn^{2+}	Zinc homeostasis	Regulates non-Zn^{2+}-containing ribosomal protein paralogs	
PerR (*B. subtilis*)	P71086	Fe^{2+}, Mn^{2+}	Peroxide stress	DNA binding is regulated by metal-catalyzed protein oxidation	Lee and Helmann JD 2006; Traore et al. 2006[b]
Mur (*R. leguminosarum*)	Y13657	Mn^{2+}, (Fe^{2+})	Manganese uptake	No structural Zn^{2+}, binds two Mn^{2+} ions	Diaz-Mireles et al. 2004, 2005
PerR (*N. gonorrhoeae*)	NGO0542	Mn^{2+}	Manganese uptake	Functionally similar to Mur	Wu et al. 2006
Nur (*S. coelicolor*)	SCO4180	Ni^{2+}	Nickel homeostasis, oxidative stress	First Ni^{2+} responsive member	Ahn et al. 2006

[a] Metal ion selectivity in vivo. Metals in parentheses are less effective corepressors
[b] References reporting a crystal structure

like regulatory system (Schneider and Hantke 1993). In contrast with Gram-negative systems, however, manganese-resistant mutants in *B. subtilis* were not derepressed for iron uptake. The identity of the *fur* gene only emerged when the genome was completed in 1997 (Kunst et al. 1997). Unexpectedly, the *B. subtilis* genome encoded three Fur homologs: now assigned Fur, Zur, and PerR (Bsat et al. 1998; Gaballa and Helmann 1998).

Early biochemical studies established that Fur binds DNA in response to available Fe^{2+} at a conserved 19 bp inverted repeat sequence designated as a Fur box (de Lorenzo et al. 1987; Griggs and Konisky 1989). Similar Fur box sequences are found preceding candidate iron uptake genes across a wide-range of bacteria: a testament to the broad distribution of Fur and Fur-like regulatory systems (Panina et al. 2001; Rodionov et al. 2004). The nature of the interaction between Fur and the 19 bp consensus site has been controversial. Studies using oligonucleotides led to the proposal that Fur recognizes tandem arrays of at least three hexamers (GATAAT) (Escolar et al. 1999). However, this model is not easy to reconcile with the dimeric state of Fur.

We proposed a revised model for interaction of Fur protein with DNA based on analysis of genes regulated by the *B. subtilis* Fur protein (Baichoo and Helmann 2002). Sequence alignment of ~ 20 Fur-regulated genes identified a conserved motif of 15 bp (a 7-1-7 inverted repeat) (Baichoo et al. 2002). Two overlapping copies of this motif generate the classic 19 bp Fur box, thereby suggesting that Fur typically binds as two pairs of dimers to opposing faces of the DNA helix. Support for this model was developed using gel mobility shift assays and a series of oligonucleotide targets (Baichoo and Helmann 2002). Subsequent studies with *E. coli* Fur are consistent with this general model (Lavrrar and McIntosh 2003). Most Fur boxes conform to the larger 19 bp consensus, and at some promoters additional dimers may also bind to create extended arrays of Fur protein. On the other hand, at some promoters the functional Fur box matches the shorter 7-1-7 consensus, but does not have obvious additional conservation corresponding to the longer 19 bp site. Interestingly, all three Fur paralogs of *B. subtilis* bind DNA consensus sequences that differ in only one or two base pairs per half-site (Fuangthong and Helmann 2003).

This revised view of the Fur box becomes important in applying bioinformatic approaches to define regulons controlled by Fur-like proteins. Searching bacterial genomes with the full 19 bp consensus is still a powerful approach to identifying candidate Fur-regulated genes and most sites that occur at levels significantly above those predicted by chance are in fact functional (Panina et al. 2001; Baichoo et al. 2002). For a genome the size of *E. coli* or *B. subtilis* (~ 4 Mb), this often corresponds to sites that match consensus at $> 14/19$ bp. The shorter consensus sequence is also useful, but false positives are more problematic, particularly in AT-rich genomes.

In most systems studied to date, regulation by Fur involves direct repression of gene expression in response to iron. However, Fur regulation can also involve additional layers of complexity. Transcriptome and proteome approaches have indicated that many genes and proteins are expressed at a lower level in cells lacking Fur (or starved for iron), suggesting a role of Fur as a positive activator of gene expression. To date, the only well-documented example of Fur acting as a direct transcriptional activator is *Helicobacter pylori* (Delany et al. 2001, 2004). In most other systems, positive regulation by Fur has turned out to be due to indirect regulation by one or more small RNAs. The best characterized example of a Fur-regulated sRNA is RyhB of *E. coli* (Masse and Arguin 2005). RyhB orthologs have been described in *Shigella* (Payne et al. 2006) and *Vibrio* (Mey et al. 2005), and analogous small RNA-based regulatory systems are present in *Pseudomonas aeruginosa* and *Bacillus subtilis* (Wilderman et al. 2004 and our unpublished data).

Fur-regulated sRNAs provide a molecular basis for understanding one of the key functions of Fur proteins: coordination of an "iron-sparing" response (Andrews et al. 2003; Masse et al. 2005; Kaplan et al. 2006). The first hints of this response emerged from the unexpected observation that *E. coli fur* mutants have decreased levels of total iron (Abdul-Tehrani et al. 1999), despite the fact that they constitutively express iron uptake functions (and are known to have elevated internal levels of "free" iron; see below). Although this observation was initially puzzling, in retrospect it makes perfect sense. When faced with insufficient iron, *E. coli* suppresses the synthesis of abundant, iron-rich enzyme complexes (Masse and Gottesman 2002; McHugh et al. 2003). For example, RyhB acts as an anti-sense RNA to suppress translation of succinate dehydrogenase and directly or indirectly downregulates other iron-containing enzymes. Some of the other targeted enzymes include NADH dehydrogenase, iron SOD, and the iron storage protein, ferritin. As a result, *E. coli* is able to remodel its proteome to more efficiently use available iron for essential functions.

This "iron-sparing" response is now appreciated as a near universal feature of the iron starvation stress response. In *Saccharomyces cerevisiae*, the iron-sparing response is coordinated by the Cth2 RNA-binding protein that targets specific mRNAs for degradation in response to iron limitation (Puig et al. 2005). In *Corynebacterium* the iron-responsive DtxR protein represses an AraC-like transcription factor called RipA (Wennerhold et al. 2005). RipA, in turn, represses iron-rich enzymes which are, therefore, positively regulated in response to available iron. The remodeling of the proteome to compensate for the limited availability of a particular metal ion may well be a general feature of metal starvation stress responses.

The molecular basis for iron-sensing by Fur is not yet clear, and the nature of the iron-sensing site is controversial. Fur from *E. coli* (Fur_{EC}) has been extensively studied and it is known that Fur_{EC} contains two metal binding sites. The structural site contains a tightly bound Zn^{2+} with coordination of $S_2(N/O)_2$, which may involve Cys92 and Cys95, and at least one histidine (Coy et al. 1994; Jacquamet et al. 1998; Althaus et al. 1999). The regulatory site contains exchangeable bound Fe^{2+}, within a hexacoordinated environment of histidines and carboxylate ligands (Adrait et al. 1999). The crystal structure of Fur from *Pseudomonas aeruginosa* (Fur_{PA}) has been resolved in complex with Zn^{2+} (Pohl et al. 2003). It consists of two domains, an N-terminal DNA binding domain and a C-terminal dimerization domain. Two zinc binding sites were identified per monomer. The putative regulatory iron binding site is located in the dimerization domain and is coordinated by the side-chains of residues His86, Asp88, Glu107, and His124. The structural zinc binding site connects the DNA-binding domain and the dimerization domain with coordination of side-chains of residues His32, Glu80, His89, and Glu100. Fur_{PA} contains only one cysteine, which corresponds to Cys92 of Fur_{EC}, and this cysteine is not involved in metal binding. These significant structural differences suggest that Fur homologs in bacteria may be structurally diverse. One striking example is Fur from *Bradyrhizobium japonicum*. Ligands of the Fur_{PA} metal binding sites are also highly conserved in *B. japonicum*. However, Fur_{BJ} with multiple point mutations in either of the two deduced metal binding sites still responds to Fe^{2+} (Friedman and O'Brian 2004). Recent functional analysis of the Fur paralog, *B. subtilis* PerR, also reveals significant differences relative to Fur_{PA}. In PerR, the ligands corresponding to the presumed Zn^{2+}-binding site in Fur_{PA} are actually involved in binding the regulatory metal ions, Fe^{2+} or Mn^{2+}, and a tightly bound structural Zn^{2+} binds to a site not present in the Fur_{PA} structure (Lee and Helmann 2006; Traore et al. 2006). Preliminary results support a similar arrangement of metal-sensing sites for *B. subtilis* Fur.

Ultimately, a careful biochemical analysis of iron-binding by Fur will be required to define the ambient levels of "free" iron in the cytosol. Although Fur_{EC} selectively senses Fe^{2+} inside cells, in vitro Fur can be activated by many other metals including Mn^{2+}, Co^{2+}, and Cu^{2+}. Fur binds Fe^{2+}, Mn^{2+}, Co^{2+}, and Cu^{2+} with dissociation constants of 55, 85, 36, and 10 μM, respectively (Hamed 1993). This suggests that in *E. coli* the ambient level of iron may be maintained in the range of 55 μM. This is consistent with measurements of intracellular iron using an iron chelator to convert chelatable iron to the EPR-active ferric species: levels of iron between 20 and 30 μM are routinely observed in wild-type cells with significantly higher levels seen in *fur* mutants and under oxidative stress conditions that lead to the dissambly of Fe-S clusters (Keyer and Imlay 1996; Woodmansee and Imlay 2002). Fur is likely specific for iron since other metal ions (with the exception in some cases of Mn^{2+}) rarely achieve levels sufficient to activate Fur. Indeed, there is thought to be essentially no free cytosolic copper in bacteria (Changela et al. 2003) so the fact that copper binds Fur with an apparent affinity somewhat higher than that for iron is irrelevant in vivo. This is an example of the principle described in the chapter by Tottey et al. in this volume: in vivo metal-responsiveness is governed not only by *affinity*, but also by *access* and *allostery*.

2.1.2 Zur: Zinc Uptake Regulator

We initially discovered Zur (Zur_{BS}) in the course of a functional characterization of the multiple Fur paralogs encoded in *B. subtilis* genome (Gaballa and Helmann 1998). Zur_{BS} represses a zinc ABC transporter encoded by the *ycdHI-yceI* operon. Zur_{BS} also represses YciC, an abundant protein originally detected in the membrane fraction of *zur* mutant cells. Indeed, characterization of the *yciC* promoter region led to the initial observation that Zur functioned as a Zn^{2+}-selective transcriptional repressor (Gaballa and Helmann 1998). We now appreciate that *yciC* is part of a complex operon: Zur_{BS} also regulates the upstream *yciA* and *yciB* genes (Gaballa et al. 2002). Although physiological studies initially suggested that the YciABC system might function as part of a low affinity transport system, recent results suggest an alternative role for these proteins. YciA has recently been found to encode an alternate enzyme for folate biosynthesis that can substitute, under zinc limitation conditions, for the Zn-dependent FolE protein (V. de Crecy-Lagard, personal communication). YciC is a putative metallochaperone of still uncertain specificity, which is postulated to allow metal insertion into the YciA protein (and perhaps other metalloenzymes). The substitution of one metalloenzyme by another, with different metal cofactor specificity, is an example of what we term a "metal substitution" response (see below).

E. coli Zur (Zur_{EC}) is the best characterized representative of the Zur subfamily (Hantke 2005). Zur_{EC} was discovered as a genetic locus that affected expression of the high-affinity zinc-uptake system ZnuABD in the presence of zinc (Patzer and Hantke 1998). Zur_{EC} represses transcription by binding to DNA sequences located between the divergently transcribed *znuCB* and *znuA* genes (Patzer and Hantke 1998; Outten and O'Halloran 2001). Zur_{EC} has two Zn^{2+} binding sites per monomer. Site A binds zinc very tightly with a coordination environment of $S_3(N/O)$, while site B contains loosely bound zinc in $S(N/O)_3$ coordination with tetrahedral geometry. Tetrahedral coordination in site B of Zur_{EC}, which is different from Fe^{2+} binding site in Fur that contains 5–6 N/O ligands, favors Zn^{2+} binding over Fe^{2+} (Outten et al. 2001). Interestingly, Zur_{EC} has an extraordinarily high zinc sensitivity in femtomolar (10^{-15}) range, suggesting that there is almost no persistent free zinc in cytoplasm (Outten and O'Halloran 2001). It remains to be determined whether this is a general feature of Zur proteins; preliminary studies with Zur_{BS} suggest a significantly lower Zn affinity with the implication that cytosolic levels of free zinc may be higher than postulated in *E. coli*.

Bioinformatic analyses have augmented our understanding of the Zur regulon in *B. subtilis* and *E. coli*, and allowed a preliminary prediction of likely zinc-regulated genes in a variety of bacterial species (Panina et al. 2003). One unexpected feature to emerge from this analysis is the finding that Zur (and zinc-dependent metalloregulators in general) represses several genes encoding homologs of ribosomal proteins. It had been noted, since the early days of bacterial genome sequencing, that most ribosomal proteins are encoded by single copy, presumably essential, genes. However, a small number of genes are duplicated in many bacterial genomes and these typically occur in two flavors originally designated C^+ and C^- (Makarova et al. 2001). The C^+ paralogs contains two CxxC motifs indicative of zinc binding, while the C^- paralogs lack many or all of the presumed zinc ligands. With the finding that the C^- paralogs in many species are associated with Zur boxes, and are therefore predicted to be expressed when zinc is limiting, the role of these alternative proteins becomes clearer (Panina et al. 2003). It is likely that cells growing in zinc-limited conditions replace Zn-containing ribosomal proteins with non-zinc binding paralogs as part of a "zinc-sparing" response. A good example of this may be the replacement of the essential S14 protein in *B. subtilis* in newly synthesized ribosomes with a non-zinc-containing paralog (H. Nanamiya, personal communication).

In some cases, this protein substitution may play an additional role we refer to as "zinc mobilization". For example, ribosomes from zinc-starved *B. subtilis* cells lose the zinc-containing r-protein L31 and instead are found associated with the C^- paralog, YtiA (Nanamiya et al. 2004). In this case, L31 is not an essential r-protein and cells lacking both L31 and YtiA are viable, although somewhat slower growing. Since L31 is surface exposed, it is postulated that this protein provides the cell with a reservoir of zinc that can be mobilized in times of need, much like iron stored in bacterioferritins. In this case, newly synthesized YtiA displaces L31 from the surface of extant ribosomes, as it has been shown to do in vitro (Akanuma et al. 2006), and the released L31 releases zinc either spontaneously of by degradation of the protein. Since ribosomes are abundant in rapidly growing cells, we estimate that more than half of total cell-associated zinc could be in the ribosome, and the ability of cells to mobilize some or all of this "stored" zinc would clearly be advantageous under zinc-limiting conditions.

2.1.3 Mur: Manganese Uptake Regulator

In *Rhizobium leguminosarum* a Fur homolog functions as a Mn^{2+}-responsive transcriptional regulator by repressing manganese transporter (encoded by *sitABCD*) in response to Mn^{2+}. Therefore, this protein was named Mur (manganese uptake regulator) (Wexler et al. 2003; Diaz-Mireles et al. 2004). In the presence of Mn^{2+}, Mur binds to an MRS (Mur-responsive sequence), which is distinct from a classical Fur box, thereby repressing the *sitABCD* operon (Diaz-Mireles et al. 2005). Interestingly, *R. leguminosarum mur* partially complements an *E. coli fur* mutant, and Mur protein binds to a canonical Fur box in an iron-dependent manner. Thus, Mur can recognize two different sequences in response to two different metals. The observation that Mur is

Fe^{2+}-responsive in *E. coli*, but Mn^{2+}-responsive in *Rhizobium*, highlights the key role of cellular milieu in controlling metal responsiveness in the cell.

Mur lacks a structural Zn^{2+} binding site, and instead binds two manganese ions per dimer (Bellini and Hemmings 2006). In vitro studies demonstrate that Mur binds to Mn^{2+} with slightly higher affinity than Fe^{2+}, Co^{2+}, Zn^{2+}, or Ni^{2+} and all bind to Mur with micromolar dissociation constants. Mur binds to MRS and a classical Fur box as one and two homodimers, independent of the metal co-repressors. Presumably non-physiological metal ions are maintained (by their cognate metalloregulatory proteins) at levels too low to be effective co-repressors in vivo.

In *Neisseria gonorrhoeae*, a Mur-like protein regulates a manganese transporter, MntABC, in a Mn^{2+}-dependent manner (Wu et al. 2006). This protein was designated PerR because, like PerR, it responds to Mn^{2+} and the null mutant is more resistant to H_2O_2 killing than the wild-type. However, this regulator may in fact be more appropriately classified as Mur since none of its regulated genes are directly involved in protection against oxidative stress. For example, catalase is under control of OxyR in *N. gonorrhoeae* (Tseng et al. 2003). The H_2O_2-resistant phenotype of the *perR* mutant may simply arise from accumulation of Mn^{2+} which is know to have antioxidant properties.

2.1.4 Nur: Nickel Uptake Regulator

Recently, a new member of the Fur family, Nur (nickel-uptake regulator) was discovered in *Streptomyces coelicolor* (Ahn et al. 2006). Nur controls nickel homeostasis in *S. coelicolor* by negatively regulating the *nikABCDE* nickel uptake system, and also represses one of two superoxide dismutase (SOD) genes. *S. coelicolor* contains two types of SOD: NiSOD (encoded by the *sodN* gene) and FeSOD (encoded by the *sodF* gene). In the presence of Ni, the expression of FeSOD is repressed by Nur binding to operator of the *sodF* gene, whereas the expression of NiSOD encoded by *sodN* is induced (by a still unknown mechanism). Nur does not appear to be an activator for *sodN* nor does it bind to the *sodN* regulatory region. An sRNA-mediated mechanism also seems unlikely since Ni^{2+} has no significant effect on the half-life of the *sodN* transcript.

Nur is highly specific for Ni^{2+} both in vivo and in vitro. In vivo, the expression of FeSOD and NiSOD responds to a very small amount of Ni (nanomolar range), but to a much higher amount of Co (micromolar range) (Chung et al. 1999). In vitro, purified Nur protein specifically responds to Ni^{2+}, while *S. coelicolor* cell extracts can be partially activated by 1 mM cobalt. Nur exhibits significant similarity to other members of the Fur family. When compared to Fur_{PA} (Pohl et al. 2003), the putative structural zinc binding is conserved with three matches out of four. In contrast, the predicted regulatory site is poorly conserved, with only with one match out of four. Structural and biochemical analyses will be required to test the role of these putative metal-binding sites in sensing of nickel ions.

2.1.5
PerR: Peroxide Regulon Repressor

PerR is one of the three Fur paralogs identified in *B. subtilis* (Herbig and Helmann 2001). PerR senses oxidative stress and the major function of the PerR regulon is to protect cells against peroxide-mediated damage. PerR, like most other Fur homologs, requires a metal ion co-repressor to bind DNA and functions to negatively regulate genes involved in oxidative stress, such as *katA* (catalase), *ahpCF* (alkylhydroperoxide reductase), and *mrgA* (encoding a Dps homolog) (Helmann et al. 2003). PerR also directly and indirectly affects metal ion homeostasis. PerR negatively regulates the ZosA P-type ATPase, proposed to import Zn ions in response to oxidative stress, and Fur (Fuangthong et al. 2002; Gaballa and Helmann 2002).

PerR senses H_2O_2 by metal-catalyzed protein oxidation (Lee and Helmann 2006). PerR can exist in two forms (PerR:Zn,Fe and PerR:Zn,Mn) but only the Fe-cofactored form is sensitive to peroxides in vivo and in vitro. Fe^{2+} in PerR plays a direct role in peroxide sensing: peroxides react with Fe^{2+} to produce OH· (the Fenton reaction) leading to the oxidation of either of two histidine ligands to the bound iron cofactor (Lee and Helmann 2006). The major mechanism of peroxide-mediated cell killing is DNA damage mediated by Fenton chemistry. Thus, the use of Fenton chemistry by PerR allows the cell's defensive response to be activated by the same set of conditions that make the cell most vulnerable: peroxides and the presence of cytosolic iron.

2.1.6
Irr: Iron-Responsive Regulator

In addition to Fur, *B. japonicum* has a second Fur-like regulator that controls heme biosynthesis genes in response to iron availability (Hamza et al. 1998). Irr is a conditionally stable protein that only accumulates in cells that are iron-limited and acts to repress heme biosynthesis. Irr itself binds two molecules of heme, which it likely acquires by direct association with the terminal enzyme of heme biosynthesis, ferrochelatase (Qi and O'Brian 2002; Yang et al. 2005). Heme-binding triggers protein degradation (Qi et al. 1999). Since Irr binds a metal-ion chelate, rather than directly interacting with iron, it is technically a heme-sensor rather than a metalloregulatory protein as defined above.

2.2 DtxR/MntR Family

This family of regulatory proteins includes members that sense Fe, Mn, or in some cases, both (Table 3). Unlike the Fur family, this particular group of regulators has proven highly amenable to structural analyses and high resolution structures are available for three different members (DtxR, IdeR, and MntR) in complex with various metal ion co-repressors and as the DNA-bound complex (White et al. 1998; Pohl et al. 1999; Wisedchaisri et al. 2004).

2.2.1 Iron Sensors: DtxR/IdeR

DtxR was originally characterized as the regulator of diphtheria toxin expression in *Corynebacterium diphtheriae* (reviewed in Holmes 2000). It was observed as early as 1936 that diphtheria toxin is produced under conditions of iron limitation (Tao et al. 1994). Since the human body has very little available iron, this coordinates the induction of toxin synthesis with infection of the host. However, regulation of toxin synthesis is not the *raison d'etre* for DtxR. Indeed, the toxin gene itself is on an integrated prophage, and DtxR and most DtxR-regulated genes are not phage-associated. Functionally, DtxR is most closely analogous to Fur in that it regulates genes for siderophore biosynthesis, siderophore uptake, and indirectly controls an iron-sparing response. In contrast with the sRNA-mediated iron-sparing response described in *E. coli*, DtxR represses the *ripA* gene encoding an AraC-family repressor protein (Wennerhold et al. 2005). RipA, in turn, represses genes encoding dispensable, iron-rich proteins (Wennerhold and Bott 2006).

DtxR is representative of a family of iron-dependent repressors found in high GC Gram-positive bacteria. The other well-characterized member of this family is the iron-dependent repressor (IdeR) from *Mycobacterium tuberculosis* (Rodriguez and Smith 2003). IdeR is encoded by an essential gene, and regulates genes involved in iron acquisition and storage, and oxidative stress responses (Gold et al. 2001). Structurally, both DtxR and IdeR are very well characterized and there are numerous crystal structures available. Each DtxR/IdeR monomer consists of two domains. The N-terminal domain contains a helix–turn–helix DNA-binding motif, two metal binding sites, and a dimerization interface. The C-terminal domain is structurally similar to eukaryotic SH3 domain (Qiu et al. 1995; Pohl et al. 1999; Feese et al. 2001). These two domains are connected by a flexible proline-rich segment. In DtxR, metal site 1 (ancillary site) includes H79, E83, and H98 from the N-terminal domain, and E170 and Q173 from the SH3-like domain, and adopts a distorted trigonal bipyramidal geometry. Metal site 2 (primary site) includes M10, C102, E105, and H106, and adopts a distorted octahedral geometry (Andrews et al. 2003). In the absence of metal ions, DtxR/IdeR exists as an in-

Table 3 DtxR family members

Protein (organism)	ID #	Corepressor(s) [a]	Function	Refs. [b]
DtxR (*C. diphtheriae*)	CAE49945	Fe^{2+}	Iron homeostasis, oxidative stress	Tao et al. 1994; Holmes 2000
IdeR (*M. tuberculosis*)	P0A672	Fe^{2+}	Iron homeostasis, oxidative stress	Rodriguez and Smith 2003
MntR (*B. subtilis*)	NP_390332	Mn^{2+}	Manganese uptake	Que and Helmann 2000; Glasfeld et al. 2003 [b]; Kliegman et al. 2006 [b]
MntR (*B. anthracis*)	Q81M27	Mn^{2+}	Annotated as MntR, but also called AntR	Sen et al. 2006
TroR (*T. pallidum*)	AAC45729	Mn^{2+} (Zn^{2+}?)	Manganese (zinc) uptake	Posey et al. 1999; Hazlett et al. 2003
SirR (*S. epidermidis*)	X99128	Mn^{2+}, Fe^{2+}	Manganese and/or iron uptake	Hill et al. 1998
ScaR (*S. gordonii*)	AAF25184	Mn^{2+}	Manganese uptake	Jakubovics et al. 2000
EfaR (*E. faecalis*)	AAK97638	Mn^{2+}	Manganese uptake	Low et al. 2003

[a] Metal ion selectivity in vivo. Metals in parentheses are less effective corepressors
[b] References reporting a crystal structure

active monomer. Binding of metal ions triggers dimerization and activates the protein (Tao et al. 1995; Chou et al. 2004).

Although DtxR and IdeR require iron as a co-repressor in vivo, several other divalent transition metal ions (including Fe^{2+}, Ni^{2+}, Co^{2+}, Mn^{2+}, and Cd^{2+}) can act as co-repressors in vitro (Pohl et al. 1997; Dussurget et al. 1999; Chou et al. 2004). DtxR can be activated by Fe^{2+} at much lower levels than Mn^{2+} (Spiering et al. 2003). Similarly, IdeR requires lower Fe^{2+} concentration for dimerization and DNA binding (Chou et al. 2004). This specificity arises, in part, from the nature of the DtxR/IdeR coordination sphere, which favors the binding of Fe^{2+}.

2.2.2 MntR: Manganese Uptake Regulator

The observation that *B. subtilis* encoded a DtxR homolog was originally puzzling since all known iron-regulated genes are controlled by a Fur protein. The function of this gene emerged when it was found that mutants are extremely sensitive to elevated levels of Mn^{2+} (and Cd^{2+}): a phenotype linked to the constitutive expression of two transporters involved in manganese uptake, MntH and MntABCD (Que and Helmann 2000). This led to the finding that MntR (manganese transport regulator) senses Mn^{2+} rather than Fe^{2+}. MntR binds to an imperfect 19 bp inverted repeat preceding both genes and hence prevents transcription (Que and Helmann 2000). MntR has been most thoroughly characterized in *B. subtilis*, but it is now known that orthologs are present in many other bacteria including *B. anthracis* (Sen et al. 2006), *Staphylococcus aureus* (Horsburgh et al. 2002; Ando et al. 2003), *Corynebacterium diphtheriae* (Schmitt 2002), *E. coli* (Patzer and Hantke 2001), and *S. enterica* (Qi and O'Brian 2002).

Overall, the structure of MntR is similar to that of DtxR, except that MntR lacks the C-terminal SH3-like domain and their metal binding sites are different (Glasfeld et al. 2003). Like DtxR, MntR can be activated by several different transition metal ions in vitro (Que and Helmann 2000). However, both are highly selective in vivo (Guedon and Helmann 2003). The origins of metal-selectivity are complex: but both *affinity* and *access* play important roles.

Presumably, amino acid changes in the metal-sensing sites of DtxR/MntR have led to alterations in the relative *affinity* for iron and manganese. There are at least two amino acids within the conserved metal binding sites that differ between the Fe^{2+}-sensing DtxR and Mn^{2+}-sensing MntR. Met10 and Cys102 in DtxR are replaced in MntR with Asp and Glu, respectively. Substitution of sulfur-containing residues in DtxR with carboxylate-containing residues in MntR presumably contributes to the altered specificity of DtxR/MntR. When these two residues in MntR were altered to the corresponding residues from DtxR, MntR specificity was relaxed and it was now able to sense both Mn^{2+} and Fe^{2+} (Guedon and Helmann 2003). Moreover, when the metal-binding residues in DtxR were replaced with those in MntR, the mutant DtxR (expressed in *B. subtilis*) was highly selective for manganese. These results suggest that the identity of the amino acids at the metal binding site is an important determinant for DtxR/MntR selectivity (Guedon and Helmann 2003).

Comparisons of DtxR and MntR also provide an instructive example of the role of cellular milieu (i.e., *access*) in metal responsiveness. When *C. diphtheriae* DtxR was expressed in *B. subtilis* it sensed both Fe^{2+} and Mn^{2+} (Guedon and Helmann 2003). This suggests that manganese ions are normally present at a higher ambient level in *B. subtilis* than in *C. diphtheriae*, where DtxR is highly selective for iron. Since Mn^{2+} levels in the *C. diphtheriae* cytosol are presumably regulated by its own MntR homolog ($MntR_{CD}$) (Schmitt 2002), this predicts that $MntR_{CD}$ binds Mn^{2+} with higher affinity than $MntR_{BS}$. As noted above, when DtxR was mutated to include amino acids typically found in MntR, it now responded only to manganese (in *B. subtilis*). However, if a *fur* mutation was introduced (leading to elevated intracellular levels of iron), this mutant protein now responded to added iron. Thus, the ability of a protein to respond to a particular metal in vivo is governed, to a significant extent, by both the biochemical *affinity* of the metalloregulator for its ligand and the ambient level of free metal ion in the cytosol (*access*).

Like DtxR, MntR is activated by the binding of two metal ions per monomer (Lieser et al. 2003; Golynskiy et al. 2005). However, in MntR the two Mn^{2+} ions bind in a binuclear complex (Glasfeld et al. 2003). The first conformation of MntR to be visualized by crystallography contained two ions at sites A and B, separated by 3.3 Å (the AB conformation). However, subsequent studies revealed that a preferred conformation, and the one presumed to be biologically relevant, includes two ions separated by 4.4 Å (the AC conformation) (Kliegman et al. 2006). Both conformations use the same set of metal binding ligands and retain almost the same geometries. However, in the AB conformation, E11 and E102 are the bridging ligands, while E99 and E102 bridge the Mn^{2+} ions in the AC conformation. It is speculated that MntR selectivity for Mn^{2+} over Fe^{2+} is due to the geometry and size of metal binding sites, and that occupancy of site A may help organize the ligands of site B/C: the binding of Mn^{2+} or Cd^{2+} at site A leads to a suitable environment for site B binding (Kliegman et al. 2006).

2.2.3
Other DtxR/MntR Family Members

Other members of the DtxR/MntR family have been described that exhibit somewhat more divergent characteristics. A Mn^{2+}-dependent DtxR homolog has been reported in *Treponema pallidum* a bacterium with no requirement for iron, and it is called TroR (Posey et al. 1999). TroR regulates the transcription of *troABCD* (encode ABC transporter), *troR*, and *gpm* (encodes the glycolytic enzyme phosphoglyceromutase). Biochemistry studies showed that TroR responds to Mn^{2+}. In vivo, TroR could repress promoter–operator with *lacZ* transcriptional fusion in *E. coli*, but metal dependence was not observed. However, later studies suggested that Zn^{2+} might be the primary corepressor of TroR in vivo (Hazlett et al. 2003). Immunoblot and RT-PCR analyses revealed that TroR requires Zn^{2+} to repress TroA. In addition, TroA (a transporter component) has the same affinity for both Zn^{2+} and Mn^{2+}. SirR in *Staphylococcus epidermidis* may be the only known DtxR homolog that responds physiologically to both Fe^{2+} and Mn^{2+}. SirR regulates expression of *sitABC* operon that encodes a putative ABC transporter of unknown specificity. The Mn^{2+} and Fe^{2+} selectivity of SirR is illustrated both by in vivo immunoblot analysis and in vitro gel shift assays (Hill et al. 1998).

2.3
NikR: Nickel Responsive Repressor

NikR is representative of a family of nickel sensors found in a variety of bacterial systems (Dosanjh and Michel 2006). Unlike other described families of bacterial metalloregulatory proteins, NikR interacts with DNA via a ribbon–helix–helix motif in which DNA contacts are made by a double-stranded, intersubunit beta sheet rather an alpha-helix (typically part of a helix–turn–helix unit) (Chivers and Sauer 1999). This mode of DNA interaction was first characterized for the Arc and MetJ repressors (Raumann et al. 1994).

The *E. coli* NikR protein is the best characterized representative of this family and mediates the nickel-dependent repression of the *nikABCDE* uptake operon. NikR is a tetramer in solution, with each tetramer containing two DNA-binding motifs (Chivers and Sauer 2002). NikR has at least two distinct Ni^{2+} binding sites (Chivers and Sauer 2002; Bloom and Zamble 2004). A very high affinity site ($K_d \sim 9$ pM) that activates DNA-binding and a second, lower affinity site ($K_d \sim 30$ nM) that further increases the extent and affinity of DNA-binding (Chivers and Sauer 2002). As expected for a Ni^{2+} complex, the coordination geometry is square planar and involves primarily His residues (Chivers and Sauer 2002). Unexpectedly, this coordination geometry changes significantly upon complexation with operator DNA (Carrington et al. 2003). The metal site in the DNA-bound complex is apparently six-coordinate and a Cys residue, seen as a ligand in unbound protein, is not longer coordinated to Ni^{2+}. This suggests that caution needs to be applied in extrapolating inferences about metal coordination derived from solution to the complexes present with DNA. To date, NikR-like regulators are only known as nickel sensors: it will be interesting to determine if the NikR protein scaffold has also been adapted to sense other metal ions. For further details on NikR and its role, the reader is referred to a recent review (Dosanjh and Michel 2006) and to the chapter by Hausinger and Zamble in this volume.

3 Metal Resistance, Efflux, and Detoxification Regulators

Bacteria must tightly control intracellular metal ion levels to avoid toxicity. Toxicity can result either from the over-accumulation of essential or beneficial metal ions, or from exposure to metals with no biological role (e.g., mercury and lead). The ability of cells to grow in the presence of high levels of metal ions (resistance) can result from either sequestration (production of metal-ion binding proteins such as ferritin or metallothionein), efflux, or detoxification. These processes are typically regulated by metalloregulatory proteins of either the MerR or ArsR families.

3.1 MerR Family

MerR is the namesake for a large and diverse family of regulators, including both metalloregulatory proteins and proteins that respond to other types of signals (reviewed in Brown et al. 2003; Hobman et al. 2005). MerR, one of the first described metalloregulatory proteins (O'Halloran and Walsh 1987), is associated with mercury resistance operons, which are often found on transposable elements such as Tn*21* and Tn*501* (Barkay et al. 2003). Members of the MerR family sense metal ion excess and activate the transcription of metal resistance genes. Typically, regulation involves promoter elements, unusually long promoter spacing, and MerR-like regulators that bind to a site between the −35 and −10 promoter elements to induce a remodeling of the promoter region DNA (Summers 1992). Well-characterized members of this family include MerR (mercury resistance regulator), CueR (copper efflux regulator), ZntR (zinc transport regulator), PbrR (lead resistance regulator), CoaR or CorR (cobalt responsive regulator), and CadR (cadmium resistance regulator) (Table 4). The proliferation of MerR family members is exemplified in the highly metal-tolerant organism *Cupriavidus metallidurans* CH34, where at least seven metal-sensing paralogs are found associated with distinct efflux and resistance systems (Mergeay et al. 2003). The large *S. coelicolor* genome may harbor as many as 23 MerR homologs (Brown et al. 2003).

The best characterized MerR proteins are from mercury resistance operons encoded on the Tn*21* and Tn*501* transposons (the two MerR proteins differ at only five amino acids). The Tn*21 mer* operon consists of five structural genes (*merTPCAD*) from which MerR itself is divergently transcribed. MerTPCAD are involved in mercury detoxification by a mechanism that, counter-intuitively, involves mercury import into the cell. MerP binds mercury in the periplasm, transfers it to MerT (a membrane bound transporter), which delivers Hg^{2+} to MerA. MerA reduces Hg^{2+} to the volatile form Hg^0, which is then able to diffuse through the cell membrane and into the environment (Barkay et al. 2003) (see chapter by Silver and Hobman in this volume).

Table 4 MerR family members

Protein	ID #	Organism	Metal inducers	Ligands	Refs.[a]
MerR	AF071413	Tn21	Hg^{2+}	C82, C117′, C126′	Summers 1992
MerR	P22853	*B. cereus*	Hg^{2+}	C79, C114′, C123′	Helmann et al., 1990
CueR	1Q05(pdb)	*E. coli*	Cu^{1+}, Ag^{1+}, Au^{1+}	C112, C120	Changela et al. 2003[a]
ZntR	1Q0A(pdb)	*E. coli*	Zn^{2+}, Cd^{2+}, Pb^{2+}	C79, C114, C124	Changela et al. 2003[a]
PbrR	YP 145623	*Cupriavidus metallidurans*	Pb^{2+}	Unknown	Borremans et al. 2001
CoaR	NP 442632	*Synechocystis* PCC 6803	Co^{2+}	C363, H364, C365	Rutherford et al. 1999
CadR	AAK48830	*Pseudomonas putida* 06909	Cd^{2+}, [Zn^{2+}, Hg^{2+}]	Unknown	Brocklehurst et al. 2003

[a] References reporting a crystal structure

Tn*21* MerR binds as a dimer to an operator sequence in the promoter for the *merTPCAD* operon, and activates transcription in response to mercuric ion. This same binding site also negatively regulates transcription from the overlapping and divergent *merR* promoter, thereby allowing for negative autoregulation. MerD may also bind the *mer* operator sequence in complex with MerR. This finding suggests that MerD serves as a co-regulator of this system and in essence resets the response by allowing MerR without Hg^{2+} bound to be made, and thereby repress transcription again (Nucifora et al. 1989; Champier et al. 2004).

MerR is distinct from most other metalloregulatory protein families in that binding of metal ion does not elicit a large change in DNA-binding affinity. Rather, metal binding triggers a conformational change in the bound protein to convert a repressor into an activator of transcription (O'Halloran et al. 1989; Frantz and O'Halloran 1990; Ansari et al. 1995). The underlying mechanism has been well-studied and involves a change in the twist of the spacer region of the promoter DNA that helps to realign the −35 and −10 promoter elements to allow productive interactions with RNA polymerase. This has been dubbed the "untwist and shout" mechanism to highlight the role that altered DNA twisting plays in promoter activation (Summers 1992). However, the precise details of the promoter conformational change may also involve changes in DNA bending or extension. These details have not been visualized in the case of MerR (there is no high resolution structure available for full-length MerR), but a DNA-bound complex of the MerR homolog BmrR allows some insights into the likely DNA structural alterations associated with promoter activation (Schumacher and Brennan 2002). When activated by ligand and bound to DNA, BmrR breaks a central AT base pair and slides the bases relative to one another, bends the DNA by $\sim 50^\circ$ away from the protein, and effects a shortening of the DNA spacer by the equivalent of nearly 2 bp (5 Å).

One of the remarkable features of MerR is that it senses a single mercuric ion per dimer. Genetic analyses indicated that three different Cys residues were required for mercury-responsive activation of transcription (Ross et al. 1989). However, the chemistry of mercuric ion dictates that it forms stable bi- and tridentate complexes, so the involvement of six cysteine thiols (per dimer) in binding a single mercuric ion presented a conundrum. In experiments with a MerR homolog from *B. cereus*, it was shown that inactive MerR proteins could form functional heterodimers (Helmann et al. 1990). With this insight as a starting point, it was possible to demonstrate that a functional, high affinity mercury binding site required one Cys from one subunit (C79) and two other Cys residues from the other subunit (C114 and C127); no other combinations of cysteine thiolates were able to restore function. Since MerR only binds one mercury per dimer, this implies a high level of negative cooperativity between the two binding sites. In principle, this could result from a protein conformational change or two symmetric sites that physically overlap. In light of subsequent structural studies of the MerR mercury-binding domain (Song et al. 2004) and of MerR homologs (such as CueR and ZntR; Changela et al. 2003), it is apparent that the two sites are actually quite distant in the protein structure and binding to one site distorts the other so that metal binding is no longer favored.

Since MerR binds Hg^{2+} at only one site per dimer, the resulting complex is asymmetric. It would be interesting to know, for example, whether MerR can activate transcription with Hg^{2+} bound only in the upstream metal-binding site (relative to the start site of transcription), the downstream site, or either site. Similar questions have been addressed with the *E. coli* cyclic AMP receptor protein (CRP) using an oriented heterodimer approach (Meibom et al. 2000). Such an experimental system would also allow a fine-scale dissection of the role of the individual amino acids that might interact directly with the RNAP (in addition to indirectly affecting RNAP activity by reconfiguration of the promoter structure). To date, the only metal-sensing MerR homologs for which full-length crystal structures are available are *E. coli* CueR and ZntR (Changela et al. 2003). The CueR and ZntR structures identify key Cys residues that form the high affinity metal-binding sites. The details of these coordination complexes have been recently reviewed (Pennella and Giedroc 2005).

3.2
ArsR Family

The ArsR/SmtB family includes several well-characterized metalloregulatory proteins. This family has been the focus of several recent reviews (Busenlehner et al. 2003; Pennella and Giedroc 2005; Tottey et al. 2005): here we briefly summarize the key features of some of the best studied members. The ability of regulators of this family to sense a variety of metals, all using a similar protein backbone, has led to a "theme and variations" model for structural adaptation (Table 5). The detailed biochemical analyses of family members have contributed significantly to our current perspectives on metalloregulator function, and contributed directly to the development of the "affinity, access, allostery" model for metal-sensing (Tottey et al. 2005) (see chapter by Tottey and Robinson in this volume).

The metalloregulatory members of this protein family bind DNA as repressors, and are induced by binding of a metal ion. Typically, these family members bind an operator sequence that contains a 12-2-12 imperfect inverted repeat upstream of regulated genes. Once bound with metal ions, a conformational change is induced that dissociates the repressor from its operator sequences, resulting in transcription of the efflux pumps and thereby allowing the removal of toxic or excess metal ions from the cell.

ArsR/SmtB family members have at least two distinct metal-sensing sites (Shi et al. 1994, 1996). Sequence alignments of some of the first-described ArsR proteins focused attention on a highly conserved ^{30}ELCVCD^{35} motif.

Table 5 ArsR family members

Protein	ID #	Organism	Metal inducer[a]	Ligands	Type I or II site[b]	Refs.[c]
ArsR	2208345A	*E. coli*	As^{3+}, Bi^{3+}, Sb^{3+}	C32,C34	Type I	Shi et al. 1994
SmtB	CAA45872	*Synechococcus* PCC 7942	Zn^{2+} [Cd^{2+}, Co^{2+}]	D104, H106, H117', E120'	Type II	VanZile et al. 2002; Eicken et al. 2003[c]
CzrA	IRIV (pdb#)	*Staphylococcus aureus*	Zn^{2+}, Co^{2+}	D84, H86, H97', H100'	Type II	Eicken et al. 2003)[c]; Pennella and Giedroc 2003
NmtR	n/a	*M. tuberculosis*	Ni^{2+}, Co^{2+}	D91, H93, H104, H107, H109& H116	Type II	Cavet et al. 2002
CmtR	n/a	*M. tuberculosis*	Cd^{2+}, Pb^{2+}	C57, C61	Type I	Wang et al. 2005
CadC	P20047	*S. aureus* pI258	Cd^{2+}, Pb^{2+}, Bi^{3+}, Zn^{2+}	C58, C60, C7, C11	Type I	Ye et al. 2005[c]
ZiaR	Q55940	*Synechocystis* PCC 6803	Zn^{2+}	C71, C73, H116	Both	Thelwell et al. 1998
BxmR	BAD11074	*Oscillatoria brevis*	Ag^{+}, Cu^{+}, Zn^{2+}, Cd^{2+}	Unknown	Unknown	Liu et al. 2004
AztR	n/a	*Anabaena* PCC 7120	Zn^{2+}, Cd^{2+}, Pb^{2+}	C21, C72, C74, H24 or H27	Type I	Liu et al. 2005

n/a Not applicable

[a] Parentheses indicate metal inducers that are less potent or are inducers in other organisms

[b] Type I sites are also named $\alpha3$ or $\alpha3$n. Type II sites are also named $\alpha5$ or $\alpha5$c. Terminology as per Ye et al. 2005

[c] References reporting a crystal structure

Mutagenesis experiments demonstrated that these Cys residues were required for sensing As^{3+} (Shi et al. 1994, 1996). Unexpectedly, Cys61 (analogus to Cys32 of ArsR) in SmtB, a Zn^{2+} sensing regulator of the SmtA metallothionein, was found to be dispensable for sensing of its cognate inducer, zinc (Turner et al. 1996). Mutational analysis revealed that His105 and His106 were necessary for sensing Zn^{2+} in vivo. Other members of the ArsR/SmtB family that are also involved in Zn homeostasis are most similar to SmtB.

These two distinct metal-sensing sites were named for their position within the tertiary structure of proteins: $\alpha 3$ and $\alpha 5$. The $\alpha 3$ site typically binds metals through cysteine residues, as seen in the conserved ELCVCD motif first observed in ArsR. The $\alpha 5$ site is composed mainly of histidines and carboxylate residues and is formed at the dimer interface (see Fig. 1). Ye and coworkers (Ye et al. 2005) introduced the terminology of type I and type II binding sites to simplify the terminology for $\alpha 3$ and $\alpha 5$ sites, respectively. In general, the $\alpha 3$ site, through cysteine residues, coordinates thiophilic metals (Cd^{2+}, Pb^{2+}, Bi^{3+}, and As^{3+}) whereas the $\alpha 5$ site is more amenable to larger coordination spheres and typically binds Zn^{2+}, Co^{2+}, Ni^{2+}, and Mn^{2+}.

Recent results suggest that not all ArsR family members can be simply classified as either type I or type II metal sensors (Table 5). For example, CzrA responds to both Zn^{2+} and Co^{2+}. Recent work has identified the type II residues responsible for creating the tetrahedral coordination geometry needed for Zn^{2+} binding in CzrA (Pennella et al. 2006). Also related to SmtB is ZiaR from the cyanobacterium *Synechocystis* that regulates a P-type ATPase, ZiaA. ZiaR mutational studies suggest that both type I and type II sites are necessary to

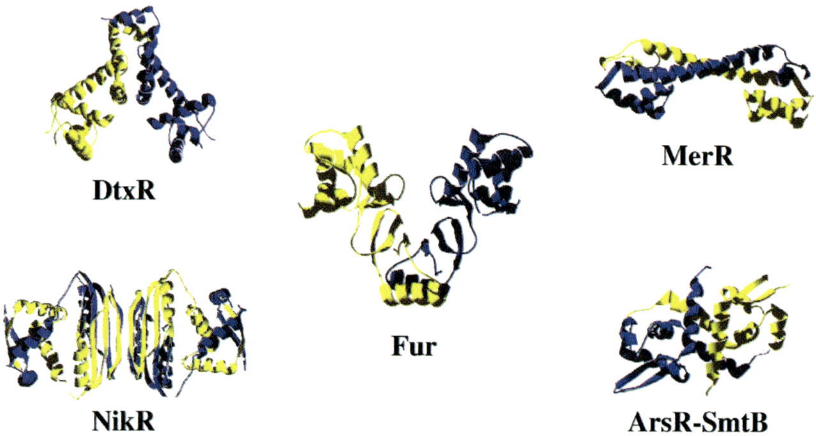

Fig. 1 Representative structures from each of the five indicated families of bacterial metalloregulatory proteins. The structures shown are *B. subtilis* MntR (a member of the DtxR/MntR family), *E. coli* ZntR (MerR family), *P. aeruginosa* Fur (Fur family), *E. coli* NikR (apoprotein), and *Synechococcus* SmtB (Zn_2 form). For each structure one subunit of the dimer is *yellow* and the other *blue*. Note that NikR is a tetramer

sense and respond to zinc (Thelwell et al. 1998). This may point to an intermediate between the ArsR-type (type I site) and SmtB-type (type II site) sensors. CadC also possess both type I and type II metal sensing sites (Ye et al. 2005). The type I site includes cysteines donated by different subunits. This creates a more complicated site than the ArsR type I site which uses two cysteines from the same subunit. CadC can also sense Zn^{2+}, although the mechanism differs significantly when compared to SmtB. To further complicate the situation, a recently described member of the ArsR/SmtB family, AztR appears to sense Zn^{2+} through a type I site (Liu et al. 2005). As these examples illustrate, it is difficult to predict metal specificity of ArsR family proteins by taking a one-dimensional approach to their characteristics (Harvie et al. 2006).

4 Emerging Themes in Metalloregulation

In this chapter we have provided a brief survey of the major families of bacterial metalloregulatory proteins, their biological roles, and current thoughts on the factors that contribute to their in vivo metal selectivity. We refer to metalloregulators as arbiters of metal sufficiency since it is these proteins that ultimately determine whether a cell has sufficient amounts, or an excess, of a given metal ion. These proteins provide a key window into the ambient levels of "free" (rapidly exchangeable) metals in the cell which, presumably, are at or near the K_d for metal-ion binding. At levels above the biochemically defined binding constants, metal ion uptake will be repressed and/or efflux will be induced. Together, the window between these response thresholds defines the biologically favored metal concentration (Finney and O'Halloran 2003).

Work in several systems has led to an appreciation for the complex interplay of biological and chemical factors that conspire to allow selective gene regulation in response to changing metal status. The three key factors have been alliteratively defined as: *affinity*, *access*, and*allostery* (Tottey et al. 2005). Clearly, metal ion *affinity* will play a key role in determining which ions trigger a biological response. However, there are numerous examples where the biologically relevant inducer (or co-repressor) binds weaker in vitro than other metal ions (often following the Irving–Williams series). This reflects the key role of *access*: metal ions can only trigger a biological response if they can accumulate in the cell at a sufficient level to interact with the metalloregulatory protein. Copper in particular often binds tightly to metalloregulatory sensory sites, yet is maintained in cells at very low levels, thereby precluding interference with sensing of other metal ions (Changela et al. 2003; Harvie et al. 2006). Finally, binding alone is not sufficient to trigger a biological response. In some cases, metal ions can bind to a metalloregulatory protein but not trigger the allosteric changes needed to alter gene expression. Perhaps the best characterized example of this phenomenon is the NmtR nickel sensor, which binds zinc in a non-effective tetracoordinate complex but responds to nickel, which interacts in a six-coordinate, biological effective complex (Cavet et al. 2002).

The biological responses controlled by metalloregulatory proteins are complex, but several themes have recently emerged. First, it is clear that metalloregulation of nutrient metal ions is often quite complex and may involve several layers of regulation, including both direct and indirect effects. In contrast, responses to metal excess, particularly to toxic metal ions and metalloids, is often quite simple and may involve a single efflux system. In general, many of the target genes for metalloregulators encode transporters involved either in uptake or efflux of metal ions (Nies 2003). Alternatively, metal ion storage proteins may also be the targets of regulation.

Analysis of the complex responses induced by limitation for nutrient metal ions has revealed responses that often lead to a large-scale remodeling of the proteome. These responses contribute to effects we refer to as metal sparing, metal substitution, and metal mobilization. The *metal-sparing* response was first defined for iron regulons in *E. coli* and *S. cerevisiae* but is likely to be a widespread feature of responses to metal limitation (Andrews et al. 2003; Masse and Arguin 2005; Kaplan et al. 2006). This metal-sparing response involves a repression (often at the translational level) of proteins that require metal ions but that are not absolutely essential for growth. It is particularly advantageous to stop the synthesis of abundant, metabolic enzymes. Since many of these enzymes are stable, even ceasing their synthesis may not have significant effects on metabolism for several generations.

A related phenomenon is *metal-substitution* wherein genes induced by metal ion limitation encode alternative isozymes that allow the cell to replace an enzyme dependent on the limiting metal ion with an enzyme dependent on another, less limiting metal ion, or by an alternative pathway that does not require metal ions. This metal substitution response is important in individual cells, but also across populations of organisms. One dramatic example is the use of a cadmium-dependent carbonic anhydrase in place of the conventional zinc-dependent enzyme in a marine diatom (Lane et al. 2005). Similarly, the use of a copper-containing plastocyanin in place of iron-containing cytochromes for photosynthesis is observed in the oceanic diatom *Thalassiosira oceanica* (Peers and Price 2006). As we have looked deeper into the transcriptional responses of *B. subtilis* to metal ion limitation, we see many more examples of apparent metal-substitution phenomena and experiments to test these ideas are currently in progress.

Finally, metal ion limitation can be expected to result in the *mobilization* of metal ions that have been stored in the cell. When iron is limiting for growth, it is likely that bacteria will mobilize iron stored in bacterioferritins and related proteins. However, the factors that enable this metal mobilization are not yet well characterized. Similarly, we have suggested that the common occurrence of non-zinc-binding ribosomal proteins as part of the Zur regulon is an example of metal mobilization from zinc "stored" in surface-accessible r-proteins (Moore and Helmann 2005).

Acknowledgements Our work on metalloregulation is supported by a grant from the National Institutes of Health (GM059323).

References

Abdul-Tehrani H, Hudson AJ, Chang YS, Timms AR, Hawkins C, Williams JM, Harrison PM, Guest JR, Andrews SC (1999) Ferritin mutants of *Escherichia coli* are iron deficient and growth impaired, and *fur* mutants are iron deficient. J Bacteriol 181:1415–1428

Adrait A, Jacquamet L, Le Pape L, Gonzalez de Peredo A, Aberdam D, Hazemann JL, Latour JM, Michaud-Soret I (1999) Spectroscopic and saturation magnetization properties of the manganese- and cobalt-substituted Fur (ferric uptake regulation) protein from *Escherichia coli*. Biochemistry 38:6248–6260

Agnoli K, Lowe CA, Farmer KL, Husnain SI, Thomas MS (2006) The ornibactin biosynthesis and transport genes of *Burkholderia cenocepacia* are regulated by an extracytoplasmic function sigma factor which is a part of the Fur regulon. J Bacteriol 188:3631–3644

Ahn BE, Cha J, Lee EJ, Han AR, Thompson CJ, Roe JH (2006) Nur, a nickel-responsive regulator of the Fur family, regulates superoxide dismutases and nickel transport in *Streptomyces coelicolor*. Mol Microbiol 59:1848–1858

Akanuma G, Nanamiya H, Natori Y, Nomura N, Kawamura F (2006) Liberation of zinc-containing L31 (RpmE) from ribosomes by its paralogous gene product, YtiA, in *Bacillus subtilis*. J Bacteriol 188:2715–2720

Alen C, Sonenshein AL (1999) *Bacillus subtilis* aconitase is an RNA-binding protein. Proc Natl Acad Sci USA 96:10412–10417

Althaus EW, Outten CE, Olson KE, Cao H, O'Halloran TV (1999) The ferric uptake regulation (Fur) repressor is a zinc metalloprotein. Biochemistry 38:6559–6569

Ando M, Manabe YC, Converse PJ, Miyazaki E, Harrison R, Murphy JR, Bishai WR (2003) Characterization of the role of the divalent metal ion-dependent transcriptional repressor MntR in the virulence of Staphylococcus aureus. Infect Immun 71:2584–2590

Andrews SC, Robinson AK, Rodriguez-Quinones F (2003) Bacterial iron homeostasis. FEMS Microbiol Rev 27:215–237

Ansari AZ, Bradner JE, O'Halloran TV (1995) DNA-bend modulation in a repressor-to-activator switching mechanism. Nature 374:371–375

Bagg A, Neilands JB (1987) Ferric uptake regulation protein acts as a repressor, employing iron(II) as a cofactor to bind the operator of an iron transport operon in *Escherichia coli*. Biochemistry 26:5471–5477

Baichoo N, Helmann JD (2002) Recognition of DNA by Fur: a reinterpretation of the Fur box consensus sequence. J Bacteriol 184:5826–5832

Baichoo N, Wang T, Ye R, Helmann JD (2002) Global analysis of the *Bacillus subtilis* Fur regulon and the iron starvation stimulon. Mol Microbiol 45:1613–1629

Barkay T, Miller SM, Summers AO (2003) Bacterial mercury resistance from atoms to ecosystems. FEMS Microbiol Rev 27:355–384

Bellini P, Hemmings AM (2006) In vitro characterization of a bacterial manganese uptake regulator of the Fur superfamily. Biochemistry 45:2686–2698

Bloom SL, Zamble DB (2004) Metal-selective DNA-binding response of *Escherichia coli* NikR. Biochemistry 43:10029–10038

Brown NL, Stoyanov JV, Kidd SP, Hobman JL (2003) The MerR family of transcriptional regulators. FEMS Microbiol Rev 27:145–163

Bsat N, Herbig A, Casillas-Martinez L, Setlow P, Helmann JD (1998) *Bacillus subtilis* contains multiple Fur homologues: identification of the iron uptake (Fur) and peroxide regulon (PerR) repressors. Mol Microbiol 29:189–198

Busenlehner LS, Pennella MA, Giedroc DP (2003) The SmtB/ArsR family of metalloregulatory transcriptional repressors: structural insights into prokaryotic metal resistance. FEMS Microbiol Rev 27:131–143

Carrington PE, Chivers PT, Al-Mjeni F, Sauer RT, Maroney MJ (2003) Nickel coordination is regulated by the DNA-bound state of NikR. Nat Struct Biol 10:126–130

Cavet JS, Meng W, Pennella MA, Appelhoff RJ, Giedroc DP, Robinson NJ (2002) A nickel-cobalt-sensing ArsR-SmtB family repressor. Contributions of cytosol and effector binding sites to metal selectivity. J Biol Chem 277:38441–38448

Champier L, Duarte V, Michaud-Soret I, Coves J (2004) Characterization of the MerD protein from *Ralstonia metallidurans* CH34: a possible role in bacterial mercury resistance by switching off the induction of the mer operon. Mol Microbiol 52:1475–1485

Changela A, Chen K, Xue Y, Holschen J, Outten CE, O'Halloran TV, Mondragon A (2003) Molecular basis of metal-ion selectivity and zeptomolar sensitivity by CueR. Science 301:1383–1387

Chen L, James LP, Helmann JD (1993) Metalloregulation in *Bacillus subtilis*: isolation and characterization of two genes differentially repressed by metal ions. J Bacteriol 175:5428–5437

Chivers PT, Sauer RT (1999) NikR is a ribbon-helix-helix DNA-binding protein. Protein Sci 8:2494–2500

Chivers PT, Sauer RT (2002) NikR repressor: high-affinity nickel binding to the C-terminal domain regulates binding to operator DNA. Chem Biol 9:1141–1148

Chou CJ, Wisedchaisri G, Monfeli RR, Oram DM, Holmes RK, Hol WG, Beeson C (2004) Functional studies of the *Mycobacterium tuberculosis* iron-dependent regulator. J Biol Chem 279:53554–53561

Chung HJ, Choi JH, Kim EJ, Cho YH, Roe JH (1999) Negative regulation of the gene for Fe-containing superoxide dismutase by an Ni-responsive factor in *Streptomyces coelicolor*. J Bacteriol 181:7381–7384

Coy M, Doyle C, Besser J, Neilands JB (1994) Site-directed mutagenesis of the ferric uptake regulation gene of *Escherichia coli*. Biometals 7:292–298

Cromie MJ, Shi Y, Latifi T, Groisman EA (2006) An RNA sensor for intracellular Mg(2+). Cell 125:71–84

de Lorenzo V, Wee S, Herrero M, Neilands JB (1987) Operator sequences of the aerobactin operon of plasmid ColV-K30 binding the ferric uptake regulation (fur) repressor. J Bacteriol 169:2624–2630

Delany I, Rappuoli R, Scarlato V (2004) Fur functions as an activator and as a repressor of putative virulence genes in *Neisseria meningitidis*. Mol Microbiol 52:1081–1090

Delany I, Spohn G, Rappuoli R, Scarlato V (2001) The Fur repressor controls transcription of iron-activated and -repressed genes in *Helicobacter pylori*. Mol Microbiol 42:1297–1309

Diaz-Mireles E, Wexler M, Sawers G, Bellini D, Todd JD, Johnston AW (2004) The Fur-like protein Mur of *Rhizobium leguminosarum* is a Mn(2+)-responsive transcriptional regulator. Microbiology 150:1447–1456

Diaz-Mireles E, Wexler M, Todd JD, Bellini D, Johnston AW, Sawers RG (2005) The manganese-responsive repressor Mur of *Rhizobium leguminosarum* is a member of the Fur-superfamily that recognizes an unusual operator sequence. Microbiology 151:4071–4078

Dosanjh NS, Michel SL (2006) Microbial nickel metalloregulation: NikRs for nickel ions. Curr Opin Chem Biol 10:123–130

Dussurget O, Timm J, Gomez M, Gold B, Yu S, Sabol SZ, Holmes RK, Jacobs WR Jr, Smith I (1999) Transcriptional control of the iron-responsive *fxbA* gene by the mycobacterial regulator IdeR. J Bacteriol 181:3402–3408

Escolar L, Perez-Martin J, de Lorenzo V (1999) Opening the iron box: transcriptional metalloregulation by the Fur protein. J Bacteriol 181:6223–6229

Feese MD, Ingason BP, Goranson-Siekierke J, Holmes RK, Hol WG (2001) Crystal structure of the iron-dependent regulator from *Mycobacterium tuberculosis* at 2.0-Å resolution reveals the Src homology domain 3-like fold and metal binding function of the third domain. J Biol Chem 276:5959–5966

Finney LA, O'Halloran TV (2003) Transition metal speciation in the cell: insights from the chemistry of metal ion receptors. Science 300:931–936

Frantz B, O'Halloran TV (1990) DNA distortion accompanies transcriptional activation by the metal-responsive gene-regulatory protein MerR. Biochemistry 29:4747–4751

Friedman YE, O'Brian MR (2004) The ferric uptake regulator (Fur) protein from *Bradyrhizobium japonicum* is an iron-responsive transcriptional repressor in vitro. J Biol Chem 279:32100–32105

Fuangthong M, Helmann JD (2003) Recognition of DNA by three ferric uptake regulator (Fur) homologs in *Bacillus subtilis*. J Bacteriol 185:6348–6357

Fuangthong M, Herbig AF, Bsat N, Helmann JD (2002) Regulation of the *Bacillus subtilis* fur and perR genes by PerR: not all members of the PerR regulon are peroxide inducible. J Bacteriol 184:3276–3286

Funahashi T, Fujiwara C, Okada M, Miyoshi S, Shinoda S, Narimatsu S, Yamamoto S (2000) Characterization of *Vibrio parahaemolyticus* manganese-resistant mutants in reference to the function of the ferric uptake regulatory protein. Microbiol Immunol 44:963–970

Gaballa A, Helmann JD (1998) Identification of a zinc-specific metalloregulatory protein, Zur, controlling zinc transport operons in *Bacillus subtilis*. J Bacteriol 180:5815–5821

Gaballa A, Helmann JD (2002) A peroxide-induced zinc uptake system plays an important role in protection against oxidative stress in *Bacillus subtilis*. Mol Microbiol 45:997–1005

Gaballa A, Wang T, Ye RW, Helmann JD (2002) Functional analysis of the *Bacillus subtilis* Zur regulon. J Bacteriol 184:6508–6514

Glasfeld A, Guedon E, Helmann JD, Brennan RG (2003) Structure of the manganese-bound manganese transport regulator of *Bacillus subtilis*. Nat Struct Biol 10:652–657

Gold B, Rodriguez GM, Marras SA, Pentecost M, Smith I (2001) The *Mycobacterium tuberculosis* IdeR is a dual functional regulator that controls transcription of genes involved in iron acquisition, iron storage and survival in macrophages. Mol Microbiol 42:851–865

Golynskiy MV, Davis TC, Helmann JD, Cohen SM (2005) Metal-induced structural organization and stabilization of the metalloregulatory protein MntR. Biochemistry 44:3380–3389

Griggs DW, Konisky J (1989) Mechanism for iron-regulated transcription of the *Escherichia coli* cir gene: metal-dependent binding of Fur protein to the promoters. J Bacteriol 171:1048–1054

Guedon E, Helmann JD (2003) Origins of metal ion selectivity in the DtxR/MntR family of metalloregulators. Mol Microbiol 48:495–506
Guedon E, Moore CM, Que Q, Wang T, Ye RW, Helmann JD (2003) The global transcriptional response of *Bacillus subtilis* to manganese involves the MntR, Fur, TnrA and sigmaB regulons. Mol Microbiol 49:1477–1491
Hamed MY (1993) Binding of the ferric uptake regulation repressor protein (Fur) to Mn(II), Fe(II), Co(II), and Cu(II) ions as co-repressors: electronic absorption, equilibrium, and ^{57}Fe Mossbauer studies. J Inorg Biochem 50:193–210
Hamza I, Chauhan S, Hassett R, O'Brian MR (1998) The bacterial Irr protein is required for coordination of heme biosynthesis with iron availability. J Biol Chem 273:21669–21674
Hantke K (1981) Regulation of ferric iron transport in *Escherichia coli* K12: isolation of a constitutive mutant. Mol Gen Genet 182:288–292
Hantke K (1987) Selection procedure for deregulated iron transport mutants (*fur*) in *Escherichia coli* K 12: Fur not only affects iron metabolism. Mol Gen Genet 210:135–139
Hantke K (2001) Iron and metal regulation in bacteria. Curr Opin Microbiol 4:172–177
Hantke K (2005) Bacterial zinc uptake and regulators. Curr Opin Microbiol 8:196–202
Harvie DR, Andreini C, Cavallaro G, Meng W, Connolly BA, Yoshida K, Fujita Y, Harwood CR, Radford DS, Tottey S, Cavet JS, Robinson NJ (2006) Predicting metals sensed by ArsR-SmtB repressors: allosteric interference by a non-effector metal. Mol Microbiol 59:1341–1356
Hazlett KR, Rusnak F, Kehres DG, Bearden SW, La Vake CJ, La Vake ME, Maguire ME, Perry RD, Radolf JD (2003) The *Treponema pallidum tro* operon encodes a multiple metal transporter, a zinc-dependent transcriptional repressor, and a semi-autonomously expressed phosphoglycerate mutase. J Biol Chem 278:20687–20694
Helmann JD, Ballard BT, Walsh CT (1990) The MerR metalloregulatory protein binds mercuric ion as a tricoordinate, metal-bridged dimer. Science 247:946–948
Helmann JD, Wu MF, Gaballa A, Kobel PA, Morshedi MM, Fawcett P, Paddon C (2003) The global transcriptional response of *Bacillus subtilis* to peroxide stress is coordinated by three transcription factors. J Bacteriol 185:243–253
Hentze MW, Muckenthaler MU, Andrews NC (2004) Balancing acts: molecular control of mammalian iron metabolism. Cell 117:285–297
Herbig AF, Helmann JD (2001) Roles of metal ions and hydrogen peroxide in modulating the interaction of the *Bacillus subtilis* PerR peroxide regulon repressor with operator DNA. Mol Microbiol 41:849–859
Hill PJ, Cockayne A, Landers P, Morrissey JA, Sims CM, Williams P (1998) SirR, a novel iron-dependent repressor in *Staphylococcus epidermidis*. Infect Immun 66:4123–4129
Hobman JL, Wilkie J, Brown NL (2005) A design for life: prokaryotic metal-binding MerR family regulators. Biometals 18:429–436
Holm RH, Kennepohl P, Solomon EI (1996) Structural and functional aspects of metal sites in biology. Chem Rev 96:2239–2314
Holmes RK (2000) Biology and molecular epidemiology of diphtheria toxin and the *tox* gene. J Infect Dis 181 Suppl 1:S156–167
Horsburgh MJ, Wharton SJ, Cox AG, Ingham E, Peacock S, Foster SJ (2002) MntR modulates expression of the PerR regulon and superoxide resistance in *Staphylococcus aureus* through control of manganese uptake. Mol Microbiol 44:1269–1286
Jacquamet L, Aberdam D, Adrait A, Hazemann JL, Latour JM, Michaud-Soret I (1998) X-ray absorption spectroscopy of a new zinc site in the Fur protein from *Escherichia coli*. Biochemistry 37:2564–2571

Kaplan J, McVey Ward D, Crisp RJ, Philpott CC (2006) Iron-dependent metabolic remodeling in *S. cerevisiae*. Biochim Biophys Acta
Keyer K, Imlay JA (1996) Superoxide accelerates DNA damage by elevating free-iron levels. Proc Natl Acad Sci USA 93:13635–13640
Kliegman JI, Griner SL, Helmann JD, Brennan RG, Glasfeld A (2006) Structural basis for the metal-selective activation of the manganese transport regulator of *Bacillus subtilis*. Biochemistry 45:3493–3505
Kojetin DJ, Thompson RJ, Benson LM, Naylor S, Waterman J, Davies KG, Opperman CH, Stephenson K, Hoch JA, Cavanagh J (2005) Structural analysis of divalent metals binding to the *Bacillus subtilis* response regulator Spo0F: the possibility for in vitro metalloregulation in the initiation of sporulation. Biometals 18:449–466
Kunst F, Ogasawara N, Moszer I, Albertini AM, Alloni G, Azevedo V, Bertero MG, Bessieres P, Bolotin A, Borchert S, Borriss R, Boursier L, Brans A, Braun M, Brignell SC, Bron S, Brouillet S, Bruschi CV, Caldwell B, Capuano V, Carter NM, Choi SK, Codani JJ, Connerton IF, Danchin A, et al. (1997) The complete genome sequence of the gram-positive bacterium *Bacillus subtilis*. Nature 390:249–256
Lane TW, Saito MA, George GN, Pickering IJ, Prince RC, Morel FM (2005) Biochemistry: a cadmium enzyme from a marine diatom. Nature 435:42
Lavrrar JL, McIntosh MA (2003) Architecture of a Fur binding site: a comparative analysis. J Bacteriol 185:2194–2202
Lee JW, Helmann JD (2006) The PerR transcription factor senses H_2O_2 by metal-catalysed histidine oxidation. Nature 440:363–367
Lieser SA, Davis TC, Helmann JD, Cohen SM (2003) DNA-binding and oligomerization studies of the manganese(II) metalloregulatory protein MntR from *Bacillus subtilis*. Biochemistry 42:12634–12642
Liu T, Golden JW, Giedroc DP (2005) A zinc(II)/lead(II)/cadmium(II)-inducible operon from the *Cyanobacterium anabaena* is regulated by AztR, an alpha3N ArsR/SmtB metalloregulator. Biochemistry 44:8673–8683
Loo CY, Mitrakul K, Voss IB, Hughes CV, Ganeshkumar N (2003) Involvement of the adc operon and manganese homeostasis in *Streptococcus gordonii* biofilm formation. J Bacteriol 185:2887–2900
Makarova KS, Ponomarev VA, Koonin EV (2001) Two C or not two C: recurrent disruption of Zn-ribbons, gene duplication, lineage-specific gene loss, and horizontal gene transfer in evolution of bacterial ribosomal proteins. Genome Biol 2:RESEARCH 0033
Masse E, Arguin M (2005) Ironing out the problem: new mechanisms of iron homeostasis. Trends Biochem Sci 30:462–468
Masse E, Gottesman S (2002) A small RNA regulates the expression of genes involved in iron metabolism in *Escherichia coli*. Proc Natl Acad Sci USA 99:4620–4625
Masse E, Vanderpool CK, Gottesman S (2005) Effect of RyhB small RNA on global iron use in *Escherichia coli*. J Bacteriol 187:6962–6971
McHugh JP, Rodriguez-Quinones F, Abdul-Tehrani H, Svistunenko DA, Poole RK, Cooper CE, Andrews SC (2003) Global iron-dependent gene regulation in *Escherichia coli*. A new mechanism for iron homeostasis. J Biol Chem 278:29478–29486
Meibom KL, Kallipolitis BH, Ebright RH, Valentin-Hansen P (2000) Identification of the subunit of cAMP receptor protein (CRP) that functionally interacts with CytR in CRP-CytR-mediated transcriptional repression. J Biol Chem 275:11951–11956
Mergeay M, Monchy S, Vallaeys T, Auquier V, Benotmane A, Bertin P, Taghavi S, Dunn J, van der Lelie D, Wattiez R (2003) *Ralstonia metallidurans*, a bacterium specifically adapted to toxic metals: towards a catalogue of metal-responsive genes. FEMS Microbiol Rev 27:385–410

Mey AR, Craig SA, Payne SM (2005) Characterization of *Vibrio cholerae* RyhB: the RyhB regulon and role of ryhB in biofilm formation. Infect Immun 73:5706–5719
Moore CM, Gaballa A, Hui M, Ye RW, Helmann JD (2005) Genetic and physiological responses of *Bacillus subtilis* to metal ion stress. Mol Microbiol 57:27–40
Moore CM, Helmann JD (2005) Metal ion homeostasis in *Bacillus subtilis*. Curr Opin Microbiol 8:188–195
Murphy ER, Dickenson A, Militello KT, Connell TD (1999) Genetic characterization of wild-type and mutant *fur* genes of *Bordetella avium*. Infect Immun 67:3160–3165
Nanamiya H, Akanuma G, Natori Y, Murayama R, Kosono S, Kudo T, Kobayashi K, Ogasawara N, Park SM, Ochi K, Kawamura F (2004) Zinc is a key factor in controlling alternation of two types of L31 protein in the *Bacillus subtilis* ribosome. Mol Microbiol 52:273–283
Nies DH (2003) Efflux-mediated heavy metal resistance in prokaryotes. FEMS Microbiol Rev 27:313–339
Nucifora G, Silver S, Misra TK (1989) Down regulation of the mercury resistance operon by the most promoter-distal gene merD. Mol Gen Genet 220:69–72
O'Halloran T, Walsh C (1987) Metalloregulatory DNA-binding protein encoded by the *merR* gene: isolation and characterization. Science 235:211–214
O'Halloran TV (1993) Transition metals in control of gene expression. Science 261:715–725
O'Halloran TV, Frantz B, Shin MK, Ralston DM, Wright JG (1989) The MerR heavy metal receptor mediates positive activation in a topologically novel transcription complex. Cell 56:119–129
Outten CE, O'Halloran TV (2001) Femtomolar sensitivity of metalloregulatory proteins controlling zinc homeostasis. Science 292:2488–2492
Outten CE, Tobin DA, Penner-Hahn JE, O'Halloran TV (2001) Characterization of the metal receptor sites in *Escherichia coli* Zur, an ultrasensitive zinc(II) metalloregulatory protein. Biochemistry 40:10417–10423
Panina EM, Mironov AA, Gelfand MS (2001) Comparative analysis of FUR regulons in gamma-proteobacteria. Nucleic Acids Res 29:5195–5206
Panina EM, Mironov AA, Gelfand MS (2003) Comparative genomics of bacterial zinc regulons: enhanced ion transport, pathogenesis, and rearrangement of ribosomal proteins. Proc Natl Acad Sci USA 100:9912–9917
Patzer SI, Hantke K (1998) The ZnuABC high-affinity zinc uptake system and its regulator Zur in *Escherichia coli*. Mol Microbiol 28:1199–1210
Patzer SI, Hantke K (2001) Dual repression by Fe(2+)-Fur and Mn(2+)-MntR of the *mntH* gene, encoding an NRAMP-like Mn(2+) transporter in *Escherichia coli*. J Bacteriol 183:4806–4813
Payne SM, Wyckoff EE, Murphy ER, Oglesby AG, Boulette ML, Davies NM (2006) Iron and pathogenesis of *Shigella*: iron acquisition in the intracellular environment. Biometals 19:173–180
Peers G, Price NM (2006) Copper-containing plastocyanin used for electron transport by an oceanic diatom. Nature 441:341–344
Pennella MA, Arunkumar AI, Giedroc DP (2006) Individual metal ligands play distinct functional roles in the zinc sensor *Staphylococcus aureus* CzrA. J Mol Biol 356:1124–1136
Pennella MA, Giedroc DP (2005) Structural determinants of metal selectivity in prokaryotic metal-responsive transcriptional regulators. Biometals 18:413–428
Pohl E, Haller JC, Mijovilovich A, Meyer-Klaucke W, Garman E, Vasil ML (2003) Architecture of a protein central to iron homeostasis: crystal structure and spectroscopic analysis of the ferric uptake regulator. Mol Microbiol 47:903–915

Pohl E, Holmes RK, Hol WG (1999) Crystal structure of the iron-dependent regulator (IdeR) from *Mycobacterium tuberculosis* shows both metal binding sites fully occupied. J Mol Biol 285:1145–1156

Pohl E, Holmes RK, Hol WG (1999) Crystal structure of a cobalt-activated diphtheria toxin repressor-DNA complex reveals a metal-binding SH3-like domain. J Mol Biol 292:653–667

Pohl E, Qui X, Must LM, Holmes RK, Hol WG (1997) Comparison of high-resolution structures of the diphtheria toxin repressor in complex with cobalt and zinc at the cation-anion binding site. Protein Sci 6:1114–1118

Posey JE, Hardham JM, Norris SJ, Gherardini FC (1999) Characterization of a manganese-dependent regulatory protein, TroR, from *Treponema pallidum*. Proc Natl Acad Sci USA 96:10887–10892

Prince RW, Cox CD, Vasil ML (1993) Coordinate regulation of siderophore and exotoxin A production: molecular cloning and sequencing of the *Pseudomonas aeruginosa fur* gene. J Bacteriol 175:2589–2598

Puig S, Askeland E, Thiele DJ (2005) Coordinated remodeling of cellular metabolism during iron deficiency through targeted mRNA degradation. Cell 120:99–110

Qi Z, Hamza I, O'Brian MR (1999) Heme is an effector molecule for iron-dependent degradation of the bacterial iron response regulator (Irr) protein. Proc Natl Acad Sci USA 96:13056–13061

Qi Z, O'Brian MR (2002) Interaction between the bacterial iron response regulator and ferrochelatase mediates genetic control of heme biosynthesis. Mol Cell 9:155–162

Qiu X, Verlinde CL, Zhang S, Schmitt MP, Holmes RK, Hol WG (1995) Three-dimensional structure of the diphtheria toxin repressor in complex with divalent cation co-repressors. Structure 3:87–100

Que Q, Helmann JD (2000) Manganese homeostasis in *Bacillus subtilis* is regulated by MntR, a bifunctional regulator related to the diphtheria toxin repressor family of proteins. Mol Microbiol 35:1454–1468

Raumann BE, Rould MA, Pabo CO, Sauer RT (1994) DNA recognition by beta-sheets in the Arc repressor-operator crystal structure. Nature 367:754–757

Rodionov DA, Dubchak I, Arkin A, Alm E, Gelfand MS (2004) Reconstruction of regulatory and metabolic pathways in metal-reducing delta-proteobacteria. Genome Biol 5:R90

Rodriguez GM, Smith I (2003) Mechanisms of iron regulation in mycobacteria: role in physiology and virulence. Mol Microbiol 47:1485–1494

Ross W, Park SJ, Summers AO (1989) Genetic analysis of transcriptional activation and repression in the Tn*21* mer operon. J Bacteriol 171:4009–4018

Schmitt MP (2002) Analysis of a DtxR-like metalloregulatory protein, MntR, from *Corynebacterium diphtheriae* that controls expression of an ABC metal transporter by an Mn(2+)-dependent mechanism. J Bacteriol 184:6882–6892

Schneider R, Hantke K (1993) Iron-hydroxamate uptake systems in *Bacillus subtilis*: identification of a lipoprotein as part of a binding protein-dependent transport system. Mol Microbiol 8:111–121

Schumacher MA, Brennan RG (2002) Structural mechanisms of multidrug recognition and regulation by bacterial multidrug transcription factors. Mol Microbiol 45:885–893

Sen KI, Sienkiewicz A, Love JF, vanderSpek JC, Fajer PG, Logan TM (2006) Mn(II) binding by the anthracis repressor from *Bacillus anthracis*. Biochemistry 45:4295–4303

Shi W, Dong J, Scott RA, Ksenzenko MY, Rosen BP (1996) The role of arsenic-thiol interactions in metalloregulation of the *ars* operon. J Biol Chem 271:9291–9297

Shi W, Wu J, Rosen BP (1994) Identification of a putative metal binding site in a new family of metalloregulatory proteins. J Biol Chem 269:19826–19829
Song L et al. (2004) Engineered single-chain, antiparallel, coiled coil mimics the MerR metal binding site. J Bacteriol 186:1861–1868
Spiering MM, Ringe D, Murphy JR, Marletta MA (2003) Metal stoichiometry and functional studies of the diphtheria toxin repressor. Proc Natl Acad Sci USA 100:3808–3813
Summers AO (1992) Untwist and shout: a heavy metal-responsive transcriptional regulator. J Bacteriol 174:3097–3101
Tao X, Schiering N, Zeng HY, Ringe D, Murphy JR (1994) Iron, DtxR, and the regulation of diphtheria toxin expression. Mol Microbiol 14:191–197
Tao X, Zeng HY, Murphy JR (1995) Transition metal ion activation of DNA binding by the diphtheria tox repressor requires the formation of stable homodimers. Proc Natl Acad Sci USA 92:6803–6807
Thelwell C, Robinson NJ, Turner-Cavet JS (1998) An SmtB-like repressor from *Synechocystis* PCC 6803 regulates a zinc exporter. Proc Natl Acad Sci USA 95:10728–10733
Thomas CE, Sparling PF (1996) Isolation and analysis of a *fur* mutant of *Neisseria gonorrhoeae*. J Bacteriol 178:4224–4232
Todd JD, Sawers G, Johnston AW (2005) Proteomic analysis reveals the wide-ranging effects of the novel, iron-responsive regulator RirA in *Rhizobium leguminosarum* bv. viciae. Mol Genet Genomics 273:197–206
Tottey S, Harvie DR, Robinson NJ (2005) Understanding how cells allocate metals using metal sensors and metallochaperones. Acc Chem Res 38:775–783
Traore DA, El Ghazouani A, Ilango S, Dupuy J, Jacquamet L, Ferrer JL, Caux-Thang C, Duarte V, Latour JM (2006) Crystal structure of the apo-PerR-Zn protein from *Bacillus subtilis*. Mol Microbiol 61:1211–1219
Tseng HJ, McEwan AG, Apicella MA, Jennings MP (2003) OxyR acts as a repressor of catalase expression in *Neisseria gonorrhoeae*. Infect Immun 71:550–556
Turner JS, Glands PD, Samson AC, Robinson NJ (1996) Zn^{2+}-sensing by the cyanobacterial metallothionein repressor SmtB: different motifs mediate metal-induced protein-DNA dissociation. Nucleic Acids Res 24:3714–3721
Wennerhold J, Bott M (2006) The DtxR regulon of *Corynebacterium glutamicum*. J Bacteriol 188:2907–2918
Wennerhold J, Krug A, Bott M (2005) The AraC-type regulator RipA represses aconitase and other iron proteins from *Corynebacterium* under iron limitation and is itself repressed by DtxR. J Biol Chem 280:40500–40508
Wexler M, Todd JD, Kolade O, Bellini D, Hemmings AM, Sawers G, Johnston AW (2003) Fur is not the global regulator of iron uptake genes in *Rhizobium leguminosarum*. Microbiology 149:1357–1365
White A, Ding X, vanderSpek JC, Murphy JR, Ringe D (1998) Structure of the metal-ion-activated diphtheria toxin repressor/tox operator complex. Nature 394:502–506
Wilderman PJ, Sowa NA, FitzGerald DJ, FitzGerald PC, Gottesman S, Ochsner UA, Vasil ML (2004) Identification of tandem duplicate regulatory small RNAs in *Pseudomonas aeruginosa* involved in iron homeostasis. Proc Natl Acad Sci USA 101:9792–9797
Wisedchaisri G, Holmes RK, Hol WG (2004) Crystal structure of an IdeR-DNA complex reveals a conformational change in activated IdeR for base-specific interactions. J Mol Biol 342:1155–1169
Woodmansee AN, Imlay JA (2002) Quantitation of intracellular free iron by electron paramagnetic resonance spectroscopy. Methods Enzymol 349:3–9

Wu HJ, Seib KL, Srikhanta YN, Kidd SP, Edwards JL, Maguire TL, Grimmond SM, Apicella MA, McEwan AG, Jennings MP (2006) PerR controls Mn-dependent resistance to oxidative stress in *Neisseria gonorrhoeae*. Mol Microbiol 60:401–416

Yang J, Ishimori K, O'Brian MR (2005) Two heme binding sites are involved in the regulated degradation of the bacterial iron response regulator (Irr) protein. J Biol Chem 280:7671–7676

Ye J, Kandegedara A, Martin P, Rosen BP (2005) Crystal structure of the *Staphylococcus aureus* pI258 CadC Cd(II)/Pb(II)/Zn(II)-responsive repressor. J Bacteriol 187:4214–4221

Microbiol Monogr (6)
D. H. Nies, S. Silver: Molecular Microbiology of Heavy Metals
DOI 10.1007/7171_2006_074/Published online: 13 February 2007

Transcriptomic Responses of Bacterial Cells to Sublethal Metal Ion Stress

Jon L. Hobman[1] (✉) · Kaneyoshi Yamamoto[2] · Taku Oshima[3]

[1]School of Biosciences, The University of Birmingham, Edgbaston, Birmingham B15 2TT, UK
J.L.Hobman@bham.ac.uk

[2]Department of Advanced Bioscience, Kinki University, Nakamachi 3327-204, 631-8505 Nara, Japan

[3]Department of Bioinformatics and Genomics, Graduate School of Information Science, Nara Institute of Science and Technology, 8916-5 Takayama, Ikoma, 630-0101 Nara, Japan

Abstract Bacterial cellular responses to metal ion stress are often measured as changes in transcription of genes involved in metal ion homeostasis, during detoxification processes or during functioning of efflux systems. Although there has been evidence for other bacterial cellular responses to metal ion stress, a view of what these responses are has been difficult to obtain. Recent measurements from genome-wide transcriptional profiling in bacteria strongly suggests that the effects of metals on cells may be very wide-ranging, and the transcriptomic responses equally wide. This chapter integrates the known biological effects of metal ion stress with data from microarray and other gene regulation studies from different bacteria responding to these stresses. Metal ion stresses elicit responses in metal ion homeostasis, oxidative stress responses, membrane stress responses, amino acid synthesis, and the expression of other metal ion import systems.

1 Introduction

A general definition of bacterial stress is difficult to formulate, but could be described as a perturbation in the cellular physiology or metabolism brought about by external factors, which if sufficiently severe (and not responded to) can result in growth limitation or death. Stress to bacteria can take many forms, some of which (e.g., heat and cold shock, acid, osmotic and oxidative stress) are well known, as are the bacterial responses to them. Metals play an essential role in microbial nutrition, but can also cause stress. Metal ion stress in bacteria can be a result of either excessive or insufficient amounts of essential metal ions being present in the cell, or can be caused by the presence of toxic metal ions in the cell. Therefore, the regulation of expression of genes whose products are specifically involved in uptake, trafficking, and/or removal of metal ions is an essential cellular function. Bacteria have to be able to both sense and respond to limiting and to excess levels of both essential and toxic metals (Moore and Helmann 2005), and appropriately control the expression of metal ion uptake and/or efflux genes in response to fluctuating levels of metal ions within the cell. In the case of essential metals, chromosomally encoded metal ion import and homeostasis systems arbitrate the levels of these metal ions in the cell. For systems that specifically confer resistance to toxic metals it is common (although not universal) to find the genes encoding these systems located on mobile genetic elements such as plasmids or transposons. The best understood metal ion response regulators are located in the cytoplasm and are becoming well characterized and understood in bacteria (see other chapters in this volume), but there is increasing evidence that two-component sensor-regulators and extracytoplasmic sensing σ factors are also part of the cellular detection systems for metal ions or the damage they induce.

2
Environment, Nutritional Requirements, and Stress

When bacteria are exposed to different challenges and fluctuations in physical conditions, they must respond rapidly, or become stressed and lose viability. These environmental challenges can include temperature fluctuations, the change from an anaerobic to aerobic environment, pH, ionic strength, water activity, osmolarity, competition for resources, and exposure to toxic chemicals and metals (Storz and Hengge-Aronis 2000). Detection of external stresses, or cellular damage resulting from them, has to be rapid so that proteins involved in removal of the metal ions, negation of the cellular effects caused by the metal ions, or repair of the damage caused can become functionally active. Against this background, bacteria also face the additional challenge that they have to acquire all of the elements they need for growth and replication, including essential metal ions, from their local environment, which may be of low nutrient status or physically and chemically hostile.

2.1
Essential Bulk Nutritional Requirements

Bacteria have absolute nutritional requirements for the bulk chemical elements C, H, O, N, P and S, which are the major component atoms in nucleic acids, proteins, and lipids and together comprise approximately 97% of an *Escherichia coli* cell dry mass (Wackett et al. 2004) H_2O is approximately 80% of total wet mass (Silver S, personal communication). Living organisms also contain inorganic cations and anions. Many of these major elemental cations are metal ions, and include Na^+, Mg^{2+}, K^+, and Ca^{2+} (Wackett et al. 2004; Hobman 2007).

2.2 Metals: Essential, Non-essential, and Toxic

Although required in smaller amounts than the bulk elements, the transition elements V, Mn, Fe, Co, Ni, Cu, Zn, Mo, and W are essential to life, and have to be acquired mostly as inorganic ions by bacteria from their local environment. These metals are required for functions that include acting as cofactors, for structural and catalytic roles in enzymes and proteins, and for stabilizing biological molecules. They are also used in electron transfer and utilization of dioxygen, osmotic balance (Bruins et al. 2000), and as components of metalloenzymes, which account for ∼ 30% of all enzymes in bacterial cells (Wackett et al. 2004). There are a number of metals that have no known positive biological role, and are commonly (but probably erroneously) referred to as "heavy metals" (Nieboer and Richardson 1980; Duffus 2002) or more accurately "toxic metals". Known toxic metals are Al, Au, Ag, Bi, Cd, Cr, Hg, Pb, Sn, and Tl. There are several metalloids that are also known to be toxic, these are: As, Sb, and Te (Mergeay et al. 1985; Nies 1999; Silver and Phung 2005). In addition, there are other metals that have no known biological role and may be toxic to bacteria: these are Be, Cs, Li, and Sr, but little is known about their biological interactions (Wackett et al. 2004; Nies 1999; Bruins et al. 2000; Silver and Phung 2005).

Cellular stress caused by metal ions can therefore be caused by: (i) insufficient levels of essential metal ions; (ii) excess levels of essential metal ions; or (iii) the presence of toxic metal ions.

Insufficient levels of essential metal ions will adversely affect cell functioning simply because there is not enough of these metal ions to perform the biological functions required of them. Paradoxically, although the lists of essential, non-essential, and toxic metals appear straightforward, many essential metals can be toxic to bacteria if their intracellular concentration becomes too high (Collins and Stotzky 1989; Hughes and Poole 1989; Lippard and Berg 1994; Nies 1999).

3 Metal Toxicity

The basis of the toxic effects of some metals and metalloids is essentially a function of their inorganic and bioinorganic chemistry, and it is always the ions and not the metal itself that is toxic. Toxicity is also dependent upon:

- Environmental/abiotic factors that affect metal ion bioavailability: pH, oxidation–reduction potential of the environment, temperature, presence of inorganic anions and cations, clay and organics, and water hardness (presence of dissolved Ca^{2+} and Mg^{2+} with HCO_3^- and CO_3^{2-}) (Collins and Stotzky 1989)

- The ability of the metal(loid) to enter the cell
- The ability of the metal(loid) to interact with ligands in the cell
- The consequences of these interactions
- The ability of the metal(loids) to participate in other harmful cellular reactions

There will be both primary and secondary cellular effects caused by toxic metal ions in bacteria. These effects will initially be caused by interactions between the toxic metal and the cellular components, but there will also be consequential effects of the primary interactions of the metal ions. These secondary effects can be the displacement of metals from biological molecules, and other effects such as metal ion-mediated oxidative stress (detailed below) resulting in both widespread and specific damage to cell components, leading to death if the damage is not repaired.

3.1 Metal Ion Coordination, Thiophilicity, and Mimicry

Several concepts have been suggested to explain why some metals are toxic to cells, and others are not. Each concept has been used to explain the degrees of interaction between different metals and ligands and the consequences of these interactions. Many cellular components, including water, protein side chains, nucleic acids, small cytoplasmic cellular components, and organic cofactors, act as ligands to metals. One concept that has been widely used to understand how metals interact with biological molecules is that the power of a metal ion to complex ligands is dependent on its charge/radius ratio. A metal cation with a high charge/radius ratio (polarizing power) will interact more strongly with a ligand than a metal cation with a lower polarizing power (Hughes and Poole 1989). The ionic size of a metal ion decreases as nuclear charge increases from left to right across the Periodic Table, and the polarizing power of metal ions increases. Metal ions with the smallest ionic size have the highest polarizing power. It follows that for a given ligand the binding preference of an individual metal cation follows (but not always invariably) the Irving–Williams series, which for the first row transition metals is $Ca^{2+} < Mg^{2+} < Mn^{2+} < Fe^{2+} < Co^{2+} < Ni^{2+} < Cu^{2+} > Zn^{2+}$ (Hughes and Poole 1989; Lippard and Berg 1994). The polarizing power of a metal ion is therefore important to the strength of interactions with a ligand, with the more polarizable the atomic structure, the "softer" the metal ion, and the stronger the interactions with ligands. Many of the most toxic metals are soft Lewis acids (Pearson 1963), which have a large and polarizable atomic structure and unshared electron pairs in their valence shell. Soft Lewis acids form stable complexes with soft Lewis bases, which have polarizable donor atoms. Soft Lewis acids prefer to coordinate to S and N groups (soft Lewis bases), such as cysteine sulfydryl (SH) groups or methionine thioether linkages, and/or histidine nitrogens, as well as oxygen groups in other biological molecules. Cysteines are structurally important in protein tertiary structure (Lippard and Berg 1994) and also coordinate essential metals in the active sites of many metalloproteins. The ligand preferences of hard and soft metal cations (acids) and bases are shown in Table 1, and indicate the preferences of metals for different ligands. A different classification of metals by Nieboer and Richardson (1980) into class A (oxygen seeking), class B (sulfur and nitrogen seeking), and intermediate (O/N/S) classifies metals in a similar but not identical way to the hard and soft acids and bases theory (Collins and Stotzky 1989). Many of the hard Lewis acid metals such as Mg^{2+}, Ca^{2+}, and Na^{2+} are essential (but not always), many of the metals required in trace amounts (such as Zn^{2+} and Cu^{2+}) are intermediate, and the soft Lewis acid metals (Ag^{+}, Au^{+}, Cd^{2+}, Hg^{2+}, Pb^{2+}) are toxic and the most thiophilic. They have a higher affinity for sulfydryls than hard and interme-

Table 1 Classification of metals

Classification of metals as Class A or B (Nieboehr and Richardson 1980; Duffus 2002)

Class	Metals
Class A (hard metals)	Li, Be, Na, Mg, Al, K, Ca, Sc, Ti, Fe(III), Rb, Sr, Y, Zr, Cs, Ba, La, Hf, Fr, Ra, Ac, Th
Borderline (intermediate metals)	V, Cr, Mn, Fe(II), Co, Ni, Cu(I), Zn, Rh, Pb(IV), Sn
Class B (soft metals)	Cu(I), Pd, Ag, Cd, Ir, Pt, Au, Hg, Tl, Pb(II)

Hard and soft Lewis acids and bases (Pearson 1963; Hughes and Poole 1989; Lippard and Berg 1994)

Metals	Ligands
Hard	
Mn^{2+}, Cr^{3+}, Na^{+}, Al^{3+}, Co^{3+}, K^{+}, Ga^{3+}, Fe^{3+}, Mg^{2+}, Ca^{2+}, Tl^{3+}	H_2O, NH_3, ROH, RNH_2, $CO_3{}^{2-}$, Cl^-, $PO_4{}^{3-}$, $ROPO_3{}^{2-}(RO)_2PO_2{}^-$, $SO_4{}^{2-}$, OH^-, $NO_3{}^-$, $CO_2{}^-$, RO^-
Intermediate	
Fe^{2+}, Co^{2+}, Ni^{2+}, Cu^{2+}, Zn^{2+}	Pyridine, $NO_2{}^-$, $SO_3{}^{2-}$, Br^-, $N_3{}^-$
Soft	
Cu^{+}, Pb^{2+}, Cd^{2+}, Tl^{+}, Hg^{2+}, Au^{+}	CN^-, RS^-, RSH, RNC, R_2S, $S_2O_3{}^{2-}H^-$, CO, $(RS)_2PO_2{}^-$, $(RO)_2P(O)S^-$

diate Lewis acids. The primary effects of toxic thiophilic metals entering the cell are that they:

- Will block essential functional groups of biological molecules
- Will displace essential metals (hard Lewis acids) with a lower affinity for their ligands from enzymes
- Will coordinate to the sulfur-containing side chains of cysteine and methionines and cross-link thiol groups, resulting in the distortion of protein structure
- Will bind to low molecular weight thiol compounds such as glutathione and cysteine, reducing the cellular pool of the reduced form of these
- Will bind to other biological molecules, altering their conformation and interfering with their function
- May participate in chemical reactions that produce products harmful to the cell

The consequences of these interactions can be both general and widespread, and the reported effects of metal ion toxicity include damage to membranes, DNA and proteins, interference in enzyme function, and impairment of essential cellular functions. Some of these effects are similar to those reported for oxidative stress (see below).

Even in bacteria that carry specific metal efflux or detoxification mechanisms, excess concentrations of metal ions reach a "tilt" point: beyond this point the bacterial homeostasis and repair mechanisms cannot effectively cope with the excess levels of the metal within the cell and the consequent damage caused by the metal. At this point cellular damage is detected experimentally by loss of cell viability in minimum inhibitory concentration (MIC) assays, or cell death time-course experiments (killing curves). For a perspective on death and survival in bacteria see Aertson and Michiels (2004). In reality, any damage and stress caused by metal ions and cellular responses to them has occurred before the cells lose viability.

3.2 Metals and Oxidative Stress: Redox Active and Redox Inactive Metals

Work carried out predominantly in eukaryotic organisms has shown that the redox-active metals Fe, Cu, Cr, and V, and some redox-inactive metals such as Pb, Cd, Hg, Ni, and As are involved in oxidative stress damage in a variety of different model organisms (see Geslin et al. 2001; Mattie and Freedman 2004; Stohs and Bagchi 1995; and Ercal et al. 2001 for reviews). Other reports from higher organisms link As- and Cr-mediated oxidative stress and apoptosis to carcinogenesis (reviewed in Shi et al. 2004).

Under usual conditions, cells produce reactive oxygen species (ROS), as by-products of their use of molecular oxygen (O_2) in respiration or nutrient oxidation. ROS are partially reduced derivatives of molecular oxygen, which in the one-electron reduced form is superoxide ($O_2^{\cdot-}$), and in the two-electron reduced form is hydrogen peroxide (H_2O_2). Both of these molecules are generated by the sequential univalent reduction of molecular oxygen by respiratory chain enzymes (Storz and Hengge-Aronis 2000; Cabiscol et al. 2000). Superoxide, hydrogen peroxide, and the hydroxyl radical (OH·) generated from hydrogen peroxide in the Fenton reaction are produced by aerobically grown cells and damage nucleic acids, proteins, and lipids. Oxidative stress responses are found in all aerobic organisms, and aim to limit the levels of ROS in the cell at levels that are not deleterious, or to repair damage caused by ROS (Cabiscol et al. 2000). The key regulators of the oxidative stress response in *E. coli* are OxyR and SoxRS (with some contribution to oxidative stress response from RpoS), which respectively respond to peroxides (OxyR-H_2O_2) or superoxide (SoxR-$O_2^{\cdot-}$). SoxR senses oxidative stress via cysteines forming a [2Fe – 2S] center which is oxidized by superoxide to the $Fe^{3+} - Fe^{3+}$ form. When the Fe – S clusters are oxidized, SoxR activates expression of SoxS from the *soxS* promoter, and SoxS regulates the expression of, amongst many genes, DNA repair enzymes and Mn superoxide dismutase (MnSOD). Tetrameric OxyR senses hydrogen peroxide through cysteine residues being oxidized to form an intramolecular disulfide bond. Oxidized OxyR binds DNA at its cognate promoters and interacts with the C-terminal domain of the alpha subunit of RNA polymerase, which stimulates transcription (Pomposiello and Demple 2001). The OxyR and SoxRS regulons have been defined in *E. coli* K-12 by transcriptomics analysis (Pomposiello et al. 2001; Zheng et al. 2001). In *Bacillus subtilis* the PerR regulator senses peroxide by reduction of hydrogen peroxide to generate a hydroxyl radical using a PerR-bound ferrous ion. Oxidation of a histidine in PerR by a hydroxyl radical leads to sensing of the presence of peroxides and derepression of the PerR regulon, which includes catalase, a DPS-like protein (DNA protection during starvation protein/DNA-binding protein from starved cells) and alkylhydroperoxide reductase (Lee and Helmann 2006).

As well as environmental shocks, such as ionizing radiation and redox cycling compounds like menadione and paraquat, transition metal ions and toxic metal ions can also cause oxidative stress (Stohs and Bagchi 1995; Ercal et al. 2001) – including metal ions that are not redox active. The best-known redox-active metals are Fe (Andrews et al. 2003) and Cu (Gaetke and Chow 2003; Rensing and Grass 2003; Solioz and Stoyanov 2003), but Cr, Co, Ti, and V are also reported to be redox-active (Kasprzak 2004; Stohs and Bagchi 1985). Redox-active metals participate with the superoxide ion and hydrogen peroxide in the Fenton/Haber–Weiss reactions, the product of which is the hydroxyl free radical (OH·). In *E. coli* the generation of superoxide by Cu has been demonstrated, as has induction of the *soxRS* regulon under Cu stress conditions (Kimura and Nishioka 1997).

The general and specific reactions that redox-active metals participate in to generate the hydroxyl free radical are shown below.

Generalized Fenton-like reaction equation:

$$M^{n+} + H_2O_2 \rightarrow M^{(n+1)+} + OH^- + OH\cdot \qquad \text{Fenton reaction}\,.$$

The oxidized metal ion can be reduced by reduction with $O_2{}^{\cdot-}$:

$$M^{(n+1)+} + O_2{}^{\cdot-} \rightarrow M^{n+} + O_2\,.$$

The balance of these two reactions is:

$$H_2O_2 + O_2{}^{\cdot-} \rightarrow O_2 + OH^- + OH\cdot \qquad \text{Haber–Weiss reaction}\,,$$

where cycling between the two oxidation states of the metal forms a redox couple and where M is the metal and n is the charge carried by the metal (Kasprzak 2004).

For iron, the Fenton and Haber–Weiss reactions are (Andrews et al. 2003):

$$\begin{aligned} &Fe^{2+} + H_2O_2 \rightarrow Fe^{3+} + OH^- + OH\cdot && \text{Fenton reaction}\\ &H_2O_2 + O_2{}^{\cdot-} \rightarrow OH^- + OH\cdot + O_2 && \text{Haber–Weiss reaction (Fe-catalyzed)}\\ &O_2{}^- + Fe^{3+} \rightarrow Fe^{2+} + O_2 && \text{Iron reduction}\,. \end{aligned}$$

For Cu^+ the Fenton-like reaction is:

$$Cu^+ + H_2O_2 \rightarrow Cu^{2+} + OH^- + OH\cdot \qquad \text{Fenton-like reaction}\,.$$

Apart from free hydroxyl radicals, metals can form other oxidants such as metallo-oxo and peroxo species as well as singlet oxygen (Kasprzak 2004). The hydroxyl radical (OH·) is highly reactive; it reacts with most biological molecules (Gaetke and Chow 2003) and damages nucleic acids, proteins, and lipids. DNA damage causes strand breaks and depurination (Ueda et al. 1998), and oxidation of bases. Damage to lipids is through lipid peroxidation, which decreases membrane fluidity, disrupts membrane proteins, and alters membrane properties. Further generation of free radicals leads to degradation of polyunsaturated fatty acids, and the formation of "toxic secondary message" molecules such as malonaldehyde and 4-hydroxyalkenals, which cause cellular damage (Cabiscol et al. 2000). Oxidation of free amino acids and amino acids in proteins is another common and damaging effect of transition metal-induced oxidative stress, with the side chains of Arg, Cys, His, Lys, and Pro residues being targets, leading to protein fragmentation and inter- and intraprotein cross-linking as well as cross-linking of proteins to DNA (Kasprzak 2004; Ercal et al. 2001; Stohs and Bagchi 1995; Shi et al. 2004). It is believed that the process of oxidative damage to lipids and proteins is caused by the hydroxyl radical removing a hydrogen from an unsaturated fatty acid to form a lipid radical, or removing a hydrogen from an amino-bearing carbon to form a protein radical (Powell 2000). Reaction of redox-active metals with biological reductants such as glutathione leads directly to generation of OH· radicals (Ueda et al. 1998).

There are several probable reasons for why both redox-active and redox-inactive metals cause oxidative stress responses:

- Redox-active metals participate in Fenton reactions, leading to generation of the hydroxyl free radical (OH·) (Stohs and Bagchi 1995)
- Metals bind to thiols (– SH), which are central components of redox signaling and control pathways, causing damage to the cell's antioxidant defence systems (Hansen et al. 2006; Imlay 2006; Ercal et al. 2001)
- Reduction of toxic metal oxyanions in cellular reactions can lead to the production of hydroxyl free radicals (Nies 1999)
- Metals with a higher affinity for metal binding sites than the metal occupying it (e.g., Fe or Cu) could displace that metal, which could be free to participate in Fenton reactions in the cytoplasm
- The tripeptide glutathione is the predominant low molecular weight thiol in most cells, and is a major contributor to protection against oxidative stress

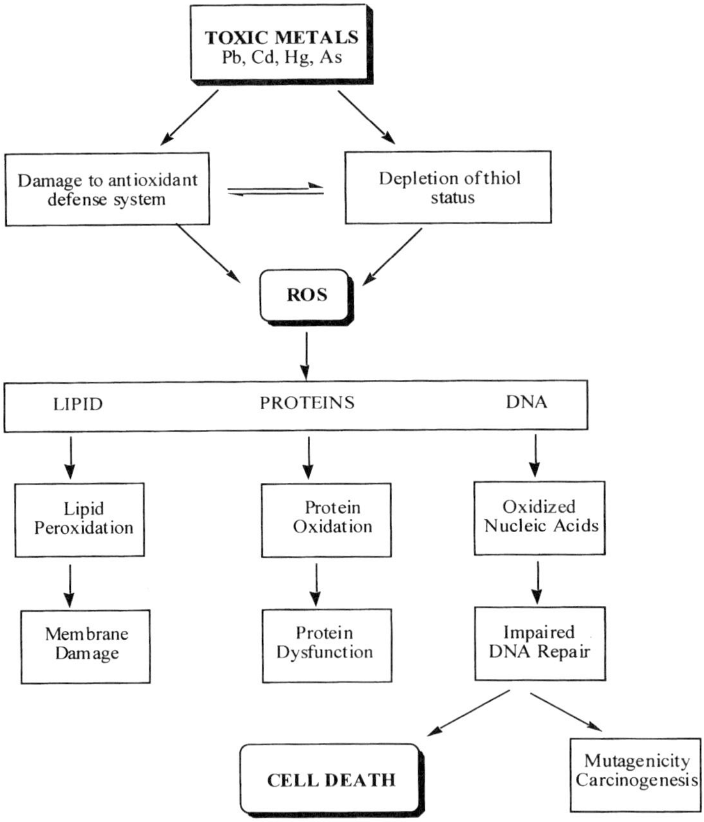

Fig. 1 Possible mechanisms of metal ion-induced oxidative stress and cellular damage (Modified from Ercal et al. 2001, with permission)

Glutathione is known to complex with thiophilic metals, and acts as an intracellular "sponge" for them, but when thiophilic metals bind to reduced glutathione, they form a complex which reacts with oxygen to form oxidized glutathione, H_2O_2, and the metal ion. This reaction both decreases the reduced glutathione pool in cells that require an NADPH-dependent reduction of the oxidized glutathione to reduce it (Nies 1999), and generates peroxide, which will participate in Fenton-like reactions to generate the hydroxyl free radical. The availability of cysteine is the major limiting factor in glutathione synthesis (Hultberg et al. 2001).

The gross manifestations of (metal-induced) oxidative stress are: (i) depletion of sulfydryls, (ii) DNA damage, (iii) lipid peroxidation, (iv) amino acid and protein oxidation, and protein cross-linking. These effects lead to cellular dysfunction and ultimately death. Figure 1 summarizes the effects of metal ion induced toxicity.

4
Bacterial Interactions with Metals: A Multilayered Approach

Bacteria will interact with metals throughout the cell, from the outer membrane, periplasm, and inner membrane (in Gram-negative bacteria) through to the cytoplasm. The interactions will be through the cell wall, porins, membrane-bound importers, the membranes themselves, and the cytoplasmic contents including proteins and cellular regulators. Bacteria need to take up essential metals (which may be insoluble or poorly soluble) from their local environment, and regulate the intracellular levels of these metal ions as well as other metal ions that have entered the cells through non-specific uptake systems. In some circumstances bacteria must also be able to sense and resist toxic metal ions that have entered the cell. For metal ion homeostasis systems in bacteria to work correctly, the bacterial cell must balance the uptake, efflux, and sequestration of metal ions, and if they cannot prevent non-essential or toxic metal ions from entering the cell, they must remove or detoxify them. Ideally, bacterial cells would be equipped with metal ion uptake and efflux systems that work uniquely on one metal, and these would be regulated by specific regulators. In practice, this is not always the case, and the chemistry of metal ions makes it such that biological systems, particularly importers, may have relatively low metal ion specificity.

4.1
Mechanisms for Importing and Controlling the Levels of Essential Metal Ions

The cell membrane passively prevents metal ions from entering the cell (Moore and Helmann 2005), but as a consequence of their requirements for essential metal ions, at higher concentrations than may be found in their local environment, bacteria carry both non-specific and specific metal ion import systems. The non-specific, high rate, and constitutively expressed importers are often driven by the chemiosmotic gradient of the bacterial cytoplasmic membrane (Fig. 2 summarizes the known chromosomally encoded *E. coli* metal ion import and export systems), but have a major disadvantage in their lack of specificity of metal ion import (see Hughes and Poole 1989). Under specific conditions such as shortage of an essential metal, bacteria can use specific, low rate, and inducible high affinity importers that can use the chemiosmotic gradient as well as ATP hydrolysis to drive metal ion uptake (Nies 1999; Silver 1996; Nies and Silver 1995). The general use of high turnover, non-specific transporters is energetically efficient, but there is a penalty to pay for this, which is that metals that are not required by cells (including toxic metals) can enter via these relatively non-specific importers (Fig. 2). Eukaryotes are also faced with similar problems of unwanted toxic metal ion import (Ballatori 2002).

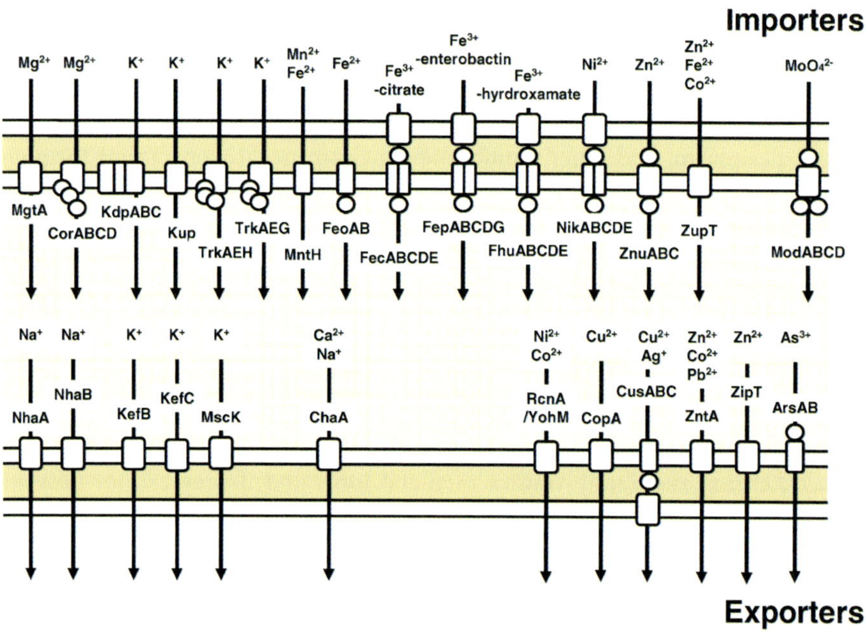

Fig. 2 Metal uptake and efflux systems in *E. coli*. The metal transport systems in *E. coli* are summarized (Earhart 1996; Silver 1996; Boos and Lucht 1996; Csonka and Epstein 1996; Makui et al. 2000; Patzer and Hantke 1998; Li et al. 2004; Rensing et al. 1997, 2000; Grass et al. 2001, 2002; Franke et al. 2001; Outten et al. 2001; Beard et al. 1997; Carlin et al. 1995; Rodrigue et al. 2005). Three pumps, the ZupT transporter (Grass et al. 2005), ChaA exporter (Shijuku et al. 2002), and ZntA exporter (Binet and Poole 2000), have a broad substrate spectrum in addition to their cognate metal

4.2
Mechanisms for Preventing Entry of, or Detoxifying/Removing Toxic Metal Ions

Unlike organic molecules, metals cannot be broken down by enzymatic action, and because of the wide ranging toxic or potentially toxic effects of some metals, strategies for dealing with metals are limited. There have been several mechanisms that have been postulated as metal resistance strategies for bacteria (Rouch et al. 1995). From the exterior of the cell through to the interior of the cell these are:

- Extracellular sequestration of the metal
- Prevention of the metal from entering the cell by reduced permeability
- Active transport of the metal away from the cell
- Intracellular sequestration of the metal by binding to proteins
- Enzymatic alteration of the metal to a less toxic form
- Reduction in sensitivity to the metal by the cellular targets

The last of these proposed mechanisms is extremely difficult for any living organism because the cellular targets for toxic metals are so apparently wide and diverse that only a fundamental re-engineering of the cellular structure would enable an organism to become less sensitive to a particular toxic metal. Therefore, for many bacteria active transport of the metal away from the cell, sequestration and/or enzymatic alteration of the metal to a less toxic form, along with repair of damage caused by the metal, are common mechanisms for responding to metal ion stress. In order to activate these systems, bacteria must be able to sense the presence of the metals, or the damage that they cause.

4.3
Mechanisms for Sensing Metal Ions and Their Cellular Effects

One of the first measurable responses of an organism to a change in the local environment or within the cell is in changes in mRNA transcription and, for reasons of economy, bacteria use the initiation of RNA-transcript formation as the key step in response regulation (Browning and Busby 2004). This is controlled by RNA polymerase, and different σ factor subunits of RNA polymerase will control the expression of different subsets of genes. Transcription factors further modulate gene expression by repressing or activating transcription in response to specific signals. Transcription factors recognize and bind to specific DNA sequences within or proximal to promoters in order to do this. There are clearly defined regulons associated with different transcription factors, and conserved DNA binding sites for those regulators. There can be multiple transcription factors positively and negatively influencing transcription from a single promoter, and many regulons overlap with each other.

The problem that all organisms have to overcome is rapidly sensing the presence of metal ions, or damage caused by them, and transmitting that "alarm" signal to produce an appropriate response through transcription initiation. It is clear that many of these response system regulons are not discrete, but instead the regulons (stimulons) overlap, which means expression of a particular gene can be controlled by a number of different regulators.

As the cellular targets of metal toxicity are so widespread and the adverse effects caused so diverse, the cell requires a variety of sensors and regulators for an efficient and rapid response to these effects. In bacteria there are several fundamentally different designs of regulators: (i) cytoplasmic regulators, (ii) two-component systems, and (iii) ECF σ factors (see Fig. 3a and b).

Gram-negative bacteria need to sense what is happening both to the cytoplasm and the extracytoplasmic compartments, and to the periplasm and outer membrane, which are collectively known as the cell envelope (Ruiz and Silhavy 2005). The conditions in the periplasmic space are different to those in the cytoplasm. The cell wall and peptidoglycan layer border the periplasm, which is viscous, contains high levels of protein, and is oxidizing, leading to the formation of disulfide bonds in proteins. These specific features of the periplasm are used to sense environmental changes, as the condition of the proteins, cell wall components, and presence of oxidizing metals can be used to monitor external conditions directly interacting with the periplasmic space. The cellular localizations of the cytoplasmic, two-component, and ECF σ factor regulator systems are shown in Fig. 2b, and represent the response regulators known to be involved in sensing metals and metal ion damage in *E. coli.*

4.3.1 Cytoplasmic Regulators

The specific metal ion response regulators, such as those belonging to the ArsR/SmtA family (Busenlehner et al. 2003; and the chapter by Bhattacharjee and Rosen in this volume), the MerR family (Permina et al. 2006; Hobman et al. 2005; Pennella and Giedroc 2005; Brown et al. 2003; and the chapter by Soosanga et al. in this volume), Fur and Zur (Pennella and Giedroc 2005; Hantke 2001; Outten et al. 2001; Patzer and Hantke 1998) and other regulators described elsewhere in this volume are located in the cytoplasm of the cell for the simple reason that, as DNA-binding transcriptional regulators, they have to bind to the promoter DNA that they are regulating. However, a major disadvantage of the toxic metal ion sensing regulatory proteins that are located in the cytoplasm is that they only sense the presence of a metal *after* it has entered the cytoplasm and has presumably caused damage. Bacteria (and eukaryotes, including plants) have other sensor systems that allow them to sense what is happening at the cell membrane and beyond the cell membrane.

4.3.2
Two-Component Regulators

Two-component sensor kinase/regulator systems are widely used sensing/regulation systems (Wolanin et al. 2002; Galperin 2006) and have a major advantage over cytoplasmically located regulators in that the sensor is spatially decoupled from the regulator and can transduce a signal across the inner membrane to a cytoplasmic regulator, which acts on the genome (Nies and Brown 1998).

There are approximately 30 known two-component sensor/regulator systems in *E. coli*, some of which cross-talk with each other (Dorel et al. 2006; Oshima et al. 2002). The classic two-component regulators work by an inner membrane-embedded histidine kinase, which acts as a sensor relaying a signal from the inner membrane/periplasm to a cognate cytoplasmic response regulator. The signal transduction occurs via phosphorylation, and the sensor is autophosphorylated on a conserved histidine residue using ATP, and transfers the phosphoryl group to a conserved aspartate residue on the response regulator, which becomes phosphorylated, and eventually then dephosphorylated. The state of phosphorylation of the regulator affects its affinity for target promoters, controlling gene expression (Dorel et al. 2006; Ruiz and Silhavy 2005). The involvement of two-component systems in the regulation of resistance or homeostasis of toxic metals has been recognized for some time, particularly their regulation of the large RND (resistance-nodulation-cell division) cation/proton antiporter efflux pumps (Mergeay et al. 2003; Nies 2004). Metal ion-specific regulators include the *E. coli* chromosomal copper sensing CusRS system (Munson et al. 2000), the plasmid-borne PcoRS copper-responsive two-component system (reviewed in Nies and Brown 1998) and the SilRS silver sensing system (Gupta et al. 1999). In *Cupriavidus metallidurans* strain CH34, there are predicted to be over 40 sensor kinases and response regulators, several of which have been characterized, such as the cobalt- and nickel-resistant *czcRS* system (van der Lelie et al. 1997; Mergeay et al. 2003).

There are also two-component systems that are recognized as important modulators of general cellular stress responses. Some of the most well-known of these in *E. coli* are CpxRA and BaeSR, which control envelope stress response, and multidrug resistance, respectively (Table 2).

4.3.3
Extracytoplasmic Function σ Factors

Alternate σ factors are important in activation of transcription of subsets of genes in response to stresses (Helmann 2002; Browning and Busby 2004). Extracytoplasmic function (ECF) σ factors respond to extracytoplasmic signals such as denaturation of proteins, or physical stresses via a conserved mechan-

Table 2 DNA recognition sites for metal ion/stress-dependent regulators in *E. coli* K-12

Regulators	Recognition sequence	Refs.
ArsR	ArsR family TCAT $(N)_7$ TTTG	Xu et al. 1996
BaeR	OmpR family TGAAGAA $(N)_4$ TGAAGAA	Yamamoto et al., unpublished
BasR	OmpR family Unknown	–
CpxR	OmpR family GTAAA $(N)_{4–8}$ GTAAA	Yamamoto and Ishihama 2006
CueR	MerR family CCTTCC $(N)_8$ GGAAGG	Outten et al. 2000
CusR	OmpR family AAAATGACAA (T/A)(T/A) TTGTCATTTT	Yamamoto and Ishihama 2005
Fur	Fur family GATAATGAT A ATCATTATC	Robison et al. 1998
EvgA	NarL family TTC(C/T)TACA $(N)_2$ TGTA(A/G)GAA	Masuda and Church 2003
HydG	NtrC family GAGTAAAAATGACTCGC $(N)_{12}$ GCGAGTCATTTTTACT	Leomhartsberger et al. 2001
KdpE	OmpR family TTTTTA(T/C)AC $(N)_2$ TTTTTA(T/C)AC	Sugiura et al. 1992
MntR	MntR family AAACATAGC $(N)_4$ GCTATGTTT	Patzer and Hantke 2001
ModE	ModE family TATAT $(N)_7$ TA(C/T)AT	Grunden et al. 1999
NikR	CopG family GTATGA $(N)_{16}$ TCATAC	Chivers and Sauer 2000
SoxR	MerR family CCTCAAGTT AACTTGAGG	Martin and Rosner 2002
SoxS	AraC family A(A/T/C)(A/G)GCAC(A/G)(A/T)(A/T)NN(A/G)(C/T)(C/T)AAA(A/C/T)N	Hidalgo et al. 1995
OmpR	OmpR family TTTACTTTTT GNAACATNTT	Harlocker et al. 1995
PhoP	OmpR family (T/G)(G/A)TT(T/G)A $(N)_5$ (T/G)(G/A)TT(T/G)A	Yamamoto et al. 2002
RcsB	NarL family TAAGAAT ATTCCTA	Hagiwara et al. 2003
RstA	OmpR family Unknown	–
YedW	OmpR family Unknown	–
ZntR	MerR family ACTCTGGAGTC GACTCCAGAGT	Brocklehurst et al. 1999
Zur	Fur family GAANTGTNATA N TATNACANTTC	Patzer and Hantke 1998

ism, and multiple ECF σ factors are found in a wide variety of bacteria (Raivio and Silhavy 2001; Helmann 2002). The *E. coli* ECF σ factor, σ^E, senses change in the periplasm, and can transduce a signal across the inner membrane to the genome, but the mode of action of σ^E regulation is different to that of two-component sensor regulator systems. The σ^E activity in *E. coli* is normally controlled by the inner membrane spanning anti-σ factor, RseA, which sequesters σ^E activity until RseA proteolysis occurs by the action of DegS, RseP, ClpA, and an unidentified protease (Dartigalongue et al. 2001). Protease activity is activated by periplasmic signals of envelope protein misfolding, which activate DegS activity, as well as being induced by high temperatures and abnormal lipopolysaccharide (LPS) (reviewed in Ruiz and Silhavy 2005). When RseA is degraded, σ^E is free of the anti-σ activity of RseA and becomes active as a σ factor, controlling the expression of over 20 promoters in *E. coli*, including four genes (*dsbC*, *fkpA*, *skp*, and *surA*) directly involved in the folding of envelope proteins (Dartigalongue et al. 2001; Ruiz and Silhavy 2005)

There are several known metal ion-responsive ECF σ factors. In *E. coli* under conditions of iron depletion, the ECF σ factor, FecI, activates the expression of *fecABCDE*, which encodes proteins for the ferric citrate transport system. In this system, FecR (inner membrane protein)–FecA (outer membrane protein) interactions and consequent FecI activation are important (Braun et al. 2003, 2006). The expression of *fecIR* is controlled by Fur and, during conditions where iron is depleted, *fecIR* transcription is derepressed. If the citrate iron is present, FecI can activate *fecABCDE* expression. In *C. metallidurans* CH34 recent evidence shows that in the nickel resistance and efflux system encoded by *cnrYHXCBAT*, CnrY and CnrX are membrane-bound proteins that may act as anti-σ factors to CnrH, which is the ECF σ factor controlling expression of the structural *cnr* genes from the promoters *cnrY*P and *cnrC*P (Grass et al. 2005).

5 Genomics and Postgenomics

There are in excess of 340 completed and annotated microbial genome sequences deposited in GenBank (NCBI Genome resources, complete microbial genomes http://www.ncbi.nih.gov/genomes/lproks.cgi). DNA sequences have been explored using bioinformatics for over 20 years and genome sequences that have been assembled, annotated, and compared are now being exploited by postgenomic technologies such as transcriptomics, proteomics, and metabolomics. The exploitation of genomic data allied with high-throughput experimental technologies is altering the way that research can be conducted into the understanding of how biological systems work (Twyman 2004; Hobman et al. 2007).

5.1 Functional Genomics–Transcriptomics as a Research Tool

Although the transcriptome (sum of all of the transcripts in a cell) does not precisely match the proteome (the sum of all of the proteins in a cell), measuring gene expression in response to changing external conditions has been a central activity of molecular biology because it is reproducible, measurable, and there are experimental tools with which to do it. DNA microarrays are an increasingly popular tool for studying the genome-wide transcriptional response of an organism to specific mutations, stresses, and changes in growth conditions (Rhodius and LaRossa 2003). In a simple transcriptomics experiment a culture of growing bacterial cells will be split, with one half of the culture exposed to an environmental stress (physical or chemical). After a short period of exposure the total RNA from cells in the control (unexposed) and test (exposed) culture are separately isolated. The total RNA extracted from each sample is converted to cDNA by reverse transcriptase and the control and test cDNAs labeled with different fluorophore dyes (the dyes Cy3 and Cy5 are popular as they excite and emit at different wavelengths to each other). The labeled cDNAs from the control and test cultures are hybridized together on a DNA microarray, and the signal intensities for each dye measured using a scanner or imager. The resulting data gives the ratio of signal from the test and control at each feature on the array. So, if the test sample gives a higher signal than the control sample at one position on the array corresponding to gene x, then gene x is up-regulated under the stress conditions, compared to the non-stress conditions.

The advantage of DNA array technology, compared to other experimental techniques used for the direct measurement of mRNA species, such as Northern Blots, reverse-transcriptase PCR (RT-PCR), and real-time or quantitative RT-PCR (qRT-PCR), is the high-throughput nature of the technique. It is effectively a massively parallel nucleic acid hybridization (Hobman et al. 2007) and can be used to simultaneously assay the transcriptional response of all ORFs in a genome. There are disadvantages to using microarrays in transcriptomics studies, which include cost, time, high noise/signal problems, and the effort required for data handling, analysis, and interpretation. However, transcriptomics is helping to identify gene targets that can be further characterized using other molecular biology techniques, as well as giving a "snapshot" of what is happening in a cell at a particular time under a particular set of conditions, and defining regulons and stimulons.

6 Transcriptomics in Metal Ion Stress

Much of the published data on understanding what happens to bacterial cells under metal ion stress has been acquired using the model bacteria *E. coli*

K-12 and *B. subtilis*, rather than specialist metal-tolerant/resistant bacteria, such as *C. metallidurans* CH34. Comparisons between the responses to metal ion stress of generalist and specialist bacteria will be of great value, because carriage of a toxic metal resistance mechanism such as those found in *C. metallidurans* CH34 may not be the exclusive or only requirement for bacteria to survive metal ion associated stress.

6.1 Transcriptomic Responses to Metal Stresses in *E. coli* K-12

MG1655 and W3110 are the two sequenced strains of *E. coli* K-12 (Blattner et al. 1997; Hayashi et al. 2006; Riley et al. 2006), and differ by only eight sequence differences (as well as two sites of DNA inversion in W3110, and 13 sites where an IS element or prophage was inserted in only one of the strains). This information is useful for comparison of data between laboratories using these two K-12 strains for their work. Published data indicates that the transcriptomic responses of liquid cultures of *E. coli* to sublethal metal ion stress are widespread, and are summarized below. Figure 3a and b summarize the Cu/Zn stress regulons, and Fig. 4 summarizes known metal ion homeostasis regulation and cross talk in *E. coli* K-12.

6.1.1 Oxidative Stress Response

OxyR and SoxRS are the major transcription factors involved in regulation of the oxidative stress response. Under copper and cadmium stress, the transcriptional profiles of *E. coli* K-12 strains were similar, but not identical, to the oxidative stress response controlled by the SoxRS regulon (as defined by Pomposiello et al. 2001) and not the OxyR regulon (as defined by Zheng et al. 2001), (Kershaw et al. 2005; Yamamoto and Ishihama 2005; Wang and Crowley 2005). The SoxRS regulon is induced primarily by superoxide-generating compounds (Nunoshiba et al. 1992). Regulation of the SoxRS regulon occurs by a two-step mechanism. First, SoxR, a MerR family transcriptional activator, is converted to the active form by the oxidation of its Fe – S centers on the SoxR dimer, by superoxide. The activated form of SoxR stimulates *soxS* transcription by DNA distortion of a MerR family-type promoter (Table 2). Increasing intracellular levels of SoxS, which is an AraC family-type regulator and binds to the asymmetric promoter sequence shown in Table 2, regulates expression of the SoxS regulon. By comparison, OxyR, a LysR family regulator, directly senses hydrogen peroxide and is then converted to the active DNA-binding form of the regulator by oxidation of an intramolecular disulfide bond. It interacts with the alpha-subunit of RNA polymerase to stimulate transcription of oxidative stress response genes (Pomposiello and Demple 2001). Amongst the genes whose expression is stimulated by copper, not all members of the SoxRS regulon are included (Kershaw et al. 2005), but statistical cut-offs used in this study excluded from the final gene list many genes with significantly changed transcriptional profiles (and highlights problems with data analysis, particularly the use of arbitrary standards, and false positives/negatives). In *E. coli*, SoxS is extremely unstable with a 2 min half-life (Griffith et al. 2004). The degradation of SoxS depends on the Lon and FtsH proteases and is inhibited by interaction with a promoter or with RNA polymerase (Griffith et al. 2004; Shah et al. 2006). These transcriptomic data support a previous analysis for *E. coli* K-12 (Kimura and Nishioka 1997) showing that, under copper stress, the SoxRS regulon is activated by the generation of superoxide, and that SodA, SodB, and DNA repair enzyme mutants were more sensitive to copper than the wild-type strain. Previous work showed that the primary targets for Cu(II) ions in membrane proteins are thiol groups, which are oxidized during the reduction of Cu(II) to Cu(I). Reoxidation of Cu(I) to Cu(II) by molecular oxygen generates the superoxide radical (Forman et al. 1980).

6.1.2 Membrane Stress Response

Under both Cu and Zn stress conditions, membrane stress-related regulons are activated. The major membrane stress response is controlled by CpxRA, a two-component regulator that senses misfolding of the NlpE and RapE proteins. CpxRA is a typical two-component signal transduction system, and CpxA, the inner-membrane protein, phosphorylates its conserved histidine residue in response to environmental signals and then transphosphorylates the conserved asparate residue in CpxR (an OmpR family regulator). The phosphorylated and activated form of CpxR (CpxR-P), activates transcription of a set of genes by binding to a conserved tandem repeat of pentanucleotide sequence, GTAAA$(N)_{4-8}$GTAAA (Table 2). CpxRA regulates the transcription of over 20 genes. Recent studies have indicated that the primary role of the CpxRA regulon is in the regulation of adherence for invasion and the formation of biofilms (Raivio 2005; Dorel et al. 2006), which are regarded as more resistant to antibiotics and metal ion stress caused by oxyanions (Harrison et al. 2005). This regulation by CpxRA was observed in both Cu and Zn stress conditions (Table 3). Induction of this regulon is one of the important regulatory effects seen in Cu and Zn stress compared with other metal stresses (Table 3). In contrast, RpoE and the BaeSR regulon, which also respond to membrane stress, are only induced by Zn stress under the conditions used in our comparative studies. This agrees with other work that showed that *rpo*E mutant cells are more sensitive to Cu and Cd stress but the effect of loss of RpoE on sensitivity to Zn stress is less (Egler et al. 2005), and that the *rpoE* regulon was induced under cadmium stress (Wang and Crowley 2005). RNA polymerase containing RpoE transcribes over 60 genes with alternative promoter sequences (Dartigalongue et al. 2001; Kabir et al. 2005; Skovierova et al. 2006). The major role of the RpoE regulon is thought to be to maintain the integrity of the bacterial envelope (Alba and Gross 2004). Recent studies have shown that the lipoproteins, whose expression is regulated by RpoE, are involved in outer membrane biogenesis (Wu et al. 2005; Ruiz et al. 2005, 2006; Onufryk et al. 2005), and the CpxRA and RpoE regulons are partially overlapping (Connoly et al. 1997). BaeSR is another regulator of the membrane stress response stimulon, with CpxRA (Raffa and Raivio 2002), and may be cooperating in the regulation of membrane stress response. BaeSR can simultaneously regulate expression of *acrD*, *mdtA*, and *spy* under indole stress conditions (Raffa and Raivio 2002; Hirakawa et al. 2005). DNase I footprinting analysis has shown the specific binding site of BasR, an OmpR family regulator, on the *acrD* and *mdtA* promoters (Hirakawa et al. 2005). BasR specifically binds to the *spy* promoter (Yamamoto, unpublished), which suggests multiple two-component regulation of genes that respond to different metal ion stresses. The DNAase I footprint data show the conserved direct repeat sequence for DNA binding of BaeR (Table 2). From these data, Zn stress appears to be more similar to membrane stress than other metal stresses are.

Each membrane stress regulon is activated by different signals. The mechanism of RpoE activation is the degradation of RseA, the anti-RpoE, by DegS and RseP (Alba and Gross 2004). This proteolysis is induced by exposure of the carboxyl terminus of DegS, caused by accumulation of misfolded outer membrane proteins in the periplasm (Alba and Gross 2004). The CpxRA regulon is induced by overproduction of lipoproteins (Snyder et al. 1995; Miyadai et al. 2004), misfolded pilus subunits (Hung et al. 2001), and alkaline pH (Danese and Silhavy 1998). However, the molecular mechanism for the activation of CpxRA is not clear.

Under conditions of Fe excess, the BasRS two-component system is activated. BasS, the sensor protein, can bind with an Fe(III) iron and transmit a signal across the inner membrane. BasRS mainly controls the expression of LPS related genes, so it is not clear why BasRS is sensing Fe. It is possible the *cusRS* (Cu) and the *hydHG* (*zraSR*, Zn, Pb) systems can also be activated with direct metal binding. In all cases, if the sensory domain of the two-component system is present in the periplasmic space, then the metal iron needs to be present in the periplasmic space for activation of the two-component system.

6.1.3 The H–NS Regulon

H – NS is one of the nucleoid proteins in *E. coli* and is involved in many stress responses, including those to osmotic stress and acid stress (Hommais et al. 2001). Under Cu and Zn stress conditions, repression of many genes related to acid and osmotic stress response is stronger than repression of other genes. In contrast, depletion of Mg and stress caused by excess Fe leads to the induction

Table 3 Effect on transcription of the metal stimulon (regulon) by other metals in *E. coli*

Gene	Effective metal[a]														
	Ba	Ca	Co	Cr	Cs	Cu	Fe(III)	Fe(II)	Li	Mn	Ni	Rb	Ru	Sn	Zn
CueR regulon															
copA	–	–	–	–	–	UP	–	–	UP*	–	–	UP*	UP*	–	–
cueo	–	–	–	–	–	UP*	–	–	–	–	–	–	–	–	–
CusR regulon															
cusS	–	–	–	–	–	–	–	–	–	–	–	–	–	–	–
cusR	–	–	–	–	–	UP	–	–	–	–	–	UP	–	–	–
cusC	–	–	UP*	–	–	UP	DOWN	–	–	–	–	–	–	–	–
cusF	–	–	–	–	–	UP	DOWN*	–	–	–	–	–	–	–	–
cusB	–	–	UP*	–	–	UP	–	–	–	–	–	–	–	DOWN*	–
cusA	DOWN*	–	–	–	–	UP*	–	–	–	–	–	–	–	–	–
CpxR regulon															
cpxP	–	–	–	–	–	UP	DOWN*	–	–	–	–	–	–	–	UP*
ebr	–	–	–	–	–	–	–	–	–	–	–	–	–	–	DOWN*
spy	–	–	–	–	–	UP*	UP*	UP*	DOWN*	–	–	–	–	–	UP
ybaJ	–	–	–	–	–	–	–	–	–	–	–	–	DOWN	–	UP*
hha	–	–	–	–	–	–	–	–	–	–	–	–	DOWN*	–	UP
yccA	–	–	UP*	–	–	UP	–	UP*	–	–	–	–	–	–	UP*
ycfS	–	–	–	–	–	UP*	–	–	–	–	–	–	–	–	–
ydeH	–	–	–	–	–	UP*	–	–	–	–	–	–	–	DOWN*	UP*
yebE	–	–	–	–	–	UP	–	–	–	–	–	–	–	–	UP
yecI	–	–	–	–	–	–	–	DOWN*	–	DOWN*	–	–	DOWN*	–	–
yqjA	–	–	–	–	–	–	–	UP*	–	–	–	–	–	–	UP*
yqjB	–	–	–	–	–	–	–	–	–	–	–	–	–	–	–
htrA	–	UP	–	–	–	UP	–	UP*	–	UP*	–	–	–	–	UP
ppiA	–	–	–	–	–	–	–	–	–	–	–	DOWN*	–	–	–
yihE	–	–	–	–	–	UP*	–	–	–	–	–	–	–	–	–
dsbA	–	–	–	–	–	–	–	–	–	–	–	–	–	–	–

Table 3 (continued)

Gene	Effective metal[a] Ba	Ca	Co	Cr	Cs	Cu	Fe(III)	Fe(II)	Li	Mn	Ni	Rb	Ru	Sn	Zn
ZntR regulon															
zntA	–	UP*	UP*	–	–	–	–	–	–	–	UP	–	–	–	UP
Zur regulon															
znuA	–	–	–	–	–	–	–	–	–	–	–	–	UP	UP	–
znuB	DOWN*	–	–	–	–	–	–	–	–	–	–	–	–	–	–
znuC	–	–	–	–	–	–	–	–	–	–	–	–	–	–	DOWN
Cysteine synthesis (CysB regulon)															
cysA	–	–	UP*	–	–	UP	–	–	UP	DOWN	–	UP*	UP	DOWN*	UP
cysC	–	–	–	–	–	–	–	–	UP	–	–	–	UP	–	UP*
cysD	–	–	–	–	–	UP	–	–	UP	DOWN	–	–	UP	–	UP
cysE	–	–	–	–	–	–	–	–	–	–	–	–	–	–	–
cysG	–	UP*	–	–	–	–	–	UP*	–	–	–	–	–	–	–
cysH	UP*	–	–	–	–	UP	–	–	UP	DOWN	–	–	UP	–	UP*
cysI	–	DOWN*	UP*	–	–	UP	–	UP*	UP	DOWN	–	–	UP	–	
cysJ	–	DOWN*	UP	–	–	UP	–	–	UP	DOWN	–	–	UP	–	UP
cysK	–	–	–	–	–	UP	–	–	–	DOWN	–	–	–	–	UP
cysM	–	–	–	–	–	–	–	–	–	–	–	–	–	–	–
cysN	–	–	UP*	–	–	UP	UP*	UP	DOWN	–	–	UP	–	UP*	
cysP	–	–	UP	–	–	UP*	–	–	UP*	DOWN*	–	UP*	–	–	UP*
cysU	–	–	UP*	–	–	UP*	–	–	UP	DOWN	–	UP*	UP	DOWN*	–
cysW	–	–	–	–	–	UP*	–	–	UP	–	–	–	–	–	–
nirB	–	–	–	–	–	–	–	–	–	–	–	–	–	–	–
nirC	–	–	–	–	–	–	–	–	–	–	–	–	–	–	–
nirD	–	–	–	–	–	–	–	–	–	–	–	–	–	–	–

Table 3 (continued)

Gene	Effective metal[a]														
	Ba	Ca	Co	Cr	Cs	Cu	Fe(III)	Fe(II)	Li	Mn	Ni	Rb	Ru	Sn	Zn
RpoE regulon															
fkpA	–	UP*	–	–	–	–	–	–	–	–	–	–	–	UP*	–
htrA/degP	–	UP	–	–	–	UP	–	UP*	–	UP*	–	–	–	–	UP
rpoE	–	UP	–	–	–	–	–	–	–	UP*	–	–	–	UP*	–
rpoH	–	UP	–	–	–	–	–	–	–	–	–	–	–	–	UP*
rseA	–	UP	–	–	–	–	–	–	–	–	–	–	–	UP	–
rseB	–	UP	–	–	–	–	–	–	–	–	–	UP*	–	–	
yggN	–	–	–	–	–	–	–	–	–	–	–	–	–	–	–
Argine synthesis (ArgR regulon)															
argA	–	UP	–	–	–	–	UP*	–	–	–	–	–	–	–	–
argB	–	UP	–	–	–	–	–	–	–	–	–	–	–	–	–
argC	–	UP	–	–	–	–	–	–	–	–	–	–	–	–	–
argD	–	UP	–	–	–	–	UP*	–	–	–	–	–	–	–	–
argE	–	UP	–	–	–	–	UP*	–	–	–	–	–	UP*	UP*	–
argF	DOWN*	UP	–	–	–	–	UP*	UP*	–	–	–	–	–	–	–
argG	–	UP	–	–	–	–	–	–	–	–	DOWN*	–	–	–	–
argH	–	UP	–	–	–	–	UP	UP*	–	UP*	–	–	–	–	–
argI	–	UP	UP*	–	UP*	–	UP	–	–	–	–	UP*	–	–	–
argK	DOWN*	–	–	–	–	–	–	–	–	–	–	–	–	–	–
artI	–	–	–	–	–	–	–	–	–	–	–	–	–	–	–
artJ	–	UP	DOWN*	–	–	–	–	–	–	–	DOWN*	–	–	–	–
artM	–	–	–	–	UP*	–	–	–	–	–	–	–	–	–	–
artP	–	UP*	–	–	–	–	–	–	–	–	–	–	–	–	–
artQ	–	–	–	–	UP*	–	–	–	–	–	–	–	–	–	–

Table 3 (continued)

Gene	Effective metal[a]														
	Ba	Ca	Co	Cr	Cs	Cu	Fe(III)	Fe(II)	Li	Mn	Ni	Rb	Ru	Sn	Zn
Histidine synthesis															
hisA	–	–	–	–	–	UP*	–	–	–	–	UP*	–	–	–	–
hisB	–	–	–	–	–	UP*	–	–	–	–	UP	–	–	–	UP*
hisC	–	UP*	–	–	–	UP	–	–	–	–	UP*	–	–	–	UP*
hisD	–	–	–	–	–	UP*	–	–	–	–	UP*	–	–	–	–
hisF	–	–	–	–	–	UP	–	–	–	–	UP	DOWN*	–	–	UP*
hisG	–	–	–	–	–	–	–	–	–	–	–	–	–	UP*	–
hisH	–	–	–	–	–	UP	–	–	–	–	–	–	–	–	–
hisI	–	–	–	–	–	UP	–	–	–	–	UP	–	–	–	–
hisM	–	–	–	–	–	–	–	–	–	–	–	–	–	–	–
hisP	–	–	–	–	–	–	–	–	–	–	–	–	–	–	–
hisQ	–	–	–	–	–	–	UP*	–	–	–	–	–	–	–	–

[a] *E. coli* BW25113 was grown in MOPS medium for 16 h. *E. coli* culture was diluted 100-fold with fresh MOPS medium and grown in a large glass tube at 37 °C with reciprocal shaking (under aerobic condition) until an OD 600 nm value of 0.2–0.3 was reached. The metal was added to the following concentrations: 100 mM LiCl, 50 mM NaCl, 25 mM $MgCl_2$, 50 mM KCl, 5 mM $CaCl_2$, 0.1 mM $CrCl_3$, 0.5 mM $MnCl_2$, 0.5 mM $FeCl_2$, 0.5 mM $FeCl_3$, 0.05 mM $CoCl_2$, 0.25 mM $NiCl_2$, 0.5 mM $CuCl_2$, 0.5 mM $ZnCl_2$, 50 mM RbCl, 1.5 mM SrCl, 100 mM $RuCl_3$, 25 mM CsCl; 1 mM $BaCl_2$, and 5 mM $SnCl_4$. After 5 min, of exposure to the metal, total RNA was prepared. Total RNA isolation, labeling, and microarray analysis were done as previously described (Oshima et al. 2002). On the DNA microarray used in this study, each probe for a gene was spotted twice. Genes were up- (UP) and down-regulated (DOWN) threefold on both duplicate spots by a metal * Threefold alteration on either duplicate spot

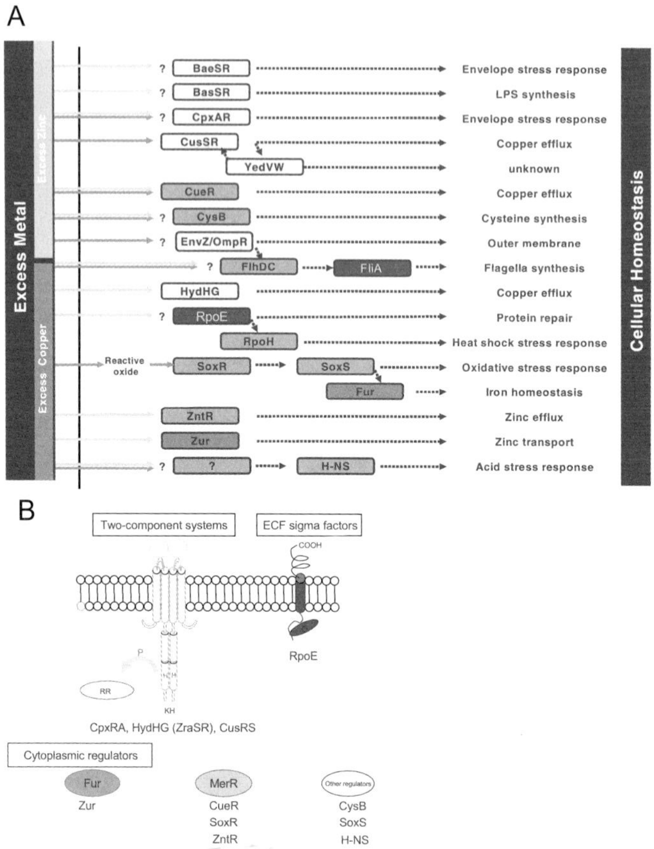

Fig. 3 **a** Copper and zinc homeostasis. The regulons stimulated by external copper and zinc, are shown (see text). These regulons include two-component systems (*pearl blue*), ECFs (*dark blue*), Mer family (*green*), Fur family (*pink*), and others (*grey*). *Blue* and *pink arrows* represent copper and zinc responses, respectively. There is cross-talk/overlap between several regulons, where one regulon regulates the expression of other regulon genes (CusSR → yedWV, EnvZ/OmpR → flhDC, RpoE → rpoH, and SoxS → fur) as a transcriptional cascade (Kershaw et al. 2005; Yamamoto and Ishihama 2005; Yamamoto and Ishihama 2005). There is also cross-talk among two-component systems, where the YedV histidine kinase phosphorylates CusR, a non-cognate response regulator (Yamamoto et al. 2005). **b** Copper and zinc homeostasis. The transcriptional regulators involved in response to copper and zinc. *Cytoplasmic regulators*: Transcriptional activators of the MerR family include ZntR (Zn response, Brocklehurst et al. 1999; Changela et al. 2003; Outten et al. 1999), CueR (Cu response, Changela et al. 2003; Outten et al. 2000; Peterson and Moller 2000; Stoyanov et al. 2001) and SoxR (superoxide response, Pomposiello and Demple 2001; Demple 1996). The transcriptional repressors Fur (Fe import response, Andrews et al. 2003) and Zur (Zn import response, Patzer and Hantke 2000). Other cytoplasmic regulators H – NS, SoxS and CysB are referenced in the text. *Two-component systems*: CpxRA (membrane stress response, de Wulf et al. 2002), HydHG (ZraSR, zinc and lead response, Leonhartsberger et al. 2001) and CusRS (periplasmic copper response, Outten et al. 2000). *ECF* sigma *factors*: RpoE periplasmic stress (Dorel et al. 2006)

of acid stress related genes, which may be controlled by the two-component systems, PhoPQ and BasRS (Hagiwara et al. 2004; Zwir et al. 2005).

6.1.4 The FlhDC Regulon

Flagellar genes are up- and down-regulated, respectively, by Cu and Zn stress. FlhDC is an activator for the regulation of flagella biosynthesis and Crp and H – NS can also activate the *flhDC* operon (Soutourina et al. 1999). Transcriptome analysis suggests that many genes in the Crp regulon are not activated by metal stress, but that some of the genes are regulated by metal stress. H – NS regulates several genes included in the Crp regulon, and the regulation is dependent on nucleoid structure and the stringent response (Johansson et al. 2000). Metal ion stress might therefore be related to the nucleoid structure or the stringent response, and can regulate several genes in an H – NS related manner, but further analysis is needed.

6.1.5 Cps/Lps Synthesis and Complex Networks of Regulation by the BasRS, PhoPQ, and Rcs Two-Component Systems

Expression of genes involved in capsular- (Cps) and lipo-polysaccharide (Lps) biosysnthesis respond to external excess Zn and Fe stress. The BasRS, PhoPQ, and Rcs two-component systems may be involved in this regulation. These regulatory networks may be closely related to others in *Salmonella* sp. (Mouslim et al. 2003). Zn and Fe can directly regulate the PhoPQ and BasRS systems in *E. coli* (Hagiwara et al. 2003, 2004). See Fig. 4 for a summary of regulators involved in responses to different metals (for EvgAS to PhoPQ, Eguchi et al. 2004). In addition, several acid resistance genes are directly regulated by PhoPQ in *Salmonella enterica* (Zwir et al. 2005) and BasRS in *E. coli* (Hagiwara et al. 2004). Thus, in some conditions, these systems are activated by several metals and positively regulate the Cps and Lps synthesis regulon and acid response regulon. It is worth noting that this regulation is observed in EB media, which contains 1.8% (wt/vol) Eiken broth (Eiken Chemical Co. Tokyo, Japan) and 0.4% glucose in 50 mM morpholinepropanesulfonic acid (MOPS) buffer (pH 7.0), and is not observed in chemostat culture conditions with synthetic media (Lee et al. 2005) and is not observed in Fe-limitation conditions with 2,2′-dypyridyl (McHugh et al. 2003).

6.1.6 The Fur and Zur Regulons

The Fur regulon is globally derepressed by Cu, Cd, and Zn addition and repressed in Fe-limitation conditions (Kershaw et al. 2005; Lee et al. 2005;

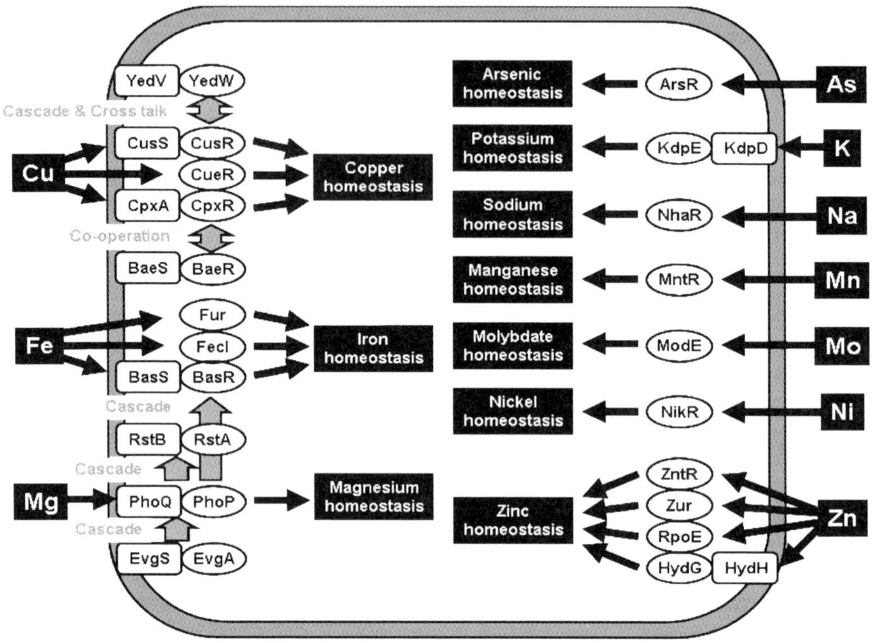

Fig. 4 Metal-ion homeostasis regulation and regulatory cross-talk in *E. coli*

McHugh et al. 2003; Wang and Crowley 2005), whilst the Zur regulon was derepressed under Cu stress (Kershaw et al. 2005). These data strongly suggest that Fe, and possibly Zn homeostasis, is closely related to adaptation to other metal stresses. McHugh et al. (2003) used Luria Bertani (LB) media and iron chelation by 2,2′-dipyridyl for their experiments. In contrast, Kershaw et al. (2005) and Lee et al. (2005) used synthetic media including Fe with excess Cu and Zn. One explanation of the induction of the Fur regulon is that it might be related to the ratio of Fe to other metals such as Cu and Zn. However, a connection between oxidative stress and Fur regulation has been identified (Zheng et al. 2001; Pomposiello et al. 2001), as well as a down-regulation of *sodB* transcription, which is consistent with a phenotypic Fe deficiency and reported regulation of Fur by SoxRS and OxyR. Two hypotheses have been put forward for why this effect should be seen. The first is that oxidative stress damages the Fur–Fe complex, leading to loss of repressor function (Zheng et al. 1999). The second hypothesis is that under oxidative stress conditions Fe^{2+} bound to Fur is oxidized to Fe^{3+}, which does not act as a corepressor, leading to derepression of Fur-regulated genes (Pomposiello et al. 2001). One other possibility is that displacement of Fe or Zn from Fur or Zur by more thiophilic metals could derepress transcription from their cognate promoters.

6.1.7
Mg-Limited Conditions

Conditions under which Mg is limited induce activity of the two-component sensor-regulator PhoPQ regulon in *E. coli* (Minagawa et al. 2003) The PhoPQ system senses external Mg to allow the bacterial cells to adapt to low Mg environments and is implicated in virulence gene expression, and resistance to some antimicrobials (Groisman 2001). PhoQ is autophosphorylated in the absence of Mg, and then transfers a phosphate to PhoP. The phosphorylated PhoP, which is an OmpR family regulator, is able to bind to a conserved direct repeat DNA sequence to regulate transcription of at least 11 genes (Kato et al. 1999; Yamamoto et al. 2002; Minagawa et al. 2003). Increases in Mg and Ca concentrations change PhoQ to an inactive form (or to a form for dephosphorylation) by direct binding to the periplasmic domain of PhoQ. High levels of Mg repress transcription of the PhoPQ regulon genes in vivo (Waldburger and Sauer 1996). The periplasmic region of PhoQ specifically binds to Mg (and Ca) in vitro (Waldburger and Sauer 1996; Garcia Vescovi et al. 1997), and several mutagenesis studies of PhoQ show that residues in the periplasmic region are required to sense low and high Mg (Regelmann et al. 2002; Marina et al. 2001; Minagawa et al. 2005), but the molecular mechanism of Mg-sensing by PhoQ is not clear. Under conditions where Zn and Fe are in excess PhoPQ regulation is also observed (Hagiwara et al. 2003).

6.1.8
Amino Acid Synthesis Pathways

Under many metal ion stress conditions, amino acid synthesis pathways are activated (Table 3). Cu and Zn stress activate the Cys synthesis pathway (Table 3). Zn and Ni can activate the His synthesis pathway (Table 3; Große and Nies, unpublished data). Wang and Crowley (2005) also noted up-regulation of the *cys* genes under cadmium stress, and this may be related to the need to replace damaged cysteine residues.

6.2
Transcriptomic Responses to Metal Stresses in *Bacillus subtilis*

Moore et al. 2005, and Moore and Helmann 2005 have studied the global and genetic responses of *B. subtilis* to metal stresses from Ag(I), Cd(II), Cu(II), Ni(II), Zn(II), and the metalloid As(V) and have determined the regulons that respond to each metal. In *B. subtilis*, there are eight metalloregulatory proteins, Fur, PerR, Zur, MntR, CueR, ArsR, AseR, and CzrA that are recognized as important in regulating the uptake and export of essential and toxic metals. Of these regulators, there are five homologs of the *E. coli* regulators, Fur, Zur, MntR, CueR, and ArsR. Three of these regulators, Fur, Zur, and MntR regulate Fe, Zn, and Mn uptake, which are essential metals. This suggests that the import systems for those essential metals are conserved in many bacteria. In addition, CueR and ArsR transcription factors regulate Cu and As resistance-related genes, and control the export of toxic metal ions. CueR is a MerR family transcriptional regulator; and MerR type transcriptional regulators not only regulate genes on the chromosome, but also on mobile genetic elements such as plasmids and transposons (summarized in, e.g., Brown et al. 2003). The ArsR regulator is encoded on a prophage-like element in the chromosome of *B. subtilis* 168, and ArsR regulators are also encoded on plasmids in several *E. coli* strains (Sato and Kobayashi 1998, and references within). This could suggest that MerR- and ArsR-family regulators and their regulons might have not only been horizontally acquired by *B. subtilis* 168, but also by *E. coli*, possibly to allow rapid and widespread acquisition of genes for resistance for many bacteria in metal polluted environments. Recently, several reports have suggested that there is a dynamic role for phage-mediated evolution of the bacterial genome (Comeau and Kisch 2005), and that there is evidence for horizontal gene transfer of P(IB)-type metal efflux ATPases (Coombs and Barkay 2004, 2005; Martinez et al. 2006)

6.3 Similarities and Differences Between *E. coli* and *B. subtilis* Metal Ion Stress Response

There are a number of interesting differences in response to metals between *E. coli* and *B. subtilis*. In *B. subtilis*, negative regulation of peroxide stress by PerR, a Fur family regulator, is derepressed under Cu, As, and Cd stress (Moore et al. 2005), but in *E. coli*, under excess Cu stress conditions, the OxyR regulon, which is induced by hydrogen peroxide, is not activated. This suggests that there are some differences in Cu homeostasis between *E. coli* and *B. subtilis*, but, the oxidative stress response is important for Cu stress adaptation in both bacteria. In contrast to *E. coli*, where Cd, Cu, and Zn derepress the Fur regulon, in *B. subtilis*, Cd, As, Ag, Zn, and Ni repress the expression of genes belonging to the Fur regulon.

In *E. coli* and *B. subtilis*, both the cysteine synthesis pathway and the histidine synthesis pathway are activated by Cu and Zn (Ag and As can also activate it in *B. subtilis*), and Zn and Ni (Ag also activates in *B. subtilis*), respectively. This evidence strongly demonstrates that activation of amino acid synthesis pathways is a common phenomenon in both *E. coli* and *B. subtilis*. Common transcriptional regulation networks are activated by several metal stresses in both *E. coli* and *B. subtilis*. The Cys synthesis pathway can also be activated by oxidative stress and sulfate limitation in *E. coli* (Zheng et al. 2001; Gyaneshwar et al. 2005) and disulfide stress in *B. subtilis* (Leichert et al. 2003).

In *E. coli*, CysB is a LysR family activator for *cys* operon expression, and recognizes *O*-acetyl-L-serine, an intermediate compound in cysteine biosynthesis, as a ligand (Kredich 1996). Most genes of the *cys* operon are induced under sulfur-limitation conditions (Gyaneshwar et al. 2005). On the other hand, *cys* operon expression in *B. subtilis* are induced under excess sulfur (Albanesi et al., 2005). In *B. subtilis*, CysL, a LysR family regulator, activates *cys* operons (Albanesi et al. 2005). Thus, the mechanism of *cys* regulation is different in *E. coli* and *B. subtilis*.

6.4 Transcriptomic Responses to Metal Stresses in Other Bacterial Species

This section summarizes the transcriptomic responses in a few other bacterial species.

6.4.1 *Caulobacter Crescentus*

Hu et al. (2005) exposed the stalked bacterium *Caulobacter crescentus* to four toxic metals Cr, Cd, Se, and U. The main transcriptomic responses to Cd and Cr stress were induction of the oxidative stress response (SodA, glutathione-*S*-transferase, thioredoxin and glutaredoxin), and the induction of expression of DNA repair enzymes. Further metal-specific responses were expression of multiple efflux pumps used to pump out Cd, whilst a sulfate transporter was down-regulated to reduce non-specific Cr uptake. A two-component regulator system appears to have been involved in the U response.

6.4.2 *Shewanella Oneidensis* MR-1

Shewanella oneidensis MR-1 is a metal ion reducing bacterium that is metabolically versatile and is of interest as a useful organism for bioremediation. The strain reduces a wide range of metals including Fe(III), Mn(IV), U(VI), Cr(VI), and Tc(VII), and can be harnessed to reduce U(VI) and Cr(VI) to less soluble and toxic forms. Genome-wide transcriptional profiling of *S. oneidensis* exposed to U(VI) and Cr(VI) stress has shown that, alongside many genes encoding proteins of unknown function, genes involved in synthesis of cytochromes and other electron transport proteins, and genes involved in membrane and cytoplasmic stress response were up-regulated (Bencheikh-Latmani et al. 2005). Sr stress in *S. oneidensis* causes induction of siderophore biosynthesis and Fe transport genes as well as lower level induction of multidrug and heavy metal efflux pumps, and sulfate ABC transporters (Brown et al. 2006)

6.4.3 *C. metallidurans* Strain CH34

C. metallidurans CH34 has long been an exemplar of a metal ion resistance specialist bacterium, because of its ability to colonize toxic metal polluted environments and because it carries genes on two large plasmids (pMOL28 and pMOL30) that confer resistance to Ag(I), Cd(II), Co(II), Cr(VI), Cu(II), Ni(II), Zn(II), Hg(II), Pb(II) (Mergeay et al. 2003), and in some cases carries multiple paralogs of these genes (Nies et al. 2006). Recent work on the pMOL30 copper resistance determinant involved cloning and sequencing the genes responsible for copper resistance, followed by transcriptomic experiments of the pMOL30 copper resistance genes, which suggests that both a time- and concentration-dependent induction of copper resistance genes is occurring (Monchy et al. 2006)

6.4.4 *Halobacterium* NRC-1

In an outstanding recently published paper, Kaur et al. (2006) described a wide-ranging systems-based study of responses of the haloarchaeal strain *Halobacterium* NRC-1 after exposure to the transition metals Mn, Fe, Co, Ni, Cu, and Zn. In this study Kaur et al. (2006) report $>20\%$ of the genome transiently changed transcription when cells were exposed to increased levels of these metals. The generalized response for each metal in this strain was an oxidative stress response, scavenging of toxic metal ions, changes in regulation of metal ion transport, increased turnover of proteins, and DNA repair.

7 Common Themes from Metal Ion Stress Response

Common themes of the bacterial metal ion stress response appear to be emerging from different experimental systems, although not all of the responses are seen in all of the systems simultaneously for each metal. These responses can be grouped as:

- Metal-specific responses (activation of efflux/detoxification systems)
- Oxidative stress responses
- Membrane stress responses
- Activation of amino acid synthesis
- Modulation of expression of metal ion import systems

The genome-wide transcriptional responses to metal ion stress strongly imply that it is not just the metal ion homeostasis systems that confer resistance to the metal, but it is in fact the activation of several different stress responses that allow bacteria to tolerate metal ion stresses. Several points come out of the genome-wide studies of responses to metals. The first is that the transcriptome responses are actually responses to the chemistry of each metal, and that many of the responses to metal ion stress may actually be responses to metal-induced oxidative stress, even for non-redox-active metals. This is probably strongly associated with the problem that aerobic bacteria face in being vulnerable to oxygen stress and Fe limitation because of their widespread use of iron-sulfur clusters, which are sensitive to oxidative damage (Imlay 2006). Consequently these cells are sensitive to any metal ion induced damage to oxidative stress response systems.

7.1 Cytoplasmic Regulators, ECF σ Factors, and Two-Component Systems are all Involved in Responses to Metal Ion Stress in *E. coli*

The key differences between the two most intensively studied prokaryotic model systems, *E. coli* and *B. subtilis*, is in differences in cell structure, with *E. coli* having a plasma membrane, a thin peptidoglycan layer, a periplasmic space and an outer membrane, whilst in *B. subtilis* there is a plasma membrane and a thick peptidoglycan layer, but no significant periplasmic space. Both *E. coli* and *B. subtilis* possess the cytoplasmic metal responsive regulators, Fur, Zur, MntR, CueR, and ArsR. There are multiple two-component systems and ECFs in both *E. coli* and *B. subtilis*, which have similar roles. For the two-component systems these roles are exclusion of organic drugs, response to environmental stresses (phosphate, glucose starvation), adaptation to anaerobic conditions, envelope stress response, and small molecule utilization (for example citrate, C4-dicarboxylate, and sugars). ECFs are involved in responses to envelope stress and iron starvation (Ogasawara N, personal communication) in both *E. coli* and *B. subtilis*. However, in *E. coli*, two-component systems and the ECF, RpoE, contribute to the regulation of metal ion homeostasis systems, but in *B. subtilis*, both two-component systems and ECFs appear not to contribute to metal ion homeostasis (with the possible exception of SigX which may contribute to iron homeostasis; Brutsche and Braun 1997).

Transcriptome analysis suggests that the membrane stress response *cpxRA*, copper stress response *cusRS*, and *cueR* regulators in *E. coli* positively regulate many genes by direct interaction with the promoter. The *cpx* and *cus* regulons may also be responding to additional independent stimuli that are caused by primary chemical interactions and reactions that copper participates in, and the secondary effects caused by those reactions. When there are very complex stresses under excess copper conditions (or from other metals), multiple two-component systems may independently respond to environmental changes caused by the metal stress. In contrast, there are no similar systems in *B. subtilis*. These periplasmic sensing systems are a specific feature of *E. coli* (and probably other Gram-negative bacteria) compared to *B. subtilis* (and other Gram-positive bacteria that lack a periplasm). This suggests that the periplasm has a role in Gram-negative bacteria in sensing metals.

8
Where Next?

Transcriptome studies show multiple effects of sublethal single metal ion stress on bacteria, and many of the gross manifestations of oxidative stress are similar to those of toxic metal stress, as are the transcriptome responses to these stresses. Although the regulation of transcription by metal responsive cytoplasmic transcription factors is becoming well understood, the importance of two-component systems and ECF σ factors in cellular response to metal ion insult is now also clear.

Transcriptome analysis will determine genome-wide transcription profiles of sublethal metal ion stress responses. However, the metal ion stress response is a complex phenomenon associated with the primary and consequential (secondary) effects of the interaction of metal ions with cellular components. This stress response is simultaneously controlled by several transcriptional regulatory systems, because of the multiple cellular effects caused by metals. In the case of Cu in *E. coli*, the main transcriptional networks that allow cells to adapt to conditions of increasing Cu concentrations are the CueR, CusRS, CpxRA, and SoxRS regulons. In addition, unknown networks activate transcription of the Cys synthesis pathway. Studies aimed at understanding the antagonistic and synergistic effects of stress caused by more than one metal may inform us of bacterial responses to "real world" situations – bacteria are unlikely to encounter reagent grade single metals in their natural environment.

Systematic studies of the responses of an organism to a range of single metals are now being published (e.g., Kaur et al. 2006). In the future, more systems biology approaches will be used to quantitatively analyze the functional interaction of cellular components (Aderem 2005) using transcriptomic, proteomic, and metabolomic data to generate a whole-organism view of cellular responses to a set of conditions. There are, however, challenges that researchers face in realizing a true systems biology understanding of cellular response to metals. One practical problem is that experimental evidence shows that there can be differences in transcriptional responses to the same metal stress, if slightly different stress conditions are used (for example, the copper stress responses described by Kershaw et al. (2005) and Yamamoto and Ishihama (2005) are not the same). Another practical problem is that metal stress transcriptomic experiments take "snapshots" of changes in gene expression at a fixed time after metals are added to the external medium. Time course experiments to measure temporal changes in gene expression may reveal a succession of changes occurring over time in response to primary and secondary cellular effects of metals.

It is also possible that there are different responses to multiple stresses that arise from subtle but significant differences in experimental conditions, for example, media (complex media or synthetic media, excess or depleted metals etc.). Therefore, to further elucidate metal stress response transcriptional networks we must understand the transcriptional profiles of multiple regulons, and the physical location on the genome of the transcription factors during this stress. Transcription factor binding can be physically mapped on the genome using ChIP-chip (Grainger et al. 2004, 2005; Wade and Struhl 2004; Herring et al. 2005) analysis allied with transcriptome studies. In this way, a picture will emerge of both the transcriptional profile and the positioning of the transcriptional regulators on promoters.

If we are to extract maximum value from transcriptome experiments, it will be very useful if the data is deposited in open databases so that other researchers have access to both the data and details of how the experiments were conducted. It is also perhaps timely to consider whether a consensus view on standardized experimental conditions (media, growth conditions, chemostat or batch culture, microarray platform, method of cDNA labeling) can be applied for genome-wide transcriptional profiling of metal ion stress. Standardized experimental methods (where practical) would facilitate cross-comparison of data, and simplify future meta-analysis of data and attempts to understand cellular responses to metal ions from a systems biology approach.

Acknowledgements JLH acknowledges funding from UK BBSRC grants EGA16107 and P15753. The collaboration between JLH and TO was funded by a BBSRC Japan Partnering Award JPA1189, and by a Japan-UK Research Cooperative Joint Program grant (from JSPS) to TO and JLH.

References

Aertson A, Michiels CW (2004) Stress and how bacteria cope with death and survival. Crit Rev Mirobiol 30:263–273

Aderem A (2005) Systems biology: Its practice and challenges. Cell 121:511–513

Alba BM, Gross CA (2004) Regulation of the *Escherichia coli* sigma-dependent envelope stress response. Mol Microbiol 52:613–619

Albanesi D, Mansilla MC, Schujman GE, de Mendoza D (2005) *Bacillus subtilis* cysteine synthetase is a global regulator of the expression of genes involved in sulfur assimilation. J Bacteriol 187:7631–7638

Andrews SC, Robinson AK, and Rodriguez-Quinones F (2003) Bacterial ion homeostasis. FEMS Microbiol Rev 27:215-237

Ballatori N (2002) Transport of toxic metals by molecular mimicry. Env Health Persp 110:689–694

Beard SJ, Hashim R, Membrillo-Hernandez J, Hughes MN, Poole RK (1997) Zinc(II) tolerance in *Escherichia coli* K-12: evidence that the *zntA* gene (o732) encodes a cation transport ATPase. Mol Microbiol 25:883–91

Bencheikh-Latmani R, Williams SM, Haucke L, Criddle CS, Wu L, Zhou J, Tebo BM (2005) Global transcriptional profiling of *Shewanella oneidensis* MR-1 during Cr(VI) and U(VI) reduction. Appl Environ Microbiol 71:7453–7460

Binet MR, Poole RK (2000) Cd(II), Pb(II) and Zn(II) ions regulate expression of the metal-transporting P-type ATPase ZntA in *Escherichia coli.* FEBS Lett 473:67–70

Blattner FR, Plunkett G, Bloch CA, Perna NT, Burland V, Riley M, Collado-Vides J, Glasner JD, Rode CK, Mayhew GF, Gregor J, Davis NW, Kirkpatrick HA, Goeden MA, Rose DJ, Mau B, Shao Y (1997) The complete genome sequence of *Escherichia coli* K-12. Science 277:1453–146

Boos W, Lucht JM (1996) Periplasmic binding protein-dependent ABC transporters. In: Niedhardt FC, Curtiss R III, Ingraham JL, Lin ECC, Brooks Low K, Magasanik B, Reznikoff WS, Riley M, Schaechter M, Umbarger HE (eds) *Escherichia coli* and *Salmonella* cellular and molecular biology, 2nd edn. ASM, Washington, DC, pp 1175–1209

Braun V, Mahren S, Ogierman M (2003) Regulation of the FecI-type ECF sigma factor by transmembrane signalling. Curr Opin Microbiol 6:173–180

Braun V, Mahren S, Sauter A (2006) Gene regulation by transmembrane signalling. BioMetals 19:103–113

Brocklehurst KR, Hobman JL, Lawley B, Blank L, Marshall SJ, Brown NL, Morby AP (1999) ZntR is a Zn(II)-responsive MerR-like transcriptional regulator of zntA in *Escherichia coli.* Mol Microbiol 31:893–902

Brown NL, Stoyanov JV, Kidd SP, Hobman JL (2003) The MerR family of transcriptional regulators. FEMS Microbiol Rev 27:145–163

Brown SD, Martin M, Deshpande S, Seal S, Huang K, Alm E, Yang Y, Wu L, Yan T, Liu X, Arkin A, Chourey K, Zhou J, Thompson DK (2006) Cellular response of *Shewanella oneidensis* to strontium stress. Appl Environ Microbiol 72:890–900

Browning DF, Busby SJW (2004) The regulation of bacterial transcription initiation. Nat Rev Microbiol 2:1–9

Bruins MR, Kapil S, Oehme FW (2000). Microbial resistance to metals in the environment. Ecotoxicol Environ Safety 45:198–207

Brutsche S, Braun V (1997) SigX of *Bacillus subtilis* replaces the ECF sigma factor *fecI* of *Escherichia coli* and is inhibited by RsiX. Mol Gen Genet 256:416–425

Busenlehner LS, Pennella MA, Giedroc DP (2003) The SmtB/ArsR family of metalloregulatory transcriptional repressors: structural insights into prokaryotic metal resistance. FEMS Microbiol Rev 27:131–143

Carlin A, Shi W, Dey S, Rosen BP (1995) The *ars* operon of *Escherichia coli* confers arsenical and antimonial resistance. J Bacteriol 177:981–6

Changela A, Chen K, Xue Y, Holschen J, Outten CE, O'Halloran TV, Mondragon A (2003) Molecular basis of metal-ion selectivity and zeptomolar sensitivity by CueR. Science 301:1383–1387

Coombs JM, Barkay T (2004) Molecular evidence for the evolution of metal homeostasis genes by lateral gene transfer in bacteria from the deep terrestrial subsurface. Appl Environ Microbiol 70:1698–1707

Coombs JM, Barkay T (2005) New findings on evolution of metal homeostasis genes: evidence from comparative genome analysis of bacteria and archaea. Appl Environ Microbiol 71:7083–7091

Csonka LN, Epstein W (1996) Osmoregulation. In: Niedhardt FC, Curtiss R III, Ingraham JL, Lin ECC, Brooks Low K, Magasanik B, Reznikoff WS, Riley M, Schaechter M, Umbarger HE (eds) *Escherichia coli* and *Salmonella* cellular and molecular biology, 2nd edn. ASM, Washington, DC, pp 1210–1223

Dartigalongue C, Missiakas D, Raina S (2001) Characterization of the *Escherichia coli* sigma E regulon. J Biol Chem 276:20866–20875

Danese PN, Silhavy TJ (1998) CpxP, a stress-combative member of the Cpx regulon. J Bacteriol 180:831–839

Demple B (1996) Redox signalling and gene control in the *Escherichia coli soxRS* oxidative stress regulon- a review. Gene 179:53–57

De Wulf P, McGuire AM, Liu X, Lin ECC (2002) Genome-wide profiling of promoter recognition by the two-component response regulator CpxR-P in *Escherichia coli.* J Biol Chem 277:26652–26661

Dorel C, Lejeune P, Rodrigue A (2006) The Cpx system of *Escherichia coli*, a strategic signalling pathway for confronting adverse conditions and for settling biofilm communities? Res Microbiol 157:306–314

Duffus JH (2002) "Heavy metals"-A meaningless term? Pure Appl Chem 74:793–807

Earhart CF (1996) Uptake and metabolism of iron and molybdenum. In: Niedhardt FC, Curtiss R III, Ingraham JL, Lin ECC, Brooks Low K, Magasanik B, Reznikoff WS, Riley M, Schaechter M, Umbarger HE (eds) *Escherichia coli* and *Salmonella* cellular and molecular biology, 2nd edn. ASM, Washington, DC, pp 1075–1090

Egler M, Grosse C, Grass G, Nies DH (2005) Role of the extracytoplasmic function protein family sigma factor RpoE in metal resistance of *Escherichia coli.* J Bacteriol 187:2297–2307

Eguchi Y, Okada T, Minagawa S, Oshima T, Mori H, Yamamoto K, Ishihama A, Utsumi R (2004) A signal transduction cascade between EvgA/EvgS and PhoP/PhoQ two-component systems of *Escherichia coli.* J Bacteriol 186:3006–3014

Ercal N, Gurer-Orhan H, Aykin-Burns N (2001) Toxic metals and oxidative stress Part I: Mechanisms involved in metal induced oxidative damage. Curr Top Med Chem 1:529–539

Forman SJ, Kumar KS, Redeker AG, Hochstein P (1980) Hemolytic anaemia in Wilson's disease: clinical findings and biochemical mechanism. Am J Hematol 9:269–276

Franke S, Grass G, Nies DH (2001) The product of the *ybdE* gene of the *Escherichia coli* chromosome is involved in detoxification of silver ions. Microbiology 147:965–972

Gaetke LM, Chow CK (2003) Copper toxicity, oxidative stress, and antioxidant nutrients. Toxicology 189:147–163

Galperin MY (2006) Structural classification of bacterial response regulators: diversity of output domains and domain combinations. J Bacteriol 188:4169–4182

Geslin C, Llanos J, Prieur D, Jeanthon C (2001) The manganese and iron superoxide dismutases protect *Escherichia coli* from heavy metal toxicity. Res Microbiol 152:901–905

Grainger DC, Hurd D, Harrison M, Holdstock J, Busby SJ (2005) Studies of the distribution of *Escherichia coli* cAMP-receptor protein and RNA polymerase along the *E. coli* chromosome. Proc Natl Acad Sci USA 102:17693–17698

Grainger DC, Overton TW, Hobman JL, Constantinidou C, Tamai E, Wade JT, Struhl K, Reppas N, Church G, Busby SJW (2004) Genomic studies with *Escherichia coli* MelR protein: applications of chromatin immunoprecipitation and microarrays. J Bacteriol 186:6938–6943

Grass G, Franke S, Taudte N, Nies DH, Kucharski LM, Maguire ME, Rensing C (2005) The metal permease ZupT from *Escherichia coli* is a transporter with a broad substrate spectrum. J Bacteriol 187:1604–1611

Grass G, Fan B, Rosen BP, Franke S, Nies DH, Rensing C (2001) ZitB (YbgR), a member of the cation diffusion facilitator family, is an additional zinc transporter in *Escherichia coli.* J Bacteriol 183:4664–4667

Grass G, Fricke B, Nies DH (2005) Control of expression of a periplasmic nickel efflux pump by periplasmic nickel concentrations. Biometals 18:437–448

Grass G, Wong MD, Rosen BP, Smith RL, Rensing C (2002) ZupT is a Zn(II) uptake system in *Escherichia coli.* J Bacteriol 184:864–866

Griffith KL, Shah IM, Wolf RE Jr (2004) Proteolytic degradation of *Escherichia coli* transcription activators SoxS and MarA as the mechanism for reversing the induction of the superoxide (SoxRS) and multiple antibiotic resistance (Mar) regulons. Mol Microbiol 51:1801–1816

Groisman EA (2001) The pleiotropic two-component regulatory system PhoP-PhoQ. J Bacteriol 183:1835–1842

Grunden AM, Shanmugam KT (1997) Molybdate transport and regulation in bacteria. Arch Microbiol 168:345–354

Gupta A, Matsui K, Lo JF, Silver S (1999) Molecular basis for resistance to silver cations in *Salmonella.* Nature Med 5:183–188

Gyaneshwar P, Paliy O, McAuliffe J, Popham DL, Jordan MI, Kustu S (2005) Sulfur and nitrogen limitation in *Escherichia coli* K-12: specific homeostatic responses. J Bacteriol 187:1074–1090

Hagiwara D, Sugiura M, Oshima T, Mori H, Aiba H, Yamashino T, Mizuno T (2003) Genome-wide analyses revealing a signaling network of the RcsC-YojN-RcsB phosphorelay system in Escherichia coli. J Bacteriol 185:5735-46

Hagiwara D, Yamashino T, Mizuno T (2004) A genome-wide view of the *Escherichia coli* BasS-BasR two-component system implicated in iron-responses. Biosci Biotechnol Biochem 68:1758–1767

Hantke K (2001) Iron and metal regulation in bacteria. Curr Opin Microbiol 4:172–177

Hansen JM, Zhang H, Jones DP (2006) Differential oxidation of thiredoxin-1, thioredoxin-2, and glutathione by metal ions. Free Radical Biol Med 40:138–145

Harrison JJ, Ceri H, Roper NJ, Badry EA, Sproule KM, Turner RJ (2005) Persister cells mediate tolerance to metal oxyanions in *Escherichia coli.* Microbiology 151:3181–3195

Hayashi K, Morooka N, Yamamoto Y, Choi S, Ohtsubo E, Baba T, Wanner BL, Mori H, Horiuchi T (2006) Highly accurate genome sequences of the *Escherichia coli* K-12 strains MG1655 and W3110. Mol Syst Biol 2:E1–E5

Helmann JD (2002) The extracytoplasmic function (ECF) sigma factors. Adv Microb Physiol 46:47–110

Herring CD, Raffaelle M, Allen TE, Kanin EI, Landick R, Ansari AZ, Palsson BO (2005) Immobilization of *Escherichia coli* RNA polymerase and location of binding sites by use of chromatin immunoprecipitation and microarrays. J Bacteriol 187:6166–6174

Hidalgo E, Bollinger JM Jr, Bradley TM, Walsh CT, Demple B (1995) Binuclear [2Fe – 2S] clusters in the *Escherichia coli* SoxR protein and role of the metal centers in transcription. J Biol Chem 270:20908–20914

Hirakawa H, Inazumi Y, Masaki T, Hirata T, Yamaguchi A (2005) Indole induces the expression of multidrug exporter genes in *Escherichia coli.* Mol Microbiol 55:1113–26

Hobman JL (2007) Molecular techniques for the study of toxic metal resistance mechanisms in bacteria. In: Crawford RL (ed) Manual of environmental microbiology, 3rd edn. ASM, Washington, DC (in press)

Hobman JL, Jones AC, and Constantinidou C (2007) An introduction to microarray technology. In: Falciani F (ed) Microarray technology. Taylor and Francis, Oxford (in press)

Hobman JL, Wilkie J, Brown NL (2005) A design for life: prokaryotic metal-binding MerR family regulators. Biometals 18:429–436

Hu P, Brodie EL, Suzuki Y, McAdams HH, Anderson GL (2005) Whole-genome transcriptional analysis of heavy metal stress in *Caulobacter crescentus*. J Bacteriol 187:8437–8449
Hughes MN, Poole RK (1989) Metals and microorganisms. Chapman and Hall, London
Hultberg B, Andersson A, Isaksson A (2001) Interaction of metals and thiols in cell damage and glutathione distribution: potentiation of mercury toxicity by dithiothreitol. Toxicology 156:93–100
Imlay JA (2003) Pathways of oxidative damage. Ann Rev Microbiol 57:395–418
Imlay JA (2006) Iron-sulphur clusters and the problem with oxygen. Mol Microbiol 59:1073–1082
Kabir MS, Yamashita D, Koyama S, Oshima T, Kurokawa K, Maeda M, Tsunedomi R, Murata M, Wada C, Mori H, Yamada M (2005) Cell lysis directed by σ^E in early stationary phase and effect of induction of the rpoE gene on global gene expression in *Escherichia coli*. Microbiology 151:2721–2735
Kasprzak KS (2002) Oxidative DNA and protein damage in metal-induced toxicity and carcinogenesis. Free Radical Biol Med 32:958–967
Kato A, Tanabe H, Utsumi R (1999) Molecular characterization of the PhoP-PhoQ two-component system in *Escherichia coli* K-12: identification of extracellular Mg^{2+}-responsive promoters. J Bacteriol 181:5516–5520
Kaur A, Pan M, Meislin M, Facciotti MT, El-Geweley R, Baliga NS (2006) A systems view of haloarchaeal strategies to withstand stress from transition metals. Genome Res 16:841–854
Kershaw CJ, Brown NL, Constantinidou C, Patel MD, Hobman JL (2005) The expression profile of *Escherichia coli* K-12 in response to minimal, optimal and excess copper concentrations. Microbiology 151:1187–1198
Kimura T, Nishioka H (1997) Intracellular generation of superoxide by copper sulphate in *Escherichia coli*. Mutat Res 389:237–242
Kredich NM (1996) Biosynthesis of cystine. In: Niedhardt FC, Curtiss R III, Ingraham JL, Lin ECC, Brooks Low K, Magasanik B, Reznikoff WS, Riley M, Schaechter M, Umbarger HE (eds) *Escherichia coli* and *Salmonella* cellular and molecular biology, 2nd edn. ASM, Washington, DC, pp 514–527
Lee JW, Helmann JD (2006) The PerR transcription factor senses H_2O_2 by metal-catalysed histidine oxidation. Nature 440:363–367
Lee LJ, Barrett JA, Poole RK (2005) Genome-wide transcriptional response of chemostat-cultured *Escherichia coli* to zinc. J Bacteriol 187:1124–1134
Leonhartsberger S, Huber A, Lottspeich F, Bock A (2001) The hydH/G Genes from *Escherichia coli* code for a zinc and lead responsive two-component regulatory system. J Mol Biol 307:93–105
Li Y, Wray R, Blount P (2004) Intragenic suppression of gain-of-function mutations in the *Escherichia coli* mechanosensitive channel, MscL. Mol Microbiol 53:485–495
Lippard SJ, Berg JM (1994) Overview of bioinorganic chemistry. In: Principles of bioinorganic chemistry. University Science Books, Mill Valley, California, pp 1–19
Lippard SJ, Berg JM (1994) Control and utilization of metal-ion concentration in cells. In: Principles of bioinorganic chemistry. University Science Books, Mill Valley, California, pp 139–173
Makui H, Roig E, Cole ST, Helmann JD, Gros P, Cellier MF (2000) Identification of the *Escherichia coli* K-12 Nramp orthologue (MntH) as a selective divalent metal ion transporter. Mol Microbiol 35:1065–1078
Marina A, Mott C, Auyzenberg A, Hendrickson WA, Waldburger CD (2001) Structural and mutational analysis of the PhoQ histidine kinase catalytic domain. Insight into the reaction mechanism. J Biol Chem 276:41182–41190

Martin RG, Rosner JL (2002) Genomics of the *marA/soxS/rob* regulon of *Escherichia coli*: identification of directly activated promoters by application of molecular genetics and informatics to microarray data. Mol Microbiol 44:1611–1624

Martinez RJ, Wang Y, Raimondo MA, Coombs JM, Barkay T, Sobecky PA (2006) Horizontal transfer of PIB-type ATPases among bacteria isolated from radionuclide- and metal-contaminated subsurface soils. Appl Environ Microbiol, pp 3111–3118

Mattie MD, Freedman JH (2004) Copper-inducible transcription: regulation by metal- and oxidative stress-responsive pathways. Am J Physiol Cell Physiol 286:C293–C301

McHugh JP, Rodriguez-Quinones F, Abdul-Tehrani H, Svistunenko DA, Poole RK, Cooper CE, Andrews SC (2003) Global iron-dependent gene regulation in *Escherichia coli*. A new mechanism for iron homeostasis. J Biol Chem 278:29478–29486

Mergeay M, Monchy S, Vallaeys T, Auquier V, Benotmane A, Bertin P, Taghavi S, Dunn J, van der Lelie D, Wattiez R (2003) *Ralstonia metallidurans*, a bacterium specially adapted to toxic metals: towards a catalogue of metal-responsive genes. FEMS Microbiol Rev 27:385–410

Mergeay M, Nies D, Schlegel HG, Gerits J, Charles P, van Gijsegem F (1985) *Alcalignes eutrophus* is a Gram-negative facultative chemolithotroph with plasmid-bound resistance to heavy metals. J Bacteriol 162:328–334

Minagawa S, Ogasawara H, Kato A, Yamamoto K, Eguchi Y, Oshima T, Mori H, Ishihama A, Utsumi R (2003) Identification and molecular characterization of the Mg^{2+} stimulon of *Escherichia coli*. J Bacteriol 185:3696–3702

Minagawa S, Okura R, Tsuchitani H, Hirao K, Yamamoto K, Utsumi R (2005) Isolation and molecular characterization of the locked-on mutant of Mg^{2+} sensor PhoQ in *Escherichia coli*. Biosci Biotechnol Biochem 69:1281–1287

Miyadai H, Tanaka-Masuda K, Matsuyama S, Tokuda H (2004) Effects of lipoprotein overproduction on the induction of DegP (HtrA) involved in quality control in the *Escherichia coli* periplasm. J Biol Chem 279:39807–39813

Monchy S, Benotmane MA, Wattiez R, van Aelst S, Auquier V, Borremans B, Mergeay M, Taghavi S, van der Lelie D, Vallaeys T (2006) Transcriptomic and proteomic analyses of the pMOL30-encoded copper resistance in *Cupriavidus metallidurans* strain CH34. Microbiology 152:1765–1776

Moore CM, Gaballa A, Hui M, Ye RW, Helmann JD (2005) Genetic and physiological responses of *Bacillus subtilis* to metal ion stress. Mol Microbiol 57:27–40

Moore CM, Helmann JD (2005) Metal ion homeostasis in *Bacillus subtilis*. Curr Opin Microbiol 8:188–195

Munson GP, Lam DL, Outten FW, O'Halloran TV (2000) Identification of a copper-responsive two-component system on the chromosome of *Escherichia coli* K-12. J Bacteriol 182:5864–5871

Nieboer E, Richardson DHS (1980) The replacement of the nondescript term "heavy metals" by a biologically and chemically significant classification of metal ions. Env Poll (Ser B) 1:3–26

Nies DH (1999) Microbial heavy-metal resistance. Appl Microbiol Biotechnol 51:730–750

Nies DH (2004) Incidence and function of sigma factors in Ralstonia metallidurans and other bacteria. Arch Microbiol 181:255–268

Nies DH, Brown NL (1998) Two-component systems in the regulation of heavy metal resistance. In: Silver S, Walden W (eds) Metal ions in gene regulation. Chapman Hall, London, pp 77–103

Nies DH, Rehbein G, Hoffmann T, Baumann C, Grosse C (2006) Paralogs of genes encoding metal resistance proteins in *Cupriavidus metallidurans* strain CH34. J Mol Micro Biotechnol 11:82–93

Nies DH, Silver S (1995) Ion efflux systems involved in bacterial metal resistances. J Indust Microbiol 14:186–199

Onufryk C, Crouch ML, Fang FC, Gross CA (2005) Characterization of six lipoproteins in the σ^E regulon. J Bacteriol 187:4552–4561

Oshima T, Aiba H, Masuda Y, Kanaya S, Sugiura M, Wanner BL, Mori H, Mizuno T (2002) Transcriptome analysis of all two-component regulatory system mutants of *Escherichia coli* K-12. Mol Microbiol 46:281–291

Outten CE, O'Halloran TV (2001) Femtomolar sensitivity of metalloregulatory proteins controlling zinc homeostasis. Science 292:2488–2492

Outten CE, Outten FW, O'Halloran TV (1999) DNA distortion mechanism for transcriptional activation by ZntR, a Zn(II)-responsive MerR homologue in *Escherichia coli.* J Biol Chem 274:37517–37524

Outten CE, Tobin DA, Penner-Hahn JE, O'Halloran TV (2001) Characterization of the metal receptor sites in *Escherichia coli* Zur, an ultrasensitive zinc(II) metalloregulatory protein. Biochemistry 40:10417–10423

Outten FW, Huffman DL, Hale JA, O'Halloran TV (2001) The independent *cue* and *cus* systems confer copper tolerance during aerobic and anaerobic growth in *Escherichia coli.* J Biol Chem 276:30670–30677

Outten FW, Outten CE, Hale JA, O'Halloran TV (2000) Transcriptional activation of an *Escherichia coli* copper efflux regulation by the chromosomal MerR homologue, CueR. J Biol Chem 275:31024–31029

Patzer SI, Hantke K (1998) The ZnuABC high-affinity zinc uptake system and its regulator Zur in *Escherichia coli.* Mol Microbiol 28:1199–210

Patzer SI, Hantke K (2001) Dual repression by Fe^{2+}-Fur and Mn^{2+}-MntR of the mntH gene, encoding an NRAMP-like Mn^{2+} transporter in *Escherichia coli.* J Bacteriol 183:4806–4813

Pearson RG (1963) Hard and soft acids and bases. J Am Chem Soc 85:3533–3539

Pennella MA, Giedroc DP (2005) Structural determinants of metal selectivity in prokaryotic metal-responsive transcriptional regulators. Biometals 18:413–428

Permina EA, Kazakov AE, Kalinina OV, Gelfand MS (2006) Comparative genomics of regulation of heavy metal resistance in eubacteria. BMC Microbiol 6:49

Pomposiello PJ, Bennik MH, Demple B (2001) Genome-wide transcriptional profiling of the *Escherichia coli* responses to superoxide stress and sodium salicylate. J Bacteriol 183:3890–3902

Pomposiello PJ, Demple B (2001) Redox-operated genetic switches: the SoxR and OxyR transcription factors. Trends Biotechnol 19:109–114

Powell SR (2000) The antioxidant properties of zinc. J Nutr 130:1447S-1454S

Raivio TL (2005) Envelope stress responses and Gram-negative bacterial pathogenesis. Mol Microbiol 56:1119–28

Raivio TL, Silhavy TJ (2001) Periplasmic stress and ECF sigma factors. Ann Rev Microbiol 55:591–624

Regelmann AG, Lesley JA, Mott C, Stokes L, Waldburger CD (2002) Mutational analysis of the *Escherichia coli* PhoQ sensor kinase: differences with the *Salmonella enterica* serovar Typhimurium PhoQ protein and in the mechanism of Mg^{2+} and Ca^{2+} sensing. J Bacteriol 184:5468–5478

Rensing C, Grass G (2003) *Escherichia coli* mechanisms of copper homeostasis in a changing environment. FEMS Microbiol Rev 27:197–213

Rensing C, Mitra B, Rosen BP (1997) The *zntA* gene of *Escherichia coli* encodes a Zn(II)-translocating P-type ATPase. Proc Natl Acad Sci USA 94:14326–14331

Rhodius VA, LaRossa RA (2003) Uses and pitfalls of microarrays for studying transcriptional regulation. Curr Opin Microbiol 6:114–119

Riley M, Abe T, Arnaud MB, Berlyn MKB, Blattner FR, Chaudhuri RR, Glasner JD, Horiuchi T, Keseler IM, Kosuge T, Mori H, Perna NT, Plunkett III G, Rudd KE, Serres MH, Thomas GH, Thomson NR, Wishart D, Wanner BL (2006) *Escherichia coli* K-12: a cooperatively developed annotation snapshot—2005. Nucleic Acids Res 34:1–9

Rodrigue A, Effantin G, Mandrand-Berthelot MA (2005) Identification of *rcnA* (*yohM*), a nickel and cobalt resistance gene in *Escherichia coli.* J Bacteriol 187:2912–2916

Rouch DA, Lee BTO, Morby AP (1995) Understanding cellular responses to toxic agents: a model for mechanism-choice in bacterial metal resistance. J Ind Microbiol 14:132–141

Ruiz N, Falcone B, Kahne D, Silhavy TJ (2005) Chemical conditionality: a genetic strategy to probe organelle assembly. Cell 121:307–317

Ruiz N, Kahne D, Silhavy TJ (2006) Advances in understanding bacterial outer-membrane biogenesis. Nat Rev Microbiol 4:57–66

Ruiz N, Silhavy TJ (2005) Sensing external stress: watchdogs of the *Escherichia coli* cell envelope. Curr Opin Microbiol 8:122–126

Sato T, Kobayashi Y (1998) The ars operon in the skin element of Bacillus subtilis confers resistance to arsenate and arsnite. J Bacteriol 180:1655–1661

Shah IM, Wolf RE Jr (2006) Inhibition of Lon-dependent degradation of the *Escherichia coli* transcription activator SoxS by interaction with "soxbox" DNA or RNA polymerase. Mol Microbiol 60:199–208

Shi H, Hudson LG, Liu KJ (2004) Oxidative stress and apoptosis in metal ion-induced carcinogenesis. Free Radical Biol Med 37:582–593

Shijuku T, Yamashino T, Ohashi H, Saito H, Kakegawa T, Ohta M, Kobayashi H (2002) Expression of *chaA*, a sodium ion extrusion system of *Escherichia coli*, is regulated by osmolarity and pH. Biochim Biophys Acta 1556:142–148

Silver S (1996) Transport of inorganic cations. In: Niedhardt FC, Curtiss R III, Ingraham JL, Lin ECC, Brooks Low K, Magasanik B, Reznikoff WS, Riley M, Schaechter M, Umbarger HE (eds) *Escherichia coli* and *Salmonella* Cellular and molecular biology, 2nd edn. ASM, Washington, DC, pp 1091–1102

Silver S, Phung LT (2005) A bacterial view of the Periodic Table: genes and proteins for toxic inorganic ions. J Industr Microbiol Biotechnol 32:587–605

Skovierova H, Rowley G, Rezuchova B, Homerova D, Lewis C, Roberts M, Kormanec J (2006) Identification of the σ^E regulon of *Salmonella enterica* serovar Typhimurium. Microbiology 152:1347–1359

Snyder WB, Davis LJ, Danese PN, Cosma CL, Silhavy TJ (1995) Overproduction of NlpE, a new outer membrane lipoprotein, suppresses the toxicity of periplasmic LacZ by activation of the Cpx signal transduction pathway. J Bacteriol 177:4216–4223

Solioz M, Stoyanov JV (2003) Copper homeostasis in *Enterococcus hirae.* FEMS Microbiol Rev 27:183–195

Stohs SJ, Bagchi D (1995) Oxidative mechanisms in the toxicity of metal ions. Free Radical Biol Med 18:321–336

Storz G, Hengge-Aronis R (2004) Bacterial stress responses. ASM, Washington

Stoyanov JV, Hobman JL, Brown NL (2001) CueR, (*ybbI*) of *Escherichia coli* is a MerR family regulator controlling expression of the copper exporter CopA. Mol Microbiol 39:502–511

Twyman RM (2004) From genomics to proteomics. In: Principles of proteomics. Bios Scientific, Oxford, pp 1–22

Ueda J-I, Takai M, Shimazu Y, Ozawa T (1998) Reactive oxygen species generated from the reaction of copper (II) complexes with biological reductants cause DNA strand scission. Arch Biochem Biophys 357:231–239

Van der Lelie D, Schwuchow T, Schwidetzky U, Wuertz S, Baeyens W, Mergeay M, Nies DH (1997) Two-component regulatory system involved in transcriptional control of heavy-metal homeostasis in *Alcaligenes eutrophus*. Mol Microbiol 23:493–503

Vescovi EG, Ayala YM, Di Cera E, Groisman EA (1997) Characterization of the bacterial sensor protein PhoQ. Evidence for distinct binding sites for Mg^{2+} and Ca^{2+}. J Biol Chem 272:1440–1443

Wackett LP, Dodge AG, Ellis LBM (2004) Microbial genomics and the Periodic Table. Appl Environ Microbiol 70:647–655

Wade JT, Struhl K (2004) Association of RNA polymerase with transcribed regions in *Escherichi coli*. Proc Natl Acad Sci USA 101:17777–17782

Waldburger CD, Sauer RT (1996) Signal detection by the PhoQ sensor-transmitter. Characterization of the sensor domain and a response-impaired mutant that identifies ligand-binding determinants. J Biol Chem 271:26630–26636

Wang A, Crowley DE (2005) Global gene expression responses to cadmium toxicity in *Escherichia coli*. J Bacteriol 187:3259–3266

Wolanin PM, Thomason PA, Stock JB (2002) Histidine protein kinases: key signal transducers outside the animal kingdom. Genome Biology 3: reviews 3013.1–3013.8

Wu T, Malinverni J, Ruiz N, Kim S, Silhavy TJ, Kahne D (2005) Identification of a multicomponent complex required for outer membrane biogenesis in *Escherichia coli*. Cell 121:235–245

Yamamoto K, Ishihama A (2005) Transcriptional response of *Escherichia coli* to external zinc. J Bacteriol 187:6333–6340

Yamamoto K, Ishihama A (2005) Transcriptional response of *Escherichia coli* to external copper. Mol Microbiol 56:215–227

Yamamoto K, Ishihama A (2006) Characterization of copper-inducible promoters regulated by CpxA/CpxR in *Escherichia coli*. Biosci Biotechnol Biochem 70:1688–1695

Yamamoto K, Ogasawara H, Fujita N, Utsumi R, Ishihama A (2002) Novel mode of transcription regulation of divergently overlapping promoters by PhoP, the regulator of two-component system sensing external magnesium availability. Mol Microbiol 45:423–438

Zheng M, Doan B, Schneider TD, Storz G (1999) OxyR and SoxRS regulation of *fur*. J Bacteriol 181:4639–4643

Zheng M, Wang X, Templeton LJ, Smulski DR, LaRossa RA, Storz G (2001) DNA microarray-mediated transcriptional profiling of the *Escherichia coli* response to hydrogen peroxide. J Bacteriol 183:4562–4570

Microbiol Monogr (6)
D. H. Nies, S. Silver: Molecular Microbiology of Heavy Metals
DOI 10.1007/7171_2006_075/Published online: 25 January 2007

Bacterial Transition Metal Homeostasis

Dietrich H. Nies

Institute for Microbiology, Martin Luther University, Kurt-Mothes-Str. 3,
06099 Halle/Saale, Germany
d.nies@mikrobiologie.uni-halle.de

Abstract Since details on metal cation transport proteins and on the allocation mechanisms for transition metals are provided elsewhere in this book, I will present aspects of transition metal homeostasis in a hopefully novel overview. We will start with a microbial look at the transition metal Periodic Table, cation speciation, and availability in the environment. This information provides rules that might govern microbial metal cation homeostasis from the outside of the cell. The fate of metal cations inside the cell is influenced by redox potentials and affinities to ligands in complex compounds. Understanding this topic requires study of interactions between metal cations and the consequences thereof. External availability and internal binding equilibria are connected by transport reactions. These lead to metal cation concentrations in cellular compartments, which are in flow equilibrium of import and export reactions. Thus, cellular cation homeostasis may be described as an interplay of transport flow backbone and competitive binding reactions. Both together provide an energy landscape for each metal cation and cellular compartment. As a recent part of the transport flow backbone in Gram-negative bacteria, efflux across the outer membrane from the periplasm to the outside has been identified. Active outer membrane efflux might indeed be taking place in Gram-negative bacteria. Thus, the periplasm is important in bacterial metal cation homeostasis.

1 Introduction

All living cells need transition metal cations. These elements share the same electronic configuration of their valence electron *s* orbitals but they differ in the occupation of their *d* orbitals. These incompletely filled *d* orbitals allow transition metals to form complex compounds, composed of a central metal cation and (mostly) four or six ligands, which are nonmetals with free electron pairs (Housecroft and Constable 2006). In living cells, these ligands in the first shell around a transition metal cation are mostly nitrogen, oxygen, or sulfur atoms. It is the formation of complex compounds (and reactions catalyzed by them) that makes transition metals so important for life. Without them, biochemistry would be impossible.

Before these cation-requiring biochemical reactions can be performed, the right metal cation has to be inserted into the catalytic site of the correct enzyme, and not the wrong one. How is this accomplished? In this chapter, we provide an overview of the series of partitioning events that result in the desired metal–enzyme complexes.

Three simple have rules seem to decide which element is used by a living cell and which is not (Nies 2004a). An ion must be biologically available to be of any use. Except for hydrogen (and partly for helium), which was there from the Big Bang 13.7 billion years ago, other elements were forged in the furnace of ancient stars and blown into space at the stars' death struggle. The synthesis routes of nuclear fusions in this hellfire and the relative stabilities of the atomic nuclei determine the elemental composition of the stellar dust at a given age of the universe (Schaifers 1984). This dust was used to form our sun 4.6 billion years ago, with some left over quickly creating the planets, including Earth. Red-hot and boiling in its infancy, heavy elements including nickel and iron sank to the middle of Earth while lighter elements formed its crust (Wood et al. 2006). This was the second element-partitioning event. The third one is determined by the solubility of the most stable ionic form of a metal ion in water. The elemental composition of seawater (Weast 1984) might serve as a standardized ecosystem to examine how much of a chemical element is available to a living cell today (Nies 2004a).

Living cells cannot influence the three first sorting events, which result in the elemental composition of seawater and other, more complicated, ecosystems. However, they control the next sorting step, the import into the cytoplasm. Since metal cations do not diffuse across the hydrophobic bilayer of a biological membrane, membrane-bound proteins or membrane-permeable carriers are needed for uptake of a metal ion. Table 1 takes a theoretical bacterial cell, assumes metal supply to this virtual bacterium by simple uptake systems ($\Delta\Psi$-driven uniport of metal cations or charge-neutral proton symport of oxyanions), and compares the amount of accu-

Table 1 Metal cation availability: a theoretical bacterium in a model ecosystem

No[a]	Element	Ionic form[b]	Concentr. in sea water[c]	MIC[d]	Radius, (pm)[e]	Concen. theoret.bacter.[f]	*E. coli*[g]
Monovalent cations: accumulation factor 46.1 (inside/outside)							
4	Sodium	Na^{+}	457 mM	≥ 50 mM	95	21 M	236 mM
8	Potassium	K^{+}	9.72 mM	≥ 50 mM	133	448 mM	161 mM, 210 mM
16	Lithium	Li^{+}	25.9 μM	≥ 50 mM	60	1.2 mM	n.d.
19	Rubidium	Rb^{+}	1.40 μM	≥ 50 mM	148	65 μM	n.d.
38	Cesium	Cs^{+}	3.76 nM	≥ 50 mM	169	170 nM	n.d.
41	Silver	Ag^{+}	2.78 nM	20 μM	126	130 nM	n.d.
60	Thallium	Tl^{+}	48.9 pM	2 mM	140	2.2 nM	n.d.
59	Gold	Au^{+}	55.8 pM	20 μM	137	2.6 nM	n.d.
Divalent cations: accumulation factor 2122 (inside/outside)							
5	Magnesium	Mg^{2+}	55.5 mM	≥ 50 mM	65	120 M	54 mM, 44 mM
7	Calcium	Ca^{2+}	9.98 mM	≥ 50 mM	99	21 M	1.5 mM, 0.3 mM
22	Barium	Ba^{2+}	218 nM	≥ 50 mM	135	460 μM	n.d.
23	Iron	Fe^{2+}/Fe^{3+}	179 nM	III≥50 mM	76	380 μM	180 μM
25	Zinc	Zn^{2+}	153 nM	1 mM	74	325 μM	270 μM
27	Nickel	Ni^{2+}	92.0 nM	1 mM	72	195 μM	5 μM
28	Copper	Cu^{2+}/Cu^{+}	47.2 nM	1 mM	69	100 μM	18 μM
31	Manganese	Mn^{2+}	36.4 nM	20 mM	80	77 μM	1.7 μM
37	Cobalt	Co^{2+}	4.58 nM	1 mM	74	10 μM	0.5 μM
45	Cadmium	Cd^{2+}	979 pM	0.5 mM	97	2 μM	n.d.
53	Mercury	Hg^{2+}	150 pM	10 μM	110	320 nM	n.d.
54	Lead	Pb^{2+}	145 pM	5 mM	120	310 nM	n.d.
58	Beryllium	Be^{2+}	66.6 pM	*≥ 50 mM*	31	140 nM	n.d.
64	Radium	Ra^{2+}	265 aM	n.d.	140	560 fM	n.d.
Divalent oxyanions: accumulation factor 46.1 (inside/outside)							
6	Sulfur	S(VI)	27.6 mM	n.d.	29	1.3 M	62 mM
26	Molybdenum	Mo(VI)	104 nM	≥ 50 mM	62	4.8 μM	8 μM
35	Uranium	U(VI)	12.6 nM	2 mM	*97*	580 nM	n.d.
47	Chromium	Cr(VI)	962 pM	VI:0.2 mM	52	44 nM	1.1 μM
48	Tungsten	W(VI)	544 pM	≥ 50 mM	68	25 nM	n.d.
Trivalent oxyanions: accumulation factor 313 (inside/outside)							
18	Phosphorus	P(V)	2.26 μM	n.d.	34	710 μM	270 mM
29	Arsenic	As(V)	40.0 nM	n.d.	47	12 μM	n.d.
30	Vanadium	V(V)	39.3 nM	*III:5 mM*	59	12 μM	77 μM
42	Antimony	Sb(V)	2.71 nM	*III:5 mM*	62	850 nM	n.d.
55	Niobium	Nb(V)	108 pM	ins	70	34 nM	n.d.
trivalent/tetravalent cations: factor 0							
21	Aluminum	Al^{3+}	371 nM	*2 mM*	50	0	n.d.
24	Indium	In^{3+}	< 174 nM	*5 mM*	81	0	n.d.
33	Tin	Sn^{4+}	25.3 nM	*5 mM*	71	0	n.d.
34	Titanium	Ti^{4+}	20.9 nM	*5 mM*	68	0	n.d.
39	Yttrium	Y^{3+}	3.37 nM	*2 mM*	93	0	n.d.
40	Cerium	Ce^{3+}	2.85 nM	n.d.	111	0	n.d.
46	Germanium	Ge^{4+}	964 pM	ins	53	0	n.d.
49	Gallium	Ga^{3+}	430 pM	*5 mM*	62	0	n.d.
51	Zirconium	Zr^{4+}	241 pM	ins	80	0	n.d.
52	Thorium	Th^{4+}	215 pM	n.d.	95	0	n.d.
56	Lanthanum	La^{3+}	86.4 pM	*2 mM*	115	0	n.d.
57	Bismuth	Bi^{3+}	81.3 pM	ins	120	0	n.d.
61	Hafnium	Hf^{4+}	44.8 pM	ins	810	0	n.d.
62	Tantalum	Tm^{3+}	13.8 pM	n.d.	95	0	n.d.
63	Proactinium	Pa^{4+}	8.66 fM	n.d.	98	0	n.d.
65	Actinium	Ac^{3+}	0.002 aM	n.d.	118	0	n.d.

Table 1 continued

[a] Elements were ranked according to their concentration in seawater (Weast 1984) as a standard ecosystem (Nies 2004a). Metals plus sulfur and phosphate as reference were taken from this publication. The sorting number is the abundance in seawater as in Nies (2004a). The major bioelements are on a green backgound, the trace elements on a yellow background, toxic-only elements on a red background, and mostly toxic elements on a gold backgound
[b] The most probable oxidation state was used to sort the metals into the groups indicated. Iron and copper were sorted as divalent cations
[c] The occurrence in seawater is given in molar concentrations for better comparison with biological levels
[d] The minimal inhibitory concentration (MIC) was determined with *E. coli* (Nies, unpublished results) in Tris-buffered mineral salts medium (Mergeay et al. 1985), starting pH 7.0, with 2 g/l sodium gluconate as the carbon source and 1 g/l yeast extract to complement for *E. coli* auxotrophies, 30 °C, 2 days. Values in italics indicate acidic conditions that might contribute to growth inhibition at the MIC concentrations. Some data were taken from Nies (1999, 2000)
[e] Weast (1984)
[f] These values indicate to what extent an ideal virtual bacterium will accumulate the respective ions if it contains only very simple uptake systems and the uptake process reaches the thermodynamic equilibrium. The accumulation factors used are indicated above the ion groups. They were derived from a uniport of one or two positive charges with a proton motive force (PMF) of 150 mV as driving force and a $\Delta\Psi$ part of 100 mV of this PMF. For the oxyanions, a charge-neutral symport with two or three protons was assumed
[g] These data give the elemental content of *E. coli* (Outten and O'Halloran 2001 or Heldal et al. 1985, or both).
n.d., not determined or no data available; ins, insoluble or unstable

mulated metals with that experimentally determined in *Escherichia coli*. This leads to interesting insights into the basic rules for the supply of cells with elements.

Half of the monovalent cations available in seawater are alkali metals, either essential macro-bioelements or not used (Nies 2004a). Na^+ is used by most organisms to form a chemiosmotic gradient, which drives uptake processes by sodium symporters. The imported Na^+ is exported again by sodium proton antiporters (Padan et al. 2001) maintaining the sodium chemiosmotic gradient. Consequently, Na^+ is not accumulated by *E. coli*, leading to a cytoplasmic Na^+ concentration that is half of that of modern seawater (Table 1). In contrast, K^+ is accumulated to an approximately similar extent in *E. coli* and the virtual bacterium. This cation is the most important monovalent cation of the cell, and is used in bacteria to maintain the correct osmotic pressure of the cytoplasm (Walderhaug et al. 1992; Altendorf and Epstein 1996).

While the remaining alkali metal cations Li^+ and Cs^+ might be either too big or too small to be effectively transported by K^+ uptake systems (and con-

sequently these metals are not toxic, Table 1), Ag^+, Rb^+, Tl^+, and—if stable enough—Au^+ have a size similar to that of K^+ and might be taken up. Theoretically, this leads to upper nanomolar (silver) or lower nanomolar (thallium, gold) concentrations in the cytoplasm, and to micromolar MIC values of these cations. A thallium-resistance phenotype as the result of a mutation in a potassium uptake system agrees with this assumption (Sensfuss et al. 1986).

Reminiscent of the situation with the alkali metals, two alkaline earth metals are macro-bioelements, with Mg^{2+} being taken up and Ca^{2+} not, while the other group members are not needed and not toxic (Table 1). There is sufficient Mg^{2+} in modern seawater so that this metal does not even have to be accumulated; the cytoplasmic concentration in *E. coli* is remarkably similar to that in seawater. Since magnesium is also the seventh most abundant element in the Earth's crust (Weast 1984), a possibility to repress Mg^{2+} uptake might be more important to bacterial cells in many ecosystems than vigorous magnesium uptake systems.

Because Mg^{2+} competitively inhibits accumulation of some divalent transition metal cations (Webb 1970; Nies and Silver 1989; Snavely et al. 1991), easy supply with these metals should require uptake systems for cations with a diameter of 75 pm. Such a system would import all the essential transition metals and provide more to the theoretical bacterium than found in *E. coli* (Table 1). However, if this transport systems allows the entry of larger divalent cations, it might also import toxic metal cations like Cd^{2+}.

Modern seawater contains more than sufficient sulfur for cellular needs. The metals that might form divalent oxyanions in the VI oxidation state are inferior in concentration to these nonmetal oxyanions, but may be accumulated by simple proton symporters to nanomolar or even micromolar cytoplasmic concentrations. While molybdenum and tungsten do not harm *E. coli*, uranate and chromate show MIC values of 2 or 0.2 mM, respectively. These values, however, are 10^5-fold above the concentrations in seawater, so that bacteria might have to bother with the toxicity of these elements only in special ecosystems. Seawater contains nearly 40 nM vanadium. However, vanadate exists only at alkaline pH values and polymerizes otherwise (Housecroft and Constable 2006). Sb(V) is also less stable than As(V), which is present in high concentrations compared to the (for a macronutrient) low concentrations of the structurally related oxyanion phosphate. Thus, the only widely occurring problem concerning trivalent oxyanions is arsenate.

Tri- and tetravalent cations form insoluble hydroxides and are generally (with the exception of Fe^{3+}) not taken up by the cells. Consequently, they are of low toxicity. Under laboratory conditions, the low pH values these ions produced in solution might also contribute substantially to the toxicity measured for these metals. Siderophore-mediated uptake of Fe^{3+} is treated in detail by Braun and Hantke (in this volume).

The bioavailability of an element is also influenced by sequestration of its cations by inorganic and organic compounds in a given ecosystem. Unshielded d-block metals do not exist in aqueous solution (Housecroft and Constable 2006). Mostly, hexaaqua ions are formed, which might lose protons due to a strong polarization of the central metal cation. This could lead to a change of the nuclearity of the complex, and to hydroxide precipitates in the long run.

Since the pOH value is connected to the pH value via pOH = 14–pH, the availability of hydroxide anions influences the solubility of metal cations. Calculated from the solubility constants (Weast 1984), the maximum soluble concentration of the divalent cations Ba^{2+}, Ca^{2+}, Mg^{2+}, Sr^{2+}, Mn^{2+}, Cd^{2+}, Pb^{2+}, Ag^{+}, and Fe^{2+} is larger than 100 mM. For other cations, this concentration decreases (at pH 7) in the following order: Co^{2+} 25 mM, Ni^{2+} 16 mM, Zn^{2+} 4.5 mM, Cu^{2+} 16 μM, Cr^{3+} 670 pM, Al^{3+} 5 pM, Hg^{2+} 3 pM, Sn^{2+} 300 fM, Fe^{3+} 60 aM, and Co^{3+} 0.25 zM. This concentration increases with decreasing pH value, for divalent cations 100-fold per pH unit and for trivalent cations 1000-fold. Vice versa, solubility decreases by the same factors with increasing pH values. Other ligands may precipitate metal cations that do not precipitate so easily as hydroxide complexes. These are phosphate for Ca^{2+} and Pb^{2+}, carbonate for Mg^{2+}, Mn^{2+}, Cd^{2+}, Ni^{2+}, Zn^{2+}, and Cu^{2+}, and chloride for Ag^{+}.

Thus, the availability, size, and charge of metals lead to simple basic "availability rules" for transition metal homeostasis: trivalent cations can mostly be ignored, except Fe^{3+} in oxic environments. Arsenate is a more dangerous piggy-back rider of the oxyanion uptake systems for phosphate than is chromate for sulfate. K^{+} uptake systems might import Rb^{+} and Ag^{+}, but the latter metal is nearly unavailable in ecosystems with high concentrations of chloride (such as seawater). Similarly, lead may not be available in phosphate-rich environments. Mg^{2+} uptake systems may be repressed in many ecosystems and, furthermore, high Mg^{2+} concentrations should competitively inhibit uptake of other divalent cations. Thus, uptake systems should exist for import of the essential divalent transition metal cations with diameter around 75 pm. Availability of Zn^{2+} and Cu^{2+}, but not so much of Ni^{2+} and Co^{2+}, depends on the pH value of the environment. Finally, carbonate produced during degradation of organic matter precipitates some divalent metals, too. This has indeed been observed for Zn^{2+}, which was precipitated during growth of a bacterium by formation of transient mixed carbonate–hydroxide complexes (Legatzki et al. 2003b).

2 Metal Cation Homeostasis as an Interplay of Transport Flow and Binding Equilibria

Once inside the cell, some transition metals can undergo redox reactions, and especially the cations form complex compounds. At least all cations should be expected to form tetra- or hexaaqua complexes.

Gram-negative bacterial cells contain the tripeptide glutathione at a concentration of about 5 mM, which is kept almost completely reduced (Kosower and Kosower 1978; Aslund and Beckwith 1999). The GSH/GSSG (reduced/oxidized glutathione) couple has a standard redox potential at pH 7.0 of $E_0' = -240$ mV (Rost and Rapoport 1964), which is close to the in vivo potential of about -260 mV in the cytoplasm (Kirlin et al. 1999). Reduced glutathione should be able to reduce all transition metal forms with a higher potential than $E_0' = -260$ mV, which corresponds at pH = 0 with $E_0 = +0.16$ V. Looking at the half-cell potentials of the elements in Table 1, Ag^{+}, Au^{+}, and Hg^{2+} could be reduced to the metallic forms (E_0 =0.8, 1.7, and 0.85 V, respectively), Fe^{3+} to Fe^{2+} (E_0 = 0.77 V), UO_4^{2-} to U^{4+}, CrO_4^{2-} to Cr^{3+} (E_0 = 0.33 and 1.35 V, respectively), and AsO_4^{3-} to As^{3+} (E_0 = 0.56 V). It should be noted that an even number of electrons is transferred in the redox reaction concerning Hg^{2+}, uranate, and arsenate, while an odd number is transferred when Ag^{+}, Au^{+}, Fe^{3+}, and chromate are being reduced. Odd electron numbers could mean creation of radicals, short living and extremely dangerous compounds for the cells. Therefore, toxic effects resulting from reduction inside the cell should be more pronounced for Ag, Au, Fe, and Cr than for Hg, U, or As.

The redox potential of the Cu^{2+}/Cu^{+} pair is E_0 = 0.15 V and thus close to the redox value of the cytoplasm. This makes copper an extremely dangerous cation once inside the cell, because of the ease of redox cycling of Cu^{2+}/Cu^{+} combined with the generation of radicals. The cell should not be able to reduce the remaining metal cations.

3
Binding

Whether reduced by the cell or not, all transition metal cations should form complex compounds when inside. Formation of these complex compounds may lead to beneficial effects, metal toxicity, or regulatory events (Fig. 1). The total of all complex compounds of any metal in a cellular compartment can be addressed as a "pool" of this individual metal cation. A metal cation pool is not static but is part of a kinetic flow equilibrium, the result of uptake of an ion into a cellular compartment and efflux from it. In Gram-negative bacterial cells four basic transport fluxes form the transport flow backbone of the cellular metal ion homeostasis: (1) uptake across the outer membrane into the periplasm, (2) uptake across the cytoplasmic membrane (CPM) into the cytoplasm, (3) efflux back from the cytoplasm into the periplasm, and (4) efflux across the outer membrane from the periplasm to the outside. Before the proteins responsible for these transport reactions are described, we will try to understand which rules govern binding of metal cations to their beneficial, toxic, or regulatory complexes.

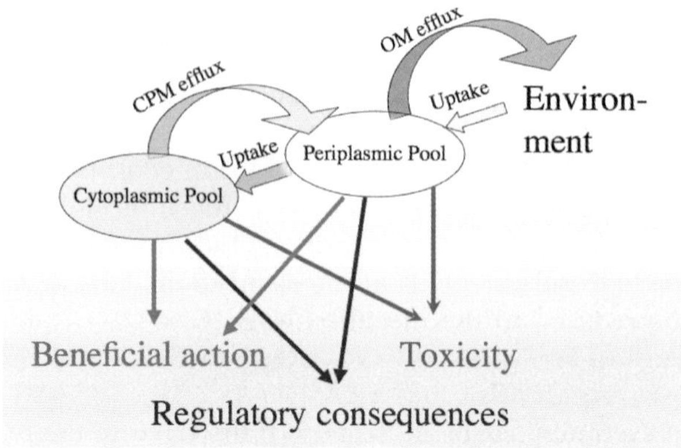

Fig. 1 Metal ion homeostasis in bacteria as an interplay of transport flow and binding equilibria. Metals from the environment are transported across the outer membrane (OM) of Gram-negative bacteria into the periplasm and further on across the cytoplasmic membrane into the cytoplasm. Uptake is antagonized by efflux back into the periplasm (or to the outside in Gram-positive bacteria) by cytoplasmic membrane efflux, which may be followed by outer membrane efflux to the outside in Gram-negative bacteria. The cytoplasmic concentration of available metals is thus in a flow equilibrium of uptake and efflux, that of the periplasm governed by two import processes (uptake across the outer membrane and cytoplasmic membrane efflux) and two export processes (uptake from the periplasm to the cytoplasm and outer membrane efflux). In both cellular compartments, metals bind to metal binding sites causing beneficial actions, toxicity, and regulatory events

3.1
The Metal Energy "Landscape" of the Cellular Compartment

To link the transport flow equilibrium that defines the size of a metal ion pool in a given cellular compartment with the binding reactions that define the composition of this cation metal pool, an energy landscape can be envisioned. This encompasses all possible metal binding sites in a compartment at a given condition and time with ligand exchange reactions leading into these binding sites. This energy landscape could be defined by hexaaqua complexes of metal cations or, if the compartment contains glutathione, their glutathionato complexes. Metal ions entering a compartment by uptake follow the kinetic path of ligand exchange reactions downhill until a suitable "valley" is reached. Filling up a valley or removing another metal cation from there might lead to metal toxicity, to beneficial effects, or to regulatory actions.

The binding sites of efflux systems can also be included as sinks, because a metal bound there should be exported. In this way, transport and binding processes are being linked and can be described by the same model. To understand metal homeostasis in bacteria, all we need to know is the identity and sizes of these subcellular pools.

3.2 Cationic Binding Forces and the Consequences

The binding energy of a metal ion to a binding site could in theory be revealed by quantum mechanical calculations of the specific metal complexes and experimentally determined by isothermal calorimetry. Ligands in the first shell of transition metal cations are mostly oxygen, nitrogen, and sulfur atoms. They donate a free electron pair to the metal cation and compete with water to recoordinate a metal cation out of the standard hexaaqua complex. While oxygen is a "hard" ligand that binds to hard (electron dense) metals, sulfur is a soft ligand searching soft metal cations (Housecroft and Constable 2006). Nitrogen is somewhat in the middle, as a hard ligand with a smaller electronegativity than oxygen.

To judge the affinities of the various metal cations to oxygen and sulfur as ligands, the solubility product of a metal sulfide was plotted against that of the respective metal hydroxide (Fig. 2). With the exception of Ag^+

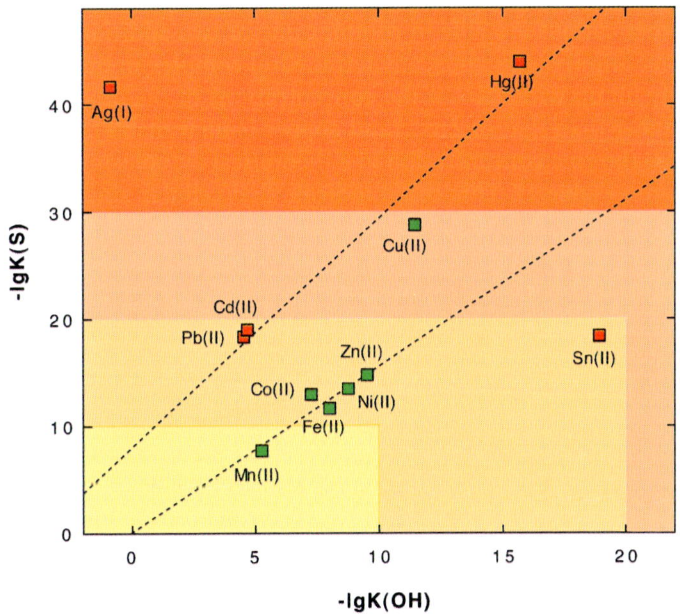

Fig. 2 Solubility product of metal sulfides and metal hydroxides as a measure of affinity to sulfur and oxygen. The negative logarithm of the solubility products of divalent heavy metal cations plus Ag^+ (Weast 1984) were plotted against each other. Essential trace elements are *green* and toxic elements *red*. With the exception of tin and the monovalent cation Ag^+, the data points are located on two lines. The essential elements except copper follow $-\lg K(S) = 1.55 \times [-\lg K(OH)] + 0.096$, the toxic elements and copper $-\lg K(S) = 2.14 \times [-\lg K(OH)] + 8.06$. The different *colors* of the background fields indicate regions of similar $K(S)$ and $K(OH)$ values

and Sn^{2+}, the connection between these solubility products could be described with two linear functions in this double logarithmical plot, one for soft metals and one for intermediate metals. This shows that within these two groups of metal cations, affinity to sulfur and oxygen changes in the same direction. In other words, if a metal cation has a higher affinity to sulfur than another metal cation of the same group, it also has a higher affinity to oxygen, probably also to nitrogen, and in general to any metal binding site provided the coordination number and geometry are not taken into account.

The affinity of the soft cations was $Hg^{2+} > Ag^{+} > Cu^{2+} > Cd^{2+} = Pb^{2+}$. Thermodynamically, these metals should end up in cytoplasmic sulfide complexes once inside the cell. With the exception of copper, the essential divalent transition metal cations exhibit similar affinities for sulfide and hydroxide with $Zn^{2+} > Ni^{2+} = Co^{2+} > Fe^{2+} > Mn^{2+}$ (Fig. 1). Thus, Zn^{2+} should be able to replace the other metals in their complexes, Co^{2+} and Ni^{2+} substitute Fe^{2+} and Mn^{2+}, and Fe^{2+} only Mn^{2+}. On the other hand, Zn^{2+} has its $3d$ orbitals completely filled, leaving only the $4s$ and $4p$ orbitals available for complex formation. Therefore, Zn^{2+} can only accommodate four ligands effectively, while the five other cations are able to complex one or two additional ligands. Due to its higher affinity to ligands, any tetrahedral complex in the cell should automatically fill up with Zn^{2+} if the concentrations of the other metal cations are not higher than that of Zn^{2+} (and copper is unavailable). It is interesting to note that the zinc concentration in a theoretical cytoplasm matches closely that of *E. coli* (Table 1). This could indicate that cellular zinc homeostasis in bacteria may indeed be simple.

Fe^{2+} can be preferentially inserted into complexes when its redox ability is used for the discrimination process. This might be the reason for the complicated redox cycling that yeast cells perform when they take up iron (Kosman 2003). Otherwise, high Co^{2+} and Ni^{2+} concentrations might interfere with an effective iron homeostasis. The concentrations of these two metals are indeed low in the theoretical cytoplasm and *E. coli* (Table 1). This leaves manganese, which is differentiated from Zn^{2+} because it can form octahedral complexes and from the other cations because its is harder than those. However, use of manganese seems to be a problem because of the possible interference of the other metal cations with manganese homeostasis. This may be the reason for the use of Mn^{2+} in very limited cases only (see the chapter by Papp, Moomaw and Maguire, in this volume), despite its central role in oxygenic photosynthesis. Maybe, a redox change between Mn(VI) and Mn(IV) can be employed to deliver manganese to its place in the photosystem II.

This rough analysis focuses and extends the rules summarized above to 15 rules of metal homeostasis in bacteria. The predictions can be experimentally addressed, or have been already:

1. In oxic environments, insoluble Fe^{3+} has to be mobilized by siderophores. Fe^{2+} can be imported by uptake systems, either specific for Fe^{2+} or unspecific for divalent cations with an ionic diameter of around 75 pm. Octahedral complexes and redox reaction could be used to differentiate Fe^{2+} against Co^{2+}, Ni^{2+}, and Zn^{2+}.
2. Cytoplasmic nickel and cobalt concentrations have to be kept low to avoid interference with iron homeostasis.
3. Cells should take up all Zn^{2+} in environments with lower Zn^{2+} concentrations than seawater; it will automatically migrate into most tetrahedral preformed complexes. Zinc toxicity should not be a problem in environments with alkaline pH values.
4. Copper is extremely dangerous; it is able to perform excessive redox cycling, radical production, and to push other cations out of their complexes. Cells can use sulfide complexes to extract and export it.
5. Manganese is difficult to use because all other transition metal cations may remove it from its complexes.
6. Cells must remove cadmium from cellular sulfide complexes and can do so by using other sulfide complexes. There is nearly no useful function for Cd^{2+}.
7. Hg^{2+} binds so tightly to thiol compounds that it is difficult to remove. However, it can be reduced to the metallic form without formation of radicals. There is no useful function for Hg^{2+}.
8. Pb^{2+} is not a problem in most ecosystems due to the low solubility of lead phosphate. However, danger might originate from imported lead phosphate. This problem could be solved in a mode similar to that for cadmium.
9. Chromate is not a problem in most environments. When it comes in, cells should reexport chromate before it can be reduced to Cr^{3+}, which leads to radical production.
10. Arsenate comes in by phosphate uptake systems but can be differentiated from phosphate by its redox potential. It can be reduced in an easy two-electron step but phosphate cannot.
11. Vanadate is no problem in most environments because it polymerizes at neutral and acidic pH values.
12. Sb(V) is far less stable than As(V) and is therefore no threat in most environments.
13. Silver may be taken up by K^+ uptake systems and accumulate in cellular sulfide compounds. Different sulfide complexes are needed to extract it from there and export it back to the outside.
14. Tl^+ and Au^+ are probably no problem in most ecosystems.
15. Cells don't have to bother with trivalent cations, they form insoluble hydroxide complexes at neutral pH values.

4 Transport Systems

Import of metals by uptake systems and export of metals by efflux systems form a kinetic equilibrium that serves as a "transport flow backbone" of cellular metal homeostasis.

4.1 Import

Since Gram-negative bacteria contain an outer membrane in addition to their cytoplasmic membrane, they possess the periplasm as an additional cellular compartment that is absent in Gram-positive bacteria. Thus, uptake of metals in Gram-negative bacterial cells is first across the outer membrane, followed by transport across the cytoplasmic membrane.

4.1.1 Import into the Periplasm

Transport of metals across the outer membrane is probably facilitated diffusion across porin proteins (Nikaido and Vaara 1985) or, in the case of iron-siderophore complexes, active transport (see the chapter by Braun and Hantke, in this volume). The major porins of *E. coli* are OmpF, OmpC, and PhoE, with OmpF being the most abundant protein of the outer membrane (Stenberg et al. 2005). OmpF and OmpC favor cations over anions (Danelon et al. 2003), and OmpF transports cations only at a neutral pH value and salt concentrations below 100 mM (Alcaraz et al. 2004). Thus, OmpF and OmpC and their orthologs in other Gram-negative bacteria should be responsible for the transport of metal cations into the periplasm.

However, transport of metals by porins may not be to steady-state equilibrium by facilitated diffusion, as long assumed. An *E. coli* mutant strain with decreased expression of OmpC exhibits increased copper sensitivity (Egler et al. 2005). This is counterintuitive because decreased expression of an outer membrane porin should lead to decreased copper uptake by the cells, thus leading to increased copper resistance. An explanation might be that cells are able to control uptake of copper into the periplasm using OmpC. If OmpC is absent, other porins like OmpF take over, but transport through these substitute pathways may not be controlled as efficiently as in the case of OmpC. More details on this explanation are given in the section below that deals with outer membrane efflux.

4.1.2
Import into the Cytoplasm

Uptake of metal cations and anions, one by one, into the cytoplasm is dealt with in other chapters of this book and would be redundant here. In general, cytoplasmic membrane uptake systems can be grouped into high rate and rather unspecific secondary transport systems that supply the basic need of a range of metal ions, and into highly substrate-specific, inducible, primary transport systems for times of need of one special ion.

The primary uptake systems are often ABC-type ATPase transporters with periplasmic substrate-binding proteins. Examples are uptake systems for manganese, iron, cobalt-containing cobalamin derivatives, nickel, zinc, sulfate, molybdate, tungstate, and phosphate. P-type ATPases can also be involved in metal cation uptake. Members of this family of ATPases can function as uptake or efflux systems, depending on the individual protein. Substrates are mostly cations from protons to lead. Examples of P-type dependent uptake ATPases import Mg^{2+}, K^{+}, Zn^{2+}, and copper. Exceptions to the rule that inducible high-affinity transport systems are ATPases are the NiCoT proteins, which are inducible, membrane-potential-dependent transport systems involved in uptake of Ni^{2+} and Co^{2+} (Eitinger et al. 2005; Rodionov et al. 2006).

Unspecific uniporters or proton symporters responsible for a general supply with a range of metals are discussed in other chapters of this book. Examples are permeases for sulfate that transport also chromate, inorganic phosphate transporters also for arsenate, magnesium uptake systems like CorA for a wide range of divalent metal cations, unspecific NRAMP (see especially the chapter by Papp, Moomaw and Maguire, in this volume), and ZIP transport proteins. Bacterial ZIP transport proteins have a broad substrate range (Grass et al. 2005a) and are responsible for uptake of divalent transition metal cations with diameters around 75 pm when magnesium uptake systems are downregulated or uptake of these cations is competitively inhibited by Mg^{2+}.

4.2
Export

In the flow equilibrium of the transport flow backbone, metal import reactions, which cannot be controlled in any case, are often balanced by regulated efflux. Again, in Gram-negative bacteria, export is first across the cytoplasmic membrane into the periplasm, followed by export across the outer membrane to the outside.

4.2.1 Cytoplasmic Membrane Efflux

The combination of high rate and unspecific with inducible high-affinity uptake systems results in an efficient supply with the metal ions needed. A disadvantage is the possible import of unwanted ions like arsenate, chromate, and Cd^{2+} or the import of too much of an otherwise nutrient metal. Cells address this problem by inducible efflux systems that decrease the cytoplasmic concentration of metal cations.

Again, details about cytoplasmic membrane efflux systems are provided in other chapters of this book. These efflux systems are primary transport ATPases or secondary membrane potential-coupled transport systems. CHR efflux proteins export chromate (Nies et al. 1998) and maybe other substrates (Nies et al. 2006), as outlined by Cervantes and Campos-Garcia (in this volume), orthologs of ArsB efflux arsenite (see the chapter by Bhattacharjee and Rosen, in this volume). The most widely distributed secondary efflux systems for transition metal cations are the cation diffusion facilitators of the CDF protein family (Nies 2003), which probably function as cation proton antiporters (Chao and Fu 2004). Many Ni^{2+}/Co^{2+} efflux proteins are related to CnrT and NreB (Grass et al. 2001; Nies 2003). Some cobalt-exporting proteins belong to the NiCoT family (Amoroso et al. 2000; Rodrigue et al. 2005), a protein family containing uptake proteins in most cases.

Primary efflux ATPases include ArsA that associates with the integral membrane protein ArsB to form an efficient export system for arsenite (see the chapter by Bhattacharjee and Rosen, in this volume), and CPx-type ATPases. These proteins belong to the superfamily of P-type ATPases, contain a conserved proline residue flanked by at least one cysteine residue, and export transition metal cations in most cases (Fagan and Saier 1994; Rensing et al. 1999; Nies 2003). Since their export activity is stimulated by thiol compounds (Rensing et al. 1999; Sharma et al. 2000), CPx-type ATPases might be able to take metals out of glutathionato complexes for export from the cytoplasm. Thus, these proteins might be important links between the transport flow backbone of the cellular metal homeostasis and the binding equilibrium.

4.2.2 Outer Membrane Efflux

The concept of an active transport from the periplasm across the outer membrane to the outside is a relatively new one. Previously, porin-mediated facilitated diffusion of low molecular weight substrates was believed to be more or less uncontrolled, which would lead to a concentration equilibrium of periplasm and outside. In such a setting, outer membrane efflux will not function and should be an energy-wasting futile cycle.

Protein complexes that could be able to catalyze an outer membrane efflux are resistance nodulation cell (RND) efflux systems. These protein complexes, composed of three different subunits, span the complete cell wall of a Gram-negative bacterium right from the cytoplasm to the outside. So, they might be able to catalyze transenvelope efflux (from the cytoplasm directly to the outside), outer membrane efflux (from the periplasm to the outside), or both processes.

RND efflux complexes are widespread among Gram-negative bacteria (Nies 2003) and, depending on the individual RND protein, they could be involved in transport of transition metal cations (heavy metal-exporting HME-RND proteins) or organic substances including antibiotics (hydrophobic and amphoteric substance-exporting HAE-RND proteins) (Tseng et al. 1999). The latter ability of RND-driven efflux systems to cause multiple drug resistance led to an increasing interest in these transport systems during the last decade, which resulted in the availability of protein structures for all three subunits of the protein complex, all done for multiple drug resistance proteins from *E. coli* or *Pseudomonas aeruginosa.*

The primary RND protein is an integral membrane protein of the cytoplasmic membrane (Fig. 3). It may take up substrates from the periplasm, cytoplasm or, in the case of hydrophobic substrates, also from the interior of the cytoplasmic membrane. Either way, RND proteins are driven by the proton-motive force (Nies 1995; Goldberg et al. 1999). The free energy released from the reimport of protons into the cytoplasm is transformed into the driving force expelling the RND substrates to the outside. This export is assumed to be through a molecular tube formed by the second component, the outer membrane factor (Paulsen et al. 1997). The third subunit is the "membrane fusion protein" (Saier et al. 1994), which stabilizes the contact between the RND and the outer membrane factor probably in a hexameric ringlike structure (Higgins et al. 2004). Membrane fusion proteins serve also as an adaptor to fit the respective outer membrane factor tube to the RND transporter (Andersen et al. 2001).

An interesting question is whether RND efflux complexes cause transenvelope efflux, outer membrane efflux, or both. The best-investigated metal-transporting RND efflux systems are CzcCBA from *Cupriavidus metallidurans* CH34 (Fig. 4) that mediates Co^{2+}, Zn^{2+}, and Cd^{2+} resistance, CnrCBA from the same bacterium that is involved in Co^{2+} and Ni^{2+} resistance, and SilCBA and CusCBA that mediate silver and copper resistance in enterobacteria (Fig. 4). The RND protein CzcA is able to transport substrates across a proteoliposomal membrane (Goldberg et al. 1999), which takes the place of the cytoplasmic membrane in the in vitro experiment. However, CzcA was unable to do so in the presence of glutathione (Nies, unpublished results). So, the transenvelope/outer membrane efflux question remained open (Nies 2003), although a theoretical calculation led to the conclusion that RND efflux systems can only function if they detoxify the periplasm in cooperation with

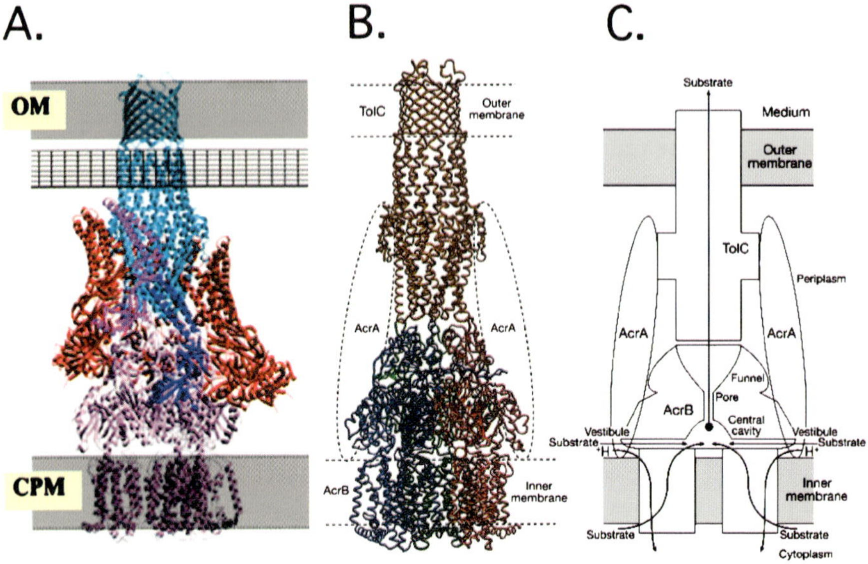

Fig. 3 Structure and possible functions of RND efflux systems. The protein complex spans the complete cell wall of a Gram-negative bacterium from the cytoplasm across the cytoplasmic membrane (CPM), the periplasm containing the bacterial peptidoglycane (bricks in panel A), and the outer membrane (OM) to the outside. The actual RND protein (resistance nodulation cell division protein family (Saier et al. 1994; Tseng et al. 1999)) resides as a trimer in the cytoplasmic membrane and extends into the periplasm (shown is AcrB from *E. coli* (Murakami et al. 2002)). Triangular openings between the individual subunits connect the interior of the RND protein with the periplasm so that substrates may be taken up from this compartment (panel C). Alternatively, substrates may also originate from the interior of the cytoplasmic membrane in the case of hydrophobic substrates or from the cytoplasm (Murakami et al. 2002, 2004). RND proteins are driven by the proton-motive force (Nies 1995; Goldberg et al. 1999). Therefore, the free energy resulting from proton import should be transformed into export of the RND-bound substrate to the outside. This export reaction is catalyzed by the outer membrane factor (protein family by Paulsen et al. (1997)). Outer membrane factors like TolC (Koronakis et al. 2000) are trimeric tubes spanning the OM by a single beta barrel and extending into the periplasm as well. They contain two openings, one to the outside, the second in their periplasmic part, and this one connects to a central opening on top of the RND protein. Finally, the membrane fusion proteins (Saier et al. 1994) probably form hexameric rings to stabilize the contact between the RND and outer membrane factor (Akama et al. 2004a,b; Higgins et al. 2004). Modified from Higgins et al. (2004) and Murakami et al. (2002)

a cytoplasmic membrane efflux protein which detoxifies the cytoplasm and delivers the substrate into the periplasm for further export (Palmer 2003). That would explain the presence of genes encoding cytoplasmic membrane efflux proteins in addition to those encoding RND efflux systems in metal cation resistance determinants (Fig. 4).

Three pieces of evidence suggest that outer membrane efflux is indeed membrane protein-dependent in bacteria. The first comes from *E. coli*. This

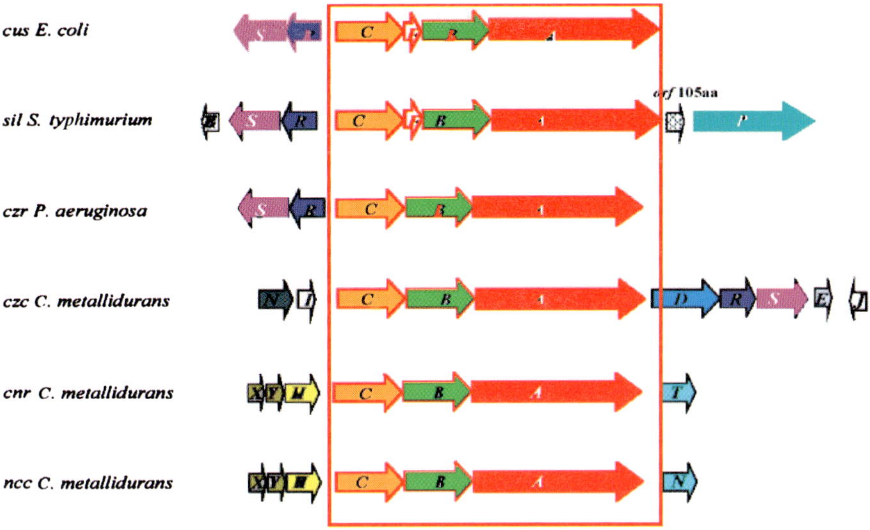

Fig. 4 Metal resistance determinants encoding RND-driven cation efflux systems. The genes (*arrows*) indicate the direction of transcription. Three or four genes in the *red rectangle* encode for an outer membrane factor protein (*C*, *orange*), a membrane fusion protein (*B*, *green*), or an RND protein (*A*, *red*). Additional genes in the vicinity of these three encode two-component regulatory systems composed of a membrane-bound histidine kinase sensor (*S*, *purple*) and an affiliated response regulator (*R*, *dark blue*). These proteins regulate expression of the resistance determinant. Alternatively, regulation may occur by an extracytoplasmic functions sigma factor (*H*, *yellow* (Missiakas and Raina 1998; Nies 2004b)) and its anti-sigma factors (*Y*, *X*, *brown*). Additional genes for transport proteins (shades of *blue*) are also part of the determinant. The four determinants at the bottom adjacent to the RND archetype *czc* are from *C. metallidurans* and *P. aeruginosa*, and mediate resistance to divalent heavy metal cations such as Zn^{2+}, Cd^{2+}, Co^{2+}, and Ni^{2+} (Nies et al. 1987, 1989; Liesegang 1993). The two determinants at the top mediate resistance to monovalent metals Cu^{+}/Cu^{2+} and Ag^{+} and are from enterobacteria (Franke et al. 2001, 2003; Silver 2003). They contain an additional gene coding for a small periplasmic metal-binding protein (*F*, *white* (Franke et al. 2003))

bacterium contains only one metal-exporting RND system, the Cus system, while some other bacteria contain more (Nies 2003), up to 12 in *C. metallidurans* (Nies et al. 2006). Deletion of the *cus* determinant led to a 50% decrease in silver resistance but no decrease in copper resistance (Franke et al. 2001). Responsible for this result was a periplasmic copper-containing copper oxidase, CueO, that complemented the action of the missing CusCBA efflux pump (Grass and Rensing 2001a,b; Outten et al. 2001; Roberts et al. 2002; Grass et al. 2004). The best explanation for this complementation of the action of an RND efflux system by a periplasmic enzyme is that both detoxify the periplasm. So, outer membrane efflux by CusCBA is part of the cellular copper tolerance system of *E. coli*.

The second source of information comes from *C. metallidurans*. This bacterium contains on its chromosomal DNA the genes for two CPx-type ATPases involved in Cd^{2+} detoxification (Nies 2003). If both genes are deleted, cadmium resistance is down to a minimal inhibitory concentration of 1 μM. *In trans* complementation with the genes for the CzcCBA efflux pump leads to some increase in cadmium resistance (50 μM), but the full resistance level of 3 mM is only reached if at least one of the two CPx exporters is able to function. Therefore, full cadmium resistance in *C. metallidurans* results from an interplay of a CPx-type ATPase efflux pump and an RND-type outer membrane efflux pump (Legatzki et al. 2003a,b). However, this evidence also allows for the existence of an additional transenvelope efflux process.

C. metallidurans also contains on its chromosome the genes for two CDF efflux proteins (Nies 2003). FieF appears to export Fe^{2+} like its ortholog from *E. coli* (Grass et al. 2005c), while DmeF has a wide substrate range for divalent transition metal cations. Deletion of *dmeF* results in a drastic decrease of Co^{2+} resistance which, however, cannot be complemented by *in trans* expression of CzcCBA or CnrCBA (Munkelt et al. 2004). Thus, in this specific situation of cobalt resistance of *C. metallidurans*, transenvelope efflux seems not to happen. Co^{2+} resistance is obligatorily a two-step detoxification process with DmeF-mediated cytoplasmic membrane efflux followed by CzcCBA- or CnrCBA-mediated outer membrane efflux.

As reviewed elsewhere (Nies 2003), substrates for the HAE-RND family of proteins (that transport organic molecules) bind to the periplasmic part of the respective RND protein. This is also true for the metal-transporting HME-RND protein CusA (Franke et al. 2003). If RND efflux systems catalyze outer membrane efflux, a futile cycle of porin-mediated outer membrane uptake and RND-mediated outer membrane efflux has to be diminished or prevented. There is evidence for this. First, full copper resistance of *E. coli* is only obtained in the presence of the outer membrane porin OmpC, and decreased expression of this porin leads to decreased copper resistance (Egler et al. 2005). As an explanation, OmpC and CusCBA form a matched pair in a way that OmpC-mediated facilitated diffusion of copper into the periplasm is coupled to Cus-mediated efflux. *E. coli* compensates for decreased expression of OmpC with increased expression of other porins that take up copper faster than Cus can efflux them. A similar close association of functions has been observed in *Neisseria gonorrhoeae* between the HAE-RND system MtrCDE and the porin PIB (Olesky et al. 2006). Thus, Gram-negative bacteria seem to maintain their periplasmic metal homeostasis as a kinetic flow equilibrium composed of porin-mediated facilitated diffusion into the periplasm and RND-driven outer membrane efflux. By using two-component regulatory systems of extracytoplasmic function sigma factors/membrane-bound anti-sigma factors to regulate expression of the RND system by the periplasmic metal concentration (Nies and Brown 1998; Grass et al. 2005b), this flow equilibrium can also be controlled at the gene level, which leads to a second-level view of bacterial metal homeostasis (see also the chapter by Hobman, Yamamoto, and Oshima, in this volume).

5
Reprise: A Glance Upon Metal Landscapes in *E. coli*

Due to the multitude of varying chemical binding sites, iron and zinc homeostasis in *E. coli* will be the most complicated to unravel. Cd^{2+} and Pb^{2+} are effluxed by the CPx-type ATPase ZntA (see the chapter by Mitra and Rensing, in this volume). It should be easier with nickel and cobalt. Ni^{2+} is an essential part of the glyoxalase I and the four classes of hydrogenases (see the chapter by Hausinger and Zamble, in this volume, and http://ecocyc.org). Uptake of Ni^{2+} and its insertion into enzymes, however, is tightly controlled and involves nickel chaperones (see the chapter by Hausinger and Zamble, in this volume). This prevents Ni^{2+} from interacting with the Fe^{2+} and Zn^{2+} metabolism, as stated above.

Interestingly, *E. coli* seems not to need inorganic Co^{2+}. There are 12 enzymes that incorporate Co^{2+} as cofactor (http://ecocyc.org); however, all of these enzymes can use other divalent metals instead of Co^{2+}. That leaves the cobalamin-dependent reactions. However, *E. coli* does not synthesize cobalamin and relies on the import of it or its precursors (Lawrence and Roth 1995). Therefore, free Co^{2+} may not be needed by the bacterium. The bioavailable form of this metal that is relatively scarce in seawater could be cobalamin and its derivatives.

A rough estimate of the copper landscape in *E. coli* was derived (Table 2). Although these calculations oversimplify, they serve to illustrate a bacterial metal cation homeostasis system until better data become available. Ninety-six percent of the cellular copper of *E. coli* is located in the periplasm (Table 2) as O'Halloran suggested (Changela et al. 2003); 25% of all copper should end up in the CuB site of the terminal cytochrome o oxidase (Abramson et al. 2000). Since the assembly of the CuA and CuB sites of cytochrome c oxidase from *Paracoccus denitrificans* may occur entirely from the periplasm (Richter and Ludwig 2003), this 25% of the *E. coli* copper pool does not require uptake of copper into the cytoplasm.

Sixty-five percent of the cellular copper of *E. coli* could be in the five copper sites of the periplasmic copper-containing copper oxidase CueO (Roberts et al. 2002), followed by 3% on the histidine kinase CusS, which controls expression of the CusCBA RND efflux pump. Other factors are periplasmic copper-containing enzymes, followed by the periplasmic copper chaperone CusF (Loftin et al. 2005) and the binding site to the RND protein CusA (Franke et al. 2003). If a periplasmic copper-insertion mechanism was uncovered, 96% of the cellular copper would not have to enter the cytoplasm.

Table 2 The copper pool of *E. coli*

Protein: binding site[a]	Number /*E. coli* cell[b]	% Binding sites[c]	Ligand atoms S	N	O	Binding energy (kJ/mol)[d]
Periplasm:						
CusA	170	0.33	4	0	0	–419
CusS	1430	2.76	0	5	0	–398
Ndh	872	1.68	3	1	0	–394
CueO:rCu	6700	12.90	2	0	3	–373
CueO:T1 Cu	6700	12.90	2	2	0	–369
SodC	442	0.85	0	4	0	–318
TynA	440	0.85	0	3	1	–293
CueO:Cu2	6700	12.90	0	3	1	–293
CueO:Cu3	6700	12.90	0	3	1	–293
CusF	< 130	< 0.25	2	1	0	–289
CyoBACD	12 800	24.70	0	3	0	–239
CueO:T2 Cu	6700	12.90	0	2	1	–214
Cytoplasm:						
CopA	330	0.64	2	1	3	–453
YhcH	440	0.85	0	2	3	–323
CueR	< 1190	< 2.30	2	0	0	–209
GSH	< 9 000 000	Not included	1	0	2	–214

[a] The known copper-binding proteins from *E. coli* and their binding sites are sorted by the cellular compartment and the estimated binding energy of copper to these sites
[b] Calculated from the proportion of the mRNA for each protein of the total mRNA pool as taken from gene array experiments (Egler et al. 2005). Control cells not treated with added copper
[c] Number of a binding site divided by the number of total binding sites (51 744), glutathione not included
[d] $(N(S)* - 104.7 + N(N)* - 79.6 + N(O)* - 54.53)$ kJ/mol. $N(S)$, $N(N)$, $N(O)$ = number of S, N, and O ligand atoms, respectively. The free energies were those of the hydroxide (O) and sulfide (S) complexes (Weast 1984); that of nitrogen was a mean values of those

However, *E. coli* might need cytoplasmic copper to synthesize its molybdenum cofactor (see the chapter by Schwarz, Hagedoorn and Fischer, in this volume). Moreover, the copper-containing enzyme YhcH has been identified that seems to reside in the cytoplasm (Teplyakov et al. 2005). Most of the cytoplasmic copper of *E. coli*, however, might be part of the regulatory protein CueR (Changela et al. 2003), followed by copper binding to the copper-exporting CPx-type ATPase CopA and YhcH (Table 2). Glutathione as binding factor is not included in this calculation. *E. coli* should contain about ten million individual protein molecules with only a few copper proteins, but about nine million glutathione molecules all able to sequester the metal cation.

It is interesting to compare the calculated binding energies of copper to these cellular sites. Copper binds strongest always to the binding sites of the export proteins: CopA in the cytoplasm and CusA in the periplasm. So, copper released by chance from one binding site or released by degradation of the protein should migrate down the energy path to the respective efflux protein. Thus, cellular copper homeostasis in *E. coli* is organized like water homeostasis in the bathroom: the sink should always be at the lowest part of the floor.

6 Outlook

In principle, but perhaps never practically, it should be possible to measure an energy landscape experimentally. That would explain the entire cellular metalome based on principles of physical chemistry. The binding energy of each metal cation to each metal-binding protein or other macromolecules can be determined by isothermal calorimetry giving enthalpy. Proteomics should reveal the amount of the metal-binding factor in the cell at a given time and condition. Atom absorption spectrophotometry or other techniques can be used to quantify the amount of metals in the cell (see the chapter by Wesenberg, Bleuel and Krauss, in this volume). Finally, fractionation of the cells and assessment of the metal content per fraction should evaluate the predictions of the energy landscape model. Moreover, we should be able to comprehend what happens when this homeostasis is being perturbed, a process leading to serious diseases in humans. So, it can be done, and moving forward from *E. coli* to the elephant should deepen our understanding of metal homeostasis of all kinds cellular compartments.

References

Abramson J, Riistama S, Larsson G, Jasaitis A, Svensson-Ek M, Laakkonen L, Puustinen A, Iwata S, Wikstrom M (2000) The structure of the ubiquinol oxidase from *Escherichia coli* and its ubiquinone binding site. Nat Struct Biol 7:910–917

Akama H, Kanemaki M, Yoshimura M, Tsukihara T, Kashiwagi T, Yoneyama H, Narita S-I, Nakagawa A, Nakae T (2004a) Crystal structure of the drug discharge outer membrane protein, OprM, of *Pseudomonas aeruginosa*. Dual modes of membrane anchoring and occluded cavity end. J Biol Chem 279:52816–52819

Akama H, Matsuura T, Kashiwagi S, Yoneyama H, Narita SI, Tsukihara T, Nakagawa A, Nakae T (2004b) Crystal structure of the membrane fusion protein, MexA, of the multidrug transporter in *Pseudomonas aeruginosa*. J Biol Chem 279:25939–25942

Alcaraz A, Nestorovich EM, Aguilella-Arzo M, Aguilella VM, Bezrukov SM (2004) Salting out the ionic selectivity of a wide channel: the asymmetry of OmpF. Biophys J 87:943–957

Altendorf K, Epstein W (1996) The Kdp-ATPase of *Escherichia coli*. In: Dalbey RE (ed) Advances in cell and molecular biology of membranes and organelles. JAI, Greenwich, UK, pp 401–418

Amoroso MJ, Schubert D, Mitscherlich P, Schumann P, Kothe E (2000) Evidence for high affinity nickel transporter genes in heavy metal resistant *Streptomyces* spec. J Basic Microbiol 40:295–301

Andersen C, Hughes C, Koronakis V (2001) Protein export and drug efflux through bacterial channel-tunnels. Curr Opin Cell Biol 13:412–416

Aslund A, Beckwith J (1999) Bridge over troubled waters: sensing stress by disulfide formation. Cell 96:751–753

Bhattacharjee H, Rosen BP (2007) Arsenic Metabolism in Prokaryotic and Eukaryotic Microbes (in this volume). Springer, Heidelberg

Braun V, Hantke K (2007) Acquisition of Iron by Bacteria (in this volume). Springer, Heidelberg

Cervantes C, Campos-Garcia J (2007) Reduction and Efflux od Chromate by Bacteria (in this volume). Springer, Heidelberg

Changela A, Chen K, Xue Y, Holschen J, Outten CE, O'Halloran TV, Mondragon A (2003) Molecular basis of metal-ion selectivity and zeptomolar sensitivity by CueR. Science 301:1383–1387

Chao Y, Fu D (2004) Kinetic study of the antiport mechanism of an *Escherichia coli* zinc transporter, ZitB. J Biol Chem 279:12043–12050

Danelon C, Suenaga A, Winterhalter M, Yamato I (2003) Molecular origin of the cation selectivity in OmpF porin: single channel conductances vs. free energy calculation. Biophys Chem 104:591–603

Egler M, Große C, Grass G, Nies DH (2005) Role of ECF sigma factor RpoE in heavy metal resistance of *Escherichia coli*. J Bacteriol 187:2297–2307

Eitinger T, Suhr J, Moore L, Smith JAC (2005) Secondary transporters for nickel and cobalt ions: theme and variations. Biometals 18:399–405

Fagan MJ, Saier MH Jr (1994) P-type ATPases of eukaryotes and bacteria: sequence comparisons and construction of phylogenetic trees. J Mol Evol 38:57–99

Franke S, Grass G, Nies DH (2001) The product of the *ybdE* gene of the *Escherichia coli* chromosome is involved in detoxification of silver ions. Microbiology 147:965–972

Franke S, Grass G, Rensing C, Nies DH (2003) Molecular analysis of the copper-transporting CusCFBA efflux system from *Escherichia coli*. J Bacteriol 185:3804–3812

Goldberg M, Pribyl T, Juhnke S, Nies DH (1999) Energetics and topology of CzcA, a cation/proton antiporter of the RND protein family. J Biol Chem 274:26065–26070

Grass G, Rensing C (2001a) CueO is a multi-copper oxidase that confers copper tolerance in *Escherichia coli*. Biochem Biophys Res Commun 286:902–908

Grass G, Rensing C (2001b) Genes involved in copper homeostasis in *Escherichia coli*. J Bacteriol 183:2145–2147

Grass G, Fan B, Rosen BP, Lemke K, Schlegel HG, Rensing C (2001) NreB from *Achromobacter xylosoxidans* 31A is a nickel-induced transporter conferring nickel resistance. J Bacteriol 183:2803–2807

Grass G, Thakali K, Klebba PE, Thieme D, Müller A, Wildner GF, Rensing C (2004) Linkage between catecholate siderophores and the multicopper oxidase CueO in *Escherichia coli*. J Bacteriol 186:5826–5833

Grass G, Franke S, Taudte N, Nies DH, Kucharski LM, Maguire ME, Rensing C (2005a) The metal permease ZupT from *Escherichia coli* is a transporter with a broad substrate spectrum. J Bacteriol 187:1604–1611

Grass G, Fricke B, Nies DH (2005b) Control of expression of a periplasmic nickel efflux pump by periplasmic nickel concentrations. Biometals 18:437–448

Grass G, Otto M, Fricke B, Haney CJ, Rensing C, Nies DH, Munkelt D (2005c) FieF (YiiP) from *Escherichia coli* mediates decreased cellular accumulation of iron and relieves iron stress. Arch Microbiol 183:9–18
Hausinger RP, Zamble D (2007) Micobial Physiology of Nickel and Cobalt (in this volume). Springer, Heidelberg
Heldal M, Norland S, Tumyr O (1985) X-ray microanalytic method for measurement of dry matter and elemental content of individual bacteria. Appl Environ Microbiol 50:1251–1257
Higgins MK, Bokma E, Koronakis E, Hughes C, Koronakis V (2004) Structure of the periplasmic component of a bacterial drug efflux pump. Proc Natl Acad Sci USA 101:9994–9999
Hobman J, Yamamoto K, Oshima T (2007) Transciptomic Responses of Bacterial Cells to sublethal Metal Ion Stress (in this volume). Springer, Heidelberg
Housecroft CE, Constable EC (2006) Chemistry, 3rd edn. Pearson Education, Essex, UK
Kirlin WG, Cai J, Thompson SA, Diaz D, Kavanagh TJ, Jones DP (1999) Glutathione redox potential in response to differentiation and enzyme inducers. Free Radic Biol Med 27:1208–1218
Koronakis V, Sharff A, Koronakis E, Luisi B, Hughes C (2000) Crystal structure of the bacterial membrane protein TolC central to multidrug efflux and protein export. Nature 405:914–919
Kosman DJ (2003) Molecular mechanisms of iron uptake in fungi. Mol Microbiol 47:1185–1197
Kosower NS, Kosower EM (1978) The glutathione status of cells. Int Rev Cytol 54:109–160
Wesenberg D, Bleuel C, Krauss GJ (2007) A Glossary of Microanalytical Approaches to Assess the Metallome (in this volume). Springer, Heidelberg
Lawrence JG, Roth JR (1995) The cobalamin (coenzyme B12) biosynthetic genes of *Escherichia coli*. J Bacteriol 177:6371–6380
Legatzki A, Anton A, Grass G, Rensing C, Nies DH (2003a) Interplay of the Czc system and two P-type ATPases in conferring metal resistance to *Ralstonia metallidurans*. J Bacteriol 185:4354–4361
Legatzki A, Franke S, Lucke S, Hoffmann T, Anton A, Neumann D, Nies DH (2003b) First step towards a quantitative model describing Czc-mediated heavy metal resistance in *Ralstonia metallidurans*. Biodegradation 14:153–168
Liesegang H, Lemke K, Siddiqui RA, Schlegel H-G (1993) Characterization of the inducible nickel and cobalt resistance determinant *cnr* from pMOL28 of *Alcaligenes eutrophus* CH34. J Bacteriol 175:767–778
Loftin IR, Franke S, Roberts SA, Weichsel A, Heroux A, Montfort WR, Rensing C, McEvoy MM (2005) A novel copper-binding fold for the periplasmic copper resistance protein CusF. Biochemistry 44:10533–10540
Mergeay M, Nies D, Schlegel HG, Gerits J, Charles P, van Gijsegem F (1985) *Alcaligenes eutrophus* CH34 is a facultative chemolithotroph with plasmid-bound resistance to heavy metals. J Bacteriol 162:328–334
Missiakas D, Raina S (1998) The extracytoplasmic function sigma factors: role and regulation. Mol Microbiol 28:1059–1066
Mitra B, Rensing C (2007) Zinc, Cadmium and Lead Resistance and Homeostasis (in this volume). Springer, Heidelberg
Munkelt D, Grass G, Nies DH (2004) The chromosomally encoded cation diffusion facilitator proteins DmeF and FieF from *Wautersia metallidurans* CH34 are transporters of broad metal specificity. J Bacteriol 186:8036–8043

Murakami S, Nakashima R, Yamashita R, Yamaguchi A (2002) Crystal structure of bacterial multidrug efflux transporter AcrB. Nature 419:587–593

Murakami S, Tamura N, Saito A, Hirata T, Yamaguchi A (2004) Extramembrane central pore of multidrug exporter AcrB in *Escherichia coli* plays an important role in drug transport. J Biol Chem 279:3743–3748

Nies DH (1995) The cobalt, zinc, and cadmium efflux system CzcABC from *Alcaligenes eutrophus* functions as a cation-proton antiporter in *Escherichia coli.* J Bacteriol 177:2707–2712

Nies DH (1999) Microbial heavy metal resistance. Appl Microbiol Biotechnol 51:730–750

Nies DH (2000) Heavy metal resistant bacteria as extremophiles: molecular physiology and biotechnological use of *Ralstonia* spec. CH34. Extremophiles 4:77–82

Nies DH (2003) Efflux-mediated heavy metal resistance in prokaryotes. FEMS Microbiol Rev 27:313–339

Nies DH (2004a) Essential and toxic effects of elements on microorganisms. In: Anke K, Ihnat M, Stoeppler M (eds) Metals and their compounds in the environment. Wiley-VCH, Weinheim (part II.1)

Nies DH (2004b) Incidence and function of sigma factors in *Ralstonia metallidurans* and other bacteria. Arch Microbiol 181:255–268

Nies DH, Brown N (1998) Two-component systems in the regulation of heavy metal resistance. In: Silver S, Walden W (eds) Metal ions in gene regulation. Chapman Hall, London/New York, pp 77–103

Nies DH, Silver S (1989) Metal ion uptake by a plasmid-free metal-sensitive *Alcaligenes eutrophus* strain. J Bacteriol 171:4073–4075

Nies DH, Mergeay M, Friedrich B, Schlegel HG (1987) Cloning of plasmid genes encoding resistance to cadmium, zinc, and cobalt in *Alcaligenes eutrophus* CH34. J Bacteriol 169:4865–4868

Nies DH, Nies A, Chu L, Silver S (1989) Expression and nucleotide sequence of a plasmid-determined divalent cation efflux system from *Alcaligenes eutrophus.* Proc Natl Acad Sci USA 86:7351–7355

Nies DH, Koch S, Wachi S, Peitzsch N, Saier MHJ (1998) CHR, a novel family of prokaryotic proton motive force-driven transporters probably containing chromate/sulfate transporters. J Bacteriol 180:5799–5802

Nies DH, Rehbein G, Hoffmann T, Baumann C, Grosse C (2006) Paralog of genes encoding metal resistance proteins in *Cupriavidus metallidurans* strain CH34. J Mol Microbiol Biotechnol 147:82–93

Nikaido H, Vaara M (1985) Molecular basis of bacterial outer membrane permeability. Microbiol Rev 49:1–32

Olesky M, Zaho S, Rosenberg RL, Nicholas RA (2006) Porin-mediated antibiotic resistance in *Neisseria gonorrhoeae*: ion, solute, and antibiotic permeation through PIB proteins with *penB* mutations: interaction OMP and RND transporter. J Bacteriol 188:2300–2308

Outten CE, O'Halloran TV (2001) Femtomolar sensitivity of metalloregulatory proteins controlling zinc homeostasis. Science 292:2488–2492

Outten FW, Huffman DL, Hale JA, O'Halloran TV (2001) The independent *cue* and *cus* systems confer copper tolerance during aerobic and anaerobic growth in *Escherichia coli.* J Biol Chem 276:30670–30677

Padan E, Venturi M, Gerchman Y, Dover N (2001) Na^+/H^+ antiporters. Biochim Biophys Acta 1505:144–157

Palmer M (2003) Efflux of cytoplasmically acting antibiotics from Gram-negative bacteria: periplasmic substrate capture by multicomponent efflux pumps inferred from their cooperative action with single-component transporters. J Bacteriol 185:5287–5289

Papp KM, Moomaw AS, Maguire M (2007) Manganese: Uptake, Biological Function and Role in Virulence (in this volume). Springer, Heidelberg

Paulsen IT, Park JH, Choi PS, Saier MH JR (1997) A family of Gram-negative bacterial outer membrane factors that function in the export of proteins, carbohydrates, drugs and heavy metals from Gram-negative bacteria. FEMS Microbiol Lett 156:1–8.

Rensing C, Ghosh M, Rosen BP (1999) Families of soft-metal-ion-transporting ATPases. J Bacteriol 181:5891–5897

Richter OM, Ludwig B (2003) Cytochrome c oxidase—structure, function, and physiology of a redox-driven molecular machine. Rev Physiol Biochem Pharmacol 147:47–74

Roberts SA, Weichsel A, Grass G, Thakali K, Hazzard JT, Tollin G, Rensing C, Montfort WR (2002) Crystal structure and electron transfer kinetics of CueO, a multicopper oxidase required for copper homeostasis in *Escherichia coli*. Proc Natl Acad Sci USA 99:2766–2771

Rodionov DA, Hebbeln P, Gelfand MS, Eitinger T (2006) Comparative and functional genomic analysis of prokaryotic nickel and cobalt uptake transporters: evidence for a novel group of ATP-binding cassette transporters. J Bacteriol 188:317–327

Rodrigue A, Effantin G, Mandrand-Berthelot MA (2005) Identification of *rcnA (yohM)*, a nickel and cobalt resistance gene in *Escherichia coli*. J Bacteriol 187:2912–2916

Rost J, Rapoport S (1964) Reduction potential of glutathione. Nature 201:185–187

Saier MH Jr, Tam R, Reizer A, Reizer J (1994) Two novel families of bacterial membrane proteins concerned with nodulation, cell division and transport. Mol Microbiol 11:841–847

Schaifers K, Traving G (1984) Meyers Handbuch Weltall. Bibliographisches Institut, Mannheim

Schwarz G, Hagedoorn P-L, Fischer K (2007) Molybdate and Tungstate: Uptake, Homeostasis, Cofactor Synthesis and Enzymes (in this volume). Springer, Heidelberg

Sensfuss C, Reh M, Schlegel HG (1986) No correlation exists between the conjugative transfer of the autotrophic character and that of plasmids in *Nocardia opaca* strains. J Gen Microbiol 132:997–1007

Sharma R, Rensing C, Rosen BP, Mitra B (2000) The ATP hydrolytic activity of purified ZntA, a Pb(II)/Cd(II)/Zn(II)-translocating ATPase from *Escherichia coli*. J Biol Chem 275:3873–3878

Silver S (2003) Bacterial silver resistance: molecular biology and uses and misuses of silver compounds. FEMS Microbiol Rev 27:341–353

Snavely MD, Miller CG, Maguire ME (1991) The *mgtB* Mg^{2+} transport locus of *Salmonella typhimurium* encodes a P-type ATPase. J Biol Chem 266:815–823

Stenberg F, Chovanec P, Maslen SL, Robinson CV, Ilag LL, von Heijne G, Daley DO (2005) Protein complexes of the *Escherichia coli* cell envelope. J Biol Chem 280:34409–34419

Teplyakov A, Obmolova G, Toedt J, Galperin MY, Gilliland GL (2005) Crystal structure of the bacterial YhcH protein indicates a role in sialic acid catabolism. J Bacteriol 187:5520–5527

Tseng T-T, Gratwick KS, Kollman J, Park D, Nies DH, Goffeau A, Saier MHJ (1999) The RND superfamily: an ancient, ubiquitous and diverse family that includes human disease and development proteins. J Mol Microbiol Biotechnol 1:107–125

Walderhaug M, Polarek J, Voelkner P, Daniel J, Hesse J, Altendorf K, Epstein W (1992) KdpD and KdpE, proteins that control expression of the *kdpABC* operon, are members of the two-component sensor–effector class of regulators. J Bacteriol 174:2152–2159
Weast RC (1984) CRC handbook of chemistry and physics, 64th edn. CRC, Boca Raton, Florida
Webb M (1970) Interrelationship between utilization of magnesium and the uptake of other bivalent cations by bacteria. Biochim Biophys Acta 222:428–439
Wood BJ, Walter MJ, Wade J (2006) Accretion of the Earth and segregation of its core. Nature 441:825–833

Microbiol Monogr (6)
D. H. Nies, S. Silver: Molecular Microbiology of Heavy Metals
DOI 10.1007/7171_2006_076/Published online: 18 January 2007

Biosensing of Heavy Metals

Hauke Harms

Department of Environmental Microbiology, UFZ Centre for Environmental Research, Permoserstraße 15, 04318 Leipzig, Germany
hauke.harms@ufz.de

Abstract Naturally occurring, regulated resistance mechanisms of bacteria against various heavy metals and metalloids have been used to construct whole-cell living biosensors or bioreporters. Molecular fusions of regulatory circuits with reporter genes encoding easily detectable reporter proteins enable bioreporters to sense metal targets, typically at concentrations in the nanomolar to micromolar range, although more sensitive sensors also exist. The biological components of extant bioreporter constructs and the target ranges and sensitivities of bioreporter constructs are presented. An outlook on developments using novel molecular interactions as triggers of the biological responses and strategies for the improvement of bioreporters is given. Application examples are presented that illustrate the capability of bioreporters to measure bioavailable fractions of the target species rather than total loads.

1 Introduction

The binding of a chemical to a biological receptor can be regarded as the consequence of the successful recognition of these two partners. As such it is potentially useful for the analysis of the chemical, provided that there is a possibility to observe the binding in a way that has advantages over the chemical analysis of the chemical. From a practical point of view, the usefulness of receptor–ligand interactions for analytical purposes depends primarily on the kind and extent of a conformational change resulting in the receptor, and on the biological role that the receptor molecule plays. Those receptor molecules that are involved in cellular signalling chains or circuits are particularly promising targets for the development of biological sensor systems. This is for two reasons: Firstly, ligand-binding to such receptors typically has a natural function in connection with the biological effect of the ligand, be it the induction of a catabolic pathway by a substrate or the up-regulation of a defence reaction by a toxicant. Secondly, the binding typically activates a functional detection chain that provides a good basis for extension by an observable reporter principle.

2
The Principle of Bioreporter Organisms

So-called whole-cell living biosensors or bioreporter organisms rely on biological recognition and, unlike many enzyme-based biosensors, make use of an extant detection chain. This has the advantage that the technical periphery for signal transduction and interpretation can be simplified and reduced or even omitted, for instance in the case of bioreporter reactions that can be interpreted with the naked eye. The fact that living organisms are used for chemical analysis has advantages such as the ease and low price of bioreporter

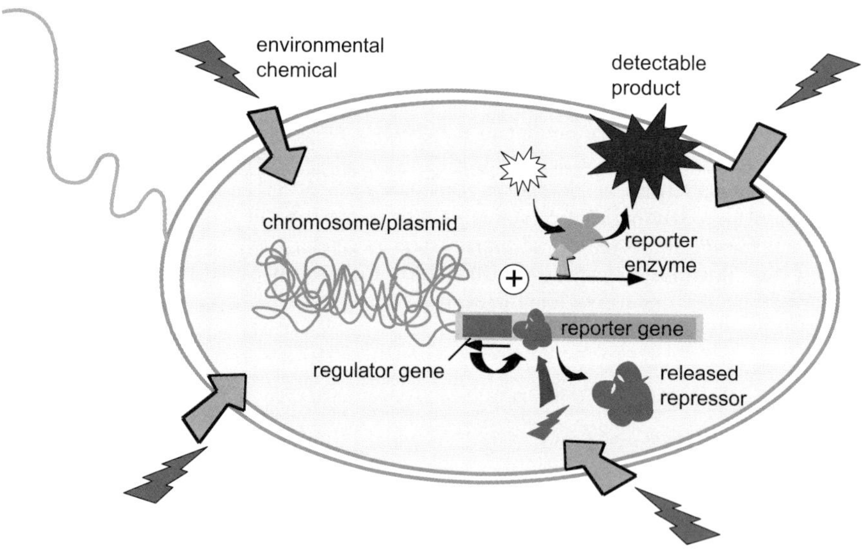

Fig. 1 Schematic representation of a class I bioreporter bacterium for an environmental chemical (drawn as a *flash*). Upon entering the cell the chemical induces the transcription of a reporter gene, in the case depicted here by causing a repressor protein to leave the promoter/operator region. The accumulation of a visible reporter enzyme product results in signal amplification. Reproduced from (Harms et al. 2006)

production by simple autocatalytic reproduction or the direct access to the analysis of bioavailable fractions of chemicals. However, there are also disadvantages connected with the required viability of fragile living beings, the restriction to physiological conditions, and the risk of unnoticed interferences within the complex regulatory network of a living cell. A schematic representation of so-called class I bioreporter bacteria (van der Meer et al. 2004) is given in Fig. 1. They are based on the interaction of a regulatory protein–ligand complex with a promoter-operator region. Upon ligand recognition the regulator protein binds to or releases the operator, thereby activating the transcription of a (typically heterologous) gene that has been chosen because it encodes a protein or enzyme that gives an easily observable signal. In the case of essential bioelements such as iron it may also be the absence of the metal species that is sensed and used to up-regulate the reporter. Bioreporters that were constructed to react to stress conditions by increased signal output (class II) or bioreporters reacting aspecifically on stress conditions with a reduced signal output (class III) are not considered here, although they have been used to analyse the effects of heavy metals and metalloids (for review see Köhler et al. 2000; Hansen and Sorensen 2001). This is the right place to mention that a sharp distinction between the bioreporter classes is sometimes impossible since, e.g. class I bioreporters for toxic chemicals are generally also subject to class III effects. Bioreporter inhibition at higher analyte concentrations is typical and restricts their use to sub-toxic concentrations.

3
Reporter Genes

Reporter genes that have been used for the construction of bioreporter bacteria have been presented in detail by Daunert et al. (2000) and Köhler et al. (2000) and are summarized in Table 1 together with some information about the characteristics that may guide their choice for different applications. When the construction of bioreporters is motivated by the wish to analyse bioavailable fractions of a metal in a natural setting, non-invasive (remote) quantification of the bioreporter signal is necessary and, typically, light-emitting proteins or enzymes are used. If easy signal interpretation is desired and spectroscopic apparatus unavailable a colour reaction may be the bioreporter signal of choice. If the goal is to localize the target element at the micro-scale a detection method of high spatial resolution may be needed, which requires that microscopic single cell analysis is possible and the small size of the reporter organisms can be exploited (Jaspers et al. 2001). In most cases quantitative results are desired and reporter genes with clear dose–signal relationships over extended concentration ranges are favoured. An important aspect is the requirement of oxygen of most light-emitting reporters, which reduces the choice of reporters for application to anoxic en-

Table 1 Reporter genes and proteins used to construct heavy metal bioreporters

Reporter protein	Genes	Detection methods	Key characteristic – advantages/drawbacks
Bacterial luciferase	*luxAB*/*luxCDABE* from *Vibrio*/*Photobacterium*	Luminometry, exposition of photographic film, microscopy	Robust bioreporter; sensitive detection; unequivocal signal for environmental analysis - no natural sources of bioluminescence in heavy-metal contaminated matrices / *luxAB* constructs require aldehyde (C9-C14) substrate; requires oxygen
Firefly luciferase	*luc*	Luminometry, exposition of photographic film, microscopy	High quantum yield; sensitive detection; unequivocal signal for environmental analysis - no natural sources of bioluminescence in heavy-metal contaminated matrices / requires luciferin as substrate, oxygen and ATP
Fluorescent proteins	*gfp* and variants	Fluorescence microscopy, steady state fluorimetry	No substrate required - non-invasive detection possible; single cell signal detection permits analyses with high spatial resolution / interference of autofluorescence of natural matrices; no signal amplification; requires oxygen for maturation
β –Galactosidase	*lacZ*	Colorimetry, fluorimetry, chemiluminometry, electrochemical	Choice of substrates for different detection methods including naked-eye detection / requires substrate addition
Ice-nucleating protein	*inaZ*	Ice-nucleation test	Sensitive / highly invasive reporter reaction

vironments. Finally, the reporter signal should be clearly distinguishable from natural signals.

4 Receptor Systems

Of even more interest in the present context is the range of biological systems, i.e. receptor proteins that have been used for the recognition of chemicals. Large parts of this volume deal with the interactions of microbial proteins of various kinds with heavy metals and metalloids and, since the early 1990s, many of these receptors have been considered or used for the construction of specific bioreporter bacteria for metal detection and quantification. Table 2 summarizes some characteristics of extant whole-cell living biosensors for heavy metals and metalloids. Despite the large total number of bioreporters that have been described to date, the range of targeted metal species is relatively small and appears to be driven by the anticipated value of resulting bioreporter products for application in environmental analysis and ecological research, but limited by the availability of information regarding the genetic mechanisms underlying bacteria–heavy metal interactions. In some cases the lack of genetic information was circumvented by random insertion (e.g. using transposons) in bacterial genomes followed by the screening of phenotypic effects of metals (e.g. Guzzo et al. 1991). Clearly the highest numbers of bioreporters were constructed to target either mercury ions or inorganic arsenic species. Targets of somewhat lesser popularity were species of Al, Cd, Cr, Cu, Fe, Ni, Pb and Zn. In addition, some bioreporters were found to have cross-reactivity with species of Sb, Bi, Co, Ag and Sn. Examination of the literature reveals that the entire wealth of extant bioreporter constructs for Hg^{2+} and AsO_2^- relies on the *mer* and *ars* resistance machineries from a small number of biological sources and varies mainly in the construction principles (e.g. plasmid-borne vs. chromosomal, design of transcriptional fusions), host organisms and reporter genes. Although reported ranges of detected concentrations have to be regarded with great caution due to very different calibration schemes, in some cases small variations such as dilution of bioreporter suspensions have led to drastically increased sensitivities (Rasmussen et al. 1997). Despite these limitations, it is clear that reported lower detection limits often reflect the concentrations at which the analytes naturally occur and/or act as toxicants or nutrients.

Table 2 Bioreporter constructs for heavy metal and metalloid analysis

Target chemical	Operator/ promoter	Source	Reporter gene	Induction time	Concentration	Refs.
$\mathbf{Al^{3+}}$, Cu^{2+}, Fe^{3+}, Ni^{2+}	*fliC*	*Escherichia coli*	*luxAB*	20 min	40–400 μM [a]	Guzzo et al. 1991, 1992
AsO_2^-, AsO_4^{3-}	*ars*	*E.coli*	*lacZ*		5–10 μ g/L, ≥ 100 μ g/L	Diario et al. 1995
AsO_2^-	*ars*	*E. coli*	*luxAB*	2 h	≥ 10 μ g/L	Cai and Dubow 1996
AsO_2^-, SbO_2^-, Cd^{2+}	*ars*	*S. aureus*	*lucFF*	2 h	≥ 100 nM, ≥ 33 nM, ≥ 330 nM	Tauriainen et al. 1997
SbO_2^-, AsO_2^-	*ars*	*E. coli*	*luxAB*	30 min, 3 h	1 fM–10 nM, 1 fM–10 nM	Ramanathan et al. 1997
SbO_2^-, AsO_2^-	*ars*	*E. coli*	*lacZ*	30 min–17 h	1 fM–10 μM, 1 nM–100 μM	Ramanathan et al. 1998
AsO_2^-, AsO_4^{3-}	*arsR*	*E. coli*	*luxAB*	1 h	50 nM–1 μM	Stocker et al. 2003
AsO_2^-, AsO_4^{3-}	*arsR*	*E. coli*	*lacZ*	1 h	100 nM–1 μM	Stocker et al. 2003
AsO_2^-, AsO_4^{3-}	*arsR*	*E. coli*	*gfp*	1 h	≥ 100 μM	Stocker et al. 2003
AsO_2^-, AsO_4^{3-}, SbO_2^-	*arsR*	*S. aureus*	*gfp*	8 h, 8 h, 8 h	≥ 0.1 μM, ≥ 0.1 μM, ≥ 0.1 μM	Liao and Ou 2005
$\mathbf{Cd^{2+}}$, Bi^{3+}, Pb^{2+}	*cad*	*S. aureus* pI258	*blaZ*	90 min	0.5–100 μM [a]	Yoon et al. 1991
Cd^{2+}	*cad*	*S. aureus, E. coli*	*luxAB*	60 min–2 h	0.5–100 μM	Corbisier et al. 1993
Cd^{2+}, Pb^{2+}, SbO_2^-, Sn^{2+}, Zn^{2+}	*cadC*	*S. aureus*	*lucFF*	2 h	≥ 3.3 nM, ≥ 33 nM, ≥ 1 nM, ≥ 100 μM, ≥ 1 μM	Tauriainen et al. 1998
$\mathbf{CrO_4^{2-}}$, $Cr_2O_7^{2-}$, Cr^{3+}	*chrA*	*Cupriavidus* metallidurans CH34	*luxCDABE* *lacZ*	60 min–2 h	1–50 nM [a]	Peitzsch et al. 1998

Table 2 continued

Target chemical	Operator/ promoter	Source	Reporter gene	Induction time	Concentration	Refs.
CrO_4^{2-}	*chrA*	*C. metallidurans* CH34 pMOL28	*luxCDABE*	3 h	2–40 μM	Corbisier et al. 1999
CrO_4^{2-}, $Cr_2O_7^{2-}$ Cr^{3+}	*chrA*	*C. metallidurans* CH34 pMOL28	*lucFF*	2 h	50 nM, 30 nM	Ivask et al. 2002
$\mathbf{Cu^{2+}}$, Cd^{2+}, Zn^{2+}	*cupC*	*C. metallidurans* CH34	*luxCDABE*	5 h	2 μM–0.1 mM	Corbisier et al. 1994
Cu^{2+}		*E. coli*	*luxCDABE*		1 μM–1 mM	Holmes et al. 1994
Cu^{2+}	*copSRA*	*C. metallidurans* CH34	*luxCDABE*	90 min	≥ 1 μM	Corbisier et al. 1999
Cu^{2+}, Ag^{+}	*copA*	*E. coli*	*luxCDABE*	60–100 min	1–30 μM, 300 nM–3 μM	Riether et al. 2001
Cu^{2+}, Ag^{+}	*copA*	*E. coli*	*lucFF*		0.7 μM–1 mM, 4–30 nM	Hakkila et al. 2004
Fe^{3+}	*pvd*		*inaZ*		Induction < 1 μM	Loper et al. 1994
Fe^{3+}	*isiAB*	*Synechococcus* sp	*luxAB*		10 nM–1 μM	Durham et al. 2002
Fe^{3+}	*fepA-fes*	*E. coli*	*luxCDABE*		30 nM	Mioni et al. 2003
Hg^{2+}	*mer*	*E. coli* Tn21	*luxAB*	5 min	50–200 nM	Condee and Summer 1992
Hg^{2+}	*merT*	*E. coli* Tn21	*luxCDABE*	100 min	0.5 nM–1 μM 10 pM	Selifonova et al. 1993, Rasmussen et al. 1997
$\mathbf{Hg^{2+}}$, Cd^{2+}	*merT*	*E. coli* Tn21	*lucFF*	60 min	0.1 fM–1 μM	Virta et al. 1995
Hg^{2+}	*mer*	*Serratia marcescens*	*lux*		0.1 nM–1 μM	Holmes et al. 1994
Hg^{2+}	*mer*	*E. coli Tn501*	*lacZ, luxCDABE, inaZ*		20–200 nM	Rouch et al. 1995

Table 2 continued

Target chemical	Operator/ promoter	Source	Reporter gene	Induction time	Concentration	Refs.
Hg^{2+}	*mer*	*E. coli* Tn21	*luxCDABE*	80 min	≥ 0.5 μ g/L	Hansen and Sorensen 2000
Hg^{2+}	*mer*	*E. coli* Tn21	*lacZYA*	4 h	≥ 0.5 μ g/L	Hansen and Sorensen 2000
Hg^{2+}	*mer*	*E. coli* Tn21	*gfp*	16 h	≥ 25 μ g/L	Hansen and Sorensen 2000
Hg^{2+}	*mer*		*luxCDABE*	90–150 min	0.05–200 nM	Pepi et al. 2006
Methyl mercury chloride	*mer*	*S. marcescens* pDU 1358	*lucFF*	2 h	0.2 nM	Ivask et al. 2001
Phenylmercury acetate					1 nM	
Dimethylmercury					10 μM	
Ni^{2+}	*celF*		*luxAB*		1–50 ppm	Guzzo et al. 1994
Ni^{2+}, Co^{2+}	*cnr*	*C. metallidurans* CH34	*lacZ*	30 min–2 h	50 μM [a]	Grass et al. 2000
Ni^{2+}, Co^{2+}	*cnr*	*C. metallidurans* CH34	*luxCDABE*	4–6 h	100 nM–60 μM 9–400 μM	Tibarzawa et al. 2001
Pb^{2+}	*pbrR*	*C. metallidurans* CH34	*luxCDABE*	3 h	500 μM–5 mM	Corbisier et al. 1999
Zn^{2+}	*smtA*	*Synechococcus* sp.	*lacZ*	2 h	12 μM	Huckle et al. 1993
Zn^{2+}, Co^{2+}, Cd^{2+}	*smtA*	*Synechococcus* sp.	*luxCDABE*	4 h	0.5–4 μM [a]	Erbe et al. 1996
Zn^{2+}	*zntA*	*E. coli*	*lucFF*	2 h	40 μM–12 mM	Ivask et al. 2002
Zn^{2+}, Cd^{2+}, Pb^{2+}, Hg^{2+}, Co^{2+}	*zntA*	*E. coli*	*luxCDABE*	60–100 min	1–30 μM, 10–300 nM 30 nM–1 μM, 1–30 μM, 3–10 μM	Riether et al. 2001

[a] Concentration measured with the analyte printed in bold

5
Bioreporters for the Determination of Bioavailability

Most researchers motivate the construction of bioreporters with their wish to measure an analyte's bioavailability as an alternative to chemical analytical methods, which typically attempt to quantify total concentrations or loads. Class I bioreporters differ from other bioreporters since they are designed to give an isolated view on bioavailability while trying to exclude secondary biological effects such as stress or toxicity. To judge how realistic this aim is, one needs to consider what a chemical's bioavailability is and what bioreporters could sense and report about it. Recently, Semple et al. (2004) defined "... the *bioavailable compound* as that which is freely available to cross an organism's cellular membrane from the medium the organism inhabits at a given time..." A difficulty lies in the fact that bioavailability is highly subjective since organisms can influence the freely available compound in their surrounding by modifying their environment, including their own extracytoplasmic structures or by actively exposing themselves. Therefore, the chemical measured by the bioreporter is not automatically representative for the bioavailability of the same chemical to other organisms one is interested in. The awareness of this fact has led Hansen and Sorensen (2000) to construct a series of biosensor vectors for mercury that can be transferred to a variety of Gram-negative bacteria having different ecological properties. However, the subjectivity of bioavailability goes beyond differences between physiologically or ecologically different bacterial strains. To be of practical use for monitoring purposes, bioreporter bacteria need to be good surrogates for toxicity targets of more concern, such as humans or fish. It is thus important to know if bacterial bioreporters would sense the same bioavailable compound.

Taking environmental arsenic contamination as an example we can easily foresee and experimentally show that any arsenite or arsenate that is bound to water-borne iron-hydroxide colloids is not freely available to the bacterium and cannot enter the cell (Harms et al. 2005). However, the same contaminated water will become acidified in the stomach of humans, leading to dissolution of iron hydroxides and the release of arsenic species, upon which potentially more arsenic is freely available for uptake by the human epithelial cells. The arsenite bioreporter bacteria would thus underestimate the toxicologically relevant load of arsenic in the sample. Changed incubation conditions in the bioreporter assay, however, can modify the bioavailable arsenic for the bacterial cell. Dissolution of the iron hydroxides can be achieved by assay acidification and subsequent complexation with EDTA, whilst keeping arsenite bioavailable to the bioreporter cells (Harms et al. 2005). Thus, the biological parameter can be modified by changing the assay conditions and thereby 'standardizing' the bacterial response to the expected human bioavailable arsenic. In contrast, the capability of a bacterium to produce effective siderophores may lead to an overestimation of metal bioavailability for other organisms.

Uptake and metabolism of target compounds by bacterial cells create a chemical flux and thus potentially change the bioavailable compound, for instance in the case of sorbed or insoluble, solid species. It is therefore recommended to construct bioreporter bacteria based on various natural isolates and to use them in combination to obtain a realistic picture of the range of bioavailabilities of an analyte. It has been suggested, for instance, that iron availability in freshwater environments can be best monitored using cyanobacterium-based bioreporters (Porta et al. 2003). If the subjectivity of bioavailability is neglected, biological detection loses its advantage over, e.g., the chemical analysis of water-dissolved fractions.

6
Potential for Bioreporter Improvement

In a recent review, van der Meer et al. (2004) have scrutinized the detection chain of bacterial bioreporters. They noticed a general lack of engineering principles used for the development of bioreporters and identified potential for improvement at various steps between the mode of exposure of the bioreporter to the target chemicals and the processing of instrumental data at the other end of the chain. In terms of biological improvements, they identified the possibility of improving the sensitivity of receptors by protein engineering. There is, for example, no obvious reason why it should be impossible to modify receptors for various metal ions in a way that they become as sensitive as some of those for Hg^{2+} (Table 2), for example. Improved sensitivity or better signal-to-noise ratios can also be achieved by eliminating components that interfere with the analyte–receptor interaction, such as efflux pumps. Diorio et al. (1995) showed that the disruption of *arsB* increased the sensitivity of an arsenite bioreporter by one to two orders of magnitude. A different approach would be to shorten the detection chain, thereby making it less susceptible to variations and interferences. A way of doing so has been presented by Deuschle et al. (2005) who designed translational fusions of two fluorogenic proteins of clearly different spectral properties with a central periplasmic binding protein (pbp). The conformational change upon ligand-binding to the pbp domain changed the separation distance between the two fluorophores, thereby influencing the frequency of fluorescence resonance energy transfer between them. Spectral analysis of the fluorescence output (the ratio between both emission wave lengths) could be used for analyte quantification. This strategy eliminated the entire genetic machinery required for the induction, transcription and translation in traditional class I bioreporters from the detection chain. Rational protein design was applied to modify the sensitivity of the fusion proteins and to change their target range (Looger et al. 2003). Wells et al. (2005) shortened the detection chain by analysing green fluorescent bioreporters for arsenic by using laser-induced confocal microscopy in combination with fluorescence correlation spectroscopy. This method directly determines the number of synthesized Gfp molecules rather than inferring the degree of expression from the fluorescence output. It thus eliminates uncertainty about the quantum yield of the fluorescing protein. An important improvement over existing bioreporters for Hg_2^+ is a construct expressing an organomercurial lyase that cleaves organic mercury compounds thereby making the mercury available to the strain's *mer*-based luminescent reporter function. Application of this reporter, together with a reporter lacking the organomercurial lyase, allows quantification of organic mercury compounds in the presence of inorganic mercury (Ivask et al. 2001).

7 Application of Whole Cell Living Biosensors for Environmental Analysis

Bioreporters are still somewhere between the proof-of-the-principle phase and real environmental application, as detailed in Harms et al. (2006). Many of the so-called applications of bioreporters are in fact attempts to validate the functioning of new constructs using real-world samples and do not contribute to decisions about contaminated site management. Some examples that go beyond the inclusion of environmental samples in the calibration exercises of new bioreporters are presented below.

Durham et al. used a luminescent *Synechococcus* sp. bioreporter to compare the bioavailability of iron in water samples from Lake Erie and Lake Huron. The reporter that used the Fur-regulated *isiAB* promoter detected iron at trace levels and reported varying iron bioavailability in Lake Huron and Lake Erie despite similar total iron concentrations in both lakes (Durham et al. 2002; Porta et al. 2003). Using a *Pseudomonas syringae* green fluorescent bioreporter for ferric iron, Joyner and Lindow (2000) visualized and quantified the large heterogeneity of iron availability on plant leaves. Iron limitation was also shown for the flowers of apples and pears (Temple et al. 2004). A bioreporter was used to demonstrate that the siderophore alcaligin E of the heavy metal-resistant strain *Cupriavidus metallidurans* CH34 influences the bioavailability and hence the toxicity of cadmium (Gilis et al. 1998). An *ars*-based luminescent bioreporter for arsenite and antimony was used to investigate the bioavailable part of high total concentrations of antimonite in various soils from mining and smelting sites. The antimonite was found to be unavailable over a large pH range and thus to be of limited environmental risk (Flynn et al. 2003). Large differences between the bioavailable fractions of four heavy metals in soil were detected with luminescent (*lucFF*) bioreporters. The bioavailabilities of Cd_2^+, Hg_2^+ and Zn_2^+ in water extracts were all below 2% whereas almost half of the chromate was bioavailable. When the bioreporters were exposed to soil water suspensions the bioavailabilities of cadmium and mercury rose 20 and 30 times, respectively, thus indicating the bioaccessibility (potential bioavailability) of the particle-bound metals as opposed to particle-bound zinc and chromium, which remained inaccessible (Ivask et al. 2002). Similar results were obtained by bioreporter measurements in soils polluted by metal smelters where water-extractable fractions of cadmium and lead were 115 and 40 times smaller than the corresponding biologically accessible fractions (Ivask et al. 2004). Further bioreporter applications include the quantification of copper in the run-off from patinated copper roofs (Karlén et al. 2002), the dynamics of Hg bioavailability in agricultural and forest soils (Rasmussen et al. 2000), and the prediction of the transfer of nickel into plants.

Examples of large scale bioreporter application to real environmental problems, with the goal of substituting or complementing chemical and/or toxicological monitoring, are still scarce but there is a recent example for such an application. Inorganic arsenic species have received enormous attention as contaminations of groundwater in large regions of the world; some well-documented examples being Bangladesh, India and Vietnam. The recent application of luminescent arsenite/arsenate bioreporter bacteria (Stocker et al. 2003) to samples from nearly 200 groundwater wells from the Red River and Mekong River deltas in Vietnam resulted in more than 90% correct bioreporter measurements, meaning that arsenic contents above and below the WHO guidelines were quite reliably distinguished (Trang et al. 2005). This can be seen as a major breakthrough in bioreporter technology, which has been facilitated by various favourable circumstances. Firstly, the arsenic problem is restricted to the two inorganic arsenic species, arsenite (As III) and arsenate (As V), which can both be detected with the same bioreporter. Secondly, the risk of interference from other analytes is low, since antimonate as the only cross-reacting inducer and toxicants other than inorganic arsenic are absent in the groundwater. Thirdly, groundwater is a homogenous, transparent medium allowing direct bioreporter application and recognition without sample preparation or analyte extraction. Fourthly, the patchiness of the arsenic contamination, the high number of potentially affected groundwater wells (ca. 10 million) and the local social and economic conditions require cheap and simple arsenic tests. Finally, for implementation of the method it is important that the arsenic contamination threatens millions of people with different forms of cancer – a problem that is sufficiently important to justify the use of genetically modified organisms outside of research laboratories. One will have to see if the case of arsenic will help to establish bioreporter measurements as an accepted method for groundwater analysis in Southeast Asia. To the author's knowledge, up to now bioreporter-based analytical methods have not been certified by environmental regulation authorities. Certification would be an important step towards their routine use in environmental analysis and monitoring.

References

Cai J, DuBow MM (1996) Expression of the *Escherichia coli* chromosomal *ars* operon. Can J Microbiol 42662–671

Condee CW, Summer AO (1992) A *mer-lux* transcriptional fusion for real-time examination of in vivo gene expression kinetics and promoter response to altered superhelicity. J Bacteriol 174:8094–8101

Corbisier P, Ji G, Nuyts G, Mergeay M, Silver S (1993) *LuxAB* gene fusions with the arsenic and cadmium resistance operons of *Staphylococcus aureus* plasmid pI258. FEMS Microbiol Lett 110:231–238

Corbisier P, Thiry E, Masolijn A, Diels L (1994) Construction and development of metal ion biosensors. In: Campbell AK, Kricka LJ, Stanley PE (eds) Bioluminescence and chemiluminescence: fundamentals and applied aspects. Wiley, Chichester, pp 151–155

Corbisier P, van der Lelie D, Borremans B, Provoost A, de Lorenzo V, Brown NL, Lloyd JR, Hobman JL, Csöregi E, Johansson G, Mattiasson B (1999) Whole cell- and protein-based biosensors for the detection of bioavailable heavy metals in environmental samples. Anal Chim Acta 387:235–244

Daunert S, Barrett G, Feliciano JS, Shetty RS, Shresta S, Smith-Spencer W (2000) Genetically engineered whole-cell sensing systems: Coupling biological recognition with reporter genes. Chem Rev 100:2705–2738

Deuschle K, Okumoto S, Fehr M, Looger LL, Kozhukh L, Frommer WB (2005) Construction and optimization of a family of genetically encoded metabolite sensors by semirational protein engineering. Prot Sci 14:2304–2314

Diorio C, Cai J, Marmor JS, Shinder R, DuBow MS (1995) An *Escherichia coli ars* operon homolog is functional in arsenic detoxification and is conserved in Gram-negative bacteria. J Bacteriol 177:2050–2056

Durham KA, Porta D, Twiss MR, McKay RML, Bullerjahn GS (2002) Construction and initial characterization of a luminescent *Synchococcus* sp. PCC 7942 Fe-dependent bioreporter. FEMS Microbiol Lett 209:215–221

Erbe JL, Adams AC, Taylor KB, Hall LM (1996) Cyanobacteria carrying and *smt-lux* transcriptional fusion as biosensor for the detection of heavy metal cations. J Ind Microbiol Biotechnol 17:80–83

Flynn HC, Meharg AA, Bowyer PK, Paton GI (2003) Antimony bioavailability in mine soils. Environ Poll 124:93–100

Gilis A, Corbisier P, Baeyens W, Taghavi S, Mergeay M, van der Lelie D (1998) Effect of the siderophore alcaligin E on the bioavailability of Cd to *Alcaligenes eutrophus* CH34. J Ind Microbiol Biotechnol 20:61–68

Grass G, Grosse C, Nies D (2000) Regulation of the *cnr* cobalt and nickel resistance determinant from *Ralstonia* sp. strain CH34. J Bacteriol 182:1390–1398

Guzzo A, Diorio C, DuBow MS (1991) Transcription of the *Escherichia coli fliC* gene is regulated by metal ions. Appl Environ Microbiol 57:2255–2259

Guzzo J, Guzzo A, DuBow MS (1992) Characterization of the effects of aluminum on luciferase biosensors for the detection of ecotoxicity. Toxicol Lett 64/65:687–693

Guzzo A, DuBow MS (1994) A *luxAB* transcriptional fusion to the cryptic *celF* gene of *Escherichia coli* displays increased luminescence in the presence of nickel. Mol Gen Genet 242:455–460

Hakkila K, Green T, Leskinen P, Ivask A, Marks R, Virta M (2004) Detection of bioavailable heavy metals in EILATox-Oregon samples using whole-cell luminescent bacterial sensors in suspension or immobilized onto fibre-optic tips. J Appl Toxicol 24:333–342

Hansen LH, Sørensen SJ (2000) Versatile biosensor vectors for detection and quantification of mercury. FEMS Microbiol Lett 193:123–127
Hansen LH, Sørensen SJ (2001) The use of whole-cell biosensors to detect and quantify compounds or conditions affecting biological systems. Microb Ecol 42:483–494
Harms H, Rime J, Leupin O, Hug SJ, van der Meer JR (2005) Influence of groundwater composition on arsenic detection by bacterial biosensors. Microchim Acta 151:217–222
Harms H, Wells M, van der Meer JR (2006) Whole-cell living biosensors – are they ready for environmental application? Appl Microbiol Biotechnol 70:273–280
Holmes DS, Dubey SK, Gangolli S (1994) Development of biosensors for the detection of mercury and copper ions. Environ Geochem Health 16:229–233
Huckle JW, Morby AP, Turner JS, Robinson NJ (1993) Isolation of a prokaryotic metallothionein locus and analysis of transcriptional control by metal ions. Mol Microbiol 7:177–187
Ivask A, Hakkila K, Virta M (2001) Detection of organomercurials with sensor bacteria. Anal Chem 73:5168–5171
Ivask A, Virta M, Kahru A (2002) Construction and use of specific luminescent recombinant bacterial sensors for the assessment of bioavailable fraction of cadmium, zinc, mercury and chromium. Soil Biol Biochem 34:1439–1447
Ivask A, Francois M, Kahru A, Dubourguier H-C, Virta M, Douay F (2004) Recombinant luminescent bacterial sensors for the measurement of bioavailability of cadmium and lead in soils polluted by metal smelters. Chemosphere 55:147–156
Jaspers MCM, Meier C, Zehnder AJB, Harms H, van der Meer JR (2001) Measuring mass transfer processes of octane with the help of an *alkS-alkB::gfp*-tagged *Escherichia coli*. Environ Microbiol 3:512–524
Joyner DC, Lindow SE (2000) Heterogeneity of iron bioavailability on plants assessed with a whole-cell GFP-based bacterial biosensor. Microbiology 146:2435–2445
Karlén C, Wallinder IO, Heijerick D, Leygraf C (2002) Runoff rates, chemical speciation and bioavailability of copper released from naturally patinated copper. Environ Poll 120:691–700
Köhler S, Belkin S, Schmid RD (2000) Reporter gene bioassays in environmental analysis. Fresenius J Anal Chem 366:769–779
Liao VHC, Ou KL (2005) Development and testing of a green fluorescent protein-based bacterial biosensor for measuring bioavailable arsenic in contaminated groundwater samples. Environ Toxicol Chem 24:1624–1631
Looger LL, Dwyer MA, Smith JJ, Hellinga HW (2003) Computational design of receptor and sensor proteins with novel functions. Nature 423:185–189
Loper JE, Lindow SE (1994) A biological sensor for iron availability to bacteria in their habitats on plant surfaces. Appl Environ Microbiol 60:1934–1941
Mioni CE, Howard AM, DeBruyn JM, Bright NG, Twiss MR, Applegate BM, Wilhelm SW (2003) Characterization and field trials of a bioluminescent bacterial reporter of iron bioavailability. Mar Chem 83:31–46
Peitzsch N, Eberz G, Nies DH (1998) *Alcaligenes eutrophus* as a bacterial chromate sensor. Appl Environ Microbiol 64:453–458
Pepi M, Reniero D, Baldi F, Barbieri P (2006) A comparison of *mer::lux* whole cell biosensors and moss, a bioindicator, for estimating mercury pollution. Water Air Soil Poll 173:163–175
Porta D, Bullerjahn GS, Durham KA, Wilhelm SW, Twiss MR, McKay RML (2003) Physiological characterization of a *Synechococcus sp.* (*Cyanophyceae*) strain PCC 7942 iron-dependent bioreporter for freshwater environments. J Phycol 39:64–73

Ramanathan S, Shi WP, Rosen BP, Daunert S (1997) Sensing antimonite and arsenite at the subattomole level with genetically engineered bioluminescent bacteria. Anal Chem 69:3380–3384

Ramanathan S, Shi W, Rosen BP, Daunert S (1998) Bacteria-based chemiluminescence sensing system using β-galactosidase under the control of the ArsR regulatory protein of the *ars* operon. Anal Chim Acta 369:189–195

Rasmussen LD, Turner RR, Barkay T (1997) Cell-density-dependent sensitivity of a *mer-lux* bioassay. Appl Environ Microbiol 63:3291–3293

Rasmussen LD, Sorensen SJ, Turner RR, Barkay T (2000) Application of a mer-lux biosensor for estimating bioavailable mercury in soil. Soil Biol Biochem 32:639–646

Riether KB, Dollars M-A, Billard P (2001) Assessment of heavy metal bioavailability using *Escherichia coli zntAp::lux* and *copAp::lux*-based biosensors. Appl Microbiol Biotechnol 57:712–716

Rouch DA, Parkhill J, Brown NL (1997) Induction of bacterial mercury- and copper-responsive promoters: Functional differences between inducible systems and implications for their use in gene-fusions for in vivo metal biosensors. J Ind Microbiol Biotechnol 14:349–353

Selifonova O, Burlage R, Barkay T (1993) Bioluminscent sensors for detection of bioavailable Hg(II) in the environment. Appl Environ Microbiol 59:3083–3090

Semple KT, Doick KJ, Jones KC, Burauel P, Craven A, Harms H (2004) Defining bioavailability and bioaccessibility of contaminated soil and sediment is complicated. Environ Sci Technol 38:228A-231A

Stocker J, Balluch D, Gsell M, Harms H, Feliciano JS, Malik KA, Daunert S, van der Meer JR (2003) Development of a set of simple bacterial biosensors for quantitative and rapid field measurements of arsenite and arsenate in potable water. Environ Sci Technol 37:4743–4750

Tauriainen S, Karp M, Chang W, Virta M (1997) Recombinant luminescent bacteria for measuring bioavailable arsenite and antimonite. Appl Environ Microbiol 63:4456–4461

Tauriainen S, Karp M, Chang W, Virta M (1998) Luminescent bacterial sensor for cadmium and lead. Biosens Bioelectron 13:931–938

Tauriainen SM, Virta MPJ, Karp MT (2000) Detecting bioavailable toxic metals and metalloids from natural water samples using luminescent sensor bacteria. Water Res 34:2661–2666

Temple TN, Stockwell VO, Loper JE, Johnson KB (2004) Bioavailability of iron to *Pseudomonas fluorescens* strain A506 on flowers of pear and apple. Phytopathol 94:1286–1293

Tibarzawa C, Corbisier P, Mench M, Bossus A, Solda P, Mergeay M, Wyns L, van der Lelie D (2001) A microbial biosensor to predict bioavailable nickel in soil and its transfer to plants. Environ Poll 113:19–26

Trang PTK, Berg M, Viet PH, Mui NV, van der Meer JR (2005) Bacterial bioassay for rapid and accurate analysis of arsenic in highly variable groundwater samples. Environ Sci Technol 39:3625–3630

Van der Meer JR, Tropel D, Jaspers M (2004) Illuminating the detection chain of bacterial bioreporters. Environ Microbiol 6:1005–1020

Virta M, Lampinen J, Karp M (1995) A luminescence-based mercury biosensor. Anal Chem 67:667–669

Wells M, Gosch M, Rigler R, Harms H, Lasser T, van der Meer JR (2005) Ultrasensitive reporter protein detection in genetically engineered bacteria. Anal Chem 77:2683–2689

Yoon KP, Misra TK, Silver S (1991) Regulation of the *cadA* cadmium resistance determinant of *Staphylococcus aureus* plasmid pI258. J Bacteriol 173:7643–7649

Microbiol Monogr (6)
D. H. Nies, S. Silver: Molecular Microbiology of Heavy Metals
DOI 10.1007/7171_2006_077/Published online: 13 February 2007

A Glossary of Microanalytical Tools to Assess the Metallome

Dirk Wesenberg · Corinna Bleuel · Gerd-Joachim Krauss (✉)

Inst. Biochemistry and Biotechnology
Div. Ecological and Plant Biochemistry,
Martin-Luther-University Halle-Wittenberg,
Kurt-Mothes-Str. 3, 06120 Halle/S., Germany
gerd-joachim.krauss@biochemtech.uni-halle.de

Dedicated to Dr. Sieglinde Menge †

Abstract Measurements of trace metals as parts of homeostatic networks for essential and nonessential metals in microbial cells need sensitive high-resolution techniques. The term "metallome" denotes metals and metalloid species within cells, encompassing both the inorganic element content and their complexes with biomolecules, especially with peptides and proteins. Elucidation of the physiological roles of metals and their bioinorganic speciation requires a set of microanalytical purification, separation, and identification methods. This chapter summarizes analytical tools useful to investigate bacterial responses to metal stress.

Abbreviations

2D	Two-dimensional
AAS	Atomic absorption spectrometry
AES	Atomic emission spectrometry
CE	Capillary electrophoresis
CEC	Capillary electrochromatography
CGE	Capillary gel electrophoresis
CZE	Capillary zone electrophoresis
EDS	Energy-dispersive X-ray spectrometry
EELS	Electron energy loss spectroscopy
EFTEM	Energy-filtered transmission electron microscopy
ELNES	Energy loss near edge structures
ESI	Electron spectroscopy imaging
	Electrospray ionization
ETAAS	Electrothermal atomic absorption spectrometry
(μ)EXAFS	(Micro) extended X-ray absorption fine structure
FAAS	Flame atomic absorption spectrometry
GC	Gas chromatography
HPLC	High-performance liquid chromatography
ICP-MS	Inductively coupled plasma mass spectrometry
IEF	Isoelectric focussing
INAA	Instrumental neutron activation analysis
IUPAC	International Union of Pure and Applied Chemistry
LA	Laser ablation
LC	Liquid chromatography
LIBS	Laser-induced breakdown spectroscopy
m/z	Mass-to-charge ratio
MALDI-TOF-MS	Matrix-assisted laser desorption/ionization time-of-flight mass spectrometry
MS	Mass spectrometry
MT	Metallothionein
NEXAFS	Near-edge X-ray absorption spectroscopy
NMR	Nuclear magnetic resonance spectroscopy
PAGE	Polyacrylamide gel electrophoresis
PC	Phytochelatins
PCR	Polymerase chain reaction
PIXE	Proton (or particle) induced X-ray emission
PLE	Pressurized liquid extraction
PSD	Postsource decay
RP	Reversed phase
SDS	Sodium dodecyl sulfate
SE	Size exclusion
SEC	Size exclusion chromatography
SEM	Scanning electron microscopy
SPME	Solid-phase microextraction
STEM	Scanning transmission electron microscopy
STXM	Soft scanning transmission X-ray microscopy
SXRD	Scanning X-ray diffraction
SXRF	Scanning X-ray fluorescence
TEM	Transmission electron microscopy

1
Introduction: The -omics Toolbox

Heavy metals pose significant environmental risks for terrestrial and aquatic ecosystems. Depending on speciation and bioavailability metals have exerted selective pressures on microorganisms. Many organisms have evolved resistance mechanisms to withstand high concentrations of such pollutants (Ehrlich 1997; Gadd 2004; Wackett et al. 2004; Krauss et al. 2006). Evolution has resulted in the development of a precise homeostatic regulatory network in cells to control the uptake, distribution, storage, and detoxification of metal ions.

Adaptation to heavy metal exposure is element specific and detoxification of one metal does not confer a general metal resistance. Moreover, different species of the same metal can be essential, harmless, or toxic. Thus, it is essential to know the chemical form of the heavy metal (species) outside and inside the cells, primarily the oxidation state and/or the organometallic nature. The speciation of an element is defined as the distribution of an element amongst defined chemical species in a system (Templeton et al. 2000).

The mechanisms for molecular selection of elements by living cells are very complex (Williams 2001). Understanding them requires both the characterization of the proteins involved and the description of the pool of non-proteinaceous molecules, some of which distribute metals to metalloproteins (Outten and O'Halloran 2001; Finney and O'Halloran 2003). Therefore, multiparallel analytical techniques are essential for an integrated view on the cell status and metabolism (Fiehn and Weckwerth 2003; Fridman and Pichersky 2005). To obtain a catalogue of the cellular metal-binding components, genome sequencing and functional genomics (transcriptomics, proteomics, and metabolomics) in addition to bioinformatic techniques are required (Eide 2001). The availability of complete sequences of many genomes nowadays offers outstanding opportunities to identify metalloproteins (Maret 2004) and to deduce geochemical signatures and phylogenetic relationships within prokaryotes (Zerkle et al. 2005).

Transcriptional profiles were described for *Escherichia coli* exposed to elevated Cu^{2+}, Zn^{2+}, Cd^{2+}, Co^{2+}, and Ni^{2+} concentrations (Brocklehurst and Morby 2000; Egler et al. 2005). Other expression profiling studies addressing metal biochemistry have used several other bacterial species such as *Bacillus subtilis* (Ye et al. 2000; Hobman et al., in this volume).

Protein profiling is of great importance for the study of the whole protein pattern in cells. Using proteomics data, the community of a microbial biofilm can be described as shown for a natural mine drainage (Ram et al. 2005). Proteomic analysis of the effect of Cd^{2+} was performed in the yeasts *Saccharomyces cerevisiae* (Vido et al. 2001) and *Schizosaccharomyces pombe* (Bae and Chen 2004). Interestingly, 80% of the proteins up-regulated in *S. pombe* were different from those in *S. cerevisiae* (López-Barea and Gómez-Ariza 2006). However, functional metalloproteome annotation requires a complete understanding of the function of proteins involved. A first attempt was described for zinc (Maret 2004). Our current understanding of protein interactions is derived from protein profiles that comprise comprehensive networks. Nevertheless it still remains incomplete. Besides computational modeling of proteomic networks (Janes and Lauffenburger 2006), detailed knowledge on the structures of individual proteins is required. The lack of amplification methods in proteomics, such as PCR in genomics, gives a need for highly sensitive analytical approaches. Herein, metals and metalloids function as markers for specific atomic instrumental devices. Various scientific fields related to biometals have been independently developed. The complete sequence of microbial genomes suggests important opportunities to identify metalloproteins and chelating peptides, allowing a detailed view on the role of essential and toxic metals.

Studies on the metallome (the entirety of metal and metalloid species within cells (Szpunar 2005)) now provide possibilities (1) to study the compartmentation and behavior of metals involved in cellular metabolism and environmental stress response (= ionome), (2) to detect metal metabolism related metabolites (= metallo-metabolome), and (3) to achieve a deep insight into metalloprotein structure and function (= metalloproteome). The function of metalloproteins in microorganisms depends on their interaction with metals, for example Cu^{2+}, Fe^{2+}, Zn^{2+}, Ni^{2+}, or $MoO_4{}^{2-}$. Other proteins

Table 1 Metallomics: microanalytical methods available for the investigation of the (bacterial) metallome

Research topics	Analytical approaches
Content and compartmentation of heavy metals in bacteria and biofilms	Atomic absorption spectrometry (AAS, ICP-AES, ICP-MS); Analytical electron microscopy (SEM- and TEM-based microanalysis: EDS, PIXE, EELS, ESI)
Chemical speciation and metabolite profiling	Hyphenated methods (LC-MS, (MALDI-MS, ESI-MS), LC-ICP-AES, LC-ICP-MS, GC-ICP-MS, LC-NMR), stabile isotope tracers
Identification of metal binding peptides and proteins (glutathione, phytochelatins, metallothioneins, metallochaperones), and metalloproteins (metalloenzymes)	Hyphenated techniques (Fig. 1), INAA, PIXE, biosensors
Structural analysis of metal-binding molecules	X-ray diffraction analysis, X-ray analysis μEXAFS, NMR, MALDI-MS, PSD-MS
Elucidation of reaction mechanisms using model complexes (bioorganic chemistry)	Microtitration, NMR, Laser and Raman spectroscopy, Fluorescence spectroscopy, Circular dichroism

or peptides (metallothioneins and phytochelatins, respectively) are expressed as a defense mechanism or act as transporters of essential or toxic ions. Many metalloproteins are used as biomarkers for environmental metal exposure. Metallomics can be regarded as speciation analysis that investigates the individual metal species (qualitative metallomics) and their concentrations (quantitative metallomics). Deciphering a metallome will inform of (1) the intracellular distribution of metals and metalloids among the compartments, (2) binding sites of heavy metals, and (3) the concentrations of the metal species present. Monitoring of changes within the metallome has recently been called "comparative metallomics" (Szpunar 2004). In order to validate metallomics data links to the genome and proteome of the cells are needed.

Advances in metallomics were reviewed recently (Gomez-Ariza et al. 2004b; Haraguchi 2004; Wind and Lehmann 2004; Gomez-Ariza et al. 2005; Szpunar 2005; Garcia et al. 2006; Lobinski et al. 2006). As a conclusion, new tools for metallomics as well as successful new technology implementations clearly reflect a shift in metallome research toward studies of dynamic aspects of metal biochemistry. New approaches in metalloproteome analysis relevant for microbial studies are discussed in this chapter. A set of techniques is available for the separation, identification, and characterization of metalloproteins and metal-binding peptides and proteins (Table 1).

2 Microchemical Analysis of Heavy Metals in the Environment

Pollution risk assessment includes the measurement of physicochemical and geochemical processes (Nieboer et al. 1999). These data are used (1) to locate possible heavy metal origins and (2) to assess their multiscale effects. Detailed historical knowledge of the sampling site is essential for considering long-term effects, as shown for highly heavy metal polluted sites in the Mansfeld Land area in Germany (Krauss et al. 2003). An important factor often neglected in toxicological and ecotoxicological assessment is the bioavailability of heavy metals.

Modern microanalytical approaches are available for determination of chemical and mineralogical features of heavy metal containing soils, sediments, and precipitates as well as microbial degradation of minerals:

- X-ray diffraction of metalliferous grains in stream sediments permits the identification of the source of metals (Schreck et al. 2005).
- Laser-induced breakdown spectroscopy (LIBS) for field experiments and sensitive simultaneous analysis of elements (Harmon et al. 2005).
- X-ray fluorescence spectroscopy (wavelength dispersive and energy dispersive) for detection of heavy metals in environmental samples (Wennrich et al. 2004; Schubert et al. 2005).

- Energy-dispersive X-ray spectrometry (EDS) attached to scanning electron microscopy (SEM) for elemental analysis, e.g., description of fungal involvement in heavy metal precipitation in a highly polluted surface water (Krauss et al. 2007).
- Laser ablation (LA-ICP-MS) to detect heavy metals in geological matrices (Günther and Heinrich 1999).
- Scanning and transmission electron microscopy (SEM and TEM) for geophysical imaging of stimulated microbial biomineralization (Williams et al. 2005).

Differential fractionation procedures according to the IUPAC are established procedures to assess heavy metal chemistry during their interaction with geological matrices and their environmental fate (Templeton et al. 2000).

3 Electron Microscopy Techniques

Electron microscopy and related microanalytical tools have significantly increased our knowledge of cellular interactions of microorganisms with heavy metals from their environment. Such investigations might be included into metallomic approaches (Table 1).

Electron microscopy studies were employed to assess types and abundances of bacteria (Konstantinidis et al. 2003) and fungi (Turnau and Kottke 2005; Krauss et al. 2007) in contaminated ecosystems. The microanalytical localization of elements on the cellular and subcellular level yields essential data on their distribution and their role in biochemical processes, and their interaction between heavy metal species. These techniques provide links between physiological, biochemical, and histological studies, which are especially crucial when studying the environmental stress response of microorganisms.

While traditional methods for speciation and quantification of metals in biofilms are based on fractionation and extraction, electron microscopy techniques such as scanning electron microscopy (STEM) coupled with energy-dispersive microscopy (EDS) are able to directly map metals in biofilms (Webb et al. 2000; Jackson and Leppard 2002; Leppard et al. 2003). However, the necessity of dehydration and sectioning can produce artifacts such as shrinkage and aggregation of particles. Soft scanning transmission X-ray microscopy (STXM), in which chemical sensitivity is accomplished through near-edge X-ray absorption spectroscopy (NEXAFS), was used to examine "real" samples in hydrated environments without fractionation and extraction. STXM can provide quantitative maps of chemical species with a spatial resolution higher than 50 nm (Lawrence et al. 2003). *Caulobacter crescentus* and Al mineral particles were observed by STXM (Yoon et al. 2004). Dynes et al. (2006) reported quantitative mapping of metals in biofilms by STXM

(Fe, Mn, Ni) including the correlation of metals with the biochemical and morphological characteristics of the biofilm.

Careful specimen preparation allows for preservation of structures and metal content and is most crucial in successful microscopy and microanalysis (Turnau and Kottke 2005; Krauss et al. 2006). Scanning electron microscopy (SEM) or transmission electron microscopy (TEM) can be applied. Whereas SEM rapidly visualizes highly resolved three-dimensional surfaces based on electron density, TEM leads to cellular ultrastructure requiring thin ($\sim$ 30–300 nm), electron beam semitransparent samples. For example, while SEM confirmed electron dense accumulations of metal sulfide to be associated with cell surfaces of *Desulfovibrio vulgaris* within sediments, TEM revealed the minerals to be sphalerite (ZnS) and mackinawite (FeS) (Williams et al. 2005). TEM is useful to localize metals inside or around (e.g., as electron-dense accumulations within ruthenium red labeled extracellular polysaccharides) bacterial cells, as shown for uranium or palladium exposed *Acidithiobacillus ferrooxidans* (Merroun et al. 2003a) or *Bacillus sphaericus* (Pollmann et al. 2006, Merroun et al. 2003b), respectively. *Desulfosporosinus* spp. precipitated uraninite crystals less than 3 nm in diameter, as shown by high-resolution TEM (Suzuki et al. 2002). As described by Hayat (1986) and Keyse et al. (1998), rastering capabilities can be added to TEM resulting in scanning transmission electron microscopy (STEM). As an example *Plectonema boryanum* (cyanophyceae) was shown to take up cadmium, cobalt, copper, mercury, nickel, and lead and sequestrate the metals in high amounts within polyphosphate bodies (Jensen et al. 1982). Energy-dispersive X-ray (EDX) spectrometers enable both SEM and TEM to rapidly detect elements (heavier than boron) qualitatively and quantitatively in the range of 0.1 wt % on flat, uniform samples. Thus, analytical metal detection can be added to the more descriptive imaging systems SEM or TEM. For a deeper discussion of problems and solutions for biological materials, see Echlin 2001 (SEM) and Goldstein et al. 2003 (SEM-based X-ray microanalysis). EDX was successfully used for speciation of deposited metals as phosphate (uranium) compounds connected to the extracellular polysaccharides of *A. ferrooxidans* (Merroun et al. 2003a) and *Myxococcus xanthus* (Gonzalez-Munoz et al. 1997), or within the intracellular uranium polyphosphate bodies (EDS spot mode with 70 Å, Merroun et al. 2003a) or on the cell wall of *A. ferrooxidans* (Merroun et al. 2003). Electron energy loss spectroscopy (EELS) enhanced TEM/STEM can detect elements at high spatial resolution (up to even single atoms) by classification of electrons (originated by primary beam) due to their energy loss traversing a (biological) sample (see Egerton 1997). Extraction of chemical bonding information near the absorption edges for each element leads to electron energy loss near edge structures (ELNES). Maps of electrons of the inner shells of elements between lithium and uranium can be generated by energy-filtered TEM (EFTEM) or electron spectroscopy imaging (ESI).

Proton (or particle) induced X-ray emission (PIXE) is basically EDS using scanned protons in the 2–4 MeV range as the excitation source (see Sect. 7.3). With microPIXE, elemental distribution maps at the spatial resolution of SEM/EDS and concentrations down to the ppm range for elements are possible (Ryan et al. 2001). Extended X-ray absorption fine structure (EXAFS; determination of chemical bonding information) has been used for the characterization of metal microbial interactions, e.g., uranium exposed *Desulfosporosinus* spp. (Suzuki et al. 2002), *A. ferrooxidans* (Merroun et al. 2003a), *Bacillus cereus*, and *B. sphaericus* (Hennig et al. 2001) and palladium exposed *B. sphaericus* (Merroun et al. 2003b; Pollmann et al. 2005). Analyzing the uraninite crystals precipitated in *Desulfosporosinus* spp. by EXAFS, Suzuki et al. (2002) found a large proportion of uranium ions on the surface of the nanoparticles, as expected for tiny particles. The cellular investigations were accompanied by detection of uranite in the surrounding sediment. Scanning X-ray fluorescence (SXRF; detection of trace elements) was used for investigation of metal quantification in *Euglena gracilis* (Shen et al. 2002) and aquatic protists (Twining et al. 2003); scanning X-ray diffraction (SXRD; examination of crystal structure) may applicable.

4
Sample Preparation

Sampling and sample preparation are the first steps to study the metallome (Fig. 1). A carefully planned procedure is essential for speciation analysis (Caruso and Montes-Bayon 2003). Sample pretreatment is the most critical stage of metal analysis, mainly when solid matrices are measured (Alonso-Rodriguez et al. 2006). Conventional sample extraction procedures using large volumes are based on solvents assisted by mechanical shaking.

Microwave-assisted extraction is an attractive preparation technique achieving different pressure, temperature, solvents, and variable time (Ackley et al. 1999). Solid-phase microextraction (SPME) uses different modes of physicochemical interactions with solid matrices. They were developed to satisfy the need for rapid sampling and sample preparation, both in field and laboratory experiments (Pawliszyn 1997). Recent developments are reviewed by Ouyang and Pawliszyn (2006). Analytical microsystems allow the miniaturization of SPME (Ríos et al. 2006). Pressurized liquid extraction (PLE = "accelerated solvent extraction") has been reviewed recently (McKiernan et al. 1999; Schantz 2006). Extraction of organometallic species by PLE characterizes a new trend for total metal extraction (Alonso-Rodriguez et al. 2006). Thus, Gómez-Ariza et al. (2004a) extracted water-soluble Se species from spiked and native yeast. Several techniques for speciation analysis of Cu, Zn, Cd, Pb, and Ni in freshwater systems are reviewed and compared with respect to their performance and to the metal species detected

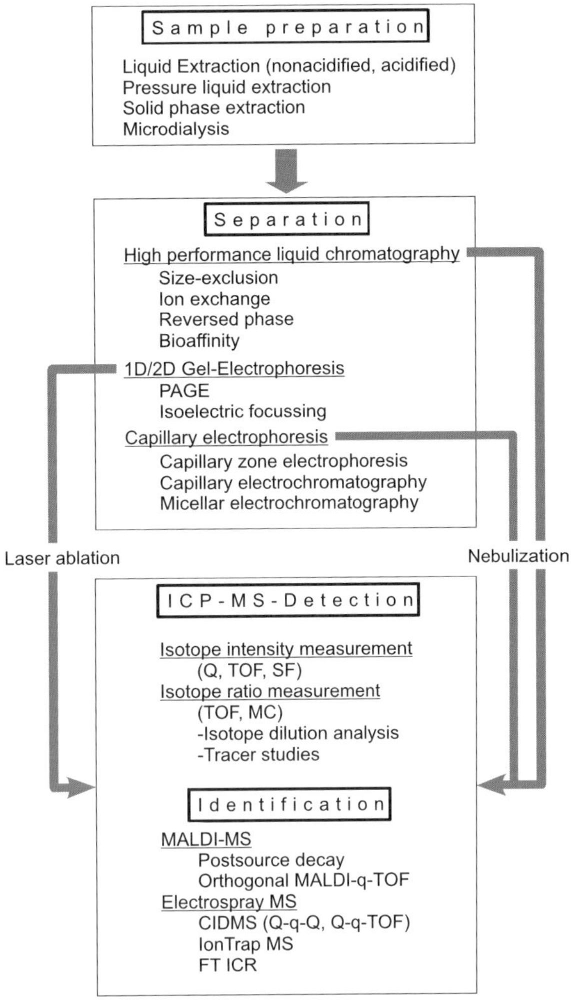

Fig. 1 Hyphenated methodology for analysis of metal-containing proteins

(Sigg et al. 2006). Microdialysis represents a new membrane-based sampling technique which is used in coupling with HPLC and capillary electrophoresis for separation of peptides and other biomolecules (Ruiz-Jimenez and de Castro 2006).

5 High-Performance Liquid Chromatography (HPLC)

Besides electrophoretic techniques, HPLC contributes to the isolation of target species from the sample ("selectivity component") of a metalloprotein analysis procedure (Gomez-Ariza et al. 2004b; Garcia et al. 2006). A guard column should be coupled to the HPLC column to protect it from the effects of lipids, carbohydrates, and particulate material. For example, an online purification and preconcentration step was integrated in a selenopeptide mapping procedure by nano-HPLC and nano-HPLC-ESI-MS/MS (Giusti et al. 2006).

A variety of separation modes have been developed expressing different mechanisms during the chromatographic separation process of proteins: normal phase, reversed phase (RP), RP ion pair, hydrophilic interaction, molecu-

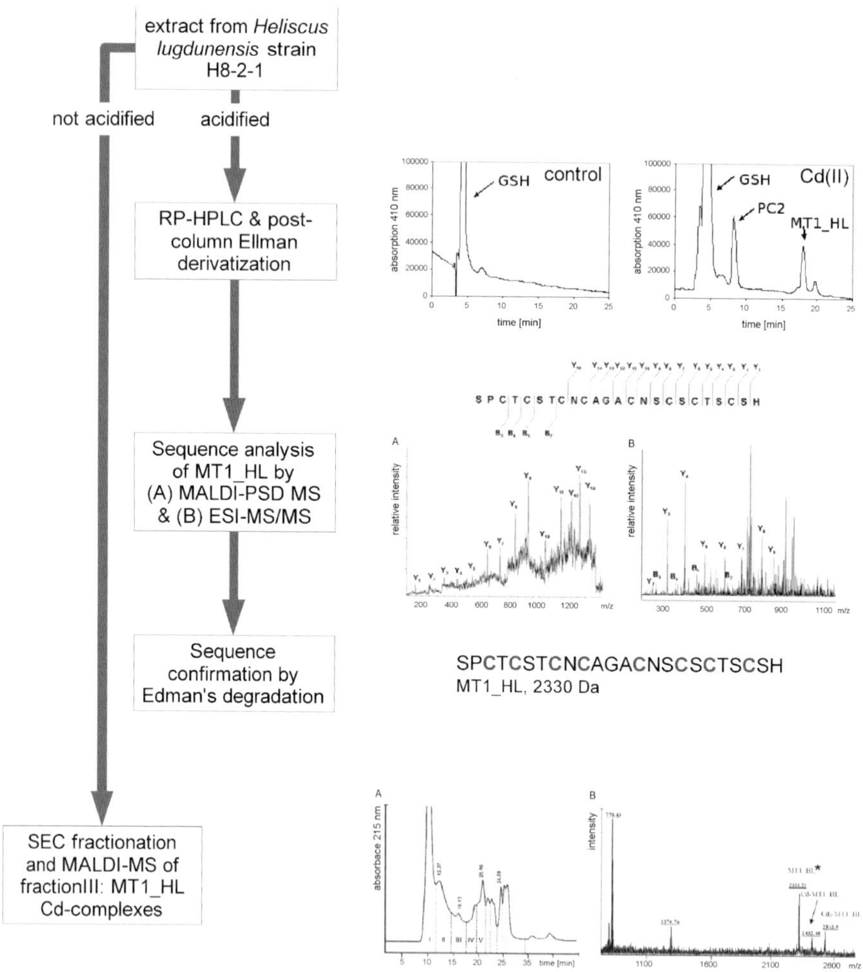

Fig. 2 Strategy for the separation and identification of metal-binding peptides and proteins from the aquatic fungus *Heliscus lugdunensis*; * all cysteines form disulfide bridges (Jaeckel et al. 2005)

lar imprint, chiral separations, ion-exchange, size exclusion (SE), and affinity chromatography (Weston 1997; Caruso and Montes-Bayon 2003) (Fig. 1). The requirements and limitations concerning the interface between HPLC and the determination technique (atomic spectrometry, mass spectrometry) vary as a function of chromatographic mode and the detection system used (Szpunar 2000).

Using higher concentrations of organic modifier makes RP-HPLC poorly compatible with inductively coupled plasma mass spectrometry (ICP-MS) and MS (Szpunar 2000), as exemplified in the separation of fungal thiol-containing peptides (Fig. 2). Microbore (1.0 mm) HPLC columns are becoming popular especially for classical RP-HPLC (Lobinski et al. 2006). The separation of intact metal complexes of proteins needs to extract the sample under nonacidified conditions and to separate it by size exclusion chromatography (SEC), as shown in the separation of metal-containing metallothioneins (Fig. 2). In the case of relatively pure samples, the resolution capacity of SEC may be sufficient to accomplish the separation of metalloproteins.

6 Electrophoretic Techniques

This section compiles an overview of electrophoretic methods as tools for metallomics.

6.1 Flat Bed Gel Electrophoresis

Polyacrylamide is a common electrophoresis phase for metallomics. Its porosity depends on the amount of acrylamide and the degree of cross-linking. Gradient pore size gels are frequently used and proteins migrate until a specific size hampers their migration. Thus, classical electrophoresis with its different modes, e.g., native polyacrylamide gel electrophoresis (PAGE; separation of intact protein metal complexes), sodium dodecyl sulfate polyacrylamide gel electrophoresis (SDS-PAGE), isoelectric focussing (IEF), and immunoelectrophoresis may be used for the characterization of metalloproteins and metal-binding proteins (Szpunar 2000; Garcia et al. 2006; López-Barea and Gómez-Ariza 2006). Two-dimensional (2D) peptide maps are commonly used in studies of structures and modifications of proteins (Andrews 1992; Hames 1998). Items of interest to the 2D electrophoretic protein gel research community are freely available at www.lecb.ncifcrf.gov/EP/. The capabilities and limitations of classical gel electrophoresis for elemental speciation and metalloproteomics were described recently (Chéry et al. 2006).

6.2 Capillary Electrophoresis (CE)

CE is a very appropriate technique for metalloprotein separation. CE leads to high-resolution separation, rapid analysis, and minimal sample and reagent consumption as reviewed by Richards and Beattie (1994). Proteins move according to the applied field and the induced electroosmotic flow (Weston 1997). Various separation modes are applicable for metalloprotein analysis (Fig. 1), e.g., capillary zone electrophoresis (CZE) using a sieving matrix for separation, capillary gel electrophoresis (CGE), and capillary electrochromatography (CEC) (Prange and Pröfrock 2005; Garcia et al. 2006). CZE is less prone (than HPLC) to artifacts such as metal exchange by interaction with the stationary phase. Because of its simplicity, low cost, and adequate capillary lifetime, separation in uncoated fused silica capillaries is widely appreciated (Schaumlöffel 2004). On the other hand, coated capillaries were suggested to reduce adsorption problems and to improve the separation efficiency (Wang and Prange 2002). CZE and CEC have been critically compared for the separation of Cd-, Cu-, and Zn-containing metalloproteins (Montes-Bayon et al. 2006). CZE is most commonly used in metalloproteomics because of its flexibility, simplicity, and good ability to connect with mass spectrometry-based detectors such as ICP-MS (Sect. 7.1) and ESI-MS (Sect. 8.2) (Prange and Pröfrock 2005).

7 Analytical Techniques for Measurement of Metals and Metalloids

Prerequisites for element species quantification are detectors, highly sensitive for elements (= "sensitivity component" of a metalloprotein analysis procedure), coupled both online and offline with separation modules (Gomez-Ariza et al. 2004b; Garcia et al. 2006). In the case of planar electrophoretic techniques, especially 2D-PAGE, metal species can be detected offline by (1) excising/digestion of peptide spots followed by ICP-MS or atomic absorption spectrometry (AAS) quantification, (2) inside the gel (PIXE), or (3) by laser ablation of the spots followed by transporting via carrier gas (pneumatic nebulization) and detection by ICP-MS (Fig. 1). The hyphenation (online coupling) between separation techniques (HPLC, CE) and metal-specific detectors (e.g., ICP-MS, Sect. 7.1) is implemented via a nebulizer (Prange and Pröfrock 2005; Lobinski et al. 2006). Such hyphenation techniques are actually the most powerful techniques for chemical speciation (Fig. 1, Table 1). New developments in atomic spectroscopy were reviewed recently by Bings et al. (2006).

7.1
Inductively Coupled Plasma Mass Spectrometry (ICP-MS)

ICP is a multielement ion source efficiently generating primarily single positively charged ions for MS from most elements. ICP-MS has a great potential for elemental analysis in biotechnology (Baranov et al. 2002) and allows for the determination of metal concentrations in the ppm to sub-ppb range when applied as a detector for HPLC or electrophoresis (Caruso et al. 2000; Caruso and Montes-Bayon 2003; Prange and Pröfrock 2005; Lobinski et al. 2006) (Fig. 1) or for capillary and nano-flow HPLC and CE (Kannamkumarath et al. 2002; Schaumlöffel 2004; Lobinski et al. 2006). The requirements of CE coupling to ICP-MS and ESI-MS are discussed by Prange and Pröfrock (2005). A comparison of CE-ICP-MS and capillary liquid chromatography hyphenated with ICP-MS was given by Montes-Bayon (2006) investigating Cd-, Cu-, and Zn-containing metalloproteins in complex matrices. ICP-MS was compared with other methods such as AAS and PIXE (Bertrand and Poirier 2005).

Using online coupled size-exclusion HPLC-ICP-MS, zinc metalloproteases from *Bacillus cereus* and *Pseudomonas aeruginosa* were separated and quantified (Leopold and Fricke 1997). Size-exclusion ICP-MS was used to measure the Fe/S and Mn/S ratios in Mn superoxide dismutase from *Anabaena* and catalase–peroxidase from *Synechocystis* (heterologously expressed in *E. coli*) (Hann et al. 2004). Both nano-HPLC-ICP-MS (75 μm column) and nano-HPLC-electrospray MS/MS were applied for mapping a selenium–yeast protein digest (Giusti et al. 2006). The selenium speciation was recently reviewed (B'Hymer and Caruso 2006). HPLC-ICP-MS has been widely used for the quantification and characterization of metallothioneins (Gomez-Ariza et al. 2005; López-Barea and Gómez-Ariza 2006), which show different isoforms and are used as stress biomarkers (Nordberg 2000; El Ghazi et al. 2003). Metallothioneins (MTs) are gene-encoded, cysteine-rich, low molecular weight (< 7 kDa), metal-binding (thiolate bond clusters) proteins in both prokaryotes and eukaryotes (Clemens et al. 2003). Their physiological functions have not yet been fully elucidated and are still a matter of debate. Nevertheless, MT analytics explain the synergetic use of different detection modes. CE-ICP-MS binding studies of Cu^{+}, Zn^{2+}, and Cd^{2+}, chelated by either native or recombinant metallothioneins, were reviewed recently (Lobinski et al. 2006). Mounicou et al. (2000) showed the parallel identification of metallothioneins by CZE-ICP-MS and CZE-ESI-MS to measure ratios of MT-Cd^{2+}, -Cu^{2+}, and -Zn^{2+} complexes.

The possibilities and limitations of 2D-PAGE for elemental speciation coupled with laser ablation (LA)-ICP-MS were discussed earlier (Chéry et al. 2003, 2006). The ablation of the protein was done with a laser beam over the gel within an electrophoretic lane. The ablated analyte is transferred into the ICP by a continuous stream of argon, and the ions are determined by MS

(Chéry et al. 2003; Szpunar 2005). Thus, simultaneous multielement in-gel screening (Cu, Zn, Fe, and P) of proteins from yeast mitochondria was carried out (Becker et al. 2004). Binet et al. (2003) characterized (Zn, Cd) binding proteins in cytoplasmic fractions from *E. coli*.

In summary, ICP-MS used in metallomics includes important benefits, as described in recent reviews (Szpunar 2004; Garcia et al. 2006; Lobinski et al. 2006):

- Availability of collision/reaction cell instruments, extending the detectable elements to phosphorus and sulfur while improving the isotope ratio measurements of such elements as selenium, chromium, vanadium, and iron
- High sensitivity (upper attogram level)
- Possibility to use the isotope-dilution techniques for internal standardization and also to monitor transformations of metal speciation which may occur during sample pretreatment/separation
- Available interfaces allowing the transfer of effluents from capillary columns (300 μm i.d.) and nanocapillary columns (< 150 μm i.d.) via direct injection or nebulization (Prange and Pröfrock 2005), resulting in sub-femtogram detection limit
- Scanning and laser ablation of 2D-PAGE for the detection of metal-containing spots

7.2 Atomic Absorption Spectrometry (AAS)

AAS methods are widely used to quantify metal ions in biological matrices (Garcia et al. 2006). Bertrand et al. (2003) compared AAS techniques with ICP-MS (Sect. 7.1) and PIXE (Sect. 7.3). During flame AAS (FAAS) analytes are transferred as a mist into a flame by a gaseous oxidant or reductant flow. Solid particles are transformed into gaseous atoms and elementary ions. The quantification limits are in the mg L^{-1} range. In electrothermal AAS (ETAAS) (employed for trace element quantification), the sample is ashed in an electrically heated graphite tube resulting in higher sensitivity (μg L^{-1} range). FAAS and ETAAS are methods for metal-binding protein quantification after chromatography or gel electrophoresis (Szpunar 2000). Offline measurement of metal species can also be carried out applying ETAAS (Scancar et al. 2000). Recently, a method for real-time ETAAS was described using a direct thermospray interface coupling CE with ETAAS (Li et al. 2005). ICP-atomic emission spectrometry (ICP-AES, Sect. 8.2) was used for determination of the Zn content in a purified bacterial metallothionein (Blindauer et al. 2001) and for the measurement of Cd metallothionein stoichiometry (Bongers et al. 1988; Capdevila et al. 2005).

7.3 Proton (Particle) Induced X-ray Emission (PIXE)

PIXE is the most mature microanalytical technique (also used in electron microscopy) based purely on spectral imaging, which allows a sensitive, multielementary, and nondestructive analysis of metalloproteins using denaturing and nondenaturing gel electrophoresis. An excellent comparison of PIXE with other methods such as AAS, ICP-AES, and ICP-MS is given by Bertrand et al. (2003). Due to the high absolute sensitivity of PIXE the metal ions concentrated in the narrow bands of an electropherogram can be successfully measured in situ. Thus, iron was determined in an iron–sulfur protein (HiPiP) from *Thiocapsa roseopersicina* after electrophoretic separation (Szökefalvi-Nagy et al. 1990). Using combined PIXE-PAGE analysis Fe and Ni were detected within hydrogenases from *Desulfovibrio gigas*, showing that these metals are localized on different polypeptides or subunits (Szökefalvi-Nagy et al. 1999). The capability of PIXE to quantitatively detect Fe in protein bands obtained by SDS-PAGE of the photosystem I complex from the cyanobacterium *Synechococcus* sp. was explored by Solis et al. (1998). PIXE was also coupled with CE for elemental analysis, too (Wittrisch et al. 1997). Detection limits as low as $\sim 10^{-5}$ M have been achieved.

7.4 Nuclear Magnetic Resonance Spectroscopy (NMR)

NMR spectroscopy is a technique used to characterize the structure of molecules as large as proteins. When exposed to a magnetic field, the atomic nuclei in molecules can alter their spin state. Dependent on resonance frequencies in the local chemical environment, measurements can be used to analyze the 3D structures of proteins. To gain insight into the functional mechanisms of metalloproteins and metal-binding proteins NMR, in conjunction with isotope labeling and computational methods, is a very valuable tool in metallomics. For example, the delivery of copper for thylakoid import into the cyanobacterium *Synechocystis* PCC6803 was observed by NMR (Banci et al. 2006). The copper metallochaperone ScAtx1 acting as intracellular mediator (Cavet et al. 2003; see the chapter by Tottey et al., in this volume) was characterized. Besides ICP-AES and ESI-MS, multinuclear NMR spectroscopy was used to clarify the structural dynamics of the cyanobacterial metallothionein SmtA from *Synechococcus* PCC7942 (Blindauer et al. 2001; Blindauer and Sadler 2005). The 3D NMR solution structure of a small metallothionein of *Neurospora crassa* (Cu_6NcMT) was reported by Cobine et al. (2004).

Hyphenation of capillary HPLC or CE to NMR spectroscopy is an intensively developing field (Webb 2005). Miniaturized techniques, such as continuous flow capillary HPLC-NMR, were developed for structural determinations of RP-HPLC separated tocopherol homologues (Krucker et al. 2004). Webb et al. (2005) reviewed some of the approaches in the evolving field of coupling NMR detection with pressure-drive and electrophoretic micro-preparations, which needs the development of small NMR detectors.

8 Metalloprotein Identification and Confirmation with Mass Spectrometry

In parallel to measurement of metal species, techniques for the identification of protein ligands are required. MS allows the determination of the molecular mass of compounds, e.g., proteins, by separation of molecular ions according to their mass-over-charge ratio (m/z) and significant structural confirmation (Garcia et al. 2006). Thereby, molecular ions are formed by inducing a gain or loss of charge (e.g. electron ejection, deprotonation, or protonation). The MS represents the "structural component" of a metalloprotein analysis procedure (Gomez-Ariza et al. 2004b; 2005). MALDI and electrospray ionization (ESI) MS together with ICP-MS are successfully used complementary methods for the acquisition of information on the metal and metalloid speciation in metallomics research (Haraguchi 2004; Wind and Lehmann 2004; Prange and Pröfrock 2005; Szpunar 2005; Lobinski et al. 2006) (Fig. 1).

8.1 Matrix-Assisted Laser Desorption/Ionization Time-of-Flight Mass Spectrometry (MALDI-TOF-MS)

MALDI-TOF-MS is less sensitive to matrix effects (salts and buffers) than ESI-MS. The technique produces mostly singly charged ions with little fragmentation and is an attractive method to determine species that would have remained undetected by ESI-MS (Chalmers and Gaskell 2000).

Samples are prepared for MALDI by embedding the analyte in a condensed matrix of small molecules at high molar excess. In MALDI-TOF-MS the molecules are protonated and desorbed into the gas phase by a laser pulse. The mass-to-charge ratio of the analyte ions is measured by their time-of-flight providing the identification of low femtomole quantities of proteins, as shown for the identification of metalloproteins (Garcia et al. 2006; Lobinski et al. 2006). The pattern of fragmentation into peptide ions provides the possibility of obtaining the amino acid sequence. MALDI-postsource decay (PSD) techniques (Spengler 1997) were compared for the characterization of selenium non-peptide species in a selenized yeast (Encinar et al. 2004). The combination of LA-ICP-MS with MALDI-Fourier transform resonance-MS allows the identification of protein phosphorylation as well as determination of the metal content of proteins from yeast mitochondria (Becker et al. 2004).

A typical application of the combined use of different extraction procedures and MALDI-MS and ESI-MS (Sect. 8.2) for identification of metal-binding proteins is the purification of a novel Cd-induced small metallothionein from an aquatic fungus (Jaeckel et al. 2005) (Fig. 2). An acidified extract was separated by RP-HPLC and MALDI-PSD-MS and ESI-MS/MS were used to sequence the MT. In order to maintain the intact Cd protein complex and to exclude acid proteolysis, the nonacidified extract was characterized by SEC and MALDI-MS, showing the binding of at least two Cd ions by the novel MT (Jaeckel et al. 2005).

8.2 Electrospray Ionization Mass Spectrometry (ESI-MS)

ESI-MS is an electrodynamic technique to generate ions directly from a solution. Different types of mass analyzers and detectors can be coupled with an ESI source to build a mass spectrometer device for the detection of the molecular mass of peptides and proteins and also for their sequencing. Applications of ESI-MS to metal speciation studies in biochemistry and metalloprotein research were reviewed (Chassaigne et al. 2000; Rosenberg 2003; Lobinski et al. 2006). Protein–metal interactions and stoichiometries in *Bacillus* species were studied by SEC-ESI-MS (Benson et al. 2003). Selenopeptides were characterized in a selenium–yeast protein digest (Giusti et al. 2006).

Mass spectrometric techniques and their synergistic use with ICP are the methods of choice for detection of metal-binding peptides. LC-ESI-MS and LC-ICP-MS were complementarily used for the characterization of bacterial copper proteins. ESI-TOF-MS confirmed both the stoichiometry of ligation and the oxidation state of the metal center in plastocyanin and the CuA domain of cytochrome *c* oxidase from the cyanobacterium *Synechocystis* (Hann et al. 2006).

The intracellular synthesis of glutathione–metal complexes is important for heavy metal detoxification. Arsenic–glutathione complexes were separated by different HPLC modes and characterized by synergistic use of ESI-MS and ICP-MS (Raab et al. 2004). An arsenic–glutathione conjugate $As(GS)_3$ was synthesized by a hypertolerant strain of *Aspergillus* spec. (Canovas et al. 2004). High concentrations of arsenic provoke a massive formation of $As(GS)_3$-filled vacuoles.

The binding of Cd^{2+} by glutathione is discussed as an important way to detoxify this metal in phytochelatin-less plants (bryophytes, Bruns et al. 2001) and fungi, e.g., *Paxillus involutus* (Courbot et al. 2004). Glutathione-mediated Cd^{2+} sequestration in *Rhizobium leguminosarum* was reported by Lima et al. (2006). ESI-MS and ICP provided complementary data for the characterization of Cd^{2+}-induced phytochelatins (PCs) in the plant *Oryza sativa* (Vacchina et al. 2000). Phytochelatins, which were determined initially in Cd^{2+}-exposed fungi (cadystins from *Schizosaccharomyces pombe*), is the most common name for glutathione-derived metal-complexing peptides with the general structure (γ-Glu-Cys)$_n$-Gly occurring in plants and fungi (Pocsi et al. 2004; Mendoza-Cózatl et al. 2005). ESI-tandem MS was used to describe the induction of PC2 in the aquatic fungus *Heliscus lugdunensis* (Jaeckel et al. 2005, Fig. 2). Applications of ESI-MS to metal speciation studies with a particular emphasis on metal complexation by PCs and MTs were reported (McSheehy and Mester 2003). ESI-MS and tandem MS have greatly contributed to unraveling the molecular microheterogeneity of glutathione and PCs (Vacchina et al. 2000; Jaeckel et al. 2005; Sarry et al. 2006).

ESI-MS represents a suitable technique for the determination of intact metallothionein–metal complexes or of metal-free MT ligands after acidification of biological extracts (Jaeckel et al. 2005) or disintegration of the complexes by postchromatographic acidification (McSheehy and Mester 2003). Figure 2 shows an example procedure. As a suitable prerequisite for studying demetalated proteins, a fast and mild method was reported to eliminate metal

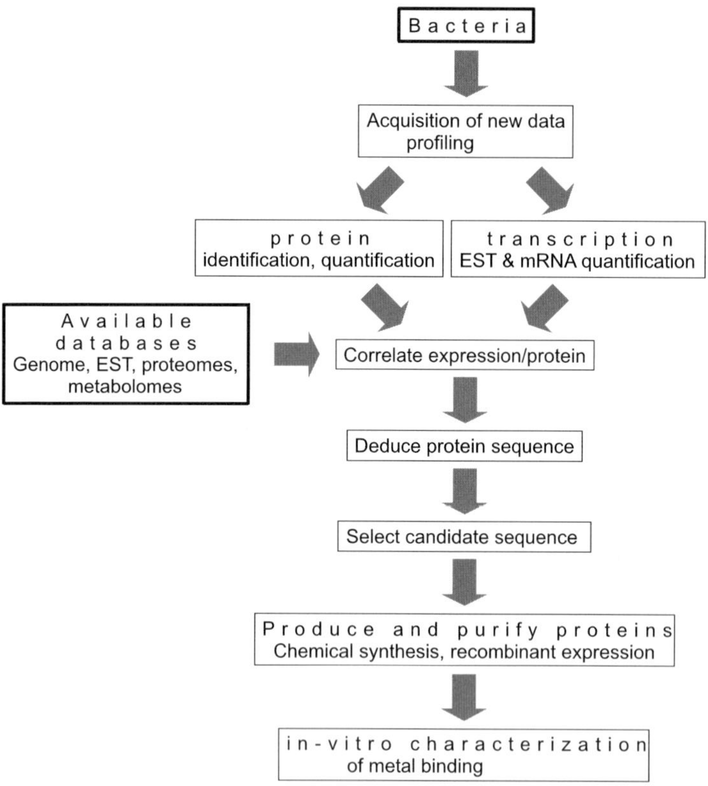

Fig. 3 Proposed scheme for identification and characterization of metal-binding proteins from bacteria

ions from proteins by HPLC using a chelating agarose-based stationary phase (Carrer et al. 2006).

Gene sequence and gene expression data can be used to correlate information about proteins and mRNAs to purify proteins as prerequisites for metal-binding assays (Fig. 3). Figure 4 shows a procedure for the production and purification of a recombinant chaperone and its characterization by ESI-MS (Urvoas et al. 2003). The methodology is important for the structural analysis of such small soluble proteins involved in intracellular transport and delivery of metals to their targets (see the chapter by Tottey et al., in this volume).

Metallothionein studies highlight the particular potential of metal speciation and protein structure analysis. Figure 5 presents an analytical case in which the bacterial metallothionein SmtA from *Synechococcus* PCC7942 was characterized (Blindauer et al. 2001; Blindauer and Sadler 2005; Tottey et al., in this volume). ICP-AES, multinuclear NMR spectroscopy, and ESI-MS were used to elucidate the structure, composition, and dynamics of bacterial MTs. The metallothionein SmtA was expressed heterologously in *E. coli* and purified. The following positive ESI-MS under acidic and nonacidic conditions resulted in 5610 amu for the Apo-SmtA and 5863 amu for the Zn_4SmtA. The experiments suggested that for SmtA half the protein molecules were apo

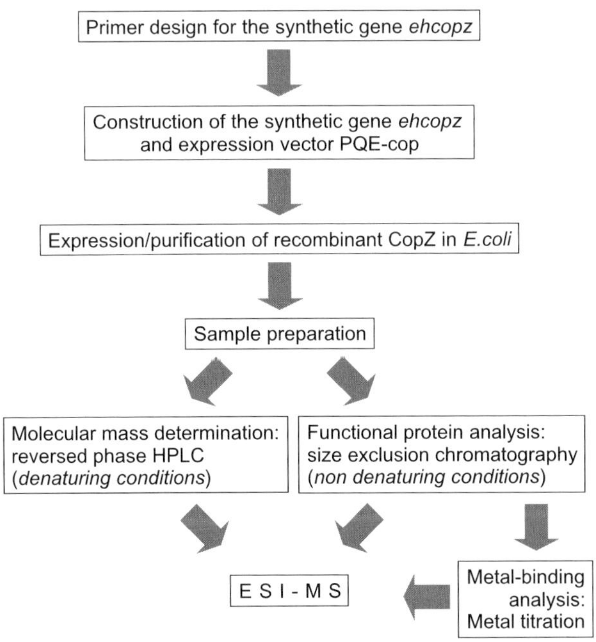

Fig. 4 Analysis of the metal-binding selectivity of the metallochaperone CopZ (7593 Da) from *Enterococcus hirae* (Urvoas et al. 2003)

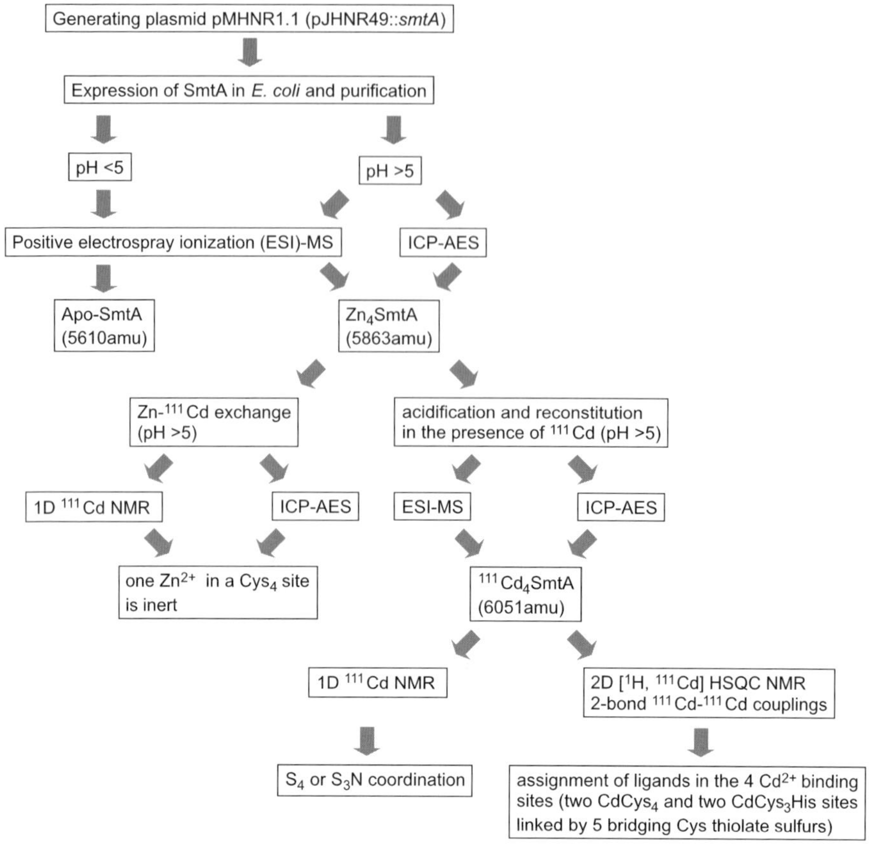

Fig. 5 Strategy for structural characterization of the bacterial metallothionein SmtA from *Synechococcus* PCC7942 (Blindauer et al. 2001; Blindauer and Sadler 2005)

and that the remainder were fully loaded. The Zn content was determined by ICP-AES. To identify the Zn ligands, the native Zn^{2+} ions of Zn_4SmtA were replaced by $^{111}Cd^{2+}$ after acidification resulting in $^{111}Cd_4$SmtA (6051 amu). One-dimensional ^{111}Cd NMR experiments indicated a S_4 or S_3N coordination. After 2D heteronuclear NMR experiments and two-bond ^{111}Cd–^{111}Cd couplings, two $CdCys_4$ and two $CdCys_3$His sites linked by five bridging Cys thiolate sulfurs could be identified as the ligands within the four $^{111}Cd^{2+}$-binding sites. In a second experiment, after Zn–^{111}Cd exchange under nonacidic conditions, one-dimensional ^{111}Cd NMR and ICP-AES revealed an inert Zn^{2+} in a Cys_4 site. The presence of a complete Zn_4 cluster is necessary for creating an inert site (Blindauer and Sadler 2005).

References

Ackley KL, B'Hymer C, Sutton KL, Caruso JA (1999) Speciation of arsenic in fish tissue using microwave-assisted extraction followed by HPLC-ICP-MS. J Anal At Spectrom 14:845–850

Alonso-Rodriguez E, Moreda-Pineiro J, Lopez-Mahia P, Muniategui-Lorenzo S, Fernandez-Fernandez E, Prada-Rodriguez D, Moreda-Pineiro A, Bermejo-Barrera A, Bermejo-Barrera P (2006) Pressurized liquid extraction of organometals and its feasibility for total metal extraction. Trends Anal Chem 25:511–519

Andrews AT (1992) Electrophoresis: theory, techniques and biochemical and clinical applications. Clarendon, Oxford

Bae W, Chen X (2004) Proteomic study for the cellular responses to Cd^{2+} in *Schizosaccharomyces pombe* through amino acid-coded mass tagging and liquid chromatography tandem mass spectrometry. Mol Cell Proteomics 3:596–607

Banci L, Bertini I, Ciofi-Baffoni S, Kandias NG, Robinson NJ, Spyroulias GA, Su XC, Tottey S, Vanarotti M (2006) The delivery of copper for thylakoid import observed by NMR. Proc Natl Acad Sci USA 103:8320–8325

Baranov VI, Quinn ZA, Bandura DR, Tanner SD (2002) The potential for elemental analysis in biotechnology. J Anal At Spectrom 17:1148–1152

Becker JS, Zoriy M, Krause-Buchholz U, Becker JS, Pickhardt C, Przybylski M, Pompe W, Rodel G (2004) In-gel screening of phosphorus and copper, zinc and iron in proteins of yeast mitochondria by LA-ICP-MS and identification of phosphorylated protein structures by MALDI-FT-ICR-MS after separation with two-dimensional gel electrophoresis. J Anal At Spectrom 19:1236–1243

Benson LM, Kumar R, Cavanagh J, Naylor S (2003) Protein–metal ion interactions, stoichiometries and relative affinities determined by on-line size exclusion gel filtration mass spectrometry. Rapid Commun Mass Spectrom 17:267–271

Bertrand M, Poirier I (2005) Photosynthetic organisms and excess of metals. Photosynthetica 43:345–353

Bertrand M, Weber G, Schoefs B (2003) Metal determination and quantification in biological material using particle-induced X-ray emission. Trends Anal Chem 22:254–262

B'Hymer C, Caruso JA (2006) Selenium speciation analysis using inductively coupled plasma-mass spectrometry. J Chromatogr A 1114:1–20

Binet MR, Ma R, McLeod CW, Poole RK (2003) Detection and characterization of zinc- and cadmium-binding proteins in *Escherichia coli* by gel electrophoresis and laser ablation-inductively coupled plasma-mass spectrometry. Anal Biochem 318:30–38

Bings NH, Bogaerts A, Broekaert JAC (2006) Atomic spectroscopy. Anal Chem 78:3917–3945

Blindauer CA, Sadler PJ (2005) How to hide zinc in a small protein. Acc Chem Res 38:62–69

Blindauer CA, Harrison MD, Parkinson JA, Robinson AK, Cavet JS, Robinson NJ, Sadler PJ (2001) A metallothionein containing a zinc finger within a four-metal cluster protects a bacterium from zinc toxicity. Proc Natl Acad Sci USA 98:9593–9598

Bongers J, Walton CD, Richardson DE, Bell JU (1988) Micromolar protein concentrations and metalloprotein stoichiometries obtained by inductively coupled plasma atomic emission spectrometric determination of sulfur. Anal Chem 60:2683–2686

Brocklehurst KR, Morby AP (2000) Metal-ion tolerance in *Escherichia coli*: analysis of transcriptional profiles by gene-array technology. Microbiology 146:2277–2282

Bruns I, Sutter K, Menge S, Neumann D, Krauss G-J (2001) Cadmium lets increase the glutathione pool in bryophytes. J Plant Physiol 158:79–89

Canovas D, Vooijs R, Schat H, de Lorenzo V (2004) The role of thiol species in the hypertolerance of *Aspergillus sp P37* to arsenic. J Biol Chem 279:51234–51240

Capdevila M, Domènech J, Pagani A, Tío L, Villarreal L, Atrian S (2005) Zn- and Cd-metallothionein recombinant species from the most diverse phyla may contain sulfide (S^{2-}) ligands. Angew Chem Int Ed 44:4618–4622

Carrer C, Stolz M, Lewitzki E, Rittmeyer C, Kolbesen BO, Grell E (2006) Removing coordinated metal ions from proteins: a fast and mild method in aqueous solution. Anal Bioanal Chem 385:1409–1413

Caruso JA, Montes-Bayon M (2003) Elemental speciation studies: new directions for trace metal analysis. Ecotoxicol Environ Saf 56:148–163

Caruso JA, Sutton KL, Ackley KL (2000) Elemental speciation: new approaches for trace element analysis, 1st edn. Elsevier, Amsterdam

Cavet JS, Borrelly GP, Robinson NJ (2003) Zn, Cu and Co in cyanobacteria: selective control of metal availability. FEMS Microbiol Rev 27:165–181

Chalmers MJ, Gaskell SJ (2000) Advances in mass spectrometry for proteome analysis. Curr Opin Biotechnol 11:384–390

Chassaigne H, Vacchina V, Lobinski R (2000) Elemental speciation analysis in biochemistry by electrospray mass spectrometry. Trends Anal Chem 19:300–313

Chéry CC, Günther D, Cornelis R, Vanhaecke F, Moens L (2003) Detection of metals in proteins by means of polyacrylamide gel electrophoresis and laser ablation-inductively coupled plasma-mass spectrometry: application to selenium. Electrophoresis 24:3305–3313

Chéry CC, Moens L, Cornelis R, Vanhaecke F (2006) Capabilities and limitations of gel electrophoresis for elemental speciation: a laboratory's experience. Pure Appl Chem 78:91–103

Clemens S, Simm C, Maier T (2003) Heavy metal-binding proteins and peptides. In: Fahnestock S, Steinbüchel A (eds) Polyamides and complex proteinaceous materials II. Wiley-VCH, Weinheim, pp 255–288

Cobine PA, McKay RT, Zangger K, Dameron CT, Armitage IM (2004) Solution structure of Cu_6 metallothionein from the fungus *Neurospora crassa*. Eur J Biochem 271:4213–4221

Courbot M, Diez L, Ruotolo R, Chalot M, Leroy P (2004) Cadmium-responsive thiols in the ectomycorrhizal fungus *Paxillus involutus*. Appl Environ Microbiol 70:7413–7417

Dynes JJ, Tyliszczak T, Araki T, Lawrence JR, Swerhone GDW, Leppard GG, Hitchcock AP (2006) Speciation and quantitative mapping of metal species in microbial biofilms using scanning transmission X-ray microscopy. Environ Sci Technol 40:1556–1565

Echlin P (2001) Biological X-ray microanalysis: the past, present practices, and future prospects. Microsc Microanal 7:211–219

Egerton R (1997) Electron energy-loss spectroscopy. Phys World 10:47–51

Egler M, Grosse C, Grass G, Nies DH (2005) Role of the extracytoplasmic function protein family sigma factor RpoE in metal resistance of *Escherichia coli*. J Bacteriol 187:2297–2307

Ehrlich HL (1997) Microbes and metals. Appl Microbiol Biotechnol 48:687–692

Eide DJ (2001) Functional genomics and metal metabolism. Genome Biol 2:reviews 1028.1021–reviews 1028.1023

El Ghazi I, Menge S, Miersch J, Chafik A, Benhra A, Elamrani MK, Krauss G-J (2003) Quantification of metallothionein-like proteins in the mussel *Mytilus galloprovincialis* using RP-HPLC fluorescence detection. Environ Sci Technol 37:5739–5744

Encinar JR, Polatajko A, Szpunar J, Lobinski R (2004) Identification of non-peptide species in selenized yeast by MALDI mass spectrometry using post-source decay and orthogonal Q-TOF detection. Analyst 129:846–849

Fiehn O, Weckwerth W (2003) Deciphering metabolic networks. Eur J Biochem 270:579–588

Finney LA, O'Halloran TV (2003) Transition metal speciation in the cell: insights from the chemistry of metal ion receptors. Science 300:931–936

Fridman E, Pichersky E (2005) Metabolomics, genomics, proteomics, and the identification of enzymes and their substrates and products. Curr Opin Plant Biol 8:242–248

Gadd GM (2004) Microbial influence on metal mobility and application for bioremediation. Geoderma 122:109–119

Garcia JS, De Magalhaes CS, Arruda MAZ (2006) Trends in metal-binding and metalloprotein analysis. Talanta 69:1–15

Giusti P, Schaumloffel D, Preud'homme H, Szpunar J, Lobinski R (2006) Selenopeptide mapping in a selenium-yeast protein digest by parallel nanoHPLC-ICP-MS and nanoHPLC-electrospray-MS/MS after on-line preconcentration. J Anal At Spectrom 21:26–32

Goldstein J, Newbury D, Joy D, Lyman C, Echlin P, Lifshin E, Sawyer L, Michael J (2003) Scanning electron microscopy and X-ray microanalysis. Plenum, New York

Gomez-Ariza JL, de la Torre MAC, Giraldez I, Morales E (2004a) Speciation analysis of selenium compounds in yeasts using pressurised liquid extraction and liquid chromatography–microwave-assisted digestion–hydride generation–atomic fluorescence spectrometry. Anal Chim Acta 524:305–314

Gomez-Ariza JL, Garcia-Barrera T, Lorenzo F, Bernal V, Villegas MJ, Oliveira V (2004b) Use of mass spectrometry techniques for the characterization of metal bound to proteins (metallomics) in biological systems. Anal Chim Acta 524:15–22

Gomez-Ariza JL, Garcia-Barrera T, Lorenzo F, Arias A (2005) Analytical characterization of bioactive metal species in the cellular domain (metallomics) to simplify environmental and biological proteomics. Int J Environ Anal Chem 85:255–266

Gonzalez-Munoz MT, Merroun ML, Ben Omar N, Arias JM (1997) Biosorption of uranium by *Myxococcus xanthus*. Int Biodeterior Biodegradation 40:107–114

Günther D, Heinrich CA (1999) Enhanced sensitivity in laser ablation-ICP mass spectrometry using helium–argon mixtures as aerosol carrier —Plenary lecture. J Anal At Spectrom 14:1363–1368

Hames BD (1998) Gel electrophoresis of proteins: a practical approach. Oxford University Press, Oxford

Hann S, Koellensperger G, Binger C, Furtmuller PG, Stingeder G (2004) SEC-ICP-DRCMS and SEC-ICP-SFMS for determination of metal–sulfur ratios in metalloproteins. J Anal At Spectrom 19:74–79

Hann S, Obinger C, Stingeder G, Paumann M, Furtmuller PG, Koellensperger G (2006) Studying metal integration in native and recombinant copper proteins by hyphenated ICP-DRC-MS and ESI-TOF-MS capabilities and limitations of the complementary techniques. J Anal At Spectrom 21:1224–1231

Haraguchi H (2004) Metallomics as integrated biometal science. J Anal At Spectrom 19:5–14

Harmon RS, De Lucia FC, Miziolek AW, McNesby KL, Walters RA, French PD (2005) Laser-induced breakdown spectroscopy (LIBS)—an emerging field-portable sensor technology for real-time, in-situ geochemical and environmental analysis. Geochemistry 5:21–28

Hayat MA (1986) Basic techniques for transmission electron microscopy. Academic Press, London

Hennig C, Panak PJ, Reich T, Rossberg A, Raff J, Selenska-Pobell S, Matz W, Bucher JJ, Bernhard G, Nitsche H (2001) EXAFS investigation of uranium(VI) complexes formed at *Bacillus cereus* and *Bacillus sphaericus* surfaces. Radiochim Acta 89:625–631

Hobman J, Yamamoto K, Oshima T (2007) Transcriptomic responses of bacterial cells to sublethal metal ion stress (in this volume). Springer, Heidelberg

Jackson TA, Leppard GG (2002) Energy-dispersive X-ray microanalysis and its applications in biogeochemical research. In: Violante A, Huang PM, Bollag J-M, Gianfreda L (eds) Soil mineral–organic matter–organism interactions and ecosystem health. Elsevier, Amsterdam

Jaeckel P, Krauss G, Menge S, Schierhorn A, Rücknagel P, Krauss G-J (2005) Cadmium induces a novel metallothionein and phytochelatin 2 in an aquatic fungus. Biochem Biophys Res Commun 333:150–155

Janes KA, Lauffenburger DA (2006) A biological approach to computational models of proteomic networks. Curr Opin Chem Biol 10:73–80

Jensen TE, Baxter M, Rachlin JW, Jani V (1982) Uptake of heavy metals by Plectonema boryanum (cyanophyceae) into cellular components, especially polyphosphate bodies: an X-ray energy dispersive study. Environ Pollut 27:119–127

Kannamkumarath SS, Wrobel K, B'Hymer C, Caruso JA (2002) Capillary electrophoresis–inductively coupled plasma–mass spectrometry: an attractive complementary technique for elemental speciation analysis. J Chromatogr A 975:245–266

Keyse RJ, Garrat-Read AJ, Goodhew PJ, Lorimer GW (1998) Introduction to scanning transmission electron microscopy. Springer, New York

Konstantinidis KT, Isaacs N, Marsh TL, Fett J, Simpson S, Long DT (2003) Microbial diversity and resistance to copper in metal-contaminated lake sediment. Microb Ecol 45:191–202

Krauss G, Sridhar KR, Jung K, Wennrich R, Ehrman J, Bärlocher F (2003) Aquatic hyphomycetes in polluted groundwater habitats of central Germany. Microb Ecol 45:329–339

Krauss GJ, Wesenberg D, Ehrman JM, Solé M, Miersch J, Krauss G (2007) Fungal responses to heavy metals. In: Novel techniques and ideas in mycology. Fungal Diversity, Hong Kong, in press

Krucker M, Lienau A, Putzbach K, Grynbaum MD, Schuger P, Albert K (2004) Hyphenation of capillary HPLC to microcoil 1H NMR spectroscopy for the determination of tocopherol homologues. Anal Chem 76:2623–2628

Lawrence JR, Swerhone GDW, Leppard GG, Araki T, Zhang X, West MM, Hitchcock AP (2003) Scanning transmission X-ray, laser scanning, and transmission electron microscopy mapping of the exopolymeric matrix of microbial biofilms. Appl Environ Microbiol 69:5543–5554

Leopold I, Fricke B (1997) Inhibition, reactivation, and determination of metal ions in membrane metalloproteases of bacterial origin using high-performance liquid chromatography coupled on-line with inductively coupled plasma mass spectrometry. Anal Biochem 252:277–285

Leppard GG, Droppo IG, West MM, Liss SN (2003) Compartmentalization of metals within the diverse colloidal matrices comprising activated sludge microbial floes. J Environ Qual 32:2100–2108

Li Y, Yan XP, Jiang Y (2005) Interfacing capillary electrophoresis and electrothermal atomic absorption spectroscopy to study metal speciation and metal–biomolecule interactions. Angew Chem Int Ed 44:6387–6391

Lima AIG, Corticeiro SC, Figueira EMDP (2006) Glutathione-mediated cadmium sequestration in *Rhizobium leguminosarum*. Enzyme Microb Technol 39:763–769

Lobinski R, Schaumlöffel D, Szpunar J (2006) Mass spectrometry in bioinorganic analytical chemistry. Mass Spectrom Rev 25:255–289
López-Barea J, Gómez-Ariza J-L (2006) Environmental proteomics and metallomics. Proteomics 6:51–62
Maret W (2004) Exploring the zinc proteome. J Anal At Spectrom 19:15–19
McKiernan JW, Creed JT, Brockhoff CA, Caruso JA, Lorenzana RM (1999) A comparison of automated and traditional methods for the extraction of arsenicals from fish. J Anal At Spectrom 14:607–613
McSheehy S, Mester Z (2003) The speciation of natural tissues by electrospray-mass spectrometry. II: Bioinduced ligands and environmental contaminants. Trends Anal Chem 22:311–326
Mendoza-Cózatl D, Loza-Tavera H, Hernández-Navarro A, Moreno-Sánchez R (2005) Sulfur assimilation and glutathione metabolism under cadmium stress in yeast, protists and plants. FEMS Microbiol Rev 29:653–671
Merroun M, Hennig C, Rossberg A, Reich T, Selenska-Pobell S (2003a) Characterization of U(VI)-*Acidithiobacillus ferrooxidans* complexes using EXAFS, transmission electron microscopy, and energy-dispersive X-ray analysis. Radiochim Acta 91:583–591
Merroun ML, Pollmann K, Raff J, Scheinost A, Selenska-Pobell S (2003b) EXAFS studies of palladium nanoclusters formed at the cells and S-layers of *Bacillus sphaericus* JG-A12. Forschungszentrum Rossendorf. FZR-Rep 400:25
Montes-Bayon M, Profrock D, Sanz-Medel A, Prange A (2006) Direct comparison of capillary electrophoresis and capillary liquid chromatography hyphenated to collision-cell inductively coupled plasma mass spectrometry for the investigation of Cd-, Cu- and Zn-containing metalloproteins. J Chromatogr A 1114:138–144
Mounicou S, Polec K, Chassaigne H, Potin-Gautier M, Lobinski R (2000) Characterization of metal complexes with metallothioneins by capillary zone electrophoresis (CZE) with ICP-MS and electrospray (ES)-MS detection. J Anal At Spectrom 15:635–642
Nieboer E, Fletcher GG, Thomassen Y (1999) Relevance of reactivity determinants to exposure assessment and biological monitoring of the elements. J Environ Monit 1:1–14
Nordberg M (2000) Trace elements and metallothionein related to geo-environment. J Trace Elem Exp Med 13:97–104
Outten CE, O'Halloran TV (2001) Femtomolar sensitivity of metalloregulatory proteins controlling zinc homeostasis. Science 292:2488–2492
Ouyang G, Pawliszyn J (2006) Recent developments in SPME for on-site analysis and monitoring. Trends Anal Chem 25:692–703
Pawliszyn J (1997) Solid phase microextraction—theory and practice. Wiley-VCH, New York, USA
Pocsi I, Prade RA, Penninckx MJ (2004) Glutathione, altruistic metabolite in fungi. Adv Microb Physiol 49:1–76
Pollmann K, Merroun M, Raff J, Fahmy K, Hennig C, Scheinost S, Selenska-Pobell S (2005) Interactions of the S-layer of *Bacillus sphaericus* JG-A12 with palladium and formation of Pd-nanoclusters. FZR Rep 419:32
Pollmann K, Raff J, Merroun M, Fahmy K, Selenska-Pobell S (2006) Metal binding by bacteria from uranium mining waste piles and its technological applications. Biotechnol Adv 24:58–68
Prange A, Pröfrock D (2005) Application of CE-ICP-MS and CE-ESI-MS in metalloproteomics: challenges, developments, and limitations. Anal Bioanal Chem 383:372–389
Raab A, Meharg AA, Jaspars M, Genney DR, Feldmann J (2004) Arsenic–glutathione complexes—their stability in solution and during separation by different HPLC modes. J Anal At Spectrom 19:183–190

Ram RJ, VerBerkmoes NC, Thelen MP, Tyson GW, Baker BJ, Blake RC, Shah M, Hettich RL, Banfield JF (2005) Community proteomics of a natural microbial biofilm. Science 308:1915–1920

Richards MP, Beattie JH (1994) Analysis of metalloproteins and metal-binding peptides by capillary electrophoresis. J Capillary Electrophor 1:196–207

Ríos A, Escarpa A, González MC, Crevillén AG (2006) Challenges of analytical microsystems. Trends Anal Chem 25:467–479

Rosenberg E (2003) The potential of organic (electrospray- and atmospheric pressure chemical ionisation) mass spectrometric techniques coupled to liquid-phase separation for speciation analysis. J Chromatogr A 1000:841–889

Ruiz-Jimenez J, de Castro MDL (2006) Coupling microdialysis to capillary electrophoresis. Trends Anal Chem 25:563–571

Ryan CG, Jamieson DN, Griffin WL, Cripps G, Szymanski R (2001) The new CSIRO-GEMOC nuclear microprobe: first results, performance and recent applications. Nucl Instrum Meth Phys Res B 181:12–19

Sarry J-E, Kuhn L, Ducruix C, Lafaye A, Junot C, Hugouvieux V, Jourdain A, Bastien O, Fievet JB, Vailhen D, Amekraz B, Moulin C, Ezan E, Garin J, Bourguignon J (2006) The early responses of *Arabidopsis thaliana* cells to cadmium exposure explored by protein and metabolite profiling analyses. Proteomics 6:2180–2198

Scancar J, Milacic R, Falnoga I, Cemazar M, Bukovec P (2000) Use of nitric acid in sample pretreatment for determination of trace elements in various biological samples by ETAAS. J Pharma Biomed Anal 22:993–1002

Schantz MM (2006) Pressurized liquid extraction in environmental analysis. Anal Bioanal Chem 386:1043–1047

Schaumlöffel D (2004) Capillary liquid separation techniques with ICP MS detection. Anal Bioanal Chem 379:351–354

Schreck P, Schubert M, Freyer M, Treutler HC, Weiss H (2005) Multi-metal contaminated stream sediment in the Mansfeld mining district: metal provenance and source detection. Geochemistry 5:51–57

Schubert M, Wennrich R, Weiß H, Schreck P, Zeller T, Otto HH, Wolfram H (2005) Formation of heavy metal bearing phases at a spring affected by the weathering of ore processing residues. Eur J Mineral 17:119–128

Shen H, Ren QG, Mi Y, Shi XF, Yao HY, Jin CZ, Huang YY, He W, Zhang J, Liu B (2002) Investigation of metal ion accumulation in *Euglena gracilis* by fluorescence methods. Nucl Instrum Meth Phys Res B 189:506–510

Sigg L, Black F, Buffle J, Cao J, Cleven R, Davison W, Galceran J, Gunkel P, Kalis E, Kistler D, Martin M, Noel S, Nur Y, Odzak N, Puy J, vanRiemsdijk W, Temminghoff E, Tercier-Waeber M-L, Toepperwien S, Town RM, Unsworth E, Warnken KW, Weng L, Xue H, Zhang H (2006) Comparison of analytical techniques for dynamic trace metal speciation in natural freshwaters. Environ Sci Technol 40:1934–1941

Solis C, Oliver A, Andrade E (1998) PIXE analysis of proteins from a photochemical center. Nucl Instrum Meth Phys Rev B 137:928–931

Spengler B (1997) Post-source decay analysis in matrix-assisted laser desorption/ionization mass spectrometry of biomolecules. J Mass Spectrom 32:1019–1036

Suzuki Y, Kelly SD, Kemner KM, Banfield JF (2002) Radionuclide contamination: nanometre-size products of uranium bioreduction. Nature 419:134

Szökefalvi-Nagy Z, Bagyinka C, Demeter I, Hollos-Nagy K, Kovacs I (1999) Speciation of metal ions in proteins by combining PIXE and thin layer electrophoresis. Fresenius J Anal Chem 363:469–473

Szökefalvi-Nagy Z, Bagyinka C, Demeter I, Kovacs KL, Quynh LH (1990) Location and quantification of metal ions in enzymes combining polyacrylamide-gel electrophoresis and particle-induced X-ray-emission. Biol Trace Elem Res 26–7:93–101
Szpunar J (2000) Bio-inorganic speciation analysis by hyphenated techniques. Analyst 125:963–988
Szpunar J (2004) Metallomics: a new frontier in analytical chemistry. Anal Bioanal Chem 378:54–56
Szpunar J (2005) Advances in analytical methodology for bioinorganic speciation analysis: metallomics, metalloproteomics and heteroatom-tagged proteomics and metabolomics. Analyst 130:442–465
Templeton DM, Ariese F, Cornelis R, Danielsson LG, Muntau H, Van Leeuwen HP, Lobinski R (2000) Guidelines for terms related to chemical speciation and fractionation of elements. Definitions, structural aspects, and methodological approaches (IUPAC Recommendations 2000). Pure Appl Chem 72:1453–1470
Tottey S, Harvie D, Robinson N (2007) Understanding how cells allocate metals (in this volume). Springer, Heidelberg
Turnau K, Kottke I (2005) Fungal activity as determined by microscale methods with special emphasis on interactions with heavy metals. In: Dighton J, White JF, Oudemans P (eds) The fungal community, its organization and role in the ecosystem. Taylor & Francis, Boca Raton, pp 287–305
Twining BS, Baines SB, Fisher NS, Maser J, Vogt S, Jacobsen C, Tovar-Sanchez A, Sanudo-Wilhelmy SA (2003) Quantifying trace elements in individual aquatic protist cells with a synchrotron X-ray fluorescence microprobe. Anal Chem 75:3806–3816
Urvoas A, Amekraz B, Moulin C, Le Clainche L, Stocklin R, Moutiez M (2003) Analysis of the metal-binding selectivity of the metallochaperone CopZ from *Enterococcus hirae* by electrospray ionization mass spectrometry. Rapid Commun Mass Spectrom 17:1889–1896
Vacchina V, Lobinski R, Oven M, Zenk MH (2000) Signal identification in size-exclusion HPLC-ICP-MS chromatograms of plant extracts by electrospray tandem mass spectrometry (ES MS/MS). J Anal At Spectrom 15:529–534
Vido K, Spector D, Lagniel G, Lopez S, Toledano MB, Labarre J (2001) A proteome analysis of the cadmium response in *Saccharomyces cerevisiae*. J Biol Chem 276:8469–8474
Wackett LP, Dodge AG, Ellis LBM (2004) Microbial genomics and the periodic table. Appl Environ Microbiol 70:647–655
Wang Z, Prange A (2002) Use of surface-modified capillaries in the separation and characterization of metallothionein isoforms by capillary electrophoresis inductively coupled plasma mass spectrometry. Anal Chem 74:626–631
Webb AG (2005) Nuclear magnetic resonance coupled microseparations. Magn Reson Chem 43:688–696
Webb SM, Leppard GG, Gaillard JF (2000) Zinc speciation in a contaminated aquatic environment: characterization of environmental particles by analytical electron microscopy. Environ Sci Technol 34:1926–1933
Wennrich R, Mattusch J, Morgenstern P, Freyer K, Treutler HC, Stark HJ, Bruggemann L, Paschke A, Daus B, Weiss H (2004) Characterization of sediments in an abandoned mining area; a case study of Mansfeld region, Germany. Environ Geol 45:818–833
Weston A, Brown PR (1997) HPLC and CE: principles and practice. Academic Press, San Diego
Williams KH, Ntarlagiannis D, Slater LD, Dohnalkova A, Hubbard SS, Banfield JF (2005) Geophysical imaging of stimulated microbial biomineralization. Environ Sci Technol 39:7592–7600

Williams RJP (2001) Chemical selection of elements by cells. Coordin Chem Rev 216:583–595

Wind M, Lehmann WD (2004) Element and molecular mass spectrometry: an emerging analytical dream team in the life sciences. J Anal At Spectrom 19:20–25

Wittrisch H, Conradi S, Rohde E, Vogt J, Vogt C (1997) Characterization of metal complexes of pharmaceutical interest by capillary electrophoresis with element sensitive detection. J Chromatogr A 781:407–416

Ye RW, Tao W, Bedzyk L, Young T, Chen M, Li L (2000) Global gene expression profiles of *Bacillus subtilis* grown under anaerobic conditions. J Bacteriol 182:4458–4465

Yoon TH, Johnson SB, Benzerara K, Doyle CS, Tyliszczak T, Shuh DK, Brown GE (2004) In situ characterization of aluminum-containing mineral-microorganism aqueous suspensions using scanning transmission X-ray microscopy. Langmuir 20:10361–10366

Zerkle AL, House CH, Brantley SL (2005) Biogeochemical signatures through time as inferred from whole microbial genomes. Am J Sci 305:467–502

Part II
Molecular Physiology of Metal-Microbe Interactions According to Groups

Microbiol Monogr (6)
D. H. Nies, S. Silver: Molecular Microbiology of Heavy Metals
DOI 10.1007/7171_2006_078/Published online: 2 February 2007

Acquisition of Iron by Bacteria

Volkmar Braun (✉) · Klaus Hantke

Mikrobiologie/Membranphysiologie, Universität Tübingen, Auf der Morgenstelle 28, 72076 Tübingen, Germany
volkmar.braun@uni-tuebingen.de

Abstract Bacteria have evolved multiple mechanisms to cope with the extreme iron limitations in their natural environments. Fe^{3+} forms insoluble hydroxy aquo complexes. The free Fe^{3+} concentration lies orders of magnitude below the concentration required for microbial growth (0.1 μM). Bacteria synthesize and secrete low-molecular-weight compounds, called siderophores, which bind Fe^{3+} with very high affinity and specificity, and host organisms of bacteria bind Fe^{3+} to proteins that serve as iron sources for bacteria. Energy-coupled transport systems bring Fe^{3+}, Fe^{3+}-siderophores, and heme across the outer membrane, the periplasm, and the cytoplasmic membrane into the bacterial cytoplasm. There, iron is released from the carrier molecules and incorporated mostly into heme and iron-sulfur proteins. Intracellular iron metabolism is poorly understood. The transport systems and the biosynthesis of the siderophores are regulated by proteins, usually by Fur in Gram-negative bacteria, and DtxR and IdeR in Gram-positive bacteria. These proteins act as transcriptional repressors when loaded with Fe^{2+}. Additional regulatory devices control siderophore biosynthesis and transport. The Fec-type of regulation is of particular interest because it involves a novel mechanism in which the ferric siderophore binds to the outer membrane transport protein and from there induces transcription of the transport and biosynthesis genes in the cytoplasm. Another recently detected device is the regulation of genes positively regulated by Fur via RhyB, a small regulatory RNA. RhyB facilitates degradation of positively regulated mRNAs, which does not occur when Fe^{2+}-Fur represses RhyB synthesis.

1 Introduction

The insolubility of Fe^{3+} and the solubility of Fe^{2+} requires different modes of iron acquisition by bacteria. The insoluble Fe^{3+} hydroxide polymer must be solubilized by iron-complexing compounds, called siderophores, which are synthesized and secreted by bacteria and fungi. In eukaryotic hosts, iron is incorporated into heme or carried by proteins, in particular hemoglobin, hemopexin, transferrin, lactoferrin, and ferritin. Bacteria have developed specific mechanisms to mobilize heme released from the heme proteins and iron released from transferrin and lactoferrin. When Fe^{3+} is provided by transferrin and lactoferrin, ionic Fe^{3+} is transported across the cytoplasmic membrane. The Fe^{3+}-siderophore complexes are transported into the bacterial cytoplasm where iron is released from the siderophores and incorporated into iron proteins, such as iron-sulfur proteins and cytochromes. The number of different siderophores a bacterial species synthesizes is lower than the number of different Fe^{3+} siderophores it takes up (it is more economical to synthesize only a few siderophores and use siderophores from other bacteria and fungi than to synthesize all siderophores for every available type of transport system). If a bacterial cell is unable to transport the siderophores secreted by other competing microbes, iron deprivation could result; therefore, the bacteria require a broad spectrum of Fe^{3+}-siderophore transport systems. These systems consist of specific outer-membrane transport proteins that are necessary because of the low concentration of Fe^{3+}-siderophores and their size, which may exceed the diameter of the pores formed by the outer-membrane porins. Synthesis of siderophores and Fe^{3+}-siderophore transport systems are regulated. Transcription of siderophore synthesis genes and Fe^{3+}-siderophore transport genes is repressed when cells contain enough iron. Transport of Fe^{3+}, Fe^{3+}-siderophores, and heme across the cytoplasmic membrane of Gram-positive and Gram-negative bacteria is catalyzed by ABC transporters, which involve a binding protein and one or two transmembrane proteins linked to a membrane-associated ATPase. Fe^{2+}, in contrast, diffuses across the outer membrane and is transported across the cytoplasmic membrane by a mechanism that differs from that of Fe^{3+} and Fe^{3+}-siderophores.

Transport of Fe^{2+} is a GTP-coupled process, directly or indirectly catalyzed by a protein, designated FeoB.

Space limitation does not allow us to discuss many pertinent papers. We will concentrate on recent findings and refer the reader in particular to the book *Iron transport in bacteria* (2004) and to the comprehensive reviews (Andrews et al. 2003; Braun et al. 1998; Cornelissen 2003; Faraldo-Gomez and Sansom 2003; Klebba 2003; Perkins-Balding et al. 2004; Postle and Kadner 2003; Prakash et al. 2003; Raymond et al. 2003; Rohde and Dyer 2003; Schalk et al. 2004; Ferguson and Deisenhofer 2004; Wandersman and Delepelaire 2004; Cartron et al. 2006).

2 Fe^{3+}-Siderophore Transport Systems in Gram-Negative and Gram-Positive Bacteria

Siderophores consist of a large variety of compounds with the function of chelating Fe^{3+}. They are classified as hydroxamates, catecholates, polycarboxylates, or mixed structures (Winkelmann et al. 1991). Fe^{3+}-siderophores are transported into bacteria by highly specific transport systems. Usually, bacteria synthesize a number of Fe^{3+}-siderophore transport systems. For example, pathogenic *Escherichia coli* may use ferrichrome, enterobactin and its dihydroxybenzoyl serine degradation products DHBS, $(DHBS)_2$, and $(DHBS)_3$; citrate; aerobactin; yersiniabactin; salmochelin; coprogen; and heme. All these siderophores except ferrichrome and coprogen are synthesized by *E. coli*. Citrate and heme, however, are not secreted in growth-promoting amounts and come from extracellular sources.

2.1 Energy-Coupled Transport Across the Outer Membrane of Gram-Negative Bacteria

Energy-coupled transport across the outer membrane has attracted much attention because there is no energy source in the outer membrane. Energy is derived from the cytoplasmic membrane. The principal questions are then: how is energy transmitted from the cytoplasmic membrane into the outer membrane and how does it affect transport? Most of the Fe^{3+}-siderophores are too large to diffuse through the pores of porins. The larger pores required for diffusion would render cells vulnerable to toxic compounds. In addition, diffusion would not suffice to support growth because the concentration of Fe^{3+}-siderophores is too low. To acquire enough iron—approximately 10^5 ions per cell per generation—high-affinity proteins extract the Fe^{3+}-siderophores from the medium by strong binding with K_d values in the nanomolar range (Scott et al. 2001).

The crystal structures of a number of outer-membrane proteins have been determined. All of these proteins form β-barrels. The β-barrels of monomeric Fe^{3+}-siderophore transporters consist of 22 antiparallel β-strands that form a pore. The pore is completely closed by a globular domain, called the cork, plug, or hatch. The β-strands are connected by surface loops of up to 37 residues and short turns of a few amino acids in the periplasm. The Fe^{3+}-siderophore binding sites at the transport proteins are located well above the cell surface. The sites involve approximately 10 residues of the plugs and loops. Release of the Fe^{3+}-siderophore from the strong binding site is an active energy-consuming process. Energy is provided by the electrochemical potential of the cytoplasmic membrane through interaction of the TonB protein with the outer membrane transport proteins. TonB is inserted in the cytoplasmic membrane and forms a complex there with two other proteins, ExbB and ExbD, which are both required for Fe^{3+}-siderophore transport. How TonB responds to the energized cytoplasmic membrane, why and how it changes conformation, how the conformational change is transmitted to the outer membrane transport proteins, and how these proteins are triggered to change their conformation are not known.

Conformational changes in the outer membrane transport proteins not only release the substrate from their binding sites. The closed pores of the Fe^{3+}-siderophore transporters must be opened so that the Fe^{3+}-siderophores can enter the periplasm. Vectorial transport into the periplasm requires either a sequence of binding sites along which the substrates move or prevention of the escape of the substrates into the medium by closure of the pores at the cell surface. Closure of the pore entry site has been demonstrated in the crystal structure of the FecA protein. Upon binding of the substrate diferric dicitrate in the external pocket, FecA loops 7 and 8 at the cell surface move 11 and 15 Å, respectively, thereby closing the gate (Ferguson et al. 2002; Yue et al. 2003). In vivo experiments have provided support of the importance of loops 7 and 8. Their deletion abolishes FecA transport activity and FecA signaling for transcription initiation of the *fecABCDE* transport genes (see Sect. 3) (Sauter and Braun 2004). In contrast, the crystal structures of FhuA do not reveal such strong loop movements upon binding of its substrate ferrichrome (Ferguson et al. 1998; Locher et al. 1998). Deletion of loops 7 and 8 of FhuA does not inactivate FhuA, but reduces ferrichrome transport 15–73%, depending on the copy number of the plasmid on which the mutant *fhuA* gene is cloned (Endriss and Braun 2004). Whether one of these mechanisms—closing of the gate, as with FecA, or leaving open the entry site, as with FhuA—reflects how the transporters function or whether both mechanisms exist remains open to question. Unfortunately, the crystal structures of other Fe^{3+}-siderophore transporters do not provide an answer because they either are bound to a metal-free siderophore, e.g., FpvA with bound pyoverdine siderophore (Cobessi et al. 2005a) and FptA with bound Fe^{3+} pyochelin (Cobessi et al. 2005b), or have a poorly defined loading, e.g.,

FepA loaded with degraded Fe^{3+}-enterobactin (Buchanan et al. 1999). FpvA loaded with pyoverdine most likely does not reflect the same structure as FpvA loaded with Fe^{3+}-pyoverdine because binding of dicitrate to FecA does not induce the structural changes in FecA induced by diferric dicitrate (Yue et al. 2003). It is known that strong movement of surface loops also occurs in FepA. Surface loop 7 of FepA can be chemically cross-linked to major outer-membrane proteins. Cross-linkage is prevented by binding of ferric enterobactin (Scott et al. 2002). Strong loop movements also occur in BtuB, the outer-membrane transporter of vitamin B_{12} that has the same basic structure as the Fe^{3+}-siderophore transporters, upon binding of vitamin B_{12}, but the entry site is not closed (Chimento et al. 2003). In BtuB, two Ca^{2+} ions bind to an aspartate cage. Such a cage is not seen in Fe^{3+}-siderophore transporters. Ca^{2+} binding partially orders surface loops, which facilitates access of vitamin B_{12} to its BtuB binding site. Although the basic design of outer-membrane transporters is very similar, the way they function differs in the details.

Binding of Fe^{3+}-siderophores involves amino acid residues in loops connecting β-strands and in the plug domains. Binding causes strong structural changes throughout the entire FhuA or FecA molecule. The loaded and unloaded states can be compared. A network of interacting amino acid residues probably transmits the structural changes from far above the cell surface to the periplasm. A number of small (1–2 Å) shifts throughout the proteins and very large changes in the portions of the proteins exposed to the periplasm occur. In both proteins, a short helix, called the switch helix, is only observed in the unliganded structures. Resolution of the switch helix results in a flexible extended conformation, so that in FecA a defined electron density is only seen from residue 95 upward (note that FecA contains an additional N-terminal sequence of 79 residues not present in FhuA, see Sect. 3). In FhuA, residue W22 is displaced by 17 Å. As a result of the structural transitions in periplasmic regions, the region of the transporters that interacts with the TonB protein, the so-called Ton box, becomes flexible, which may facilitate binding to TonB. Consistent with this interpretation are the observations that in the crystal structure of FpvA loaded with iron-free pyoverdine, the Ton box is seen, whereas in FptA loaded with Fe^{3+}-pyochelin, the Ton box is not seen. In BtuB, the Ton box becomes disordered upon binding of vitamin B_{12}.

Very recently, crystal structures of TonB fragments extending from residue 158 to 235 and from 148 to 239, linked to FhuA (Pawelek et al. 2006) and BtuB (Shultis et al. 2006), respectively, were determined. Both structures delineate the interaction of TonB with the transporters (Fig. 1). Previous results demonstrated that certain single amino acid replacements in the Ton box of FhuA and BtuB inactivate the transporters. Amino acid replacements in Gln160 of TonB partially restore the activities of the FhuA and BtuB Ton box mutants. This was taken as evidence for the interaction of these two regions. This conclusion was supported by spontaneous in vivo formation of disulfide bridges between cysteine residues introduced in the Ton box of BtuB

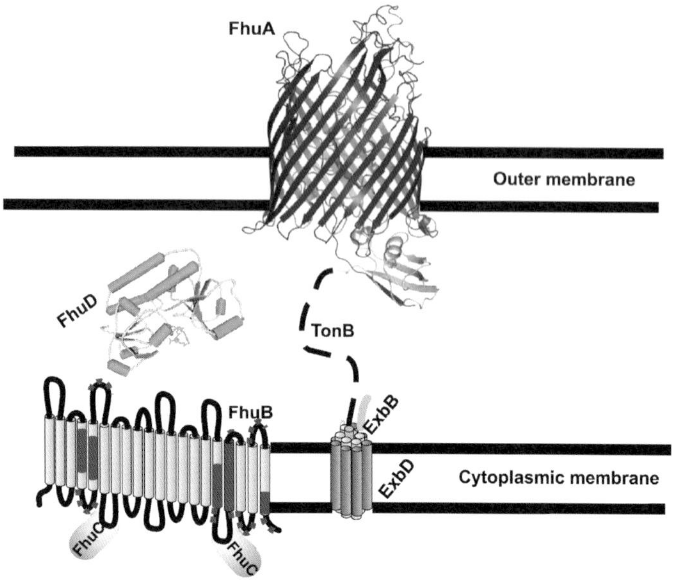

Fig. 1 Model of ferrichrome transport across the outer membrane and the cytoplasmic membrane of *Escherichia coli*. The known crystal structures of FhuA and FhuD, both loaded with substrates, and the TonB fragment (*in yellow*) that binds to the Ton box of FhuA are illustrated. The structure of the TonB-ExbB-ExbD complex is not known and only the transmembrane portions are shown in the cytoplasmic membrane. The structure of the TonB segment that connects the C-terminal crystal form to the transmembrane portion is also not known and is drawn as an *interrupted line*. FhuB, FhuC, and FhuD form the ABC transporter that catalyzes ferrichrome transport across the cytoplasmic membrane. Part of the figure has been kindly provided by (Pawelek et al. 2006)

and FecA, and cysteine residues introduced in region 160 of TonB. In the FhuA crystal structure, the Ton box (residues 6–13) is not seen because it is flexible.

In the new crystal structure, the TonB fragment associates with the FhuA Ton box by forming a parallel β interaction with the β3-strand of TonB. The β3-strand is part of a three-stranded β-sheet including the β1- and β2-strands (Fig. 1). An earlier NMR structure of TonB(152–239) reveals the same three β-strands and in addition a fourth β-strand, in anti-parallel orientation, which is replaced by the β-strand of the FhuA TonB box (Peacock et al. 2005). Furthermore, the crystal structure of another TonB fragment, TonB(148–239), shows a fold similar to that of TonB(152–239), in which one β-strand of the three-stranded β-sheet of one monomer forms a fourth anti-parallel β-strand with the three-stranded β-sheet of another monomer (Ködding et al. 2005). Gln160 of TonB is not seen in the FhuA-TonB(158–235) structure, but can be oriented such that it forms a hydrogen bond with Thr12. In the BtuB-TonB(153–233) structure, Gln160 is seen and interacts with Asp6, Leu8, and

Val10 of the BtuB Ton box (residues 6–12). The new crystal structures confirm the earlier results of genetic suppressor analyses and cysteine cross-linking experiments.

There are numerous additional interactions between the TonB fragments and the transporters. In both structures, the TonB fragments occupy approximately half of the periplasmic surface area of the transporters. Of particular interest with regard to function is the position of the TonB α1-helix close to the plug domain, which allows interaction of TonB Arg166 with Glu56 of the FhuA plug. This interaction might be important for the dislocation of the plug to open the pore. In BtuB-TonB(153–233), Arg158 of TonB interacts with Asp6 of the BtuB Ton box.

As mentioned above, FpvA was isolated with bound iron-free pyoverdine. Fluorescence of pyoverdine allows the binding to and release from FpvA to be studied with fluorescence techniques. Metal-loaded pyoverdine, in contrast, is not fluorescent. The affinity for pyoverdine is ten times lower than the affinity for iron-loaded pyoverdine. Metal-loaded pyoverdine binds in minutes, whereas pyoverdine binds in an hour. Binding of both compounds is not affected by the proton motive force and TonB, but dissociation of pyoverdine from FpvA is accelerated—in a *tonB*$^+$ strain, dissociation takes 1 min, in a *tonB* mutant, it takes hours (Clement et al. 2004). Since iron-poor growth conditions are prevalent, a surplus of unloaded siderophores may actually occur; therefore, binding of metal-free siderophores to the transporters may be the rule. This phenomenon has been observed with binding of dicitrate to FecA (Yue et al. 2003), pyochelin to FptA, and deferri-ferrichrome to FhuA (Hoegy et al. 2005). However, whether the binding of a siderophore prior to loading with a metal-loaded siderophore is advantageous is not clear since, for example, the iron-pyoverdine transport rate is the same regardless of whether FpvA is preloaded with pyoverdine or not.

Salmochelin is a recently discovered, frequently occurring siderophore of *Salmonella enterica* and related enterobacteria. It is a derivative of enterobactin, with glucose linked to the C5 position of two catecholate residues (Bister et al. 2004). Fe^{3+}-salmochelin is transported across the outer membrane by the IroN transporter. In *E. coli*, linear Fe^{3+}-salmochelin hydrolysis products are probably taken up by the Fe^{3+}-enterobactin FepA transporter and the Cir and Fiu transporters because they transport linear Fe^{3+}-enterobactin hydrolysis products.

A virulence-associated adhesin receptor has been found in different pathogenic *E. coli* strains (Tarr et al. 2000). Recently it was shown that this protein is a TonB-dependent catecholate siderophore receptor (Leveille et al. 2006), as one could assume from its sequence (Fig. 2). This is the first siderophore receptor that has an additional function in adherence to epithelial cells.

Heme and heme contained in hemoglobin, the hemoglobin-haptoglobin complex, and hemopexin are abundant sources of iron for bacteria from

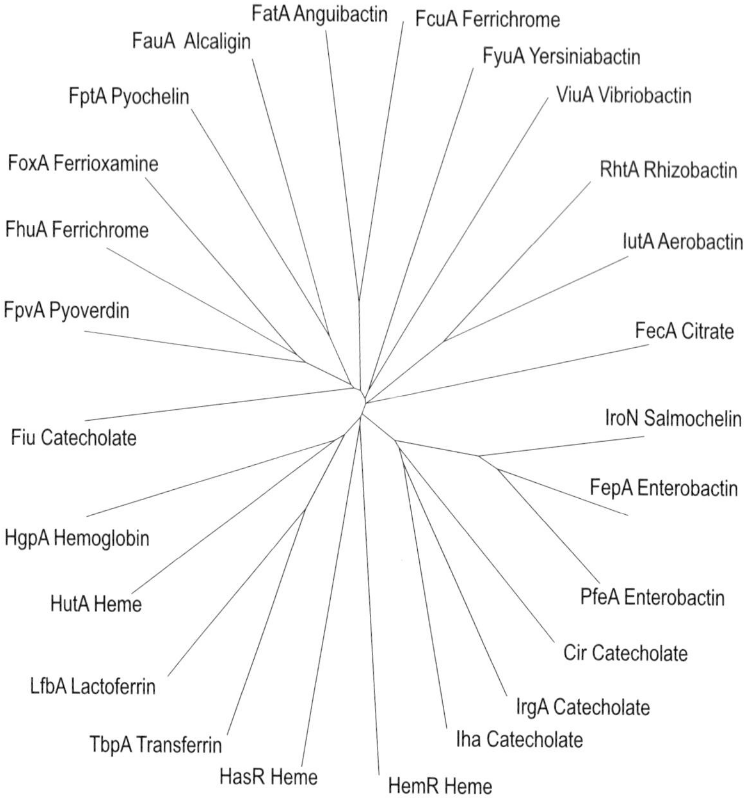

Fig. 2 Phylogram of selected outer membrane proteins that transport Fe^{3+}-siderophores, Fe^{3+} provided by host transferrin and lactoferrin, and heme and hemoglobin. Substrate specificity of the receptors is often (but not always) predictable within one group (Page 2005)

their host. Gram-negative bacteria contain receptor proteins at the cell surface for all these heme compounds. Most of the receptors are specific for heme, some for hemoglobin, and others for hemoglobin-haptoglobin and hemopexin. These receptors are mostly found in strains of *Vibrio cholerae*, *Yersinia*, *Neisseria*, *Haemophilus influenzae*, *E. coli*, and *Shigella*. In *Serratia marcescens*, the secreted HasA hemophore protein releases heme from hemoglobin and donates heme to the outer membrane receptor HasR. Heme is then transported across the outer membrane by a TonB–ExbB–ExbD-dependent mechanism. Structural transitions in HasR presumably induced by energized TonB cause release of empty HasA from HasR and heme uptake into the periplasm (Letoffe et al. 2004). The K_d of heme binding to HasA is approximately 1 μM and that of HasA binding to HasR is 5 nM, which indicates that HasA functions at very low heme or hemoglobin concentrations (Wandersman and Delepelaire 2004). The predicted HasR structure consists of a β-barrel, a plug domain, and an N-terminal extension that is important for transcription initiation by heme-loaded HasR (see Sect. 3). HasR lacking the N-terminal extension (residues 1–91) retains transport activity. HasR lacking the extension and the plug (residues 11–192) is transport and induction incompetent, but heme specifically diffuses through it, as deduced from the requirement for His603. In a His603Ala replacement mutant, diffusion is abolished and heme transport through the complete mutant HasR is abolished (Letoffe et al. 2005). A substrate-specific diffusion channel through the β-barrel of FhuA (Braun et al. 2003) and of FecA (M. Ogierman, A Sauter, V. Braun, unpublished results) has not been observed.

Since the concentration of free Fe^{3+} ions in virtually all aerobic environments is below the growth-promoting concentration, Fe^{3+} must be delivered to bacteria in eukaryotic hosts by proteins, e.g., transferrin and lactoferrin, to be transported as Fe^{3+} ion into bacterial cells (reviewed in Cornelissen 2003; Perkins-Balding et al. 2004; Rohde and Dyer 2003). The delivering proteins do not enter the bacteria. In Gram-negative bacteria, they instead bind to highly specific receptor proteins. These receptors, in contrast to siderophore and heme receptors, consist of two proteins: TbpA and TbpB for transferrin, and LbpA and LbpB for lactoferrin. Such two-component receptors are not unique. They also occur for heme proteins in *Neisseria* and *Haemophilus influenzae*. These receptors bind only human and primate transferrins and lactoferrins, even though other vertebrate transferrins and lactoferrins are homologous to the primate proteins. Since these bacteria infect only humans, the receptor specificity reflects the host. Both the A and B components bind the host iron proteins with binding constants in the nanomolar range. The A components represent the transmembrane transporters, and the B components either increase the transport activity (*Neisseria gonorrhoeae*) or are essential for transport (*Neisseria meningitidis*), as shown, e.g., for transferrin-mediated iron transport. The B component is a lipoprotein. Most of it is exposed on the cell surface to which it is anchored by the lipid of the murein lipoprotein type. Fe^{3+} delivered by transferrin and lactoferrin is transported across the outer membrane by members of the *Neisseriaceae*, *Pasteurellaceae*, and *Moraxellaceae*. Iron supply by transferrin and lactoferrin is essential for the virulence of these bacteria. How iron is released from transferrin and lactoferrin by the receptors and how iron is transported across the outer membrane are not known. Energy input via TonB, ExbB, and ExbD is not required for binding of the iron proteins to the receptors but is required to release unloaded transferrin from gonococcal TbpA. Whether also the transfer of Fe^{3+} to TbpA requires energy is not known. Modelling of *N. meningitidis* TbpA on the basis of the homologous FhuA sequence and crystal structure arrives at a structure similar to FhuA consisting of a β-barrel with 22 antiparallel β-strands and a plug domain. Much larger surface loops account for the larger size of TbpA than FhuA. The model predicts a narrow channel through the entire length of TbpA through which Fe^{3+} could diffuse. The channel opening has a diameter of approximately 8 Å and constricts to 5 Å at its narrowest site (Oakhill et al. 2005). Since Fe^{3+} is insoluble, it is likely that it moves through the channel along amino acid residues, as has been demonstrated for the facilitated diffusion of maltose through the LamB protein channel (Van Gelder et al. 2002). Figure 2 shows a sequence comparison of selected outer membrane iron transporters.

2.2
Transport of Fe^{3+}-Siderophores, Heme, and Fe^{3+} Across the Cytoplasmic Membrane of Gram-Negative Bacteria

Gram-positive and Gram-negative bacteria transport Fe^{3+}-siderophores, heme, and Fe^{3+} across the cytoplasmic membrane by ABC transporters. The ABC transporters are composed of binding proteins found in soluble form in the periplasm of Gram-negative bacteria and fixed to the surface of the cytoplasmic membrane by a lipid anchor of the murein-lipoprotein type in Gram-positive bacteria. Removal of the lipid anchor does not inactivate the binding proteins (Kempf et al. 1997) because they are bound by additional forces to the transmembrane transport proteins. Crystal structures of many binding proteins have been determined. The binding proteins are composed of two flexible lobes that move towards each other upon binding of the substrate. The structure of the iron-binding protein FbpB of *Neisseria gonorrhoeae* and of HitA (also designated hFbp) of *Haemophilus influenzae* agree with the general structure of binding proteins. Surprisingly, movement of the two lobes in the ferric hydroxamate binding protein FhuD is restricted by a rigid helix that connects the two lobes (Fig. 1). The ferric hydroxamate binding site is at the surface of the molecule and not deep inside, as observed in most of the binding proteins when the two lobes have closed. Surface exposure results in a flexible structure of the antibiotic moiety of albomycin, which binds with the ferric hydroxamate portion to FhuD (Clarke et al. 2002).

Transport across the cytoplasmic membrane was studied by isolating point mutants and deletion mutants, by complementation of mutants with isolated wild-type genes, by substrate binding to the binding proteins, and by determination of the crystal structure of binding proteins. The crystal structure of the vitamin B_{12} ABC transporter BtuCD (Locher et al. 2002) and the separately determined crystal structure of the related vitamin B_{12} binding protein BtuF (Borths et al. 2002) suggest that the periplasmic BtuF protein is positioned above the cytoplasmic membrane permease BtuC through two glutamate residues in BtuF and arginine residues in BtuC. These interacting residues are also found in a number of other bacterial iron ABC transporters, including the ferric citrate transporter. Replacement of one or both glutamate residues and some of the arginine residues by alanine and cysteine residues reduces, but does not abolish citrate-mediated iron transport (V. Braun and C. Herrmann, unpublished results). This result indicates that the predicted fixation of the monomeric binding proteins to the permease dimers through salt bridges is probably strengthened by additional interactions between the proteins. In the few cases studied, the binding proteins specifically recognize the ferric siderophores and iron (Mietzner et al. 1998; Köster 2001; Anderson et al. 2004; Shouldice et al. 2003; Dhungana et al. 2005; Pramanik and Braun 2006). ATP binding and ATP hydrolysis by the proteins bound to the permeases at the inner side of the cytoplasmic membrane presumably trigger structural changes in the permeases such that the pore between the interface of the two permease subunits is opened. Comparison of the crystal structure of the Rad50 DNA repair enzyme with bound ATP and the ATP-free BtuD protein suggests that binding of two ATP molecules at sites in the interface between the two BtuD ATPase subunits may cause structural changes in BtuD that are conveyed to BtuC such that the BtuC pore is opened (Locher et al. 2002).

Strong conformational changes have been demonstrated in the MalK ATPase of the maltose transport system of *E. coli*. In the ATP-loaded state, the substrate-loaded binding protein tightly binds to the permease and triggers ATP hydrolysis by the ATPase. Hydrolysis of ATP to ADP results in the opening of the two nucleotide binding domains (the interface between two molecules moves apart). Opening of the contact between the two nucleotide binding domains converts the ATPase to the resting state (Lu et al. 2005). After transmission of the substrates from the binding protein to the permease, the binding protein may stay bound to the permease to avoid release of the substrate to the periplasm. In addition, or alternatively, the permease might cycle between two states, one in which the pore is opened to the periplasm for accepting the substrate from the binding protein, and one where the entry is closed and simultaneously the exit to the cytoplasm is opened to release the substrate into the cytosol. Such a mechanism would guarantee a vectorial translocation through the permease. Tight structural coupling of the ATPase to the permease may convey the structural changes in the ATPase upon binding of Mg^{2+}-ATP and its hydrolysis. Whether the permeases bind the substrates or form a pore through which the substrates diffuse without recognition by the permeases is not known. Since the substrate specificity of the transport systems coincides with the substrate specificity of the binding proteins, it is likely that substrate binding to the permeases, if it occurs, plays no major role in transport.

Bacteria may encode more than one gene cluster for heme transport. These clusters in addition frequently contain genes not involved in heme transport, but somehow involved in intracellular heme utilization. For example, *Vibrio cholerae* synthesizes three distinct outer membrane receptors (HutA, HutR, and HasR), two TonB systems, and an apparent ABC transporter to transfer heme across the inner membrane. The cluster encoding the heme transport system located on the small chromosome reads *hutD hutC hutB exbD1 exbB1 tonB1 hutW hutX hutZ*. HutD is an ATPase, HutC is an inner-membrane permease, and HutB is a binding protein. The *exbB*, *exbD*, and *tonB* genes have the same transcription polarity as the genes encoding the transport system across the inner membrane, but none of the outer membrane transporters whose activity requires the TonB, ExbB, and ExbD proteins is encoded in this locus. The functions of the HutWXZ proteins are not clear. HutZ enhances heme utilization as an iron source, but is not essential (Wyckoff et al. 2004). HutZ might serve as a heme carrier and storage protein in the cytoplasm. The low solubility of heme may require heme-binding proteins that carry heme from sites of heme uptake and heme biosynthesis to sites of heme incorporation into cytochromes. No heme oxygenases have been identified in enterobacteria, but they occur in *Staphylococcus aureus*, *Corynebacterium diphtheriae*, *Neisseria meningitidis*, and *Pseudomonas aeruginosa* and are required for iron release from heme. Genes homologous to *hutWXZ* occur in other organisms, but their functions are unknown. ShuS of *Shigella dysenteriae* is another protein encoded in a heme transport locus, but is not involved in heme transport (Wyckoff et al. 2005). It is also not essential for heme utilization at low heme concentrations, but it is required for growth at high heme concentrations. ShuS binds heme in a non-toxic form and may serve additional functions in intracellular heme trafficking, as does HutZ.

2.3 Transport of Fe^{3+}-Siderophores, Heme, and Fe^{3+} Across the Cytoplasmic Membrane of Gram-Positive Bacteria

In Gram-positive bacteria, Fe^{3+}-siderophore and heme transport occurs only across the cytoplasmic membrane (Figs. 3 and 4). The few transport systems characterized in detail are ABC transporters (Clancy et al. 2006; Jin et al. 2006; Pramanik and Braun 2006; Schneider and Hantke 1993; Sebulsky et al. 2004). Although transport ATPases are usually components of a single ABC transporter, the FhuC ATPase of the *Staphylococcus aureus* Fe^{3+}-hydroxamate transport system functions also in Fe^{3+}-staphylobactin transport determined by the SirA binding protein and the SirB and SirC permease components. In *Listeria monocytogenes*, the K_m values determined for the binding of ferric siderophores to binding proteins (1–10 nM) (Jin et al. 2006) are much lower than the K_m values found for Gram-negative binding proteins (1 μM) (Rohrbach et al. 1995).

Although Gram-positive bacteria lack an outer membrane and substrates are thought to diffuse through the hydrophilic cell wall, diffusion may be facilitated by proteins, as evidenced in *S. aureus* in which surface-exposed proteins contribute to heme uptake. Isd-mediated heme transport is catalyzed by six proteins, three of which are surface-exposed. IsdA, IsdB, and IsdC are anchored to the cell wall murein, and the proteins IsdE (lipoprotein), IsdD (permease), and IsdF (ATPase) form an ABC transporter in the cytoplasmic membrane (Marraffini et al. 2006). The genes encoding these proteins, in add-

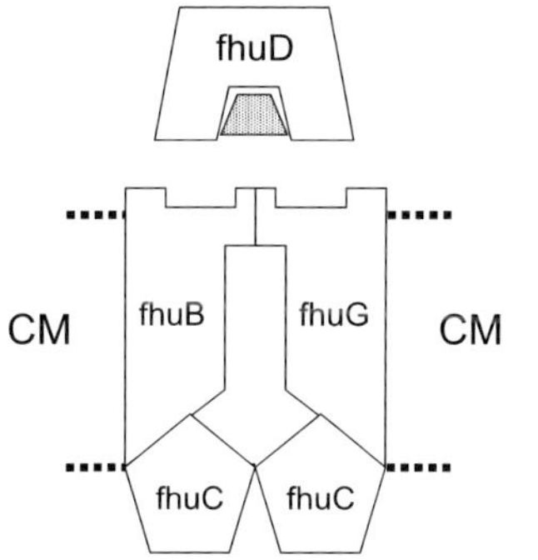

Fig. 3 Model of the ferrichrome and ferrioxamine transport system of *Streptococcus pneumoniae*. This system also transports the structurally related antibiotics albomycin and salmycin (Pramanik and Braun 2006)

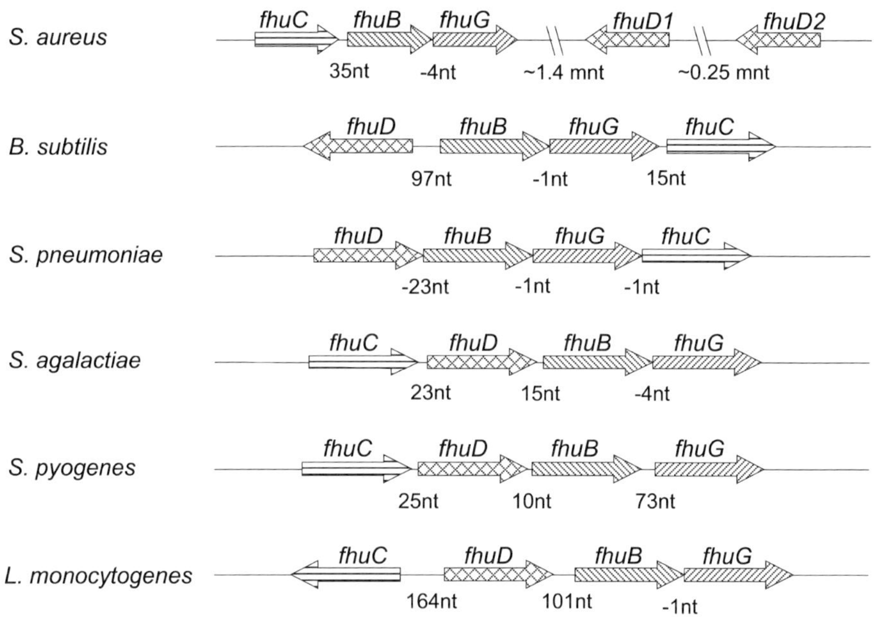

Fig. 4 Comparison of experimentally characterized systems that transport Fe^{3+}-siderophores across the cytoplasmic membrane of Gram-positive bacteria

ition to the genes encoding sortase B (*srtB*), which links IsdC to the murein, and IsdG, a cytoplasmic heme oxygenase that degrades heme and releases iron, are adjacent and regulated by iron (Fig. 5). It is proposed that IsdB at the cell surface binds hemoglobin, releases heme from the heme protein, and delivers it to IsdA further inside the cell wall. From there, heme is translocated to IsdC within the cell wall, subsequently to the IsdE binding protein

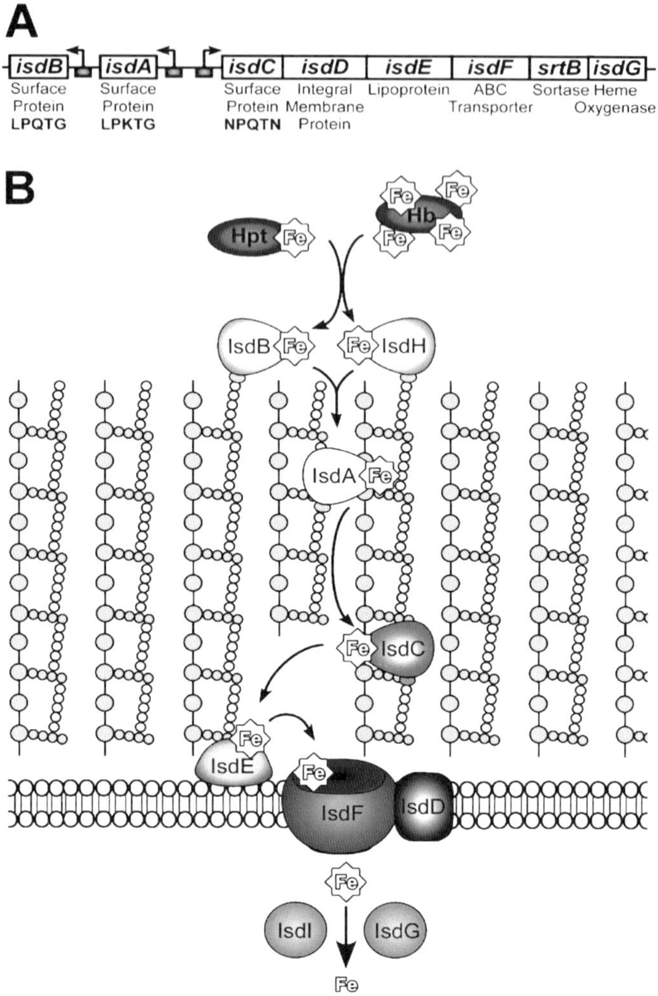

Fig. 5 **A** Genetic organization of the genes encoding the Isd proteins of *Staphylococcus aureus* and **B** model of heme uptake into *S. aureus* mediated by the Isd proteins (Marraffini et al. 2006)

anchored to the cytoplasmic membrane, and from there across the cytoplasmic membrane via IsdD and IsdF (ATP binding protein). Such an involvement of proteins in facilitated diffusion of heme across the Gram-positive cell wall may be a more general mechanism. For example, group A streptococci contain a surface-exposed protein, Shp, that binds heme and transfers heme to the HtsA lipoprotein, which is part of a heme ABC transporter (Liu and Lei 2005). Uptake through the Gram-positive cell wall is formally (but not mechanistically) reminiscent of the protein-mediated heme transport across the outer membrane of Gram-negative bacteria.

3
Outer Membrane Transporters also Function as Signal Generators and Signal Transmitters

Studies on the regulation of the ferric citrate transport system of *E. coli* unravelled a novel mechanism of transcriptional control. Ferric citrate induces transcription of the *fecABCDE* transport genes without entering the cell (Braun and Mahren 2005). Binding of ferric citrate, or rather diferric dicitrate as shown by the substrate-loaded FecA crystal structure (Ferguson et al. 2002; Yue et al. 2003), to the outer membrane FecA protein induces a signal that is transmitted across the outer membrane by FecA. FecA functions as a signal receiver and as a signal transmitter. The signal is transmitted across the cytoplasmic membrane by the FecR protein. FecR is a transmembrane protein with the N-terminus in the cytoplasm and the C-terminus in the periplasm. Compared to outer-membrane transporters without regulatory activity, mature FecA (without signal peptide) contains an N-terminal extension of 79 residues (signaling region), which is required for transcription regulation but not for transport (Kim et al. 1997). The signaling region is located in the periplasm (Kim et al. 1997) and not seen in the crystal structure because it is flexible. In the periplasm, the N-proximal region of FecA interacts with the C-proximal region of FecR (Enz et al. 2000, 2003a). Single amino acid replacements in the FecA signaling region located on one side of the signaling structure (Fig. 6) reduce transcription initiation (Breidenstein et al. 2006). In this same region, two FecA mutations are located that suppress three mutations in the C-proximal end of FecR (Enz et al. 2003a). These data suggest that the mutations are located in the FecA interface to FecR. The C-proximal region of FecR forms heptad repeats that are conserved in FecR homologs of *Nitrosomonas europeae*, *Xanthomonas campestris*, *Pseudomonas* species, and *Bordetella* species (Braun and Mahren 2005).

In the cytoplasm, FecR activates the FecI sigma factor, which recruits the RNA polymerase and directs it to the promoter upstream of the *fecA* gene, which in turn controls transcription of *fecA* and the downstream *fecBCDE* genes. The cytoplasmic $FecR_{1-85}$ fragment is sufficient to induce *fec* transport gene transcription, which requires no binding of diferric dicitrate to FecA (Stiefel et al. 2001). Among FecR homologs, the $FecR_{1-85}$ fragment contains three highly conserved tryptophan residues that are important for FecR activity, as shown by the inactivation of FecR by randomly generated mutations in which the arginine replaces tryptophan.

FecI belongs to the group 4 sigma factors, also designated extracytoplasmic function (ECF) sigma factors of the σ^{70} family. These can be subdivided, from the N- to the C-terminus, into regions 1.1, 1.2, 2.1, 2.2, 2.3, 2.4, 3, 4.1, and 4.2. Region 2.2 is the most conserved region and forms the largest interface with the RNA polymerase. Amino acid replacements in this region lead to a strong reduction in binding of FecI to the β' subunit of the RNA

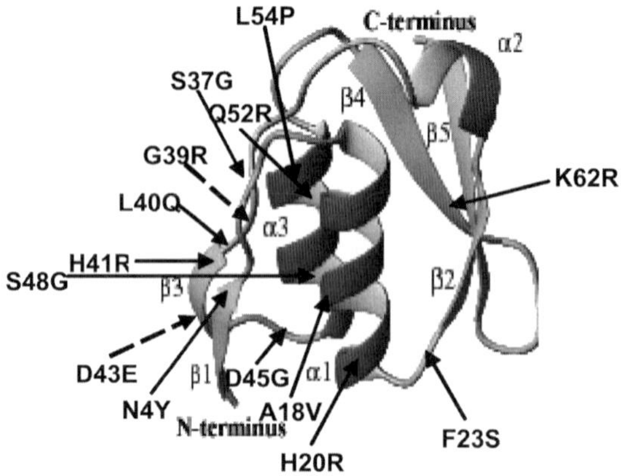

Fig. 6 Signaling domain of FecA determined by nuclear magnetic resonance spectroscopy (Garcia-Herrero and Vogel 2005). The positions of amino acid replacements that partially inactivate the signaling activity of FecA and of two mutation sites that suppress FecR mutations (*interrupted arrows*) are indicated

polymerase and FecI activity. Mutations in region 4 display a reduced FecR binding and FecI activity (Braun and Mahren 2005).

The promoter upstream of *fecA* encompasses the well-conserved – 35 region and the less-conserved – 10 region, and in addition a region extending up to the + 11 site (Angerer et al. 1995; Enz et al. 1995, 2003b). The *fecABCDE* genes are not only regulated by FecIR but also by iron via the Fur repressor protein (Angerer and Braun 1998) (see Sect. 5.1). *fecIR* transcription is only controlled by Fe^{2+}-Fur. Transcriptional control is well designed in that cells first recognize iron limitation, which results in the synthesis of the FecIR regulatory proteins. The cells then determine whether diferric dicitrate is in the growth medium by binding the inducer to FecA, which is always present in low amounts. If this is the case, the signal cascade is initiated from the cell surface into the cytoplasm, where FecI is activated. Citrate-mediated iron transport occurs as long as there is insufficient iron in the cytoplasm. If a sufficient level is attained, surplus Fe^{2+} binds to Fur, which in turn represses transcription of *fecIR* and *fecABCDE*.

Most of the sequenced *E. coli* genomes do not encode a *fec* system. This indicates that *E. coli* K-12 acquired the *fec* system through horizontal gene transfer. The *fec* genes are flanked by insertion elements that serve to transpose DNA fragments. In the closely related *Klebsiella pneumoniae*, only one of three studied strains contains a functional ferric citrate transport system. A *K. pneumoniae* virulence plasmid encodes a complete FecI protein and truncated FecA and FecR proteins. This FecA protein lacks the signaling sequence required for the initiation of *fec* transport gene transcription. The genomes of *Shigella* species (Luck et al. 2001) and *Photorhabdus luminescens* encode complete and functional *fec* transport and regulatory systems. Two

Enterobacter aerogenes strains lack a FecA protein, but synthesize functional FecIR proteins (Luck et al. 2001; Mahren et al. 2005).

Regulatory systems of the *E. coli* Fec type are found in various Gram-negative bacteria. The PupB outer membrane transporter of *Pseudomonas putida* binds the Fe^{3+}-siderophore pseudobactin, which induces transcription of the *pupB* gene via PupI and PupR. These two proteins are homologous to FecI and FecR, respectively. Fe^{3+}-pseudobactin is required for induction in the presence of PupR. Deletion of the *pupR* gene results in constitutive *pupB* transcription by PupI, which indicates that PupR functions as an anti-sigma factor of PupI (Koster et al. 1994).

In *Pseudomonas aeruginosa*, binding of Fe^{3+}-pyoverdine to the outer membrane transporter FpvA induces transcription of the *fpvA* gene and a number of other virulence-related genes. FpvA contains an N-terminal extension, which when removed abolishes induction but not transport (Shen et al. 2002). Transport is not required for induction as certain transport-inactive mutants in the plug and the barrel domains still induce (James et al. 2005). The FpvR protein in the cytoplasmic membrane negatively regulates the activity of two sigma factors: FpvI, which is required for *fpvA* transcription, and PvdS, which is required for the transcription of genes that encode enzymes for pyoverdine synthesis, exotoxin A, and the PrpL endoprotease (Beare et al. 2003). The cytoplasmic N-proximal 67 residues of FpvR bind to region 4 of FpvI, and mutants in these regions are compromised in the interaction of FpvR with FpvI and with PvdS (Redly and Poole 2005). Since the interaction-deficient FpvI and PvdS proteins occur in lower amounts, the wild-type proteins may be stabilized by binding to FpvR. Alanine scanning mutagenesis of PvdS revealed regions 2.1 and 2.2 as important for binding RNA polymerase and regions 2.3, 2.4, and 4.2 as important for directing RNA polymerase to promoter DNA (Wilson and Lamont 2006). At least ten additional outer membrane proteins of *P. aeruginosa* regulate the transcription of genes in response to substrate binding. Synthesis of the receptors FiuA for ferrichrome transport and FoxA for Fe^{3+}-ferrioxamine transport is induced by their cognate Fe^{3+}-siderophores. The regulatory proteins FiuR/FoxR and FiuI/FoxI are encoded by genes adjacent to the *fiuA* and *foxA* genes, respectively (Llamas et al. 2006).

Another well-studied signaling system consists of the HasA hemophore and the HasR receptor involved in heme transport in *Serratia marcescens* (see Sect. 2.1). Transcription of the *hasR* gene is induced by heme-loaded HasA. Heme transfer from HasA to HasR is necessary for induction, as shown by a HasA mutant that binds heme but does not transfer heme to HasR, and a HasR mutant that does not bind heme, both of which are induction incompetent (Cwerman et al. 2006). However, free heme and a HasA mutant that does not bind heme induce *hasR* transcription. Binding of free heme and HasA without heme to HasR are both required for induction. Since the HasA mutant blocks heme transport, transport and induction occur independently.

Bordetella pertussis, *Bordetella bronchiseptica*, and *Bordetella avium* bind heme to outer membrane heme transporters without the involvement of a hemophore. Binding initiates transcription of heme transport systems via the two regulatory proteins HurI and RhuR, which function equivalently to FecI and FecR (Kirby et al. 2004; Vanderpool and Armstrong 2004).

4
Uptake of Iron Without Siderophores

In addition to uptake of iron via siderophores, cells are also able to import ionic iron by several different uptake systems.

4.1
Feo-Mediated Ferrous Iron Transport

One of the first Fe^{2+} transport systems was detected (Hantke 1987a) in *E. coli* and characterized (Kammler et al. 1993). The operon consists of three genes, *feoABC*, which encode a putative regulator of activity (FeoA), a transporter in the membrane (FeoB), and a putative transcriptional regulator (FeoC). The FeoB protein has a unique domain with GTPase activity (Marlovits et al. 2002). This domain is highly similar to the large family of eukaryotic G proteins that transduce signals from outside across the cytoplasmic membrane.

FeoB homologues have been found in roughly half of the 300 completely sequenced bacterial genomes in the databases. However, only in a few cases has FeoB-dependent iron transport been demonstrated. For example, the *feoB* gene of a *Synechocystis* sp. strain was mutated and shown to be responsible for Fe^{2+} uptake (Katoh et al. 2001).

In the pathogen *Helicobacter pylori*, FeoB plays a role in iron supply and virulence (Velayudhan et al. 2000; Waidner et al. 2002). However, microarray studies indicate only very low *feoB* mRNA levels under iron-restricting conditions (Merrell et al. 2003). Similarly, in a quantitative real-time reverse transcriptase PCR assay, only small amounts of the *feoB* transcript were observed (Boonjakuakul et al. 2005). Unfortunately, the discrepancy between the importance of FeoB for virulence and the low *feoB* mRNA levels were not discussed. Fe^{2+} uptake in *H. pylori* is inhibited by FCCP, DCCD, and vanadate (Velayudhan et al. 2000). Since the iron uptake measured was FeoB-dependent, it was concluded that FeoB iron transport is ATP dependent, which contradicts the demonstrated GTPase activity of FeoB (Marlovits et al. 2002). However, since uptake into whole cells was measured, other processes in iron metabolism may be ATP-dependent and hinder uptake of iron.

Interestingly, a FeoB homologue contributes to the high iron supply needed for magnetosome synthesis in *Magnetospirillum magnetotacticum*

(Suzuki et al. 2006). *M. magnetotacticum* lives under microoxic conditions where iron(II) and iron(III) are available.

In *Leptospira biflexa*, a saprophytic relative of pathogenic *Leptospira* strains, a FeoB homologue is important for iron supply, even though these strains are recorded as aerobic bacteria (Louvel et al. 2005).

Porphyromonas gingivalis is one of several bacteria that contain two homologues of FeoB. Mutants generated in the two homologues *feoB* genes and transport experiments have shown that FeoB1 is responsible for iron uptake, whereas FeoB2 is involved in manganese uptake (Dashper et al. 2005). This result again shows that metal transporters could change their metal specificity during evolution and cautions conclusions of functional identity based on protein similarity. An increasing number of examples of proteins handling metal ions are arising that show that only slight modifications can change their metal-binding specificity. In contrast to the *P. gingivalis* FeoB1 protein, the FeoB protein of *Campylobacter jejuni* does not have iron uptake activity (Raphael and Joens 2003). In this case it would be advisable to test the uptake of other divalent cations. However, in an alignment of FeoB proteins (Cartron et al. 2006) the *C. jejuni* FeoB did not group with the manganese-specific FeoB2 of *P. gingivalis*.

In *Listeria monocytogenes*, the *feoAB* genes do not seem to be important for the iron supply of the pathogen. A FeoAB mutant is not attenuated in a mouse model and grows as well as the parent strain in plate assays (Jin et al. 2006). However, a Fe^{2+} uptake experiment is still necessary to show that this transport system contributes to the iron supply of this organism.

In conclusion, those systems in which no clear-cut FeoB-dependent iron uptake has been demonstrated must be experimentally examined before one can be sure that the FeoB homologue is an iron transporter.

4.2
Ferrous Iron and Manganese Uptake by the Sit-ABC Transporter

In *Salmonella enterica*, the binding-protein-dependent ABC transporter SitABCD is an iron-uptake system, as shown by complementation of an enterobactin-deficient strain of *E. coli* (Zhou et al. 1999). The *sitABCD* operon is regulated similarly to the manganese transporter gene *mntH* and SitABCD catalyzes Mn^{2+} uptake—even somewhat better than Fe^{2+} uptake. Kehres et al. (2002) concluded that a major task of this transporter is Mn^{2+} uptake under slightly alkaline conditions because Fe^{2+} in sufficient concentrations is not expected under natural conditions. *sitABCD* is regulated by the iron regulator Fur and by the manganese-dependent regulator MntR. SitABCD is necessary for full virulence of *S. enterica* (Boyer et al. 2002). SitABCD homologues are found in enteroinvasive and avian pathogenic *E. coli* strains (Boyer et al. 2002) and in most *Shigella* strains (Runyen-Janecky et al. 2003). Experiments with an avian pathogenic *E. coli* strain indicate that this transport system also promotes iron-dependent growth (Sabri et al. 2006).

In *Yersinia pestis*, the homologous genes *yfeABCD* are responsible for transport of iron and manganese. Also in this case, the genes contribute to the virulence of the strain (Bearden and Perry 1999).

4.3
ZupT, a Housekeeping Divalent Metal Transporter?

ZupT is a member of the ZIP (ZRT, IRT-like protein) protein family, which was initially characterized as an iron or zinc transporter in eukaryotes. The first observations in *E. coli* demonstrated zinc uptake in a *znu* mutant defective in high-affinity zinc uptake (Grass et al. 2002). Later it was shown that this transporter is astonishingly unspecific in allowing also uptake of Fe^{2+} and possibly also Mn^{2+} and other non-physiological divalent metals (Grass et al. 2005). Since this transporter is neither regulated by zinc nor by iron, it seems to be a housekeeping transporter for the supply of both metals. Its metal affinity seems to be relatively low since a phenotype was only observed when high-affinity divalent metal transporters were mutated.

4.4
Ferric Iron Transport Systems Sfu/Fbp/Hit/Yfu

In contrast to the Sit transport system, the Sfu/Fbp/Hit/Yfu binding-protein-dependent ABC transporter allows the uptake of ferric iron. The SfuABC transporter was detected in *Serratia marcescens* and is able to complement growth of an *E. coli* mutant unable to produce siderophores on an iron-poor medium (Zimmermann et al. 1989; Angerer et al. 1990). No siderophore that can deliver Fe^{3+} to the periplasmic binding protein has been found. The homologous transporter FbpABC from *Neisseria meningitidis* transports the Fe^{3+} released from transferrin or lactoferrin by specialized TonB-dependent outer membrane receptors. The source of Fe^{3+} for the Sfu transport system is still unknown. The situation for *Yersinia enterocolitica* YfuABCD (Saken et al. 2000) and *Yersinia pestis* YfuABC (Gong et al. 2001) is similar. The YfuD protein shows similarity to a lysine exporter of *Corynebacterium* and is not encoded in other homologous operons of this transport system. Further studies are required to demonstrate its participation in iron uptake. Evidence that Fe^{3+} and not Fe^{2+} is transported comes from binding of Fe^{3+} to isolated periplasmic binding proteins, crystal structures with bound Fe^{3+}, and transfer of Fe^{3+} from transferrin to binding proteins (Anderson et al. 2004; Bruns et al. 1997; Krewulak et al. 2005; Mietzner et al. 1998).

5
Three Types of General Iron Regulators: Fur, DtxR, and RirA

Many bacteria have several iron-uptake systems, which help the cells to adapt to changing environments. Since the uptake of too much iron can be toxic and the production of futile proteins and siderophores is uneconomical, their synthesis is tightly regulated. Interestingly, three general iron regulators have been described in different bacterial species. The iron regulator Fur was first characterized in *E. coli* (Hantke 2001) and was later found in many Gram-negative bacteria and in some Gram-positive bacteria. DtxR was described as the regulator of the diphtheria toxin, which is produced under low-iron conditions by the pathogen *Corynebacterium diphtheriae* (Boyd et al. 1990). This regulator was later found to be a general iron regulator in many Gram-positive bacteria. Recently, it was found that in certain *Rhizobia*, neither Fur-like nor DtxR-like repressors regulate iron-uptake genes. The repressor identified is called RirA (Todd et al. 2002) and is distantly related to a regulator of FeS cluster synthesis.

5.1
The General Iron Regulator Fur

A crystal structure of the Fur protein of *Pseudomonas aeruginosa* has been solved (Pohl et al. 2003) and many features of this protein have been reviewed (Rudolph et al. 2006; Hantke 2001). Therefore, only some major characters of Fur and some recent developments regarding its role as a global regulator will be discussed.

The Fur protein is a repressor with a DNA-binding helix-turn-helix motif in the N-terminal domain. The C-terminal domain contains two metal-binding sites, a regulatory Fe^{2+}-binding site, and a structurally important tetrahedral zinc-binding site. One zinc coordination site, a histidine in the N-terminal domain, helps to link the two domains. Recent structural studies with *E. coli* Fur showed that zinc binding promotes dimerization of the protein (Pecqueur et al. 2006). In addition, this zinc-binding site is clearly different from the zinc-binding site observed in *P. aeruginosa* Fur. There are additional data that indicate that neither the zinc-binding site nor the presence of zinc is a general feature of all Fur-like proteins (Lee and Helmann 2006). Metal binding in the C-terminal domain of Fur induces the formation of an additional helix near the N-terminus of the protein (Pecqueur et al. 2006). A major task of this global regulator in many different bacteria is the regulation of iron uptake and iron homeostasis, which is often achieved by direct binding of Fe-Fur dimers to promoters of genes responsible for iron uptake. The binding site, the 19-bp palindromic Fur box, is very similar in different bacteria. The crystal structure and other data indicate that two Fur dimers most likely bind to two sites of the Fur box (Baichoo and Helmann 2002; de Lorenzo et al. 2004).

Fur in certain bacteria may also influence the expression of genes for the oxidative stress response, the acid stress response, pathogenicity, motility, and biofilm formation. In many of the latter cases, the regulation is indirect via regulation of more specialized regulators. These regulators may belong to the ECF sigma factors (see Sect. 3). In addition, there are specific AraC-like regulators for the uptake of pyochelin and alcaligin (Beaumont et al. 1998; Heinrichs and Poole 1996), two-component regulatory systems, e.g., PfeRS for enterobactin uptake in *Pseudomonas* (Dean et al. 1996), and small RNA antisense regulators, e.g., RhyB. A LysR-type regulator together with Fur regulates the expression of the enterobactin receptor IrgA in *Vibrio cholerae* (Watnick et al. 1998).

It is interesting to note that the Fur protein family consists not only of iron regulators, but also of a group of homologues that regulate zinc-uptake systems in a zinc-dependent manner, Zur (Hantke 2001).

The Fur protein acts as a repressor in vitro (de Lorenzo et al. 1987) and in certain cases also in vivo with Mn^{2+} as a corepressor (Hantke 1987b), possibly because of the similarities between Mn^{2+} and Fe^{2+}. Therefore, it is not surprising that the Fur-like protein Mur regulates a manganese-uptake system with manganese as corepressor (Diaz-Mireles et al. 2005). In this strain of *Rhizobium*, iron uptake is regulated by RirA.

5.1.1
The Irr Protein, a Member of the Fur Family, Responds to Heme

The Irr protein, first characterized in *Bradyrhizobium japonicum*, is a member of the Fur protein family with very peculiar properties. Under low-iron conditions, Irr inhibits heme synthesis. When there is enough iron for the ferrochelatase, heme is synthesized. The Irr protein binds heme and is rapidly degraded in the cell (Qi et al. 1999). The status of heme biosynthesis seems to be the regulatory signal for recognizing the iron status of the cell. Recent studies have shown that Irr also has both direct activating or direct repressing activities on genes involved in iron homeostasis (Yang et al. 2006). In contrast to many other Fur-like proteins, Irr activates certain iron-uptake genes, which leads to a low iron level in *irr* mutants, whereas a mutant unable to synthesize heme has a high internal iron level owing to the stability of the Irr protein (Yang et al. 2006).

5.1.2
RhyB – Another Level of Regulation via RNA

Characterization of *E. coli* Fur mutants revealed as early as 1987 that these strains are unable to grow on acetate and succinate as sole carbon source

(Hantke 1987b). Genes positively regulated by Fur and iron are indirectly regulated by a small RNA (Masse and Gottesman 2002). The responsible antisense RNA, called RyhB in *E. coli*, is repressed by iron-Fur. Under iron-limiting conditions, the 90-nt RyhB RNA is produced and bound to the protein Hfq, which catalyzes the binding of the antisense RNA to certain complementary regions of certain mRNAs (Geissmann and Touati 2004). The RNAs are degraded by RNaseE or RNase III (Afonyushkin et al. 2005). The genes regulated by this mechanism often encode iron-containing proteins, e.g., *sdhCDAB* (succinate dehydrogenase), *sodB* (Fe-superoxide dismutase) and iron-storage proteins. A homologue of RyhB, twice the size of RyhB, has been found in *Vibrio cholerae* (Mey et al. 2005). This RNA has additional functions compared to those of RyhB in *E. coli*, including the regulation of genes for motility, chemotaxis, and biofilm formation. In this context, it is interesting to note that in contrast to *E. coli*, *V. cholerae* has several chemotaxis genes with a predicted Fur box, which indicates direct regulation by Fur.

A more specialized regulation by an antisense RNA has been observed in *Vibrio anguillarum*. RNAα regulates together with Fe-Fur the synthesis of the siderophore anguibactin (Crosa 1997).

5.2
DtxR, the Iron Regulator of High-GC Gram-Positive Bacteria

DtxR was detected as the chromosomally encoded iron-dependent repressor of diphtheria toxin production from a corynebacteriophage (Boyd et al. 1990). Soon afterwards, DtxR-like regulators were found in many Gram-positive bacteria and more general functions in iron regulation were identified. These proteins have many functional similarities with Fur. Iron uptake and iron homeostasis genes are regulated directly by repression, and oxidative stress genes are regulated indirectly (Rodriguez et al. 2002). Also, the overall structure and the DNA binding of DtxR (Pohl et al. 1999) and Fur (Baichoo and Helmann 2002; Pohl et al. 2003) show similarities, which were not expected from the highly divergent amino acid sequences.

A distantly related homologue of DtxR is MntR, the manganese-dependent repressor of manganese transport systems in *E. coli* and *S. enterica* (Patzer and Hantke 2001).

5.3
RirA, a Global Iron Regulator Found in Some Rhizobia

The iron regulation of several iron-uptake systems of Rhizobia proved to be enigmatic because *fur* mutants did not show derepression of iron-regulated genes (Wexler et al. 2003) even though the cloned rhizobial Fur protein repressed certain iron-regulated genes in *E. coli*. This Fur protein was then shown to regulate the manganese transporter SitABCD and was consequently renamed Mur (Diaz-Mireles et al. 2005). Further studies led to the identification of another global iron regulator, RirA (Todd et al. 2002). This protein is distantly related (ca. 30% identity) to IscR, which regulates genes responsible for Fe – S cluster biosynthesis in *E. coli*. IscR is a repressor that is only active when loaded with an Fe-S cluster. Three well-conserved cysteines in the C-terminus of RirA may also bind an Fe – S cluster. The regulatory signal might be similar to the eukaryotic aconitase-like IRP iron regulators. With these latter regulators, the presence or absence of the Fe – S cluster is the signal for the iron level in the cell, which determines the regulatory activity of the protein. However, in *Rhizobium leguminosarum*, this regulation seems to be complicated by the second global regulator Irr (see Sect. 5.1.1).

Acknowledgements We thank Karen A. Brune (Marburg) for editing the manuscript and Avijit Pramanik for drawings. The work of the authors was supported by the Deutsche Forschungsgemeinschaft and the Fonds der Chemischen Industrie.

References

Afonyushkin T, Vecerek B, Moll I, Blasi U, Kaberdin VR (2005) Both RNase E and RNase III control the stability of *sodB* mRNA upon translational inhibition by the small regulatory RNA RyhB. Nucleic Acids Res 33:1678–1689

Anderson DS, Adhikari P, Nowalk AJ, Chen CY, Mietzner TA (2004) The hFbpABC transporter from *Haemophilus influenzae* functions as a binding-protein-dependent ABC transporter with high specificity and affinity for ferric iron. J Bacteriol 186:6220–6229

Andrews SC, Robinson AK, Rodriguez-Quinones F (2003) Bacterial iron homeostasis. FEMS Microbiol Rev 27:215–237

Angerer A, Braun V (1998) Iron regulates transcription of the *Escherichia coli* ferric citrate transport genes directly and through the transcription initiation proteins. Arch Microbiol 169:483–490

Angerer A, Enz S, Ochs M, Braun V (1995) Transcriptional regulation of ferric citrate transport in *Escherichia coli* K-12. FecI belongs to a new subfamily of sigma 70-type factors that respond to extracytoplasmic stimuli. Mol Microbiol 18:163–174

Angerer A, Gaisser S, Braun V (1990) Nucleotide sequences of the sfuA, sfuB, and sfuC genes of *Serratia marcescens* suggest a periplasmic-binding-protein-dependent iron transport mechanism. J Bacteriol 172:572–578

Baichoo N, Helmann JD (2002) Recognition of DNA by Fur: a reinterpretation of the Fur box consensus sequence. J Bacteriol 184:5826–5832

Bearden SW, Perry RD (1999) The Yfe system of *Yersinia pestis* transports iron and manganese and is required for full virulence of plague. Mol Microbiol 32:403–414

Beare PA, For RJ, Martin LW, Lamont IL (2003) Siderophore-mediated cell signalling in *Pseudomonas aeruginosa*: divergent pathways regulate virulence factor production and siderophore receptor synthesis. Mol Microbiol 47:195–207

Beaumont FC, Kang HY, Brickman TJ, Armstrong SK (1998) Identification and characterization of *alcR*, a gene encoding an AraC-like regulator of alcaligin siderophore biosynthesis and transport in *Bordetella pertussis* and *Bordetella bronchiseptica*. J Bacteriol 180:862–870

Bister B, Bischoff D, Nicholson GJ, Valdebenito M, Schneider K, Winkelmann G, Hantke K, Süssmuth RD (2004) The structure of salmochelins: C-glucosylated enterobactins of *Salmonella enterica*. Biometals 17:471–481

Boonjakuakul JK, Canfield DR, Solnick JV (2005) Comparison of *Helicobacter pylori* virulence gene expression in vitro and in the Rhesus macaque. Infect Immun 73:4895–4904

Borths EL, Locher KP, Lee AT, Rees DC (2002) The structure of *Escherichia coli* BtuF and binding to its cognate ATP binding cassette transporter. Proc Natl Acad Sci USA 99:16642–16647

Boyd J, Oza MN, Murphy JR (1990) Molecular cloning and DNA sequence analysis of a diphtheria tox iron-dependent regulatory element (dtxR) from *Corynebacterium diphtheriae*. Proc Natl Acad Sci USA 87:5968–5972

Boyer E, Bergevin I, Malo D, Gros P, Cellier MF (2002) Acquisition of Mn(II) in addition to Fe(II) is required for full virulence of *Salmonella enterica* serovar Typhimurium. Infect Immun 70:6032–6042

Braun M, Endriss F, Killmann H, Braun V (2003) In vivo reconstitution of the FhuA transport protein of *Escherichia coli* K-12. J Bacteriol 185:5508–5518

Braun V, Hantke K, Köster W (1998) Bacterial iron transport: mechanisms, genetics, and regulation. Met Ions Biol Syst 35:67–145

Braun V, Mahren S (2005) Transmembrane transcriptional control (surface signalling) of the *Escherichia coli* Fec type. FEMS Microbiol Rev 29:673–684

Breidenstein E, Mahren S, Braun V (2006) Residues involved in FecR binding are localized on one side of the FecA signaling domain in *Escherichia coli*. J Bacteriol 188:6440–6442

Bruns CM, Nowalk AJ, Arvai AS, McTigue MA, Vaughan KG, Mietzner TA, McRee DE (1997) Structure of *Haemophilus influenzae* Fe(+3)-binding protein reveals convergent evolution within a superfamily. Nat Struct Biol 4:919–924

Buchanan SK, Smith BS, Venkatramani L, Xia D, Esser L, Palnitkar M, Chakraborty R, van der HD, Deisenhofer J (1999) Crystal structure of the outer membrane active transporter FepA from *Escherichia coli*. Nat Struct Biol 6:56–63

Cartron ML, Maddocks S, Gillingham P, Craven CJ, Andrews SC (2006) Feo – transport of ferrous iron into bacteria. Biometals 19:143–157

Chimento DP, Kadner RJ, Wiener MC (2003) The *Escherichia coli* outer membrane cobalamin transporter BtuB: structural analysis of calcium and substrate binding, and identification of orthologous transporters by sequence/structure conservation. J Mol Biol 332:999–1014

Clancy A, Loar JW, Speziali CD, Oberg M, Heinrichs DE, Rubens CE (2006) Evidence for siderophore-dependent iron acquisition in group B *Streptococcus*. Mol Microbiol 59:707–721

Clarke TE, Rohrbach MR, Tari LW, Vogel HJ, Koster W (2002) Ferric hydroxamate binding protein FhuD from *Escherichia coli*: mutants in conserved and non-conserved regions. Biometals 15:121–131

Cobessi D, Celia H, Folschweiller N, Schalk IJ, Abdallah MA, Pattus F (2005a) The crystal structure of the pyoverdine outer membrane receptor FpvA from *Pseudomonas aeruginosa* at 3.6 angstroms resolution. J Mol Biol 347:121–134

Cobessi D, Celia H, Pattus F (2005b) Crystal structure at high resolution of ferric-pyochelin and its membrane receptor FptA from *Pseudomonas aeruginosa*. J Mol Biol 352:893–904

Cornelissen CN (2003) Transferrin-iron uptake by Gram-negative bacteria. Front Biosci 8:d836–d847

Crosa JH (1997) Signal transduction and transcriptional and posttranscriptional control of iron-regulated genes in bacteria. Microbiol Mol Biol Rev 61:319–336

Cwerman H, Wandersman C, Biville F (2006) Heme and a five-amino-acid hemophore region form the bipartite stimulus triggering the has signaling cascade. J Bacteriol 188:3357–3364

Dashper SG, Butler CA, Lissel JP, Paolini RA, Hoffmann B, Veith PD, O'Brien-Simpson NM, Snelgrove SL, Tsiros JT, Reynolds EC (2005) A novel *Porphyromonas gingivalis* FeoB plays a role in manganese accumulation. J Biol Chem 280:28095–28102

de Lorenzo V, Wee S, Herrero M, Neilands JB (1987) Operator sequences of the aerobactin operon of plasmid ColV-K30 binding the ferric uptake regulation (fur) repressor. J Bacteriol 169:2624–2630

de Lorenzo V, Prez-Martin J, Escolar L, Pesole G, Bertoni G (2004) Mode of binding of the Fur protein to target DNA: negative regulation of iron-controlled gene expression. In: Crosa JH, Mey AR, Payne SM (eds) Iron transport in bacteria. ASM Press, Washington, DC, pp 185–196

Dean CR, Neshat S, Poole K (1996) PfeR, an enterobactin-responsive activator of ferric enterobactin receptor gene expression in *Pseudomonas aeruginosa*. J Bacteriol 178:5361–5369

Dhungana S, Anderson DS, Mietzner TA, Crumbliss AL (2005) Kinetics of iron release from ferric binding protein (FbpA): mechanistic implications in bacterial periplasm-to-cytosol Fe^{3+} transport. Biochemistry 44:9606–9618

Diaz-Mireles E, Wexler M, Todd JD, Bellini D, Johnston AW, Sawers RG (2005) The manganese-responsive repressor Mur of *Rhizobium leguminosarum* is a member of the Fur-superfamily that recognizes an unusual operator sequence. Microbiology 151:4071–4078

Endriss F, Braun V (2004) Loop deletions indicate regions important for FhuA transport and receptor functions in *Escherichia coli*. J Bacteriol 186:4818–4823

Enz S, Brand H, Orellana C, Mahren S, Braun V (2003a) Sites of interaction between the FecA and FecR signal transduction proteins of ferric citrate transport in *Escherichia coli* K-12. J Bacteriol 185:3745–3752

Enz S, Braun V, Crosa JH (1995) Transcription of the region encoding the ferric dicitrate-transport system in *Escherichia coli*: similarity between promoters for *fecA* and for extracytoplasmic function sigma factors. Gene 163:13–18

Enz S, Mahren S, Menzel C, Braun V (2003b) Analysis of the ferric citrate transport gene promoter of *Escherichia coli*. J Bacteriol 185:2387–2391

Enz S, Mahren S, Stroeher UH, Braun V (2000) Surface signaling in ferric citrate transport gene induction: interaction of the FecA, FecR, and FecI regulatory proteins. J Bacteriol 182:637–646

Faraldo-Gomez JD, Sansom MS (2003) Acquisition of siderophores in Gram-negative bacteria. Nat Rev Mol Cell Biol 4:105–116

Ferguson AD, Chakraborty R, Smith BS, Esser L, van der HD, Deisenhofer J (2002) Structural basis of gating by the outer membrane transporter FecA. Science 295:1715–1719

Ferguson AD, Deisenhofer J (2004) Metal import through microbial membranes. Cell 116:15–24

Ferguson AD, Hofmann E, Coulton JW, Diederichs K, Welte W (1998) Siderophore-mediated iron transport: crystal structure of FhuA with bound lipopolysaccharide. Science 282:2215–2220

Garcia-Herrero A, Vogel HJ (2005) Nuclear magnetic resonance solution structure of the periplasmic signalling domain of the TonB-dependent outer membrane transporter FecA from *Escherichia coli*. Mol Microbiol 58:1226–1237

Geissmann TA, Touati D (2004) Hfq, a new chaperoning role: binding to messenger RNA determines access for small RNA regulator. EMBO J 23:396–405

Gong S, Bearden SW, Geoffroy VA, Fetherston JD, Perry RD (2001) Characterization of the *Yersinia pestis* Yfu ABC inorganic iron transport system. Infect Immun 69:2829–2837
Grass G, Franke S, Taudte N, Nies DH, Kucharski LM, Maguire ME, Rensing C (2005) The metal permease ZupT from *Escherichia coli* is a transporter with a broad substrate spectrum. J Bacteriol 187:1604–1611
Grass G, Wong MD, Rosen BP, Smith RL, Rensing C (2002) ZupT is a Zn(II) uptake system in *Escherichia coli*. J Bacteriol 184:864–866
Hantke K (1987a) Ferrous iron transport mutants in *Escherichia coli* K-12. FEMS Microbiol Lett 44:53–57
Hantke K (1987b) Selection procedure for deregulated iron transport mutants (*fur*) in *Escherichia coli* K-12: fur not only affects iron metabolism. Mol Gen Genet 210:135–139
Hantke K (2001) Iron and metal regulation in bacteria. Curr Opin Microbiol 4:172–177
Heinrichs DE, Poole K (1996) PchR, a regulator of ferripyochelin receptor gene (fptA) expression in *Pseudomonas aeruginosa*, functions both as an activator and as a repressor. J Bacteriol 178:2586–2592
Hoegy F, Celia H, Mislin GL, Vincent M, Gallay J, Schalk IJ (2005) Binding of iron-free siderophore, a common feature of siderophore outer membrane transporters of *Escherichia coli* and *Pseudomonas aeruginosa*. J Biol Chem 280:20222–20230
James HE, Beare PA, Martin LW, Lamont IL (2005) Mutational analysis of a bifunctional ferrisiderophore receptor and signal-transducing protein from *Pseudomonas aeruginosa*. J Bacteriol 187:4514–4520
Jin B, Newton SM, Shao Y, Jiang X, Charbit A, Klebba PE (2006) Iron acquisition systems for ferric hydroxamates, haemin and haemoglobin in *Listeria monocytogenes*. Mol Microbiol 59:1185–1198
Kammler M, Schön C, Hantke K (1993) Characterization of the ferrous iron uptake system of *Escherichia coli*. J Bacteriol 175:6212–6219
Katoh H, Hagino N, Grossman AR, Ogawa T (2001) Genes essential to iron transport in the cyanobacterium *Synechocystis* sp. strain PCC 6803. J Bacteriol 183:2779–2784
Kehres DG, Janakiraman A, Slauch JM, Maguire ME (2002) SitABCD is the alkaline Mn(2+) transporter of *Salmonella enterica* serovar Typhimurium. J Bacteriol 184:3159–3166
Kempf B, Gade J, Bremer E (1997) Lipoprotein from the osmoregulated ABC transport system OpuA of *Bacillus subtilis*: purification of the glycine betaine binding protein and characterization of a functional lipidless mutant. J Bacteriol 179:6213–6220
Kim I, Stiefel A, Plantor S, Angerer A, Braun V (1997) Transcription induction of the ferric citrate transport genes via the N-terminus of the FecA outer membrane protein, the Ton system and the electrochemical potential of the cytoplasmic membrane. Mol Microbiol 23:333–344
Kirby AE, King ND, Connell TD (2004) RhuR, an extracytoplasmic function sigma factor activator, is essential for heme-dependent expression of the outer membrane heme and hemoprotein receptor of *Bordetella avium*. Infect Immun 72:896–907
Klebba PE (2003) Three paradoxes of ferric enterobactin uptake. Front Biosci 8:s1422–s1436
Ködding J, Killig F, Polzer P, Howard SP, Diedrichs K, Welte W (2005) Crystal structure of a 92-residue C-terminal fragment of TonB from Escherichia coli reveals significant conformational changes compared to structures of smaller TonB fragments. J Biol Chem 280:3022–3028
Koster M, van Klompenburg W, Bitter W, Leong J, Weisbeek P (1994) Role for the outer membrane ferric siderophore receptor PupB in signal transduction across the bacterial cell envelope. EMBO J 13:2805–2813

Köster W (2001) ABC transporter-mediated uptake of iron, siderophores, heme and vitamin B12. Res Microbiol 152:291–301

Krewulak KD, Shepherd CM, Vogel HJ (2005) Molecular dynamics simulations of the periplasmic ferric-hydroxamate binding protein FhuD. Biometals 18:375–386

Letoffe S, Delepelaire P, Wandersman C (2004) Free and hemophore-bound heme acquisitions through the outer membrane receptor HasR have different requirements for the TonB-ExbB-ExbD complex. J Bacteriol 186:4067–4074

Letoffe S, Wecker K, Delepierre M, Delepelaire P, Wandersman C (2005) Activities of the *Serratia marcescens* heme receptor HasR and isolated plug and beta-barrel domains: the beta-barrel forms a heme-specific channel. J Bacteriol 187:4637–4645

Leveille S, Caza M, Johnson JR, Clabots C, Sabri M, Dozois CM (2006) Iha from an *Escherichia coli* urinary tract infection outbreak clonal group A strain is expressed in vivo in the mouse urinary tract and functions as a catecholate siderophore receptor. Infect Immun 74:3427–3436

Liu M, Lei B (2005) Heme transfer from streptococcal cell surface protein Shp to HtsA of transporter HtsABC. Infect Immun 73:5086–5092

Llamas MA, Sparrius M, Kloet R, Jimenez CR, Vandenbroucke-Grauls C, Bitter W (2006) The heterologous siderophores ferrioxamine B and ferrichrome activate signaling pathways in *Pseudomonas aeruginosa*. J Bacteriol 188:1882–1891

Locher KP, Lee AT, Rees DC (2002) The *E. coli* BtuCD structure: a framework for ABC transporter architecture and mechanism. Science 296:1091–1098

Locher KP, Rees B, Koebnik R, Mitschler A, Moulinier L, Rosenbusch JP, Moras D (1998) Transmembrane signaling across the ligand-gated FhuA receptor: crystal structures of free and ferrichrome-bound states reveal allosteric changes. Cell 95:771–778

Louvel H, Saint GI, Picardeau M (2005) Isolation and characterization of FecA- and FeoB-mediated iron acquisition systems of the spirochete *Leptospira biflexa* by random insertional mutagenesis. J Bacteriol 187:3249–3254

Lu G, Westbrooks JM, Davidson AL, Chen J (2005) ATP hydrolysis is required to reset the ATP-binding cassette dimer into the resting-state conformation. Proc Natl Acad Sci USA 102:17969–17974

Luck SN, Turner SA, Rajakumar K, Sakellaris H, Adler B (2001) Ferric dicitrate transport system (Fec) of *Shigella flexneri*2a YSH6000 is encoded on a novel pathogenicity island carrying multiple antibiotic resistance genes. Infect Immun 69:6012–6021

Mahren S, Schnell H, Braun V (2005) Occurrence and regulation of the ferric citrate transport system in *Escherichia coli* B, *Klebsiella pneumoniae*, *Enterobacter aerogenes*, and *Photorhabdus luminescens*. Arch Microbiol 184:175–186

Marlovits TC, Haase W, Herrmann C, Aller SG, Unger VM (2002) The membrane protein FeoB contains an intramolecular G protein essential for Fe(II) uptake in bacteria. Proc Natl Acad Sci USA 99:16243–16248

Marraffini LA, Dedent AC, Schneewind O (2006) Sortases and the art of anchoring proteins to the envelopes of Gram-positive bacteria. Microbiol Mol Biol Rev 70:192–221

Masse E, Gottesman S (2002) A small RNA regulates the expression of genes involved in iron metabolism in *Escherichia coli*. Proc Natl Acad Sci USA 99:4620–4625

Merrell DS, Thompson LJ, Kim CC, Mitchell H, Tompkins LS, Lee A, Falkow S (2003) Growth phase-dependent response of *Helicobacter pylori* to iron starvation. Infect Immun 71:6510–6525

Mey AR, Wyckoff EE, Kanukurthy V, Fisher CR, Payne SM (2005) Iron and fur regulation in *Vibrio cholerae* and the role of *fur* in virulence. Infect Immun 73:8167–8178

Mietzner TA, Tencza SB, Adhikari P, Vaughan KG, Nowalk AJ (1998) Fe(III) periplasm-to-cytosol transporters of Gram-negative pathogens. Curr Top Microbiol Immunol 225:113–135

Oakhill JS, Sutton BJ, Gorringe AR, Evans RW (2005) Homology modelling of transferrin-binding protein A from *Neisseria meningitidis*. Protein Eng Des Sel 18:221–228

Page RDM (2005) Inferring evolutionary relationships. In: Baxevanis AD, Davison DB, Paghe RDM, Petsko GA, Stein LD, Stormo GD (eds) Current protocols in bioinformatics, Chap. 6. Wiley, New York

Patzer SI, Hantke K (2001) Dual repression by Fe(2+)-Fur and Mn(2+)-MntR of the mntH gene, encoding an NRAMP-like Mn(2+) transporter in *Escherichia coli*. J Bacteriol 183:4806–4813

Pawelek PD, Croteau N, Ng-Thow-Hing C, Khursigara CM, Moiseeva N, Allaire M, Coulton JW (2006) Structure of TonB in complex with FhuA, *E. coli* outer membrane receptor. Science 312:1399–1402

Peacock RS, Weljie AM, Peter Howard S, Price FD, Vogel HJ (2005) The solution structure of the C-terminal domain of TonB and interaction studies with TonB box peptides. J Mol Biol 345:1185–1197

Pecqueur L, D'Autreaux B, Dupuy J, Nicolet Y, Jacquamet L, Brutscher B, Michaud-Soret I, Bersch B (2006) Structural changes of *E. coli* Fur during metal dependent dimerization and activation explored by NMR and X-ray crystallography. J Biol Chem 281:21286–21295

Perkins-Balding D, Ratliff-Griffin M, Stojiljkovic I (2004) Iron transport systems in *Neisseria meningitidis*. Microbiol Mol Biol Rev 68:154–171

Pohl E, Haller JC, Mijovilovich A, Meyer-Klaucke W, Garman E, Vasil ML (2003) Architecture of a protein central to iron homeostasis: crystal structure and spectroscopic analysis of the ferric uptake regulator. Mol Microbiol 47:903–915

Pohl E, Holmes RK, Hol WG (1999) Crystal structure of a cobalt-activated diphtheria toxin repressor-DNA complex reveals a metal-binding SH3-like domain. J Mol Biol 292:653–667

Postle K, Kadner RJ (2003) Touch and go: tying TonB to transport. Mol Microbiol 49:869–882

Prakash S, Cooper G, Singhi S, Saier MH Jr (2003) The ion transporter superfamily. Biochim Biophys Acta 1618:79–92

Pramanik A, Braun V (2006) Albomycin uptake via a ferric hydroxamate transport system of *Streptococcus pneumoniae* R6. J Bacteriol 188:3878–3886

Qi Z, Hamza I, O'Brian MR (1999) Heme is an effector molecule for iron-dependent degradation of the bacterial iron response regulator (Irr) protein. Proc Natl Acad Sci USA 96:13056–13061

Raphael BH, Joens LA (2003) FeoB is not required for ferrous iron uptake in *Campylobacter jejuni*. Can J Microbiol 49:727–731

Raymond KN, Dertz EA, Kim SS (2003) Enterobactin: an archetype for microbial iron transport. Proc Natl Acad Sci USA 100:3584–3588

Redly GA, Poole K (2005) FpvIR control of *fpvA* ferric pyoverdine receptor gene expression in *Pseudomonas aeruginosa*: demonstration of an interaction between FpvI and FpvR and identification of mutations in each compromising this interaction. J Bacteriol 187:5648–5657

Rodriguez GM, Voskuil MI, Gold B, Schoolnik GK, Smith I (2002) *ideR*, an essential gene in *Mycobacterium tuberculosis*: role of IdeR in iron-dependent gene expression, iron metabolism, and oxidative stress response. Infect Immun 70:3371–3381

Rohde KH, Dyer DW (2003) Mechanisms of iron acquisition by the human pathogens *Neisseria meningitidis* and *Neisseria gonorrhoeae*. Front Biosci 8:d1186–d1218

Rohrbach MR, Braun V, Köster W (1995) Ferrichrome transport in *Escherichia coli* K-12: altered substrate specificity of mutated periplasmic FhuD and interaction of FhuD with the integral membrane protein FhuB. J Bacteriol 177:7186–7193

Rudolph G, Hennecke H, Fischer HM (2006) Beyond the Fur paradigm: iron-controlled gene expression in rhizobia. FEMS Microbiol Rev 30:631–648

Runyen-Janecky LJ, Reeves SA, Gonzales EG, Payne SM (2003) Contribution of the *Shigella flexneri* Sit, Iuc, and Feo iron acquisition systems to iron acquisition in vitro and in cultured cells. Infect Immun 71:1919–1928

Sabri M, Leveille S, Dozois CM (2006) A SitABCD homologue from an avian pathogenic *Escherichia coli* strain mediates transport of iron and manganese and resistance to hydrogen peroxide. Microbiology 152:745–758

Saken E, Rakin A, Heesemann J (2000) Molecular characterization of a novel siderophore-independent iron transport system in *Yersinia*. Int J Med Microbiol 290:51–60

Sauter A, Braun V (2004) Defined inactive FecA derivatives mutated in functional domains of the outer membrane transport and signaling protein of *Escherichia coli* K-12. J Bacteriol 186:5303–5310

Schalk IJ, Yue WW, Buchanan SK (2004) Recognition of iron-free siderophores by TonB-dependent iron transporters. Mol Microbiol 54:14–22

Schneider R, Hantke K (1993) Iron-hydroxamate uptake systems in *Bacillus subtilis*: identification of a lipoprotein as a part of a binding protein-dependent transport system. Mol Microbiol 8:111–121

Scott DC, Cao Z, Qi Z, Bauler M, Igo JD, Newton SM, Klebba PE (2001) Exchangeability of N-termini in the ligand-gated porins of *Escherichia coli*. J Biol Chem 276:13025–13033

Scott DC, Newton SM, Klebba PE (2002) Surface loop motion in FepA. J Bacteriol 184:4906–4911

Sebulsky MT, Speziali CD, Shilton BH, Edgell DR, Heinrichs DE (2004) FhuD1, a ferric hydroxamate-binding lipoprotein in *Staphylococcus aureus*: a case of gene duplication and lateral transfer. J Biol Chem 279:53152–53159

Shen J, Meldrum A, Poole K (2002) FpvA receptor involvement in pyoverdine biosynthesis in *Pseudomonas aeruginosa*. J Bacteriol 184:3268–3275

Shouldice SR, Dougan DR, Skene RJ, Tari LW, McRee DE, Yu RH, Schryvers AB (2003) High-resolution structure of an alternate form of the ferric ion-binding protein from *Haemophilus influenzae*. J Biol Chem 278:11513–11519

Shultis DD, Purdy MD, Banchs CN, Wiener MC (2006) Outer membrane active transport: structure of the BtuB:TonB complex. Science 312:1396–1399

Stiefel A, Mahren S, Ochs M, Schindler PT, Enz S, Braun V (2001) Control of the ferric citrate transport system of *Escherichia coli*: mutations in region 2.1 of the FecI extracytoplasmic-function sigma factor suppress mutations in the FecR transmembrane regulatory protein. J Bacteriol 183:162–170

Suzuki T, Okamura Y, Calugay RJ, Takeyama H, Matsunaga T (2006) Global gene expression analysis of iron-inducible genes in *Magnetospirillum magneticum* AMB-1. J Bacteriol 188:2275–2279

Tarr PI, Bilge SS, Vary JC Jr., Jelacic S, Habeeb RL, Ward TR, Baylor MR, Besser TE (2000) Iha: a novel *Escherichia coli* O157: H7 adherence-conferring molecule encoded on a recently acquired chromosomal island of conserved structure. Infect Immun 68:1400–1407

Todd JD, Wexler M, Sawers G, Yeoman KH, Poole PS, Johnston AW (2002) RirA, an iron-responsive regulator in the symbiotic bacterium *Rhizobium leguminosarum*. Microbiology 148:4059–4071

Van Gelder P, Dumas F, Bartoldus I, Saint N, Prilipov A, Winterhalter M, Wang Y, Philippsen A, Rosenbusch JP, Schirmer T (2002) Sugar transport through maltoporin of *Escherichia coli*: role of the greasy slide. J Bacteriol 184:2994–2999

Vanderpool CK, Armstrong SK (2004) Integration of environmental signals controls expression of *Bordetella* heme utilization genes. J Bacteriol 186:938–948

Velayudhan J, Hughes NJ, McColm AA, Bagshaw J, Clayton CL, Andrews SC, Kelly DJ (2000) Iron acquisition and virulence in *Helicobacter pylori*: a major role for FeoB, a high-affinity ferrous iron transporter. Mol Microbiol 37:274–286

Waidner B, Greiner S, Odenbreit S, Kavermann H, Velayudhan J, Stahler F, Guhl J, Bisse E, van Vliet AH, Andrews SC, Kusters JG, Kelly DJ, Haas R, Kist M, Bereswill S (2002) Essential role of ferritin Pfr in *Helicobacter pylori* iron metabolism and gastric colonization. Infect Immun 70:3923–3929

Wandersman C, Delepelaire P (2004) Bacterial iron sources: from siderophores to hemophores. Annu Rev Microbiol 58:611–647

Watnick PI, Butterton JR, Calderwood SB (1998) The interaction of the *Vibrio cholerae* transcription factors, Fur and IrgB, with the overlapping promoters of two virulence genes, *irgA* and *irgB*. Gene 209:65–70

Wexler M, Todd JD, Kolade O, Bellini D, Hemmings AM, Sawers G, Johnston AW (2003) Fur is not the global regulator of iron uptake genes in *Rhizobium leguminosarum*. Microbiology 149:1357–1365

Wilson MJ, Lamont IL (2006) Mutational analysis of an extracytoplasmic-function sigma factor to investigate its interactions with RNA polymerase and DNA. J Bacteriol 188:1935–1942

Winkelmann et al. (1991) CRC handbook of microbial iron chelates. CRC Press, Boca Raton, FL

Wyckoff EE, Lopreato GF, Tipton KA, Payne SM (2005) *Shigella dysenteriae* ShuS promotes utilization of heme as an iron source and protects against heme toxicity. J Bacteriol 187:5658–5664

Wyckoff EE, Schmitt M, Wilks A, Payne SM (2004) HutZ is required for efficient heme utilization in *Vibrio cholerae*. J Bacteriol 186:4142–4151

Yang J, Sangwan I, Lindemann A, Hauser F, Hennecke H, Fischer HM, O'Brian MR (2006) *Bradyrhizobium japonicum* senses iron through the status of haem to regulate iron homeostasis and metabolism. Mol Microbiol 60:427–437

Yue WW, Grizot S, Buchanan SK (2003) Structural evidence for iron-free citrate and ferric citrate binding to the TonB-dependent outer membrane transporter FecA. J Mol Biol 332:353–368

Zhou D, Hardt WD, Galan JE (1999) *Salmonella typhimurium* encodes a putative iron transport system within the centisome 63 pathogenicity island. Infect Immun 67:1974–1981

Zimmermann L, Angerer A, Braun V (1989) Mechanistically novel iron(III) transport system in *Serratia marcescens*. J Bacteriol 171:238–243

Microbiol Monogr (6)
D. H. Nies, S. Silver: Molecular Microbiology of Heavy Metals
DOI 10.1007/7171_2006_079/Published online: 18 January 2007

New Transport Deals for Old Iron

Gregor Grass

Institute for Biology/Microbiology, Martin-Luther-University, Kurt-Mothes-Str. 3, 06099 Halle/Saale, Germany
g.grass@mikrobiologie.uni-halle.de

Abstract Maintaining iron homeostasis is a necessity for almost all organisms. Microorganisms such as *Escherichia coli* possess several systems for iron acquisition and storage. In recent years further systems have been discovered. These systems comprise the first characterized bacterial ZIP transporter, ZupT. ZupT is a transporter with broad substrate specificity and beside iron and zinc ZupT also transports cobalt or probably other divalent metal cations. Another novel bacterial iron transporter, EfeU, was recently found in *E. coli* and *Bacillus subtilis*. These EfeU permeases are the first characterized bacterial members of the OFeT-family of iron transporters that are well studied in yeast and in other lower eukaryotes.

Enterobactin, the primary catecholate-type siderophore from *E. coli* and other bacteria, is secreted from the cell in a two-step mechanism, functionally connecting the major facilitator protein EntS and an efflux-complex comprising the outer membrane exit channel protein TolC. Our knowledge of iron-transport systems was extended by the identification and characterization of an iron-efflux transporter, FieF, from *E. coli*. FieF is a member of the largest subfamily of cation diffusion facilitators (CDF). CDF proteins were previously known to be involved in detoxification of divalent transition metal cations such as Zn(II) or Cd(II) but probably participate in efflux of ferrous iron as well.

1
Introduction

For investigations on several general life processes the γ-proteobacterium *Escherichia coli* is broadly regarded as the primary prokaryotic model-organism. Studies with *E. coli* also contributed significantly to our knowledge on homeostasis of essential transition metals. Interestingly, *E. coli* and other enteric bacteria possess most of the mechanisms of metal-homeostasis that can also be found in higher organisms. Yet, the advantage of the *E. coli* model is that it conjoins full genetic manipulability with simplicity of the experimental system, which is comprised of only two compartments confined by two membranes.

The plethora of systems for iron acquisition reflects the supreme importance of this element for almost all microorganisms. In *E. coli* there are, depending on the strain investigated, more than ten iron-uptake routes known. These encompass the siderophore pathways through the Fec, Fep, Cir, Fiu, Fhu, Ybt/Fyu, Iuc/Iut, or Iro proteins (reviewed by Braun 2003, also see Braun and Hantke, in this volume) or Chu for hemophore uptake (Torres and Payne 1997; also recently reviewed by Wandersman and Delepelaire 2004). In all these instances specific ferric-iron chelate-receptors of the outer membrane and ABC transporters for transport across the inner membrane are involved. Recently, an additional system of this composition, Fit, was identified in clinical *E. coli* isolate i484 (Ouyang and Isaacson 2006) but *fit* genes can be found in the genomes of other pathogenic strains as well.

While much work has been done on ferric-siderophore uptake into bacteria during the last few decades relatively little is known about how indigenous apo-siderophores leave the cell after biosynthesis. This piece of work is not devoted to ferric-siderophore uptake but it will briefly recapitulate what was elucidated for siderophore secretion within the last couple of years.

In addition to the ferric-siderophore transporters several proteins implicated in non-siderophore bound iron uptake were identified in *E. coli*. To differentiate these principal speciations of iron, a new terminology for the latter, elemental iron uptake, was introduced (Ollinger et al. 2006). A number of pathogenic *E. coli* strains take up elemental iron through ABC-transporters lacking an associated outer membrane receptor. These are the putative ferric-iron transporters AfuABC of enterohemorrhagic (EHEC) *E. coli* O157:H7 or the iron/manganese transporter SitABCD of uropathogenic (UPEC) *E. coli* CFT073 (Sabri et al. 2006). All known iron transport systems of *E. coli* and the chronology of their discovery are depicted in Fig. 1.

In contrast to the ATP-driven ferric-siderophore ABC-transporters another integral membrane protein involved in iron uptake, FeoB, possesses a N-terminal cytoplasmic domain that functions as GTPase (Marlovits et al. 2002). FeoB proteins are frequently found to be associated with iron uptake in different bacterial species (Kammler et al. 1993; Velayudhan et al. 2000; Robey and Cianciotto 2002). However, it is not known, whether GTP-hydrolysis actually drives iron transport or even whether FeoB is an iron transporter *per se* or not. Still, FeoB contributes to ferrous iron uptake in one way or another because *E. coli* strains expressing *feo* were shown to have increased cellular iron concentrations (Kammler et al. 1993; Grass et al. 2005a).

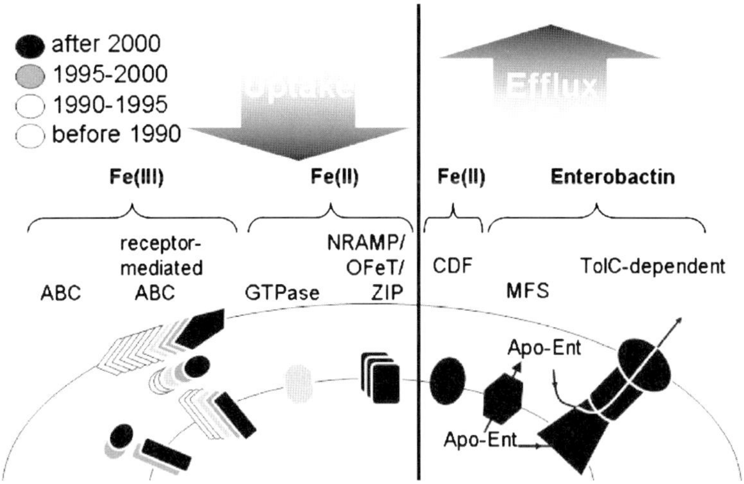

Fig. 1 Chronology of discovery of systems involved in iron transport from *E. coli*. Shown are the different mechanisms of iron-transport of *E. coli* or (where direct evidence for *E. coli* is lacking) other enterobacteria with the time-periods of their first description. Depending on the respective transporter iron may be transported elementally (as a ferrous or ferric cation) or as a ferric-siderophore or -hemophore.
ABC-receptor-mediated, chelates: ChuA (hemin, Torres and Payne 1997), Ybt/Fyu (ferric-yersiniabactin, Heesemann et al. 1993) (from *Yersinia enterolytica*), Fec (ferric-citrate, Wagegg and Braun 1981), Fep (ferric-enterobactin, Wookey and Rosenberg 1978), Fiu, Cir (ferric-enterobactin degradation catecholates, Hantke 1990), FhuA (ferric-hydroxamates, Coulton and Braun 1979), FhuE (ferric-rhodotorulic acid or -coprogen, Hantke 1983), IutA (ferric-aerobatic, de Lorenzo et al. 1986),
Iro (ferric-salmochelin, Bäumler et al. 1996), (from *Salmonella typhi*), Fit (unidentified ferric-siderophore, Ouyang and Isaacson 2006)
Receptor-less ABC-mediated, elemental: Sit (Fe(III), Zhou et al. 1999) (from *Salmonella typhimurium*), (Fe(III), Sabri et al. 2006), (from *E. coli*), Afu/Yfe (Fe(II), or Mn(II), Saken et al. 2000) (from *Yersinia enterocolitica*)
GTPase: FeoB (Fe(II) directly or indirectly, Kammler et al. 1993)
Non-ATP/GTP- driven: MntH (Mn(II), Fe(II) and others, Makui et al. 2000), ZupT (Fe(II), Zn(II), Co(II) and others, Grass et al. 2005a), EfeU (probably Fe(II)), (Grosse et al. 2006)
Cation diffusion facilitator (CDF): FieF (Fe(II), Grass et al. 2005b)
Major facilitator: EntS: (apo-enterobactin, Furrer et al. 2002)
Outer membrane exit duct: TolC: (apo-enterobactin and antibiotics, Bleuel et al. 2005)

The Mg(II) permease CorA (Hantke 1997) or the NRAMP (natural resistance associated macrophage protein) MntH were also thought to be responsible for iron uptake into *E. coli*. While recent data suggest that MntH does transport iron (Kehres et al. 2000), CorA probably does not (Papp and Maguire 2004). However, the affinity of MntH for Mn(II) is much higher than for Fe(II) (Kehres et al. 2000), thus, Mn(II) is probably the physiological substrate of MntH. NRAMPs are vital permeases of eukaryotes and it is a worth while approach to survey bacteria for transporters that bear resemblance to characterized eukaryotic counterparts. In doing so ZupT from *E. coli* was identified as the first bacterial member of the ZIP-family (ZRT- and IRT-like proteins) (Grass et al. 2002, 2005a). ZIP-transporters are paramount for iron uptake, for instance into plant roots (Guerinot 2000) or for intracellular metal-trafficking in animals and man (Eide 2004). In a similar comparative approach, YcdN (EfeU), the first bacterial member of the OFeT (oxidase-dependent ferrous iron transporters) from *E. coli* was characterized. OFeT transporters are known as major iron-uptake permeases in lower eukaryotes such as fungi and algae but no member has been investigated in prokaryotes until recently. All these recent discoveries of additional iron-transport systems strongly underscore that even for the old workhorse *E. coli* we better be prepared for surprises regarding novel mechanisms of metal-homeostasis.

2
The ZIP Protein ZupT – an Uptake Transporter for Ferrous Iron and Other Divalent Metal Cations

Amongst the metal-homeostasis mechanisms, ZIP-transporters are key uptake transporters into eukaryotic cells. In yeasts, plants and animals including man, different ZIP-transporters are implicated in Fe(II), Zn(II), Mn(II), Co(II) or Cd(II) translocation across a number of membranes and several expression studies of ZIP-genes were conducted. In contrast to the situation in eukaryotes only very few working groups study bacterial ZIP-transporters (Grass et al. 2002; Kwon et al. 2004) and ZupT, the ZIP-transporter from *E. coli*, was only recently identified as an additional iron-uptake route (Grass et al. 2005a). ZupT from *E. coli* is still the only bacterial family member characterized so far. Initially ZupT was identified as a Zn(II) uptake transporter (Grass et al. 2002) in addition to the primary ABC-type transport system Znu. A rather broad substrate spectrum is common in ZIP-transporters, so it was unsurprising that the substrate spectrum of ZupT was extended to Fe(II) and probably to other divalent metal cations as well (Grass et al. 2005a). Thus, ZupT is also an additional iron-uptake system of *E. coli*.

While the first molecular characterization of ZIP-transporters was carried out in 1996 (Eide et al. 1996, 1996a) we still have only vague ideas as to what drives ZIP-mediated metal transport. For some eukaryotic ZIP-transporters kinetic parameters were determined. Metal uptake was concentration dependent and saturable. Only a few studies addressed the question how the ZIP-mediated transport-process is energized. Contradictory data is available as to whether ZIP-transporters utilize physiological energy. ATP-binding domains within ZIP-polypeptides were not detected (Zhao and Eide 1996a) and ATP-deprivation did not diminish ZIP-transport (Gaither and Eide 2001). Function of the ZIP-transporters Zrt1 and Zrt2 from yeast were found to be energy-dependent (Zhao and Eide 1996a, b) whereas human hZIP1 and hZIP2 were found to be energy-independent and treatment with several different electron transport/oxidative phosphorylation inhibitors had no influence on transport (Gaither and Eide 2000, 2001). Similarly, the presence of an uncoupler of oxidative phosphorylation had no effect on ZupT-mediated metal uptake. Furthermore, *E. coli* cells expressing *zupT* exhibited no increase in metal transport in the presence of HCO_3^- (Taudte and Grass, unpublished observations). In contrast hZIP2-activity was stimulated by HCO_3^- indicating a Zn(II)/HCO_3^- symport mechanism (Gaither and Eide 2000). Interestingly, different pH values influenced the substrate transported. For IRT1 Zn(II) uptake was observed only at low pH (pH 4.2) (Korshunova et al. 1999) while Fe(II) and other substrates were transported at pH 6.1 (Eide et al. 1996).

Assuming that ZIP-transport does not need an additional energy source, is metal accumulation through ZupT simply uniport? Previously it was suggested that for uptake metal cations might simply follow the concentration gradient from the periplasm to the cytoplasm (Outten and O'Halloran 2001; Eide 2004). Clearly, more work is needed to clarify this issue.

Interestingly, in contrast to its eukaryotic counterparts that are frequently expressed under metal-limiting conditions *zupT*-expression in *E. coli* is not regulated in a metal-dependent manner (Grass et al. 2005a). Also, the *zupT* gene seems to be expressed constitutively and on a low level (Grass et al. 2005a). This is unlike other iron-uptake transporter genes from *E. coli* such as *feoB* or *mntH* that are regulated by iron and Fur, the global regulator of iron homeostasis. However, this might not hold true for other bacterial *zupT* genes, for example from the β-proteobacterium, *Cupriavidus metallidurans*. Expression of its ZIP-gene was found to respond positively to metal-deprivation (Scherer and Nies, unpublished observations). In this organism the ZupT orthologue possesses a histidine-rich region between III and IV similar to its eukaryotic counterparts. It remains to be seen, if there is a functional (regulatory?) relevance of this feature in ZIP-proteins.

3 A Non-Chelate Iron-Uptake System, EfeU (YcdN), from Pathogenic E. Coli Strains

Homologues of the high-affinity ferric iron permease Ftr1p are widespread in fungi and other lower eukaryotes such as algae. In yeasts Ftr1p ferric iron transporters and Fet3p-like multi copper oxidases constitute a pathway necessary for iron uptake. In the current model of Fet3p/Ftr1p-mediated iron uptake ferric iron [Fe(III)] is reduced first to ferrous iron [Fe(II)] by the membrane-bound extracellular iron reductase Fre. Fe(II) is re-oxidized by membrane-bound extracellular Fet3p to Fe(III) and finally transferred to Ftr1p for uptake. For this process close interaction of Ftr1p and Fet3p is indispensable for both function and correct localization of the protein complex (Askwith et al. 1994; Askwith and Kaplan 1997).

Interestingly, Ftr1p homologues can also be found as putative proteins in many sequenced bacteria and this family was named OFeT (oxidase dependent Fe(II) transporters, TC.9.A.10.1) as a subgroup of the larger iron/lead transporter (ILT) superfamily (Saier et al. 2006). Genome sequence projects have revealed that some *E. coli* or *Shigella* strains such as K12 suffered a frame-shift mutation close to the 5'-end of a putative FTR-homologous gene, *ycdN*, resulting in separation of a short (b1016) and a longer (b1017) peptide. Previously, it was suggested that this defective *ycdN* gene is part of an operon, *ycdNOB* (McHugh et al. 2003). The authors speculated that YcdNOB functions as a ferrous iron transporter in combination with a periplasmic iron oxidase or reductase. However, neither YcdO nor YcdB is a multi copper oxidase as in the Ftr1p/Fet3p system.

Interestingly, in genomes of several *E. coli* strains full-size *ycdN* genes are encoded. Therefore, it is most likely that functional YcdN(OB)-dependent iron-uptake systems exist in these organisms. Consequently, it was recently demonstrated that the iron-permease YcdN from the probiotic *E. coli* strain Nissle 1917 (also known as EcN or O6:K5:H1) is encoded as part of a Fur-regulated tricistronic operon. For this reason YcdN was renamed EfeU (elemental ferrous iron uptake) (Grosse et al. 2006). An *E. coli* mutant strain devoid of all known iron-uptake systems that expressed *ycdN(efeU)OB* not only reached higher cell densities in medium supplemented with iron chelator but also exhibited increased iron uptake compared with an *ycdNOB*-free control. In vitro ferrous iron permeation through EfeU established this protein as a novel iron-uptake system from *E. coli* (Grosse et al. 2006) and, because EfeU-orthologues are widespread, from other bacteria as well. Indeed, in *Bacillus subtilis* the YwbL protein is part of the YcdNOB-orthologous system, YwbLMN, which was shown to be necessary for growth in a defined medium under conditions of low iron when citrate as the siderophore was missing (Ollinger et al. 2006). Further work is needed to elucidate the function of the other components of the Ycd/Ywb-system, YcdO/YwbM and YcdB/YwbN, and the interplay with their OFeT-permeases. One of these proteins, YcdB, is a periplasmic hemoprotein translocated by the twin arginine translocation pathway. YcdB exerts peroxidase activity in vitro but the actual substrate of YcdB remains unknown (Sturm et al. 2006). A still open question is what is the connection between the peroxidase activity of YcdB and iron uptake by YcdN, if any?

Interestingly, some uropathogenic *E. coli* strains such as F11 or UTI89 and many other bacteria encode a second paralog of ILT transporters (TC.9.A.10.2). In these systems the transporter gene is not part of a tricistronic but probably of a dicistronic transcript. In contrast to OFeT permeases these transporters exhibit a long N-terminal extension that is probably located in the periplasm and that terminates with an additional transmembrane helix. The product of the second gene possesses a Tpd domain (COG3470), which is present in uncharacterized periplasmic proteins, probably involved in high-affinity Fe(II) transport. Preliminary experimental data suggests that these paralogous systems also function as iron-uptake systems when expressed in *E. coli* K12 (Koch and Grass, unpublished data).

4 Two-Step Transenvelope Efflux of Apo-Enterobactin

In contrast to the multitude of publications reporting on ferric-siderophore uptake into bacterial cells, relatively little is known as to how these crucial iron-chelating compounds in their apo-form are secreted after intracellular biosynthesis. For *E. coli* the first report on enterobactin translocation across the cytoplasmic membrane described the involvement of an ATP-independent transporter (Furrer et al. 2002). This EntS is a member of the large Major Facilitator Superfamily (MFS) of transport proteins (Fig. 1). In the meantime further members of this family from other bacteria were found to exert similar functions as EntS, namely the translocation of indigenous apo-siderophores across the cytoplasmic membrane (reviewed in Grass 2006). For Gram-negative bacteria this left the question, how apo-siderophores that are too big (>600 Da) to diffuse freely through porins in the outer membrane, arrive at the surrounding medium. Only there can they finally sequester iron for subsequent recognition and uptake through the well-studied TonB-dependent receptors.

As a first step it was recently demonstrated that secretion of the cyclic tricatecholate siderophore apo-enterobactin from *E. coli* is accomplished by the outer membrane efflux duct TolC. TolC is involved in efflux of a wide variety of substances including antibiotics and dyes. The trimeric TolC protein only functions in cooperation with active transport proteins of several families and with additional membrane fusion proteins that stabilize the resulting tripartite efflux complex (Koronakis 2003). So what are the partners of TolC in apo-enterobactin secretion?

Interestingly, it was found earlier that hemophores are secreted from *Serratia marcescens* by an efflux-ATPase-dependent transport complex comprising a close homologue of TolC (Binet and Wandersman 1996). Disappointingly, in *E. coli* K12 there is no such efflux-ATPase that could exert a similar function in apo-enterobactin secretion. Therefore, the partner(s) of TolC for apo-enterobactin transport remain unknown (Fig. 1). The current model of apo-enterobactin secretion pictures a two-step mechanism for enterobactin efflux (Bleuel et al. 2005; Grass 2006). In a first step apo-enterobactin is transported by the EntS protein from the cytoplasm to the periplasm and from there it is translocated to the outside by an efflux complex that encompasses TolC as the outer membrane efflux duct (Bleuel et al. 2005).

5
Iron Efflux by the CDF Proteins

Whereas Fe(III) is quite insoluble at physiological pH, under acidic conditions solubility rises. Also with decreasing oxygen tension, the concentration of iron as soluble Fe(II) increases. In bacteria, several mechanisms for defense against the unfavorable effects of intracellular iron overload are known, including tight regulation of iron uptake, sequestration of iron within iron storage proteins or detoxification of reactive oxygen species generated by the Fenton reaction. Although Fur-regulated iron-uptake systems should be repressed under iron-surplus conditions (Hantke 1984), transporters with a broader substrate spectrum such as ZupT might not be. Thus, under conditions of elevated iron concentrations in the environment, increased influx of Fe(II) into the cell could lead to a situation that requires additional, effective means of iron detoxification.

Only little is known about Fe(II)-efflux systems in bacteria and in other organisms. Cation diffusion facilitators (CDF) (Paulsen and Saier 1997) play a key role in metal homeostasis in humans, animals, plants and bacteria and CDF-proteins were mainly described as transporters of divalent transition metal cations. Contradictory data exist as to whether the mechanism of transport is electro-neutral antiport of the metal cation against H^+/K^+ (Guffanti et al. 2002) or whether the ratio of $Me(II)/H^+$ is 1:1 (Chao and Fu 2004).

In yeast two members of the CDF-family were implicated in mobilization of iron from mitochondrial pools (Li and Kaplan 1997) but these CDF-proteins are probably not involved in removal of surplus iron. In contrast, for four bacterial CDF-proteins experimental evidence for iron efflux exist: the FieF protein from *Escherichia coli* (Grass et al. 2005b), *C. metallidurans* CH34 (Munkelt et al. 2004) and the MamB and MamM proteins from *Magnetospirillum gryphiswaldense* (Grünberg et al. 2001). The latter two most likely transport ferrous iron from the cytoplasm into magnetosomes, thus, remove cytoplasmic iron.

E. coli harbors two members of the CDF-family, ZitB and FieF (also known as YiiP). ZitB was shown to be a zinc-efflux pump (Grass et al. 2001; Lee et al. 2002; Anton et al. 2004), whereas FieF was identified as an iron-efflux transporter (Grass et al. 2005b). In vitro, FieF was characterized additionally as a Zn(II)-transporter and the FieF protein was able to bind several metal cations including Fe(II) (Wei and Fu 2005). However, up to now there is no indication of a physiological function for FieF other than Fe(II)-efflux. Kinetic parameters support this notion because the k_m and V_{max} values of FieF for Fe(II) were in the same order of magnitude as those of the zinc transporting CDF protein ZitB for Zn(II). Conversely, transport parameters of FieF for Zn(II) suggested that Zn(II) was not the preferred substrate (Otto and Grass, unpublished results). The Fe(II)-transport parameters of FieF are thus similar to that of ZitB for Zn(II) indicative for Fe(II) being the primary substrate of FieF.

6
Outlook

Why do bacteria such as *E. coli* have so many iron- and cation-transporters? This might be both an adaptation to rapidly changing environments and survival in a host. Generally the ability to acquire Mn(II) and Fe(II) favors bacterial resistance to environmental stresses and stress during host colonization (Bearden and Perry 1999).

Possibly, *E. coli* strains that have the possibility to produce different iron-transport systems have advanced capabilities to respond to environmental stimuli such as pH changes while passing a host animal's gastrointestinal tract (Valdebenito et al. 2005). However, this has not been experimentally proven in situ but recent results point in that direction (Valdebenito et al. 2006). In this anerobic environment of the gut ferrous iron is plentiful but ferric iron is scarce (Valdebenito et al. 2005). Therefore, it would be a great step-forward for integrating iron homeostasis if we knew whether ferrous iron permeases such as FeoB, MntH, or ZupT were better suited for iron provision than siderophores under such conditions. For now the available data remain inconclusive. Although it was demonstrated that a *feoB* mutant of *E. coli* was deficient in its capability to colonize the murine intestine (Stojiljkovic et al. 1993), a *feoB* mutant of *Salmonella typhimurium* was not attenuated for mouse infection (Tsolis et al. 1996). Nonetheless, colonization does not equal infection and for *S. typhimurium* it was observed that a wild-type strain out-competed its *feoB*-lacking mutant in mixed gut colonization (Tsolis et al. 1996). For the process of infection of host organisms, the bacterium is also subject to many different and continually changing stimuli and environments. Consequently, metal-demand and gene expression of invading bacteria must also change during that period.

This might also be true for iron efflux. In general the gut is an anerobic environment and thus, ferrous iron is considered safe. However, certain host proteins of mucosal epithelia such as dual oxidase Duox provide the main source of reactive oxygen species within the intestines for microbiotic defense. Under high iron conditions Fenton-stress is unquestionably induced (Ha et al. 2005). Iron-efflux transporters such as FieF or other (CDF) proteins are ideal candidates to counter such environmental hardship by reducing the intracellular iron concentration and thus Fenton-stress.

Consequently, we can expect that new iron transporters can still be found within the bacterial kingdom. Yet, at the same time we should make first steps in conceptualizing the accumulating knowledge on the individual systems and put it in a larger perspective. What is the contribution of these transporters or acquisition systems to the whole and under which conditions do single transporter systems become particular important?

Acknowledgements The author thanks Dietrich H. Nies for his continued cooperation, Cornelia Große and Judith Scherer for carefully reading the manuscript. Work in the author's laboratory was supported by the DFG ("Deutsche Forschungsgemeinschaft") research grants GR2061/1-2 and Ni262-8.

References

Anton A et al. (2004) Characteristics of zinc transport by two bacterial cation diffusion facilitators from *Ralstonia metallidurans* CH34 and *Escherichia coli*. J Bacteriol 186:7499–7507

Askwith C et al. (1994) The FET3 gene of *S. cerevisiae* encodes a multicopper oxidase required for ferrous iron uptake. Cell 76:403–410

Askwith C, Kaplan J (1997) An oxidase-permease-based iron transport system in *Schizosaccharomyces pombe* and its expression in *Saccharomyces cerevisiae*. J Biol Chem 272:401–405

Bäumler AJ, Tsolis RM, van der Velden AW, Stojiljkovic I, Anic S, Heffron F (1996) Identification of a new iron regulated locus of *Salmonella typhi*. Gene 183:207–213

Bearden SW, Perry RD (1999) The Yfe system of *Yersinia pestis* transports iron and manganese and is required for full virulence of plague. Mol Microbiol 32:403–414

Binet R, Wandersman C (1996) Cloning of the *Serratia marcescens hasF* gene encoding the HasABC exporter outer membrane component: a TolC analogue. Mol Microbiol 22:265–273

Bleuel C et al. (2005) TolC is involved in enterobactin efflux across the outer membrane of *E. coli*. J Bacteriol 187:6701–6707

Braun V (2003) Iron uptake by *Escherichia coli*. Front Biosci 8:s1409–1421

Chao Y, Fu D (2004) Kinetic study of the antiport mechanism of an *Escherichia coli* zinc transporter, ZitB. J Biol Chem 279:12043–12050

Coulton JW, Braun V (1979) Protein II influences ferrichrome-iron transport in *Escherichia coli* K12. J Gen Microbiol 110:211–220

de Lorenzo V, Bindereif A, Paw BH, Neilands JB (1986) Aerobactin biosynthesis and transport genes of plasmid ColV-K30 in *Escherichia coli* K-12. J Bacteriol 165:570–578

Eide D, Broderius M, Fett J, Guerinot ML (1996) A novel iron-regulated metal transporter from plants identified by functional expression in yeast. Proc Natl Acad Sci USA 93:5624–5628

Eide DJ (2004) The SLC39 family of metal ion transporters. Pflugers Arch 447:796–800

Furrer JL, Sanders DN, Hook-Barnard IG, McIntosh MA (2002) Export of the siderophore enterobactin in *Escherichia coli:* involvement of a 43 kDa membrane exporter. Mol Microbiol 44:1225–1234

Gaither LA, Eide DJ (2000) Functional expression of the human hZIP2 zinc transporter. J Biol Chem 275:5560–5564

Gaither LA, Eide DJ (2001) The human ZIP1 transporter mediates zinc uptake in human K562 erythroleukemia cells. J Biol Chem 276:22258–22264

Grass G (2006) Iron transport in *Escherichia coli*: All has not been said and done. Biometals 19:159–172

Grass G, Fan B, Rosen BP, Franke S, Nies DH, Rensing C (2001) ZitB (YbgR), a member of the cation diffusion facilitator family, is an additional zinc transporter in *Escherichia coli*. J Bacteriol 183:4664–4667

Grass G et al. (2005a) The metal permease ZupT from *Escherichia coli* is a transporter with a broad substrate spectrum. J Bacteriol 187:1604–1611

Grass G et al. (2005b) FieF (YiiP) from *Escherichia coli* mediates decreased cellular accumulation of iron and relieves iron stress. Arch Microbiol 183:9–18

Grass G, Wong MD, Rosen BP, Smith RL, Rensing C (2002) ZupT is a Zn(II) uptake system in *Escherichia coli*. J Bacteriol 184:864–866

Grosse C, Scherer J, Koch D, Otto M, Taudte N, Grass G (2006) A new ferrous iron-uptake transporter, EfeU (YcdN), from *Escherichia coli*. Mol Microbiol 62:120–131

Grünberg K, Wawer C, Tebo BM, Schüler D (2001) A large gene cluster encoding several magnetosome proteins is conserved in different species of magnetotactic bacteria. Appl Environ Microbiol 67:4573–4582

Guerinot ML (2000) The ZIP family of metal transporters. Biochim Biophys Acta 1465:190–198

Guffanti AA, Wei Y, Rood SV, Krulwich TA (2002) An antiport mechanism for a member of the cation diffusion facilitator family: divalent cations efflux in exchange for K^+ and H^+. Mol Microbiol 45:145–153

Ha EM, Oh CT, Bae YS, Lee WJ (2005) A direct role for dual oxidase in *Drosophila* gut immunity. Science 310:847–850

Hantke K (1983) Identification of an iron uptake system specific for coprogen and rhodotorulic acid in *Escherichia coli* K12. Mol Gen Genet 191:301–306

Hantke K (1984) Cloning of the repressor protein gene of iron-regulated systems in *Escherichia coli* K12. Mol Gen Genet 197:337–341

Hantke K (1990) Dihydroxybenzoylserine–a siderophore for *E. coli*. FEMS Microbiol Lett 55:5–8

Hantke K (1997) Ferrous iron uptake by a magnesium transport system is toxic for *Escherichia coli* and *Salmonella typhimurium*. J Bacteriol 179:6201–6204

Heesemann J et al. (1993) Virulence of *Yersinia enterocolitica* is closely associated with siderophore production, expression of an iron-repressible outer membrane polypeptide of 65,000 Da and pesticin sensitivity. Mol Microbiol 8:397–408

Kammler M, Schön C, Hantke K (1993) Characterization of the ferrous iron uptake system of *Escherichia coli*. J Bacteriol 175:6212–6219

Kehres DG, Zaharik ML, Finlay BB, Maguire ME (2000) The NRAMP proteins of *Salmonella typhimurium* and *Escherichia coli* are selective manganese transporters involved in the response to reactive oxygen. Mol Microbiol 36:1085–1100

Koronakis V (2003) TolC–the bacterial exit duct for proteins and drugs. FEBS Lett 555:66–71

Korshunova YO, Eide D, Clark WG, Guerinot ML, Pakrasi HB (1999) The IRT1 protein from *Arabidopsis thaliana* is a metal transporter with a broad substrate range. Plant Mol Biol 40:37–44

Kwon SJ, Petri R, DeBoer AL, Schmidt-Dannert C (2004) A high-throughput screen for porphyrin metal chelatases: application to the directed evolution of ferrochelatases for metalloporphyrin biosynthesis. Chembiochem 5:1069–1074

Lee SM et al. (2002) Functional analysis of the *Escherichia coli* zinc transporter ZitB. FEMS Microbiol Lett 215:273–278

Li L, Kaplan J (1997) Characterization of two homologous yeast genes that encode mitochondrial iron transporters. J Biol Chem 272:28485–28493

Makui H, Roig E, Cole ST, Helmann JD, Gros P, Cellier MF (2000) Identification of the *Escherichia coli* K-12 Nramp orthologue (MntH) as a selective divalent metal ion transporter. Mol Microbiol 35:1065–1078

Marlovits TC, Haase W, Herrmann C, Aller SG, Unger VM (2002) The membrane protein FeoB contains an intramolecular G protein essential for Fe(II) uptake in bacteria. Proc Natl Acad Sci USA 99:16243–16248

McHugh JP et al. (2003) Global iron-dependent gene regulation in *Escherichia coli*. A new mechanism for iron homeostasis. J Biol Chem 278:29478–29486
Munkelt D, Grass G, Nies DH (2004) The chromosomally encoded cation diffusion facilitator proteins DmeF and FieF from *Wautersia metallidurans* CH34 are transporters of broad metal specificity. J Bacteriol 186:8036–8043
Ollinger J, Song K-B, Antelmann H, Hecker M, Helmann JD (2006) Role of the Fur regulon in iron transport in *Bacillus subtilis*. J Bacteriol 188:3664–3673
Outten CE, O'Halloran TV (2001) Femtomolar sensitivity of metalloregulatory proteins controlling zinc homeostasis. Science 292:2488–2492
Ouyang Z, Isaacson R (2006) Identification and characterization of a novel ABC iron transport system, *fit*, in *Escherichia coli*. Infect Immun 74:6949–6956
Papp KM, Maguire ME (2004) The CorA Mg^{2+} transporter does not transport Fe^{2+}. J Bacteriol 186:7653–7658
Paulsen IT, Saier MH Jr (1997) A novel family of ubiquitous heavy metal ion transport proteins. J Membr Biol 156:99–103
Robey M, Cianciotto NP (2002) *Legionella pneumophila feoAB* promotes ferrous iron uptake and intracellular infection. Infect Immun 70:5659–5669
Sabri M, Leveille S, Dozois CM (2006) A SitABCD homologue from an avian pathogenic *Escherichia coli* strain mediates transport of iron and manganese and resistance to hydrogen peroxide. Microbiology 152:745–758
Saier MH Jr, Tran CV, Barabote RD (2006) TCDB: the Transporter Classification Database for membrane transport protein analyses and information. Nucleic Acids Res 34:D181–186
Saken E, Rakin A, Heesemann J (2000) Molecular characterization of a novel siderophore-independent iron transport system in *Yersinia*. Int J Med Microbiol 290:51–60
Stojiljkovic I, Cobeljic M, Hantke K (1993) *Escherichia coli* K-12 ferrous iron uptake mutants are impaired in their ability to colonize the mouse intestine. FEMS Microbiol Lett 108:111–115
Sturm A, Schierhorn A, Lindenstrauss U, Lilie H, Bruser T (2006) YcdB from *Escherichia coli* reveals a novel class of Tat-dependently translocated hemoproteins. J Biol Chem 281:13972–13978
Torres AG, Payne SM (1997) Haem iron-transport system in enterohaemorrhagic *Escherichia coli* O157:H7. Mol Microbiol 23:825–833
Tsolis RM, Baumler AJ, Heffron F, Stojiljkovic I (1996) Contribution of TonB- and Feo-mediated iron uptake to growth of *Salmonella typhimurium* in the mouse. Infect Immun 64:4549–4556
Valdebenito M, Bister B, Reissbrodt R, Hantke K, Winkelmann G (2005) The detection of salmochelin and yersiniabactin in uropathogenic *Escherichia coli* strains by a novel hydrolysis-fluorescence-detection (HFD) method. Int J Med Microbiol 295:99–107
Valdebenito M, Crumbliss AL, Winkelmann G, Hantke K (2006) Environmental factors influence the production of enterobactin, salmochelin, aerobactin, and yersiniabactin in *Escherichia coli* strain Nissle 1917. Int J Med Microbiol 296:513–520
Velayudhan J et al. (2000) Iron acquisition and virulence in *Helicobacter pylori*: a major role for FeoB, a high-affinity ferrous iron transporter. Mol Microbiol 37:274–286
Wagegg W, Braun V (1981) Ferric citrate transport in *Escherichia coli* requires outer membrane receptor protein FecA. J Bacteriol 145:156–163
Wandersman C, Delepelaire P (2004) Bacterial iron sources: from siderophores to hemophores. Annu Rev Microbiol 58:611–647
Wei Y, Fu D (2005) Selective metal binding to a membrane-embedded aspartate in the *Escherichia coli* metal transporter YiiP (FieF). J Biol Chem 280:33716–33724

Wookey P, Rosenberg H (1978) Involvement of inner and outer membrane components in the transport of iron and in colicin B action in *Escherichia coli.* J Bacteriol 133:661–666
Zhao H, Eide D (1996a) The yeast *ZRT1* gene encodes the zinc transporter protein of a high-affinity uptake system induced by zinc limitation. Proc Natl Acad Sci USA 93:2454–2458
Zhao H, Eide D (1996b) The *ZRT2* gene encodes the low affinity zinc transporter in *Saccharomyces cerevisiae.* J. Biol Chem 271:23203–23210
Zhou D, Hardt WD, Galan JE (1999) *Salmonella typhimurium* encodes a putative iron transport system within the centisome 63 pathogenicity island. Infect Immun 67:1974–1981

Microbiol Monogr (6)
D. H. Nies, S. Silver: Molecular Microbiology of Heavy Metals
DOI 10.1007/7171_2006_080/Published online: 13 February 2007

Manganese: Uptake, Biological Function, and Role in Virulence

Krisztina M. Papp-Wallace · Andrea S. Moomaw · Michael E. Maguire (✉)

Department of Pharmacology, Case School of Medicine,
Case Western Reserve University, 10900 Euclid Avenue, Cleveland, OH 44106-4965, USA
mem6@cwru.edu

Abstract Recent data have demonstrated that bacterial homologs of eukaryotic Nramp transporters as well as members of the LraI family of proteins are both highly selective Mn^{2+} transporters. Mutation of these transporters in several pathogenic bacterial species causes decreased virulence in a variety of model systems. This implies that the Mn^{2+} ions are required for one or more processes essential for bacterial virulence. However, Mn^{2+}

has few known enzymatic roles compared to other divalent cations. This review will describe what is currently known about the two classes of prokaryotic Mn^{2+} transporters, how each is regulated and the virulence deficits that arise when they are mutated. Finally, possible enzymatic roles for Mn^{2+} will be outlined, and their potential for a role in virulence discussed.

1 Introduction

Manganese is transported by biological systems solely as the divalent cation. Virtually all bacteria express one or both of two major classes of Mn^{2+} transport systems: the MntH (Nramp) H^+-divalent cation transporters and ABC ATPase Mn^{2+} transporters (in Gram-positive organisms, the ABC ATPases belong to the lipoprotein receptor antigen (LraI) class of cell surface proteins). A much smaller number of bacterial species such as *Lactobacillus* appear to carry a Mn^{2+}-transporting P-type ATPase. Eukaryotes also appear to have both Nramp and ABC ATPase class Mn^{2+} transporters though the latter class has not been characterized extensively. Although both types of transporters can mediate flux of multiple transition metal divalent cations, most of the bacterial Nramp transporters and many of the ABC transporters are highly selective for Mn^{2+} over other cations. Mutation of these selective Mn^{2+} transporters in several bacterial species causes decreased virulence in a variety of models of infection. The obvious conclusion is that Mn^{2+} ion is required for one or more processes essential for bacterial virulence. This review describes the properties of prokaryotic Mn^{2+} transporters, their cation selectivity and the regulation of their expression. After a discussion of the virulence deficits that arise when the transporters are mutated, possible enzymatic roles for Mn^{2+} relevant to pathogenesis will be outlined.

2 Mn^{2+} Transport Systems

Transport systems for manganese belong to several protein families, Nramp-, ABC-transporters and P-type ATPases.

2.1 MntH (Nramp) Mn^{2+} Transporters

The Nramp (**n**atural **r**esistance **a**ssociated **m**acrophage **p**rotein) transporter is a broad spectrum H^+-coupled transition metal divalent cation antiporter, transporting Fe^{2+}, Mn^{2+}, and Zn^{2+}, initially described in macrophages of mammals (Goswami et al. 2001; Vidal et al. 1993). Nramp homologs are present in all branches of eukaryotic cells and most bacteria. They appear to be absent in the Archaea, halophiles and thermophilic bacteria based on current genomic sequences. The Nramp1 gene was first discovered 30 years ago in mammals as the *ity* locus (Plant and Glynn 1974). Mutation of this locus in mice results in increased susceptibility to *Salmonella enterica* serovar Typhimurium infection (Plant and Glynn 1979). This same locus was identified independently as conferring susceptibility to other pathogens including *Leishmania donovani* (*lsh* locus) and *Mycobacterium bovis* (*bcg* locus) (Brown et al. 1982; Gros et al. 1983; Plant et al. 1982). Mutation of Nramp1 results in a failure of the macrophage to kill intracellular bacteria (Glynn et al. 1982; Vidal et al. 1995b; Zwilling and Hilburger 1994). A second Nramp exists in eukaryotes, Nramp2 (Gruenheid et al. 1995; Vidal et al. 1995a). However, Nramp2 is not required for resistance to infection. It is a broad spectrum H^+-coupled transition metal divalent cation symporter found in the intestine. Nramp2 commonly transports Fe^{2+} and Mn^{2+} (Canonne-Hergaux et al. 1999; Fleming et al. 1997; Forbes and Gros 2003; Garrick et al. 2006a,b; Gunshin et al. 1997; Sacher et al. 2001). Mutation of Nramp2 results in microcytic anemia due to an iron deficiency (Fleming et al. 1997, 1998) as well as symptoms of Mn^{2+} deficiency (Chua and Morgan 1997).

2.1.1
The Bacterial Nramp, MntH

About 20 years after the discovery of Nramps in eukaryotes, bacteria were found to have an Nramp ortholog as well. Designated MntH (for **Mn**$^{2+}$ **t**ransport, **H**$^+$-dependent), it couples the transport of manganese to H^+ (Kehres et al. 2000; Makui et al. 2000; Que and Helmann 2000). MntH is prevalent in bacteria (Cellier et al. 2001; Richer et al. 2003). MntH is generally about 450 amino acids in length. By standard fusion protein topology experiments, it has been shown to have 11 transmembrane segments with a cytoplasmic N-terminus in *E. coli* (Courville et al. 2004) (Fig. 1).

2.1.2
Transport Properties of MntH

MntH has been studied in *Mycobacterium tuberculosis*, *Mycobacterium leprae*, *Staphylococcus aureus*, *Bacillus subtilis*, *Lactobacillus plantarum*, *E. coli*, and *S.* Typhimurium; transport properties are best characterized in *E. coli* and *S.* Typhimurium (Agranoff et al. 1999; Groot et al. 2005; Horsburgh et al. 2002; Kehres et al. 2000; Makui et al. 2000; Que and Helmann 2000; Reeve et al. 2002). The *M. tuberculosis* homolog is a pH dependent transporter of Zn^{2+} and Fe^{2+}. Mn^{2+} and Cu^{2+} may also be transported, and both cations inhibit the uptake of $^{65}Zn^{2+}$ (Agranoff et al. 1999). Growth assays imply that

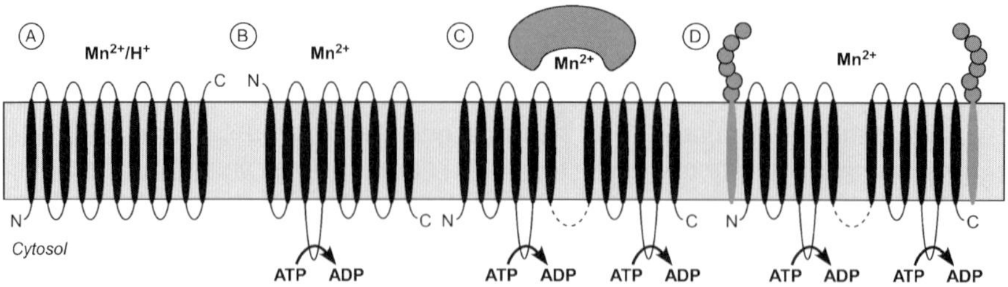

Fig. 1 *General properties of* Mn^{2+} *transporters.* The general protein properties of the various classes of Mn^{2+} transporters are shown. (A) MntH (Nramp) transporters of Gram-negative and Gram-positive bacteria are Mn^{2+}/H^+ symporters. The topology shown is based on that of *E. coli* {Courville, 2004 19543/id}. (B) MntA of the Gram-positive *L. plantarum* is a P-type ATPase {Hao, 1999 12063/id}. Its topology is uncertain as various computer algorithms predict 7–9 TM segments. (C) Putative structure of Mn^{2+}-transporting ABC ATPase systems of Gram-negative bacteria. In Gram-negative ABC ATPases, the ligand (in this case Mn^{2+}) is delivered to the transporter by a periplasmic binding protein. In general, ABC ATPases contain two homologous membrane domains of 5 or 6 TM segments each of which can be encoded as a single protein or as two proteins, hence the dotted line connecting the two halves of the protein. Further, each membrane domain is associated with a nucleotide binding domain. ATP hydrolysis is highly cooperative indicating that the two nucleotide binding domains interact. In some ABC ATPase transporters, the nucleotide binding domain is encoded as a separate protein (not shown). (D) Putative structure of Mn^{2+}-transporting ABC ATPase systems of Gram-positive bacteria. The overall structure of the transporter is similar to that of the Gram-negative bacteria with two homologous membrane domains and two nucleotide binding domains. As with Gram-negative bacteria, each domain may be encoded separately from the other and the membrane domains may or may not be fused. The major difference lies in the metal binding lipoprotein found in the Gram-positive bacteria compared to the soluble periplasmic binding protein found in the Gram-negative bacteria

MntH from *M. leprae* is also a divalent cation transporter (Reeve et al. 2002). Expression of *mntH* and *mntABCD* in *S. aureus* results in an increase in the cellular content of Mn^{2+} as determined by mass spectrometry; however, the content of Ca^{2+}, Cu^{2+}, Fe^{2+}, Mg^{2+}, Mo^{2+}, Ni^{2+}, and Zn^{2+} is not increased (Horsburgh et al. 2002). Thus, *mntH* of *S. aureus* is most likely a selective transporter of Mn^{2+}. Overexpression of *mntH* in *E. coli* results in the accumulation of Mn^{2+} as well as Fe^{2+} (Makui et al. 2000). Interestingly, *L. plantarum* has at least five Mn^{2+} transporters, three of which are MntH homologs (Groot et al. 2005), and the organism is capable of accumulating millimolar levels of Mn^{2+} (Archibald 1986; Archibald and Duong 1984). The importance of Mn^{2+} transporters in *L. plantarum* is presumably related to its unusual biology in that it does not require iron, unlike virtually all other known organisms.

Direct transport studies in *E. coli* and *S.* Typhimurium have detailed the transport capabilities of MntH (Kehres et al. 2000). The $K_{0.5}$ for Mn^{2+} uptake is 0.1 μM and 0.5–1.0 μM in *S.* Typhimurium and *E. coli*, respectively[1]. For both bacteria, the $K_{0.5}$ for iron is 100 μM, demonstrating that Mn^{2+} is the primary cation transported. Like the *M. tuberculosis* Nramp, MntH transport is influenced by pH. V_{max} for Mn^{2+} is sensitive to pH, increasing 3-fold in acidic conditions. However, $K_{0.5}$ for Mn^{2+} is not affected by pH. Conversely, the $K_{0.5}$ for Fe^{2+} decreases by 5–10 fold under acidic conditions.

2.2 The Bacterial ATP Binding Cassette ATPase Transporters

ABC ATPase transporters require the hydrolysis of ATP for transport and can be divided into two distinct but related classes. The first, primarily in Gram-positive bacteria, consists of the ABC transporters of the LraI family, which were initially described as putative adhesion proteins (Jenkinson 1994). However, many function exclusively as divalent cation transporters, frequently for Mn^{2+} (Kolenbrander et al. 1998). This class of ABC transporter is made up of an extracellular cation binding lipoprotein and an integral membrane domain required for cation transport that carries a cytoplasmic ATP binding domain. The second class, primarily in Gram-negative species, are typical ABC transporters, with a soluble periplasmic cation binding domain instead of a lipoprotein (Fig. 1).

2.2.1 ABC ATPase Transporters of the Gram-Positive Bacteria

ABC ATPase transporters have been identified in multiple species of *Streptococcus* (e.g., PsaA, FimA, AdcCBA), and *Staphylococcus* (SitABC, MntABCD) as well as *L. plantarum* (MtsC), *Corynebacterium diphtheriae* (MntABCD), *B. anthracis* (MntA), *B. subtilis* (MntABCD), and *Enterococcus faecalis* (EfaCBA). Their ability to transport Mn^{2+} has been inferred from growth, metal accumulation, and transcriptional studies (Dintilhac et al. 1997; Dintilhac and Claverys 1997; Gat et al. 2005; Groot et al. 2005; Hill et al. 1998; Horsburgh et al. 2002; Loo et al. 2003; Low et al. 2003; Oetjen et al. 2002; Que and Helmann 2000; Schmitt 2002). Specific characterization of the transport properties of ABC transporters is lacking except for limited characterization of ScaA of *S. gordonii*, MtsABC of *S. pyogenes* and SloC of *S. mutans*. *S. gordonii* ScaA has a $K_{0.5}$ for Mn^{2+} of 0.1–0.3 μM (Kolenbrander et al. 1998). Mn^{2+} inhibition of $^{55}Fe^{2+}$ transport is greater than Fe^{2+} inhibition of $^{54}Mn^{2+}$ uptake for *S. pyogenes* MtsABC, suggesting that Mn^{2+} is the favored substrate (Janulczyk et al. 2003), but actual affinities were not determined. Similarly, Mn^{2+} inhibits $^{55}Fe^{2+}$ transport by *S. mutans* SloC better than Fe^{2+} indicating

[1] Because all transport studies noted in this review were performed in intact cells rather than with purified reconstituted transport systems, the terms K_m or K_d are not appropriate. Instead, $K_{0.5}$ is used to indicate the concentration at which half-maximal transport occurs.

that Mn^{2+} is likely preferred over Fe^{2+} (Paik et al. 2003). Thus, despite indications of transport selectivity for Mn^{2+}, no ATPase transporter of this class in Gram-positive bacteria has been carefully characterized.

2.2.2 ABC ATPase Transporters of the Gram-Negative Bacteria

Potential Mn^{2+}-selective ABC ATPases in the Gram-negative bacteria are somewhat better characterized than in Gram-positive bacteria. *Synechocystis sp.* PCC 6803 , *S.* Typhimurium, *Shigella flexneri*, *Yersinia pestis*, *Porphyromonas gingivalis*, *Sinorhizobium meliloti*, and *Rhizobium leguminosarum* all possess an apparent Mn^{2+}-selective ABC transporter. Mn^{2+} transport by SitABCD of *S. flexneri*, YfeABCD of *Y. pestis*, FeoB2 of *P. gingivalis*, MntABCD/SitABCD of *S. meliloti*, and SitABCD of *R. leguminosarum* has been inferred from growth and expression studies (Bearden and Perry 1999; Chao et al. 2004; Dashper et al. 2005; Diaz-Mireles et al. 2004; Platero et al. 2003; Runyen-Janecky et al. 2003). Direct transport has been measured for ScaCBA of *Synechocystis* and SitABCD of *S.* Typhimurium. The $K_{0.5}$ of ScaCBA for Mn^{2+} transport is 1.0–3.0 μM with a V_{max} of 3–8 pmol/min/10^8 cells (Bartsevich and Pakrasi 1995, 1996). In *Salmonella*, the $K_{0.5}$ of SitABCD for Mn^{2+} transport is significantly lower at 0.1 μM and is independent of pH (Kehres et al. 2002b). Fe^{2+} is also transported, but at a $K_{0.5}$ of 3.0 to 50 μM due to pH dependencies. The V_{max} for Mn^{2+} uptake increases markedly in alkaline conditions, opposite to the results obtained for *S.* Typhimurium MntH under similar experimental parameters and suggesting that the two Mn^{2+}-selective transporters may have different physiological roles.

2.3 A Single P-type ATPase

MntA from *L. plantarum* falls into a third class, a P-type ATPase (Fig. 1), where the hydrolysis of ATP is coupled to transport within a single polypeptide chain (Hao et al. 1999a,b). The ATP dependent phosphorylation of an aspartyl residue is needed for cation binding and transport. Cd^{2+} uptake by MntA also occurs but can be inhibited only 30% by Mn^{2+}. Addition of Mn^{2+} to the growth medium represses the expression of *mntA*.

2.4 Cation Selectivities Remain Undefined

Cation selectivity and thus physiological function of the large majority of these transporters has not been adequately determined. Biologically relevant cations transported have usually been inferred from phenotypic or indirect assays or homology, as noted above. This lack of characterization is a serious problem in the study of these transporters. Homology allows one to infer that a protein is likely a cation transporter. Homology does not, however, allow inference about cation selectivity with any degree of confidence. For example, Nramp2 was presumed to solely transport Fe^{2+} due to iron deficiencies seen in animals with a mutant locus (Fleming et al. 1997). However, after direct transport measurements, Nramp2 was found to transport virtually all transition metal divalent cations (Gunshin et al. 1997). Examples such as this one occur frequently in the literature due to a lack of analysis of a broad range of cations on transport properties. In many cases, these observations result in inaccurate classification of the biologically relevant cations that are transported.

3 Regulation of Mn^{2+} Transport

Regulators that belong to several protein families are involved in homeostasis of manganese and other metals (see also the work by Helmann et al. 2006, in this volume).

3.1 MntR

MntR is best described in *B. subtilis* where it was originally identified (Que and Helmann 2000). MntR is a DtxR homolog. DtxR is the diphtheria toxin repressor, which is regulated by Fe^{2+} concentrations and is found in *C. diphtheriae* (Boyd et al. 1990). These proteins are metalloregulators which function as transcription factors regulating gene expression in response to metal concentration (see Table 1). Unlike DtxR, MntR is regulated by Mn^{2+} concentrations (Que and Helmann 2000). MntR is typically found as a dimer even in the absence of metal, whereas metal is required for dimerization of DtxR (Lieser et al. 2003; Tao et al. 1995). The crystal structures of the two proteins indicate that the metal binding sites in DtxR and MntR are very different (Glasfeld et al. 2003; Qiu et al. 1996). A distance of 9 Å separates the two bound metals of DtxR whereas MntR from *B. subtilis* appears to have three metal binding sites denoted as A, B, and C. In the "AB" conformation, they are separated by 3.3 Å but in the "AC" conformation they are separated by 4.4 Å (Glasfeld et al. 2003; Kliegman et al. 2006). Site A is always occupied by Mn^{2+}, but sites B or C may be bound by Mn^{2+} depending on the experimental conditions thus altering the structure of the MntR (Kliegman et al. 2006). MntR undergoes a dramatic conformation change upon binding Mn^{2+}, which stabilizes the protein (Golynskiy et al. 2005). Moreover, MntR′s affinity for DNA, specifically to a 16 bp palindrome (5′-TTTGCCTTAAGGAAAC-3′), is increased in the presence of Mn^{2+} (Lieser et al.

Table 1 Transcription factors regulating Mn^{2+} transport

Regulator	Organism	Transporters regulated	Cation response	Refs.
DtxR-like Regulators				
MntR	Many Gram-positive and Gram-negative bacteria	*mntH*, *sitABCD* and *sitABCD*-like transporters, *feoB2*	Mn^{2+}	(Dashper et al. 2005; Horsburgh et al. 2002; Ikeda et al. 2005; Kehres et al. 2002a; Patzer and Hantke 2001; Runyen-Janecky et al. 2003, 2006; Schmitt 2002; Sen et al. 2006; Zaharik et al. 2004)
SloR	*Streptococcus mutans*	*sloABC*	Mn^{2+}	(Paik et al. 2003)
ScaR	*S. gordonii*	*scaCBA*	Mn^{2+}	(Jakubovics et al. 2000, 2002)
AdcR	*S. gordonii*	*adcCBA*	Mn^{2+}	(Loo et al. 2003)
SirR	*S. epidermidis*	*sitABC*	Mn^{2+} and Fe^{2+}	(Hill et al. 1998)
EfaR	*E. faecalis*	*efaCBA*	Mn^{2+}	(Low et al. 2003)
MtsR	Group A Streptococci		Mn^{2+} and Fe^{2+}	(Bates et al. 2005)
Two-Component Kinase-Response Regulator				
ManS/R	*Synechocystis* sp. PCC6803	*mntABC*	Mn^{2+}	(Ogawa et al. 2002)
Fur-like Regulators				
Fur	*S.* Typhimurium, *E. coli*, *Y. pestis*, *S. meliloti* *S. flexneri*	*mntH*, *sitABCD* and *sitABCD*-like transporters	Fe^{2+}	(Bearden and Perry 1999; Ikeda et al. 2005; Kehres et al. 2002a; Patzer and Hantke 2001; Platero et al. 2004, 2003; Runyen-Janecky et al. 2003; Runyen-Janecky and Payne 2002)
Mur (Fur-like)	*R. leguminosarum*	*sitABCD*	Mn^{2+}	(Diaz-Mireles et al. 2004; Diaz-Mireles et al. 2005)
Other				
RfrA	*Synechocystis* sp. PCC6803	Unknown	Mn^{2+}	(Chandler et al. 2003)

2003; Que and Helmann 2000). In *B. subtilis*, MntR differentially regulates the expression of two Mn^{2+} transporters, *mntH* and *mntABCD* (Que and Helmann 2000). Under limiting Mn^{2+} conditions, non-metal bound MntR activates expression of *mntABCD*, whereas *mntH* is expressed independent of MntR. Under excess Mn^{2+} conditions, *mntABCD* and *mntH* are both repressed by Mn^{2+} bound MntR.

MntR regulates expression of transporters in *E. coli*, *S.* Typhimurium, *B. subtilis*, *S. aureus*, *C. diphtheriae*, *S. flexneri*, and *P. gingivalis* and many other species (Table 1) (Dashper et al. 2005; Horsburgh et al. 2002; Ikeda et al. 2005; Kehres et al. 2002a; Patzer and Hantke 2001; Runyen-Janecky et al. 2003; Schmitt 2002; Zaharik et al. 2004). Both *mntH* and *sitABCD* expression from *E. coli* and *S.* Typhimurium is inhibited by excess Mn^{2+} through the binding of MntR to a 18 bp palindrome with a 4 bp spacer (5′-AAACATAGC$_{CAAA}$GCTATGTTT-3′) found in both promoters (Ikeda et al. 2005; Kehres et al. 2002a; Patzer and Hantke 2001). As in *B. subtilis*, MntR was found to differentially regulate expression of *mntH* and *mntABC* from *S. aureus* (Horsburgh et al. 2002). MntR negatively regulates expression of *mntABC*, but positively regulates expression of *mntH*. In *C. diphtheriae* MntR represses *mntABCD* in response to excess Mn^{2+} by directly interacting with the *mntABCD* promoter (Schmitt 2002). Mn^{2+} was found to inhibit *sitABCD* expression in *S. flexneri*, and a MntR box is present in its promoter (Runyen-Janecky et al. 2003). *feoB2* is repressed by the FeoA-MntR hybrid protein in *P. gingivalis* (Dashper et al. 2005). Finally, a MntR-like protein, AntR exists in *B. anthracis* and can act as a Mn^{2+} binding repressor (Sen et al. 2006). Thus, MntR appears to be the primary transcription factor involved in Mn^{2+}-dependent gene expression throughout prokaryotes.

3.2 PerR and OxyR

PerR and OxyR are hydrogen peroxide sensors that control expression of the peroxide regulon in response to peroxide stress . Unlike OxyR, PerR is also regulated by metals (Table 1) (Dornan and Hupp 2001). In *B. subtilis*, PerR is affected by the binding of either Fe^{2+} or Mn^{2+}. Mn^{2+} appears to inhibit expression of the PerR regulon whereas Fe^{2+} stimulates expression via PerR. When Fe^{2+} is bound during exposure to H_2O_2, rapid oxidation of PerR occurs inactivating the protein and derepressing the PerR regulon (Lee and Helmann 2006a,b). When Mn^{2+} is bound, this does not occur. PerR and OxyR regulate the expression of some Mn^{2+} transporters. PerR directly regulates expression of *mntABC* but not *mntH* in *S. aureus*, *mntABC* in *Neisseria gonorrhoeae*, or *mtsABC* in *S. pyogenes* (Horsburgh et al. 2001a,b; Ricci et al. 2002; Wu et al. 2006). Moreover, regulation of PerR is affected by Mn^{2+} transport (Horsburgh et al. 2002). OxyR regulates the expression of *mntH* in *S.* Typhimurium and *E. coli* (Ikeda et al. 2005; Kehres et al. 2002a; Patzer and Hantke 2001).

Sensitivity to H_2O_2 is altered by expression of Mn^{2+} transporters, but the mechanism is unknown and does not appear to depend completely on expression of the Mn^{2+}-dependent superoxide dismutase, SodA.

3.3 Other Metalloregulators

Numerous other factors are known to regulate Mn^{2+} transport including SloR, ScaR, AdcR, SirR, EfaR, MtsR, Fur, Mur, ManR/S, and RfrA. Many, though not all, belong to the DtxR family (Table 1). SloR is a metal-dependent regulator in *S. mutans* that inhibits the expression of *sloABC* in response to excess Mn^{2+} (Paik et al. 2003). *S. gordonii* ScaR is a DtxR-like protein which regulates expression of *scaCBA* in response to Mn^{2+} (Jakubovics et al. 2000, 2002). Interestingly, a second ATPase operon in *S. gordonii*, *adcCBA*, is regulated by a separate factor AdcR in a Mn^{2+} dependent manner (Loo et al. 2003). Expression of *sitABC* from *S. epidermidis* is inhibited by SirR, a DtxR-like protein, in response to both Fe^{2+} and Mn^{2+} (Hill et al. 1998). EfaR, another DtxR-like protein, regulates expression of *efaCBA* in *E. faecalis* in response to Mn^{2+} (Low et al. 2003). In group A streptococci, MtsR is a DtxR-like Fe^{2+} and Mn^{2+} dependent regulator (Bates et al. 2005). A two-component regulatory system named ManS/R regulates expression of *mntABC* in *Synechocystis* sp. PCC 6803 (Ogawa et al. 2002). RfrA is a suppressor of Mn^{2+} transport in the same organism but is unrelated to either DtxR or Fur. The transporter it regulates has not been identified (Chandler et al. 2003).

Fur, a Fe^{2+}-sensitive metalloregulator, regulates expression of both *S.* Typhimurium and *E. coli mntH* and *sitABCD* as well as *sitABCD*-like transporters in *Y. pestis*, *S. meliloti* and *S. flexneri* (Bearden and Perry 1999; Ikeda et al. 2005; Kehres et al. 2002a; Patzer and Hantke 2001; Platero et al. 2004, 2003; Runyen-Janecky et al. 2003; Runyen-Janecky and Payne 2002). Mur, a Fur-like protein from *R. leguminosarum*, represses *sitABCD* in response to excess Mn^{2+} (Diaz-Mireles et al. 2004). It is unknown why a Fe^{2+}-responsive transcription factor would regulate Mn^{2+} transport.

4 Role of Mn^{2+} Transport in Virulence

Mn^{2+} transport in both Gram-positive and Gram-negative bacteria has been linked to virulence. The reason(s) for which Mn^{2+} transport is essential for virulence is unknown in all cases, and it is unclear at this time whether similar or different mechanisms exist in different bacteria. Generally, Mn^{2+} transport appears to be relevant during intracellular interactions of the bacterium and the macrophage, for adhesion, and/or protection from oxidative stress.

4.1
S. Typhimurium

The relationship between manganese transport and pathogenesis has been most completely studied in *S.* Typhimurium. *S.* Typhimurium causes gastroenteritis in humans and typhoid fever in mice. Upon ingestion of live bacteria, the bacteria actively invade epithelial cells in the intestine near the Peyer's patches. The bacteria will then enter underlying macrophages, which express the transition metal divalent cation transporter, Nramp1. The infected macrophages disseminate throughout the body. While inside the macrophage, the bacteria reside in a specialized organelle, the *Salmonella* containing vacuole (SCV). Studies of bacterial gene expression and survival in the SCV as well as mouse virulence studies reveal that the bacteria monitor the host cell Nramp status. Altogether, these data strongly indicate that Mn^{2+} transport is important while the bacteria are inside the SCV.

Comparing $Nramp1^{+/+}$ to $Nramp1^{-/-}$ IFNγ induced RAW264.7 mouse macrophages, expression of both of *S.* Typhimurium Mn^{2+} transporters, *mntH* and *sitABCD*, is 10–20 fold higher in the $Nramp^{+/+}$ background starting around 9 hours post invasion (Boyer et al. 2002; Kehres et al. 2000; Zaharik et al. 2002a,b 2004; Zaharik 2003; Zaharik and Finlay 2004). Survival of wild-type bacteria is significantly decreased by the presence of a functional Nramp1 in IFNγ activated RAW264.7 macrophage-like cells. However, in IFNγ activated $Nramp1^{+/+}$ and $Nramp^{-/-}$ peritoneal macrophages from C57BL/6 mice, there is no survival defect of wild-type bacteria up to 18 hours post invasion. Survival studies with mutants of *mntH* and *sitABCD* in the RAW264.7 macrophage-like cell line compared to the primary macrophages are also different. Little or no decrease in survival was observed in the $Nramp1^{+/+}$ versus $Nramp1^{-/-}$ RAW264.7 cells for the *mntH* and *sitABCD* single and double mutants compared to wild type. In $Nramp1^{-/-}$ primary macrophages, all three mutants survived equally to wild type. In contrast, the *sitABCD* mutant and the *mntH sitABCD* double mutant survived very poorly in the $Nramp1^{+/+}$ primary macrophages.

Animal studies using *S.* Typhimurium have been conducted in a variety of mouse strains using various methods of inoculation. The results further support the importance of Mn^{2+} transport in virulence. *mntH* and *sitABCD* mutants are somewhat attenuated in their abilities to infect BALB/c mice ($Nramp1^{-/-}$) upon oral inoculation compared to wild type (Janakiraman and Slauch 2000; Kehres et al. 2000). In C3H mice ($Nramp^{+/+}$), both *mntH* and *sitABCD* single mutants are markedly attenuated upon oral inoculation (Kehres and Maguire, unpublished observations). Intravenous inoculation of congenic 129/Sv $Nramp1^{+/+}$ and 129/Sv$Nramp1^{-/-}$ mice with different mutants of *mntH*, *sitABCD*, and *feoB* indicated that *sitABCD* and *feoB* are necessary for virulence in 129/Sv $Nramp1^{-/-}$ mice (Boyer et al. 2002). The dose of wild-type bacteria given to 129/Sv $Nramp1^{+/+}$ mice was insufficient to cause death in this study so no conclusions could be made about the role of *mntH* or *sitABCD* in the presence of Nramp1$^{+/+}$. A subsequent study using oral inoculation of congenic C57/BL/6 Nramp1$^{+/+}$ and Nramp1$^{-/-}$ mice with strains carrying mutations in *mntH* and *sitABCD* revealed that the Nramp1$^{-/-}$ mice were susceptible to infection whether or not the bacterial strain could express a Mn^{2+} transporter (Zaharik et al. 2004). However, the Nramp1$^{+/+}$ mice were killed only by wild-type bacteria. *mntH* and *sitABCD* mutants are significantly attenuated, and the *mntH sitABCD* double mutant was avirulent. The overall conclusion from these studies in macrophages and mice is that if the animal host cell possesses a functional Nramp1, the Mn^{2+} transporter status of the bacterium is crucial for *S.* Typhimurium survival.

4.2 Other Gram-Negative Bacteria

Studies have also been conducted on the role of Mn^{2+} transport in the virulence of *Y. pestis* and *S. flexneri*; both are intracellular pathogens that infect macrophages (Bearden and Perry 1999; Lucchini et al. 2005). The LD_{50} for the *Y. pestis yfeABCD* mutant is increased 100-fold compared to wild type after subcutaneous inoculation of NIH/Swiss Webster mice (Nramp1$^{+/+}$). *sitABCD* of *S. flexneri* is upregulated upon infection of HeLa (human epithelial) and U937 (human macrophage-like) cells. However, the *sitA* mutant of *S. flexneri* grows as well as wild type on Henle (human epithelial) cells (Runyen-Janecky et al. 2003).

4.3 *Mycobacterium Tuberculosis*

Much like the Gram-negative bacteria described above, the Gram-positive *M. tuberculosis* is also an intracellular pathogen targeting primarily macrophages. There are no differences in survival between wild type or a Nramp mutant of *M. tuberculosis* after invasion of Nramp1$^{+/+}$ and Nramp1$^{-/-}$ bone-marrow derived mouse macrophages (Boechat et al. 2002). These results are similar to those obtained in *S.* Typhimurium (Boyer et al. 2002; Zaharik et al. 2002b, 2004; Zaharik 2003; Zaharik and Finlay 2004). Moreover, there was no difference in bacterial numbers in the lung and spleen of a BALB/c mice (Nramp1$^{-/-}$) intravenously inoculated with wild type or the Nramp mutant of *M. tuberculosis*. Thus, the general literature consensus is that neither mammalian Nramp1 nor the *M. tuberculosis* Nramp is important for virulence. This conclusion is at best incomplete since, unfortunately, no experiments have been performed in Nramp1$^{+/+}$ mice or congenic Nramp1 mice. These experiments are critical, since as discussed for *S.* Typhimurium above, the Nramp1 status of the host is an important factor during bacterial infection.

4.4 Oral Streptococci

The LraI transport protein family is widespread in oral streptococci such as *S. parasanguis*, *S. mutans*, *S. gordonii*, *S. pyogenes*, and *S. pneumoniae*. These proteins were originally identified for their roles in virulence by their ability to aid in adherence to oral cavities. However, many of these LraI family proteins appear to function as Mn^{2+} transporters. *sloC* and *sloA* mutants of *S. mutans* are attenuated in a rat model for endocarditis (Paik et al. 2003). However, the *sloC* mutant and wild type colonize and proliferate similarly in a gnotobiotic rat model of caries (Kitten et al. 2000). Mutation of *mtsABC* from *S. pyogenes* results in a 40% decrease in death in BALB/c mice ($Nramp1^{-/-}$) compared to wild type *S. pyogenes* (Janulczyk et al. 2003). Upon intranasal or intraperitoneal inoculation of mice, the *psaA* mutant of *S. pneumoniae* is attenuated compared to wild type (Berry and Paton 1996). In addition *psaA*, *psaB*, and *psaC* mutants are attenuated in several mouse models for infection (Marra et al. 2002). Moreover, the *psaA* mutant adheres one tenth as well as the wild type to A549 human lung alveolar carcinoma cells. Overall, the LraI family of Mn^{2+} transporters appears important for virulence in the streptococci.

4.5 Other Gram-Positive Bacteria

E. faecalis, *S. aureus* and *B. anthracis* are also involved in pathogenesis. An *efaA* mutant is attenuated in outbred ICR mice (Nramp1 status unknown) (Low et al. 2003; Singh et al. 1998). A *mntH mntA* mutant of *S. aureus* is attenuated in an abscess model for infection in mice (Horsburgh et al. 2002). Moreover, constitutive repression of *mntABC* by a DtxR E175K mutant results in attenuated virulence in a skin abscess model (Ando et al. 2003). *B. anthracis mntA* mutant cells cannot form spores in RAW264.7 macrophage-like cells. Moreover, lysis of the RAW264.7 cells is delayed (Gat et al. 2005). These phenotypes can be rescued by the addition of 5 μM extracellular Mn^{2+}. The *mntA* mutant is greatly attenuated in a guinea pig model for *B. anthracis* infection; however, the classic virulence proteins of *B. anthracis* are not altered.

5 Mn^{2+} Dependent Enzymes

If Mn^{2+} transport is required for virulence and other functions of bacteria, it follows that one or more Mn^{2+}-dependent enzymes are required for these same functions. However, few Mn^{2+}-dependent enzymes are known and even fewer have been studied in the context of virulence. The following discussion of some potentially important Mn^{2+}-dependent enzymes serves as an illustration both of how little is known about Mn^{2+}-dependent enzymes and Mn^{2+} homeostasis and of opportunities for new and likely areas of investigation.

5.1 Manganese Superoxide Dismutase (Mn-SOD)

Prokaryotes express the enzyme superoxide dismutase as a means of detoxifying reactive oxygen species and preventing cellular damage. SodA is the manganese-containing form of this enzyme, first identified and purified from *E. coli* B (Enkvetchakul et al. 2004) but since found to be wide spread in prokaryotes. *sodA* is the only Mn^{2+}-dependent enzyme known to have an effect on virulence. Increased expression of SodA greatly increases the survival of *S.* Typhimurium and *S. flexneri* in macrophages (Franzon et al. 1990; Tsolis et al. 1995). Tsolis and colleagues showed that mutation of *sodA* in *S.* Typhimurium does not affect virulence in BALB/c mice (Tsolis et al. 1995). However, these mice are Nramp1$^{-/-}$. Since the effect of mutation of Mn^{2+} transporters depends on the Nramp1 status of the mouse, it follows that this could likewise affect the ability of Mn^{2+}-dependent enzymes to contribute to virulence. In contrast to the result in *S.* Typhimurium, oral infection of BALB/c mice or of rabbits with a *sodA* null mutant of *Y. enterocolitica* showed strong attenuation compared to wild type (Najdenski et al. 2004; Roggenkamp et al. 1997).

5.2 Protein Kinases and Phosphatases

Phosphorylation is one of the most common means of post-translational modification used by both eukaryotic and prokaryotic organisms in signal transduction pathways. In recent years several protein kinases and phosphatases have been identified in bacteria that phosphorylate/dephosphorylate Ser, Thr or Tyr residues in addition to the canonical two-component phosphorylation of His and Asp residues. The dependence of some of these protein O-phosphatases on manganese has been recently reviewed (Shi 2004). While several phosphatases and a few kinases will be discussed here, it should be noted that in many cases the physiological functions and substrates are not known. Additionally, metal ion preference is not always determined, although it has been for the enzymes presented here.

Perhaps the best characterized protein phosphatases in bacteria are PrpA and PrpB from *E. coli* and *S.* Typhimurium. PrpA and PrpB are each able to dephosphorylate Ser, Thr or Tyr residues. In *E. coli* these phosphatases have been shown to work with the CpxR–CpxA system to modulate cellular response to heat shock by regulating transcription of *htrA* (Missiakas and Raina 1997). PrpA and PrpB in *S.* Typhimurium required Mn^{2+} for catalytic activity and showed very little to no activity when other divalent cations were tested. The K_a of PrpA for Mn^{2+} was determined to be 65 μM, while the K_a for PrpB was 1 μM (Shi et al. 2001). There are many other prokaryotes whose protein phosphatases/kinases are dependent on manganese including those of *Pseudomonas aeruginosa* (Mukhopadhyay et al. 1999), the cyanobacterium *Microcystis aeruginosa* (Shi et al. 1999), *S. pneumoniae* (Morona et al. 2002), *S. agalactiae* (Rajagopal et al. 2003), *Sulfolobus solfataricus* (Lower and Kennelly 2002, 2003), *Listeria monocytogenes* (Archambaud et al. 2005), *Myxococcus xanthus* (Udo et al. 1997), and *M. tuberculosis* (Boitel et al. 2003; Chaba et al. 2002). In *S. agalactiae* virulence is attenuated in a neonatal rat sepsis model when mutations are made in the kinase Stk1 and the phosphatase Stp1 (Rajagopal et al. 2003). The role of the Ser/Thr/Tyr Mn^{2+}-dependent protein kinases and phosphatases is an area that remains to be investigated.

5.3 Stringent Response

Many bacteria normally maintain low levels of the nucleotide guanosine 3′-diphosphate 5′-diphosphate (ppGpp). Under stringent growth conditions such as amino acid starvation the degradation of ppGpp is decreased, increasing the concentration of the 'stringent factor' in the cells (Gentry et al. 1993). ppGpp is degraded to GDP by the *spoT* gene product (Gentry and Cashel 1996; Sarubbi et al. 1989). Mn^{2+} is a required co-factor for this degradation (Heinemeyer and Richter 1978; Polakis et al. 1973; Sy 1977), but the role of ppGpp and *spoT* in virulence is unknown.

5.4 Cyclic Diguanylate

3′-5′-Cyclic diguanylate is a novel global second messenger recently described in bacteria (Simm et al. 2005). Proteins containing GGDEF (GlyGlyAspGluPhe) motifs synthesize cyclic diguanylate while a family containing an EAL (GluAlaLeu) motif have phosphodiesterase activity and degrade the cyclic diguanylate. The latter activity is dependent on Mn^{2+} (Schmidt et al. 2005; Tamayo et al. 2005). Evidence has been published for cyclic diguanylate involvement in biofilm formation in *S.* Typhimurium, *P. aeruginosa*, and *Vibrio cholerae*, regulation of virulence gene expression in *V. cholerae*, and regulation of sessility and motility in *E. coli*, *S.* Typhimurium and *P. aeruginosa* (Garcia et al. 2004; Hickman et al. 2005; Simm et al. 2005; Tischler and Camilli 2004, 2005).

6
Conclusions

Two major classes of Mn^{2+} transporters, membrane potential driven MntH/Nramp and ABC-type ATPases, have been identified in prokaryotes. Many of these transporters can transport Fe^{2+}, Zn^{2+}, Cd^{2+} and often other transition metal divalent cations. Since, in most microbial species, relative affinities of the transported cations have not been determined, the physiologically relevant cation(s) for each transporter is unknown. Mn^{2+} transport in most species is regulated by a Mn^{2+}-specific transcription factor MntR and often by other factors including Fur, PerR and OxyR. Mutation of these Mn^{2+} transport systems in a variety of organisms clearly attenuates virulence in animal models of infection. The degree of attenuation appears dependent on the presence of the Nramp1 homolog in the host animal. Finally, overall Mn^{2+} homeostasis is poorly understood at present, since the roles of individual Mn^{2+}-dependent enzymes have yet to be elucidated.

References

Agranoff D, Monahan IM, Mangan JA, Butcher PD, Krishna S (1999) *Mycobacterium tuberculosis* expresses a novel pH-dependent divalent cation transporter belonging to the Nramp family. J Exp Med 190:717–724

Ando M, Manabe YC, Converse PJ, Miyazaki E, Harrison R, Murphy JR, Bishai WR (2003) Characterization of the role of the divalent metal ion-dependent transcriptional repressor MntR in the virulence of *Staphylococcus aureus*. Infect Immun 71:2584–2590

Archambaud C, Gouin E, Pizarro-Cerda J, Cossart P, Dussurget O (2005) Translation elongation factor EF-Tu is a target for Stp, a serine-threonine phosphatase involved in virulence of *Listeria monocytogenes*. Molec Microbiol 56:383–396

Archibald F (1986) Manganese: its acquisition by and function in the lactic acid bacteria. Crit Rev Microbiol 13:63–109

Archibald FS, Duong MN (1984) Manganese acquisition by *Lactobacillus plantarum*. J Bacteriol 158:1–8

Bartsevich VV, Pakrasi HB (1995) Molecular identification of an ABC transporter complex for manganese: analysis of a cyanobacterial mutant strain impaired in the photosynthetic oxygen evolution process. EMBO J 14:1845–1853

Bartsevich VV, Pakrasi HB (1996) Manganese transport in the cyanobacterium *Synechocystis* sp. PCC 6803. J Biol Chem 271:26057–26061

Bates CS, Toukoki C, Neely MN, Eichenbaum Z (2005) Characterization of MtsR, a new metal regulator in group A *Streptococcus*, involved in iron acquisition and virulence. Infect Immun 73:5743–5753

Bearden SW, Perry RD (1999) The Yfe system of *Yersinia pestis* transports iron and manganese and is required for full virulence of plague. Mol Microbiol 32:403–414

Berry AM, Paton JC (1996) Sequence heterogeneity of PsaA, a 37-kilodalton putative adhesin essential for virulence of *Streptococcus pneumoniae*. Infect Immun 64:5255–5262

Boechat N, Lagier-Roger B, Petit S, Bordat Y, Rauzier J, Hance AJ, Gicquel B, Reyrat JM (2002) Disruption of the gene homologous to mammalian Nramp1 in *Mycobacterium tuberculosis* does not affect virulence in mice. Infect Immun 70:4124–4131
Boitel B, Ortiz-Lombardia M, Duran R, Pompeo F, Cole ST, Cervenansky C, Alzari PM (2003) PknB kinase activity is regulated by phosphorylation in two Thr residues and dephosphorylation by PstP, the cognate phospho-Ser/Thr phosphatase, in *Mycobacterium tuberculosis*. Mol Microbiol 49:1493–1508
Boyd J, Oza MN, Murphy JR (1990) Molecular cloning and DNA sequence analysis of a diphtheria toxin iron-dependent regulatory element (*dtxR*) from *Corynebacterium*. Proc Natl Acad Sci USA 87:5968–5972
Boyer E, Bergevin I, Malo D, Gros P, Cellier MF (2002) Acquisition of Mn(II) in addition to Fe(II) is required for full virulence of *Salmonella enterica* serovar Typhimurium. Infect Immun 70:6032–6042
Brown IN, Glynn AA, Plant J (1982) Inbred mouse strain resistance to *Mycobacterium lepraemurium* follows the *Ity/Lsh* pattern. Immunology 47:149–156
Canonne-Hergaux F, Gruenheid S, Ponka P, Gros P (1999) Cellular and subcellular localization of the Nramp2 iron transporter in the intestinal brush border and regulation by dietary iron. Blood 93:4406–4417
Cellier MF, Bergevin I, Boyer E, Richer E (2001) Polyphyletic origins of bacterial Nramp transporters. Trends Genet 17:365–370
Chaba R, Raje M, Chakraborti PK (2002) Evidence that a eukaryotic-type serine/threonine protein kinase from *Mycobacterium tuberculosis* regulates morphological changes associated with cell division. Eur J Biochem 269:1078–1085
Chandler LE, Bartsevich VV, Pakrasi HB (2003) Regulation of manganese uptake in *Synechocystis* sp. 6803 by RfrA, a member of a novel family of proteins containing a repeated five-residues domain. Biochemistry 42:5508–5514
Chao TC, Becker A, Buhrmester J, Puhler A, Weidner S (2004) The *Sinorhizobium meliloti fur* gene regulates, with dependence on Mn(II), transcription of the *sitABCD* operon, encoding a metal-type transporter. J Bacteriol 186:3609–3620
Chua AC, Morgan EH (1997) Manganese metabolism is impaired in the Belgrade laboratory rat. J Comp Physiol B 167:361–369
Courville P, Chaloupka R, Veyrier F, Cellier MF (2004) Determination of transmembrane topology of the *Escherichia coli* natural resistance-associated macrophage protein (Nramp) ortholog. J Biol Chem 279:3318–3326
Dashper SG, Butler CA, Lissel JP, Paolini RA, Hoffmann B, Veith PD, O'Brien-Simpson NM, Snelgrove SL, Tsiros JT, Reynolds EC (2005) A novel *Porphyromonas gingivalis* FeoB plays a role in manganese accumulation. J Biol Chem 280:28095–28102
Diaz-Mireles E, Wexler M, Sawers G, Bellini D, Todd JD, Johnston AW (2004) The Fur-like protein Mur of *Rhizobium leguminosarum* is a Mn^{2+}-responsive transcriptional regulator. Microbiology 150:1447–1456
Diaz-Mireles E, Wexler M, Todd JD, Bellini D, Johnston AW, Sawers RG (2005) The manganese-responsive repressor Mur of *Rhizobium leguminosarum* is a member of the Fur-superfamily that recognizes an unusual operator sequence. Microbiology 151:4071–4078
Dintilhac A, Alloing G, Granadel C, Claverys JP (1997) Competence and virulence of *Streptococcus pneumoniae*: Adc and PsaA mutants exhibit a requirement for Zn and Mn resulting from inactivation of putative ABC metal permeases. Mol Microbiol 25:727–739
Dintilhac A, Claverys JP (1997) The *adc* locus, which affects competence for genetic transformation in *Streptococcus pneumoniae*, encodes an ABC transporter with a pu-

tative lipoprotein homologous to a family of streptococcal adhesins. Res Microbiol 148:119–131

Dornan D, Hupp TR (2001) Inhibition of p53-dependent transcription by BOX-I phosphopeptide mimetics that bind to p300. EMBO Rep 2:139–144

Enkvetchakul D, Bhattacharyya J, Jeliazkova I, Groesbeck DK, Cukras CA, Nichols CG (2004) Functional characterization of a prokaryotic Kir channel. J Biol Chem 279:47076–47080

Fleming MD, Romano MA, Su MA, Garrick LM, Garrick MD, Andrews NC (1998) Nramp2 is mutated in the anemic Belgrade (*b*) rat: evidence of a role for Nramp2 in endosomal iron transport. Proc Natl Acad Sci USA 95:1148–1153

Fleming MD, Trenor CC, Su MA, Foernzler D, Beier DR, Dietrich WF, Andrews NC (1997) Microcytic anaemia mice have a mutation in Nramp2, a candidate iron transporter gene. Nat Genet 16:383–386

Forbes JR, Gros P (2003) Iron, manganese, and cobalt transport by Nramp1 (Slc11a1) and Nramp2 (Slc11a2) expressed at the plasma membrane. Blood 102:1884–1892

Franzon VL, Arondel J, Sansonetti PJ (1990) Contribution of superoxide dismutase and catalase activities to *Shigella flexneri* pathogenesis. Infect Immun 58:529–535

Garcia B, Latasa C, Solano C, Garcia-del PF, Gamazo C, Lasa I (2004) Role of the GGDEF protein family in *Salmonella* cellulose biosynthesis and biofilm formation. Mol Microbiol 54:264–277

Garrick MD, Kuo HC, Vargas F, Singleton S, Zhao L, Smith JJ, Paradkar P, Roth JA, Garrick LM (2006a) Comparison of mammalian cell lines expressing distinct isoforms of divalent metal transporter 1 in a tetracycline-regulated fashion. Biochem J 398:539–546

Garrick MD, Singleton ST, Vargas F, Kuo HC, Zhao L, Knopfel M, Davidson T, Costa M, Paradkar P, Roth JA, Garrick LM (2006b) DMT1: which metals does it transport? Biol Res 39:79–85

Gat O, Mendelson I, Chitlaru T, Ariel N, Altboum Z, Levy H, Weiss S, Grosfeld H, Cohen S, Shafferman A (2005) The solute-binding component of a putative Mn(II) ABC transporter (MntA) is a novel *Bacillus anthracis* virulence determinant. Molec Microbiol 58:533–551

Gentry DR, Cashel M (1996) Mutational analysis of the *Escherichia coli spoT* gene identifies distinct but overlapping regions involved in ppGpp synthesis and degradation. Mol Microbiol 19:1373–1384

Gentry DR, Hernandez VJ, Nguyen LH, Jensen DB, Cashel M (1993) Synthesis of the stationary-phase sigma factor s^S is positively regulated by ppGpp. J Bacteriol 175:7982–7989

Glasfeld A, Guedon E, Helmann JD, Brennan RG (2003) Structure of the manganese-bound manganese transport regulator of *Bacillus subtilis*. Nat Struct Biol 10:652–657

Glynn AA, Bradley DJ, Blackwell JM, Plant JE (1982) Genetics of resistance to infection. Lancet 2:151

Golynskiy MV, Davis TC, Helmann JD, Cohen SM (2005) Metal-induced structural organization and stabilization of the metalloregulatory protein MntR. Biochemistry 44:3380–3389

Goswami T, Bhattacharjee A, Babal P, Searle S, Moore E, Li M, Blackwell JM (2001) Natural-resistance-associated macrophage protein 1 is an H^+/bivalent cation antiporter. Biochem J 354:511–519

Groot MN, Klaassens E, De Vos WM, Delcour J, Hols P, Kleerebezem M (2005) Genome-based in silico detection of putative manganese transport systems in *Lactobacillus plantarum* and their genetic analysis. Microbiology 151:1229–1238

Gros P, Skamene E, Forget A (1983) Cellular mechanisms of genetically controlled host resistance to *Mycobacterium bovis* (BCG). J Immunol 131:1966–1972

Gruenheid S, Cellier M, Vidal S, Gros P (1995) Identification and characterization of a second mouse Nramp gene. Genomics 25:514–525

Gunshin H, Mackenzie B, Berger UV, Gunshin Y, Romero MF, Boron WF, Nussberger S, Gollan JL, Hediger MA (1997) Cloning and characterization of a mammalian proton-coupled metal-ion transporter. Nature 388:482–488

Hao Z, Chen S, Wilson DB (1999a) Cloning, expression and characterization of cadmium and manganese uptake genes from *Lactobacillus plantarum*. Appl Environ Microbiol 65:4746–4752

Hao Z, Reiske HR, Wilson DB (1999b) Characterization of cadmium uptake in *Lactobacillus plantarum* and isolation of cadmium and manganese uptake mutants. Appl Environ Microbiol 65:4741–4745

Heinemeyer EA, Richter D (1978) Mechanism of the in vitro breakdown of guanosine 5′-diphosphate 3′-diphosphate in *Escherichia coli*. Proc Natl Acad Sci USA 75:4180–4183

Hickman JW, Tifrea DF, Harwood CS (2005) A chemosensory system that regulates biofilm formation through modulation of cyclic diguanylate levels. Proc Natl Acad Sci USA 102:14422–14427

Hill PJ, Cockayne A, Landers P, Morrissey JA, Sims CM, Williams P (1998) SirR, a novel iron-dependent repressor in *Staphylococcus epidermidis*. Infect Immun 66:4123–4129

Horsburgh MJ, Clements MO, Crossley H, Ingham E, Foster SJ (2001a) PerR controls oxidative stress resistance and iron storage proteins and is required for virulence in *Staphylococcus aureus*. Infect Immun 69:3744–3754

Horsburgh MJ, Ingham E, Foster SJ (2001b) In *Staphylococcus aureus*, *fur* is an interactive regulator with PerR, contributes to virulence, and Is necessary for oxidative stress resistance through positive regulation of catalase and iron homeostasis. J Bacteriol 183:468–475

Horsburgh MJ, Wharton SJ, Cox AG, Ingham E, Peacock S, Foster SJ (2002) MntR modulates expression of the PerR regulon and superoxide resistance in *Staphylococcus aureus* through control of manganese uptake. Mol Microbiol 44:1269–1286

Ikeda JS, Janakiraman A, Kehres DG, Maguire ME, Slauch JM (2005) Transcriptional regulation of *sitABCD* of *Salmonella enterica* serovar Typhimurium by MntR and Fur. J Bacteriol 187:912–922

Jakubovics NS, Smith AW, Jenkinson HF (2000) Expression of the virulence-related Sca (Mn^{2+}) permease in *Streptococcus gordonii* is regulated by a diphtheria toxin metallorepressor-like protein ScaR. Mol Microbiol 38:140–153

Jakubovics NS, Smith AW, Jenkinson HF (2002) Oxidative stress tolerance is manganese (Mn^{2+}) regulated in *Streptococcus gordonii*. Microbiology 148:3255–3263

Janakiraman A, Slauch JM (2000) The putative iron transport system SitABCD encoded on SPI1 is required for full virulence of *Salmonella typhimurium*. Mol Microbiol 35:1146–1155

Janulczyk R, Ricci S, Bjorck L (2003) MtsABC is important for manganese and iron transport, oxidative stress resistance, and virulence of *Streptococcus pyogenes*. Infect Immun 71:2656–2664

Jenkinson HF (1994) Cell surface protein receptors in oral streptococci. FEMS Microbiol Lett 121:133–140

Kehres DG, Janakiraman A, Slauch JM, Maguire ME (2002a) Regulation of *Salmonella enterica* serovar Typhimurium *mntH* transcription by H_2O_2, Fe^{2+}, and Mn^{2+}. J Bacteriol 184:3151–3158

Kehres DG, Janakiraman A, Slauch JM, Maguire ME (2002b) SitABCD Is the alkaline Mn^{2+} transporter of *Salmonella enterica* serovar Typhimurium. J Bacteriol 184:3159–3166
Kehres DG, Zaharik ML, Finlay BB, Maguire ME (2000) The NRAMP proteins of *Salmonella typhimurium* and *Escherichia coli* are selective manganese transporters involved in the response to reactive oxygen. Molec Microbiol 36:1085–1100
Kitten T, Munro CL, Michalek SM, Macrina FL (2000) Genetic characterization of a *Streptococcus mutans* LraI family operon and role in virulence. Infect Immun 68:4441–4451
Kliegman JI, Griner SL, Helmann JD, Brennan RG, Glasfeld A (2006) Structural basis for the metal-selective activation of the manganese transport regulator of *Bacillus subtilis*. Biochemistry 45:3493–3505
Kolenbrander PE, Andersen RN, Baker RA, Jenkinson HF (1998) The adhesion-associated *sca* operon in *Streptococcus gordonii* encodes an inducible high-affinity ABC transporter for Mn^{2+} uptake. J Bacteriol 180:290–295
Lee JW, Helmann JD (2006a) Biochemical characterization of the structural Zn^{2+} site in the *Bacillus subtilis* peroxide sensor PerR. J Biol Chem 281:23567–23578
Lee JW, Helmann JD (2006b) The PerR transcription factor senses H_2O_2 by metal-catalysed histidine oxidation. Nature 440:363–367
Lieser SA, Davis TC, Helmann JD, Cohen SM (2003) DNA-binding and oligomerization studies of the manganese(II) metalloregulatory protein MntR from *Bacillus subtilis*. Biochemistry 42:12634–12642
Loo CY, Mitrakul K, Voss IB, Hughes CV, Ganeshkumar N (2003) Involvement of the *adc* operon and manganese homeostasis in *Streptococcus gordonii* biofilm formation. J Bacteriol 185:2887
Low YL, Jakubovics NS, Flatman JC, Jenkinson HF, Smith AW (2003) Manganese-dependent regulation of the endocarditis-associated virulence factor EfaA of *Enterococcus faecalis*. J Med Microbiol 52:113–119
Lower BH, Kennelly PJ (2002) The membrane-associated protein-serine/threonine kinase from *Sulfolobus solfataricus* is a glycoprotein. J Bacteriol 184:2614–2619
Lower BH, Kennelly PJ (2003) Open reading frame sso2387 from the archaeon *Sulfolobus solfataricus* encodes a polypeptide with protein-serine kinase activity. J Bacteriol 185:3436–3445
Lucchini S, Liu H, Jin Q, Hinton JC, Yu J (2005) Transcriptional adaptation of *Shigella flexneri* during infection of macrophages and epithelial cells: insights into the strategies of a cytosolic bacterial pathogen. Infect Immun 73:88–102
Makui H, Roig E, Cole ST, Helmann JD, Gros P, Cellier MF (2000) Identification of the *Escherichia coli* K-12 NRAMP orthologue (MntH) as a selective divalent metal ion transporter. Mol Microbiol 35:1065–1078
Marra A, Lawson S, Asundi JS, Brigham D, Hromockyj AE (2002) *In vivo* characterization of the *psa* genes from *Streptococcus pneumoniae* in multiple models of infection. Microbiology 148:1483–1491
Missiakas D, Raina S (1997) Signal transduction pathways in response to protein misfolding in the extracytoplasmic compartments of *E. coli*: role of two new phosphoprotein phosphatases PrpA and PrpB. EMBO J 16:1670–1685
Morona JK, Morona R, Miller DC, Paton JC (2002) *Streptococcus pneumoniae* capsule biosynthesis protein CpsB is a novel manganese-dependent phosphotyrosine-protein phosphatase. J Bacteriol 184:577–583
Mukhopadhyay S, Kapatral V, Xu W, Chakrabarty AM (1999) Characterization of a Hank's type serine/threonine kinase and serine/threonine phosphoprotein phosphatase in *Pseudomonas aeruginosa*. J Bacteriol 181:6615–6622

Najdenski HM, Golkocheva EN, Vesselinova AM, Russmann H (2004) Comparison of the course of infection of virulent *Yersinia enterocolitica* serotype O:8 with an isogenic *sodA* mutant in the peroral rabbit model. Int J Med Microbiol 294:383–393

Oetjen J, Fives-Taylor P, Froeliger EH (2002) The divergently transcribed *Streptococcus parasanguis* virulence-associated *fimA* operon encoding an Mn^{2+}-responsive metal transporter and *pepO* encoding a zinc metallopeptidase are not coordinately regulated. Infect Immun 70:5706–5714

Ogawa T, Bao DH, Katoh H, Shibata M, Pakrasi HB, Bhattacharyya-Pakrasi M (2002) A two-component signal transduction pathway regulates manganese homeostasis in *Synechocystis* sp. 6803, a photosynthetic organism. J Biol Chem 277:28981–28986

Paik S, Brown A, Munro CL, Cornelissen CN, Kitten T (2003) The *sloABCR* operon of *Streptococcus mutans* encodes an Mn and Fe transport system required for endocarditis virulence and its Mn-dependent repressor. J Bacteriol 185:5967–5975

Patzer SI, Hantke K (2001) Dual repression by Fe^{2+}-Fur and Mn^{2+}-MntR of the *mntH* gene, encoding an NRAMP-like Mn^{2+} transporter in *Escherichia coli*. J Bacteriol 183:4806–4813

Plant J, Glynn AA (1974) Natural resistance to *Salmonella* infection, delayed hypersensitivity and *Ir* genes in different strains of mice. Nature 248:345–347

Plant J, Glynn AA (1979) Locating *Salmonella* resistance gene on mouse chromosome 1. Clin Exp Immunol 37:1–6

Plant JE, Blackwell JM, O'Brien AD, Bradley DJ, Glynn AA (1982) Are the *Lsh* and *Ity* disease resistance genes at one locus on mouse chromosome 1? Nature 297:510–511

Platero R, Peixoto L, O'Brian MR, Fabiano E (2004) Fur is involved in manganese-dependent regulation of *mntA* (*sitA*) expression in *Sinorhizobium meliloti*. Appl Environ Microbiol 70:4349–4355

Platero RA, Jaureguy M, Battistoni FJ, Fabiano ER (2003) Mutations in *sitB* and *sitD* genes affect manganese-growth requirements in *Sinorhizobium meliloti*. FEMS Microbiol Lett 218:65–70

Polakis SE, Guchhait RB, Lane MD (1973) Stringent control of fatty acid synthesis in *Escherichia coli*. Possible regulation of acetyl coenzyme A carboxylase by ppGpp. J Biol Chem 248:7957–7966

Qiu X, Pohl E, Holmes RK, Hol WG (1996) High-resolution structure of the diphtheria toxin repressor complexed with cobalt and manganese reveals an SH3-like third domain and suggests a possible role of phosphate as co-corepressor. Biochemistry 35:12292–12302

Que Q, Helmann JD (2000) Manganese homeostasis in *Bacillus subtilis* is regulated by MntR, a bifunctional regulator related to the diphtheria toxin repressor family of proteins. Mol Microbiol 35:1454–1468

Rajagopal L, Clancy A, Rubens CE (2003) A eukaryotic type serine/threonine kinase and phosphatase in *Streptococcus agalactiae* reversibly phosphorylate an inorganic pyrophosphatase and affect growth, cell segregation, and virulence. J Biol Chem 278:14429–14441

Reeve I, Hummell D, Nelson N, Voss J (2002) Overexpression, purification, and site-directed spin labeling of the Nramp metal transporter from *Mycobacterium leprae*. Proc Natl Acad Sci USA 99:8608–8613

Ricci S, Janulczyk R, Bjorck L (2002) The regulator PerR is involved in oxidative stress response and iron homeostasis and is necessary for full virulence of *Streptococcus pyogenes*. Infect Immun 70:4968–4976

Richer E, Courville P, Bergevin I, Cellier MF (2003) Horizontal gene transfer of "prototype" Nramp in bacteria. J Mol Evol 57:363–376

Roggenkamp A, Bittner T, Leitritz L, Sing A, Heesemann J (1997) Contribution of the Mn-cofactored superoxide dismutase (SodA) to the virulence of *Yersinia enterocolitica* serotype O8. Infect Immun 65:4705–4710

Runyen-Janecky LJ, Payne SM (2002) Identification of chromosomal *Shigella flexneri* genes induced by the eukaryotic intracellular environment. Infect Immun 70:4379–4388

Runyen-Janecky LJ, Reeves SA, Gonzales EG, Payne SM (2003) Contribution of the *Shigella flexneri* Sit, Iuc, and Feo iron acquisition systems to iron acquisition in vitro and in cultured cells. Infect Immun 71:1919–1928

Runyen-Janecky L, Dazenski E, Hawkins S, Warner L (2006) Role and Regulation of the Shigella flexneri Sit and MntH Systems. Infect Immun 74:4666–4672

Sacher A, Cohen A, Nelson N (2001) Properties of the mammalian and yeast metal-ion transporters DCT1 and Smf1p expressed in *Xenopus laevis* oocytes. J Exp Biol 204:1053–1061

Sarubbi E, Rudd KE, Xiao H, Ikehara K, Kalman M, Cashel M (1989) Characterization of the spoT gene of *Escherichia coli*. J Biol Chem 264:15074–15082

Schmidt AJ, Ryjenkov DA, Gomelsky M (2005) The ubiquitous protein domain EAL is a cyclic diguanylate-specific phosphodiesterase: enzymatically active and inactive EAL domains. J Bacteriol 187:4774–4781

Schmitt MP (2002) Analysis of a DtxR-like metalloregulatory protein, MntR, from *Corynebacterium diphtheriae* that controls expression of an ABC metal transporter by an Mn^{2+}-dependent mechanism. J Bacteriol 184:6882–6892

Sen KI, Sienkiewicz A, Love JF, vanderSpek JC, Fajer PG, Logan TM (2006) Mn(II) binding by the anthracis repressor from *Bacillus anthracis*. Biochemistry 45:4295–4303

Shi L (2004) Manganese-dependent protein O-phosphatases in prokaryotes and their biological functions. Front Biosci 9:1382–1397

Shi L, Carmichael WW, Kennelly PJ (1999) Cyanobacterial PPP family protein phosphatases possess multifunctional capabilities and are resistant to microcystin-LR. J Biol Chem 274:10039–10046

Shi L, Kehres DG, Maguire ME (2001) The PPP-family protein phosphatases PrpA and PrpB of *Salmonella enterica* serovar Typhimurium possess distinct biochemical properties. J Bacteriol 183:7053–7057

Simm R, Fetherston JD, Kader A, Romling U, Perry RD (2005) Phenotypic convergence mediated by GGDEF-domain-containing proteins. J Bacteriol 187:6816–6823

Singh KV, Coque TM, Weinstock GM, Murray BE (1998) In vivo testing of an *Enterococcus faecalis efaA* mutant and use of efaA homologs for species identification. FEMS Immunol Med Microbiol 21:323–331

Sy J (1977) In vitro degradation of guanosine 5′-diphosphate, 3′-diphosphate. Proc Natl Acad Sci USA 74:5529–5533

Tamayo R, Tischler AD, Camilli A (2005) The EAL domain protein VieA is a cyclic diguanylate phosphodiesterase. J Biol Chem 280:33324–33330

Tao X, Zeng HY, Murphy JR (1995) Transition metal ion activation of DNA binding by the diphtheria tox repressor requires the formation of stable homodimers. Proc Natl Acad Sci USA 92:6803–6807

Tischler AD, Camilli A (2004) Cyclic diguanylate (c-di-GMP) regulates *Vibrio cholerae* biofilm formation. Mol Microbiol 53:857–869

Tischler AD, Camilli A (2005) Cyclic diguanylate regulates *Vibrio cholerae* virulence gene expression. Infect Immun 73:5873–5882

Tsolis RM, Baumler AJ, Heffron F (1995) Role of *Salmonella typhimurium* Mn-superoxide dismutase (SodA) in protection against early killing by J774 macrophages. Infect Immun 63:1739–1744

Udo H, Inouye M, Inouye S (1997) Biochemical characterization of Pkn2, a protein Ser/Thr kinase from *Myxococcus xanthus*, a Gram-negative developmental bacterium. FEBS Lett 400:188–192

Vidal S, Belouchi AM, Cellier M, Beatty B, Gros P (1995a) Cloning and characterization of a second human NRAMP gene on chromosome 12q13. Mamm Genome 6:224–230

Vidal S, Gros P, Skamene E (1995b) Natural resistance to infection with intracellular parasites: molecular genetics identifies Nramp1 as the Bcg/Ity/Lsh locus. J Leukoc Biol 58:382–390

Vidal SM, Malo D, Vogan K, Skamene E, Gros P (1993) Natural resistance to infection with intracellular parasites: isolation of a candidate for Bcg. Cell 73:469–485

Wu HJ, Seib KL, Srikhanta YN, Kidd SP, Edwards JL, Maguire TL, Grimmond SM, Apicella MA, McEwan AG, Jennings MP (2006) PerR controls Mn-dependent resistance to oxidative stress in *Neisseria gonorrhoeae*. Mol Microbiol 60:401–416

Zaharik ML (2003) Host-Pathogen Interactions: The Impact of Nramp1 on *Salmonella Enterica* Serovar Typhimurium Virulence Gene Expression, Ph.D. Thesis, Univ. British Columbia

Zaharik ML, Finlay BB (2004) Mn^{2+} and bacterial pathogenesis. Front Biosci 9:1035–1042

Zaharik ML, Gruenheid S, Perrin AJ, Finlay BB (2002a) Delivery of dangerous goods: type III secretion in enteric pathogens. Int J Med Microbiol 291:593–603

Zaharik ML, Vallance BA, Puente JL, Gros P, Finlay BB (2002b) Host-pathogen interactions: Host resistance factor Nramp1 up-regulates the expression of *Salmonella* pathogenicity island-2 virulence genes. Proc Natl Acad Sci USA 99:15705–15710

Zaharik ML, Cullen VL, Fung AM, Libby SJ, Kujat Choy SL, Coburn B, Kehres DG, Maguire ME, Fang FC, Finlay BB (2004) The *Salmonella enterica* serovar Typhimurium divalent cation transport systems MntH and SitABCD are essential for virulence in an $Nramp1^{G169}$ murine typhoid model. Infect Immun 72:5522–5525

Zwilling BS, Hilburger ME (1994) Macrophage resistance genes: Bcg/Ity/Lsh. Immunol Ser 60:233–245

Microbiol Monogr (6)
D. H. Nies, S. Silver: Molecular Microbiology of Heavy Metals
DOI 10.1007/7171_2006_081/Published online: 24 February 2007

How Bacteria Handle Copper

David Magnani · Marc Solioz (✉)

Dept. of Clinical Pharmacology, University of Berne,
Murtenstrasse 35, 3010 Bern, Switzerland
marc.solioz@ikp.unibe.ch

Abstract Copper in biological systems presents a formidable problem: it is essential for life, yet highly reactive and a potential source of cell damage. Tight control of copper is thus a cellular necessity. To meet this challenge, cells have evolved pumps for transmembranous transport, chaperones for intracellular routing, oxidases and reductases to change the oxidation state of copper, and regulators to control gene expression in response to copper. These systems are complemented by specific mechanisms for the insertion of copper into enzymes. Copper homeostasis has evolved early in evolution and some components have been conserved from bacteria to humans. This has allowed researchers to apply knowledge across phyla and even involving human copper homeostatic diseases to elucidate the fundamental mechanism of cellular copper homeostasis. After an introduction to the properties of copper and its role in biological systems, some of the best studied bacterial systems for copper homeostasis will be discussed.

1 Introduction: Copper and Life

Copper has been known since prehistoric times. Metallic copper was available in the Middle East around 3500 B.C. It was obtained by reduction of its ores with charcoal. The discovery, some 500 years later, that the addition of tin to copper produced a much harder metal, established the Bronze Age. Copper has continued to this day to be an important metal to human kind. The abundance of copper in the earth's crust amounts to 68 ppm. It occurs mainly as the sulfide, oxide, or carbonate. Its major ores are copper pyrite (chalcopyrite, $CuFeS_2$), copper glance (chalcocite, Cu_2S), cuprite (Cu_2O), and malachite ($Cu_2CO_3(OH)_2$) (Tylecote 1992).

In the primordial, anaerobic word, copper was in the Cu(I) state in the form of water-insoluble sulphides. The ensuing oxygen evolution by microorganisms, a process which started less than 3×10^9 years ago, was a dramatic event for most living organisms. It could be considered to be an early, irreversible pollution of the earth, to which most living organisms adapted by acquiring an oxidative metabolism. While enzymes involved in anaerobic metabolism were designed to operate in the lower portion of the redox spectrum, the arrival of dioxygen created the need for a new redox active metal that could attain higher redox potentials. The oxidation of insoluble Cu(I) led to soluble and thus more bioavailable Cu(II), which was ideally suited to exploit the oxidizing power of dioxygen (Crichton and Pierre 2001). Copper is thus a modern bioelement (Kaim and Rall 1996). Concomitant with the arrival of oxygen, multi-cellular organisms developed.

Today, over 30 types of copper-containing proteins are known, prominent examples being lysyl oxidase (involved in the crosslinking of collagen) tyrosinase (required for melanin synthesis) dopamine β-hydroxylase of the catecholamine pathway, cytochrome *c* oxidase, the terminal electron acceptor of the respiratory chain, and superoxide dismutase, required for defense against oxidative damage. Another class of copper proteins, such as plastocyanins or azurines, act as electron carriers. In redox enzymes, copper serves as an electron acceptor/donor by alternating between the redox states Cu(I) and Cu(II) (Karlin 1993). Depending on the type of coordination of the copper to the protein, the redox potential can vary over the range + 200 to + 800 mV.

The redox properties of copper can, on the other hand, also cause cellular damage. A number or mechanisms have been suggested. Reactive hydroxyl radicals can be generated in a Fenton-type reaction:

$$Cu^+ + H_2O_2 = Cu^{2+} + OH^- + OH^\cdot . \quad (1)$$

The extremely reactive hydroxyl radical can participate in a number of reactions detrimental to cellular molecules, such as the oxidation of proteins and lipids (Yoshida et al. 1993). Copper can also lead to depletion of sulfhydryls, such as in cystein or glutathione, in a cycle between reactions 2 and 3:

$$2Cu^{2+} + 2RSH = 2Cu^{+} + RSSR + 2H^{+} \tag{2}$$

$$2Cu^{+} + 2H^{+} + O_2 = 2Cu^{2+} + H_2O_2\,. \tag{3}$$

The hydrogen peroxide thus generated can participate in reaction (1) and lead to further generation of toxic hydroxyl radicals. It is not clear to what extent reactions 1–3 cause copper toxicity. Cells try to keep H_2O_2 at very low levels and reaction (1) may not be the chief toxic mechanism, although this has been frequently claimed. Alternative routes of copper toxicity include the occupation of zinc or other metal sites in proteins by competition, or unspecific binding to proteins, lipids, and nucleic acids.

The toxic effect of copper is utilized in agriculture for the control of bacterial and fungal diseases (Cha and Cooksey 1991). This has in fact lead to the first thorough investigation of bacterial copper resistance, which is described in a later section.

The same property that makes copper a valuable biometal, namely the redox cycling between copper(I) and copper(II), also complicates experimental work. Copper(I) is the preferred form of copper for handling by the cell. Transporters like copper ATPases or eukaryotic CtrI-type transporters appear to transport copper(I) and copper chaperones bind copper(I) for delivery to cuproenzymes (Finney and O'Halloran 2003). While free copper(II) ions are stable in neutral, aqueous solutions exposed to the atmosphere, free copper(I) ions can only be maintained at very acidic pH or in complexed form. Highly efficient chelators of copper(I) in vitro are *o*-phenanthroline, bathophenanthroline sulfonate or 8-hydroxyquinoline (reported formation constants for copper-phenanthroline complexes are 21, irrespective of the ligands on the phenanthroline dipyrimidine ring system (Bell et al. 1991)). Copper(I) complexes which are stable in air can be formed with acetonitrile, CN^-, Tris-buffer and other complexing agents (McPhail and Goodman 1984). Phenanthrolines complex Cu(I) (and Fe(II)) so strongly that they effectively raise the redox potential to a point at which any reducing equivalent can support the reduction of the higher valency metal ion. This reaction can be counteracted by high concentrations (20 mM) of citrate or lactate, which preferentially bind the oxidized form of copper. Another difficulty is the interaction of copper(I) or (II) with all biological molecules, buffer substances, standard reducing agents etc. For example if 2 mM copper is added to complex bacterial growth media, the concentration of free, bioavailable copper remains unknown. This makes it nearly impossible to compare experimental findings by different research groups in terms of copper concentration.

The use of copper(I) rather than copper(II) by copper transporters and chaperones may have thermodynamic reasons. The lower valency states of transition metals are more exchange labile, e.g., the water exchange rate for Fe(II) is $3 \times 10^6\ s^{-1}$ compared to $3 \times 10^3\ s^{-1}$ for Fe(III). Since most copper in extracellular fluid or bacterial media is normally complexed to organic molecules, reduction of complexed Cu(II) would strongly facilitate the displacement of the ligand to which the metal ion is bound in the medium, making the metal more bioavailable.

These physico-chemical aspects have led to difficulties in experimentation with copper in biological systems which may have contributed to the delay in studying the homeostasis of this biologically important metal.

2 The State of Cytoplasmic Copper

Since copper is both essential and toxic for cells, they need to tightly control copper availability. To this end, bacteria use several strategies: access of copper to the cell is restricted by extracellular sequestration, relative impermeability of outer and inner bacterial membranes to copper, metallothionein-like copper scavenging proteins in the cytoplasm, and active extrusion of copper from the cell.

The cytoplasmic condition of cells is reducing. Glutathione (GSH) is present in millimolar concentrations and probably acts as the major reducing agent in prokaryotic and eukaryotic cells (McLaggan et al. 1990; Samuni et al. 1981). Other reducing compounds include cysteine or ascorbate. These reduce copper(II) to copper(I) and, in addition, avidly bind to copper. Thus, under normal conditions, cytoplasmic copper is in the reduced Cu^+-form and largely complexed by glutathione or other small molecules. Increased levels of GSH were observed in hepatoma cells treated with copper and inhibition of GSH synthesis with buthionine sulfoximine reduced the incorporation of copper into metallothioneins (Freedman and Peisach 1989b). It was also shown in vitro that Cu^+-GSH could mediate Cu^+-transfer into metal depleted metallothionein or copper-free Cu,Zn-superoxide dismutase (Ascone et al. 1993; Ciriolo et al. 1990; Ferreira et al. 1993). In vivo pulse-chase experiments with radioactive Cu^+ revealed that Cu^+ could also be transferred in the reverse direction from metallothionein to GSH and then to Cu,Zn-superoxide dismutase (Freedman and Peisach 1989a). These observations favor the view that GSH plays a key role in eukaryotic copper metabolism. However, inhibition of glutathione synthase in *E. coli* failed to alter the response of cells to copper, so it remains unclear if GSH plays as an important role in copper detoxification by prokaryotes as it does in higher cells.

In eukaryotic cells, resistance to copper toxicity is also influenced by the level of glutathione-peroxidase. An increased level of this enzyme was observed in copper-treated hepatoma cells, allowing efficient accommodation of increased cellular hydrogen peroxide concentrations caused by the oxidation of Cu^+-GSH or Cu^+-metallothionein (Freedman et al. 1989). To what extent such a mechanism also operates in bacteria is again not known. In the following section, the best characterized bacterial copper-resistance systems (*Escherichia coli* and *Enterococcus hirae*) will be discussed in detail. Information about copper homeostasis in *Synechocystis* can be found in the chapter by Tottey, Harvie and Robinson.

3 Copper Homeostasis in *Escherichia coli*

In *Escherichia coli* multiple systems evolved for the safe handling of copper (Rensing and Grass 2003). A central component in the copper homeostasis of *E. coli* is the integral inner membrane copper-transporting P-type ATPase CopA, which ensures removal of excess Cu(I) from the cytoplasm. In the periplasmic space, both the multi-copper oxidase CueO and the multicomponent copper efflux system CusCFBA are responsible for the control of copper levels. Additionally, various *E. coli* strains harbor plasmid-encoded genes that increase considerably the copper tolerance, allowing growth in copper-rich environments (Rouch et al. 1985).

3.1 CopA Copper ATPase

The CopA copper-transporting ATPase is the core element of copper homeostasis in *E. coli* both under aerobic and anaerobic conditions (Rensing and Grass 2003) and its expression is regulated by CueR, a transcription activator induced by Cu(I) and Ag(I) (Outten et al. 2000; Petersen and Moller 2000; Stoyanov et al. 2001). The enzyme is an integral inner membrane protein with eight hydropathic transmembrane α-helices and shares common features with other P-type ATPases (Fig. 1). It has conserved ATP binding (GDGIN), phosphorylation (DKTGT), and phosphatase (TGE) domains (Solioz and Vulpe 1996). The cytoplasmic N-terminus harbors two metal-binding CxxC motifs. Interestingly, these motifs are not required for the function (Fan et al. 2001) and apparently do not confer metal-specificity (Rensing and Grass 2003). *E. coli* Δ*copA* that has a disrupted *copA* gene is less resistant to copper than the wild-type, and complementation with the gene on a plasmid restores tolerance (Rensing et al. 2000). Accumulation of ^{64}Cu in everted membrane vesicles derived from a strain overexpressing CopA suggests copper transporting activity. Uptake is inhibited by vanadate, a specific inhibitor of P-type ATPases. DTT, a strong reductant, was required, otherwise no transport could be detected. This suggests that copper is pumped in the form of Cu(I) out of the cell (Rensing et al. 2000). Additionally, it was also shown that CopA is able to transport silver (Stoyanov et al. 2003).

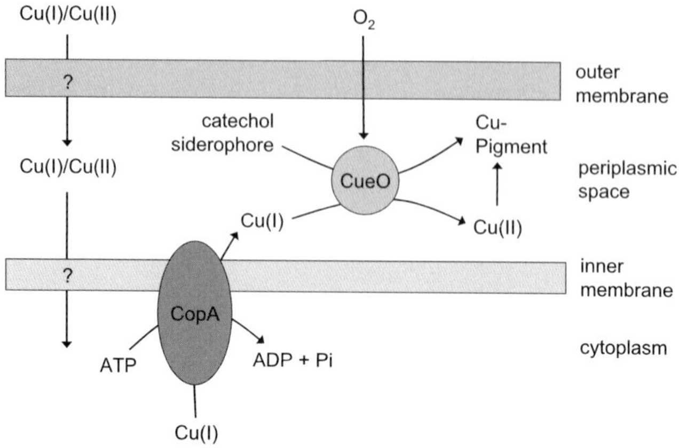

Fig. 1 The CopA ATPase and CueO oxidase of *E. coli*. CopA is an ATP-driven copper pump that expels copper(I) from the cytoplasm into the periplasmic space. There it can get oxidized by the multi-copper oxidase CueO. This enzyme can also oxidize catechol siderophores and the resulting pigments can sequester copper

3.2
CueR Copper-Responsive Regulator

CueR (for copper export regulator) belongs to the MerR-like transcription activators with an N-terminal helix-turn-helix DNA binding region, a C-terminal copper-binding region and a dimerization domain. Expression of *copA* and *cueO* (see below) are both regulated by CueR, but these genes are not organized in an operon. MerR-like regulators activate suboptimal σ^{70}-dependent promoters, in which the spacing between the – 35 and – 10 elements recognized by the σ factor is greater than the optimal 17 ± 1 base pairs. Therefore the conformation of the DNA has to be changed in order to allow access of the RNA polymerase for transcription initiation. The model suggests that there is a structural change of the MerR-like regulator upon binding of the effector molecule and this leads to a distortion of the DNA (Brown et al. 2003).

CueR is also activated by silver and, interestingly, by gold (Stoyanov et al. 2003; Stoyanov and Brown 2003). The basis of the ultrasensitivity of this regulator was further investigated by structural analysis. The sensitivity of CueR to free Cu(I) was determined to be 10^{-21} molar (zeptomolar), which corresponds to less than one copper ion per cell (Rae et al. 1999). The X-ray crystal structure revealed that the copper ion is buried in a solvent-inaccessible site in a loop at the dimer interface and has only two coordinating ligands: the sulfur atoms of the conserved Cys-112 and Cys-120. Mutation of these two residues makes CueR unresponsive to copper, silver and gold, which underscores the importance of these two residues (Stoyanov et al. 2001). The unusual metal-receptor site restricts the metal to a linear, two-coordinate geometry and uses helix-dipole and hydrogen-bonding interactions to enhance metal binding (Changela et al. 2003) (see also the chapters by Tottey, Harvie and Robinson, and by Helman, Soosanga and Gabriel for a discussions of CueR and MerR-like regulators).

3.3
CueO Periplasmic Multi-Copper Oxidase

CueO is a multi-copper oxidase involved in copper detoxification. Expression of *cueO* (copper efflux oxidase) is regulated by CueR and together with the *copA* constitutes the Cue system. It has been shown that the *cueO* promoter region shares the same features as the *copA* promoter. A copper-dependent increase of β-galactosidase activity of *cueO* promoter-*lacZ* fusions suggested that *cueO* expression is induced by copper (Outten et al. 2000). Disruption of *cueO* rendered cells more sensitive to copper, which supports the assumption that CueO is involved in copper homeostasis (Grass and Rensing 2001). It was also shown that CueO protects alkaline phosphatase from copper-induced damage. The spectroscopic properties indicate that CueO is a multi-copper oxidase. Atomic absorption spectroscopy revealed four copper atoms per molecule.

In vitro assays show comparable properties CueO to related multi-copper oxidases such as Fet3 or laccases. CueO has phenol oxidase and ferroxidase activities. There is also evidence that CueO catalyzes the oxidation of enterobactin in the presence of copper. Enterobactin is a catechol siderophore of *E. coli*. A mixture of copper and enterobactin proved to be very toxic and lead to reduced survival of *E. coli* in the presence of copper. The addition of purified CueO had a positive effect on cell survival (Grass et al. 2004). Oxidized enterobactin can also sequester copper, contributing to the detoxification mechanism. CueO seems to also have cuprous oxidase activity (Singh et al. 2004). The activity is dependent on oxygen, thus CueO works only in aerobic conditions. Therefore, *E. coli* also relies on another copper homeostatic system—the *cusCFBA* operon—which makes cells more resistant to copper under anaerobic conditions (Outten et al. 2001).

The NADH dehydrogenase-2 (NDH-2) of the *Escherichia coli* respiratory chain has been reported to also possess cupric-reductase activity. NDH-2 deficient strains were more sensitive in their growth to high or low copper concentrations than wild-type cells. This suggests that NDH-2 helps to diminish the damaging effects of copper and/or oxidative stress on the respiratory chain and contributes to copper detoxification in *E. coli* (Rodriguez-Montelongo et al. 2006).

3.4 Cus Copper Efflux System

The Cus system of *E. coli* is a chromosomally encoded copper- and silver-resistance system consisting of a CusCBA proton-cation antiporter complex under the control of the CusRS two-component regulatory system (Fig. 2). Initial studies of the Cus (Ylc) system showed that this determinant mediates silver resistance (Franke et al. 2001). The *cusCBA* genes belong to a family of homologous transport complexes involved in the export of metal ions, xenobiotics and drugs (Nies 2003). *cusA* encodes an inner membrane protein belonging to the resistance nodulation cell division family (RND) of proteins. It is assumed that CusA functions as a secondary transporter energized by proton-substrate antiport and is responsible for the substrate specificity. *cusB* codes for a membrane fusion protein (MFP) which serves as an adaptor to link CusA with CusC, which is an outer membrane factor (OMF, Fig. 2) (Rensing and Grass 2003).

CusF is a 10-kDa periplasmic protein. Yeast two hybrid experiments suggested that CusF interacts with CusB and CusC and may function as a periplasmic copper chaperone. Complementation of a *cusF* null mutant with this mutated CusF protein did not restore copper tolerance of the cells to wild-type levels, emphasizing a function of CusF in copper tolerance (Franke et al. 2003). CusF with bound copper is pink and exhibits an absorption maximum at 510 nm. This unusual feature is a consequence of its unusual structure: CusF forms a five-stranded β-barrel, classified as an OB-fold, which is a unique topology for a copper-binding protein. NMR chemical shift mapping experiments suggest that one Cu(I) is bound by conserved residues His-36,

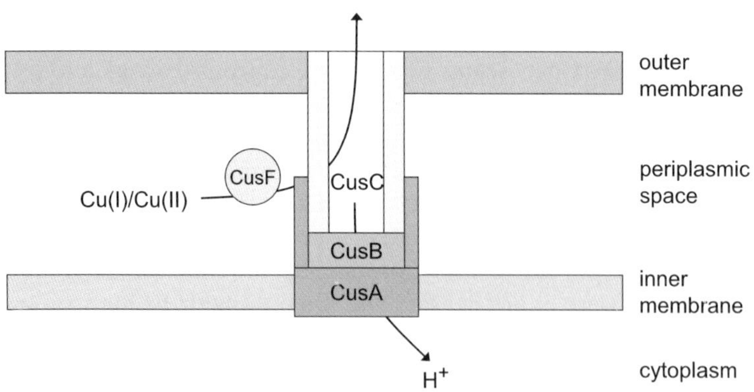

Fig. 2 The Cus system of *E. coli*. The Cus-complex consists of the inner membrane pump CusA, an integral membrane protein with 12 predicted transmembranous helices and two cytoplasmic loop, the periplasmic protein CusB, and the trimeric outer membrane protein CusC, which forms a channel bridging the periplasmic space. Entry of copper probably occurs from the periplasm via the copper chaperone CusF

Met-47, and Met-49 located in β-strands 2 and 3. These residues are clustered at one end of the β-barrel, and their side chains are oriented towards the interior of the barrel (Loftin et al. 2005). CusF thus represents a novel type of periplasmic copper chaperone.

Recently, the structure of OMF TolC, which is related to CusC, has been elucidated. It is a trimer that forms a trans-periplasmic channel embedded in the outer membrane (Koronakis et al. 2000). This suggests a model where CusA, B, and C interact to form a channel spanning the periplasm and connecting the cytoplasm to the extracellular space (Fig. 2). CusCBA belongs to a family of related transport systems that only occur in Gram-negative bacteria and that are involved in the export of metal ions, xenobiotics, and drugs. Such systems transporting various other metals have been characterized in *Pseudomonas*, *Cupriavidus*, *Synechococcus*, *Salmonella* and *E. coli* and are discussed in detail in Chap. 4.

Transcription of *cusCFBA* is regulated by the *cusRS* regulatory system that activates transcription of *cusCFBA* in the presence of copper (Munson et al. 2000). The *cusRS* operon is divergent to the *cusCFBA* operon and encodes a copper-responsive two-component signal transduction system (Franke et al. 2003). CusR is homologous to phosphate receiver response regulators and CusS is homologous to sensor histidine kinases. The closest homologs to CusRS are other two-component copper-regulatory systems: CusR has 61% sequence identity to both PcoR of *E. coli* and CopR of *P. syringae*, whereas CusS has 42% identity to CopS of *E. coli* and 38% identity to PcoS of *P. syringae*. The divergent cusCFB and cusRS operons are activated by the binding of CusR to a unique CusR box between the two promoters.

Recent experimental evidence suggests that the Cus system transports copper from the periplasmic space, rather than from the cytoplasm, to the extracellular environment (Franke et al. 2003). This is supported by the observation that related transporters such as AcrB or MexB can transport substrates that do not cross the cytoplasmic membrane and are thus restricted to the periplasmic space (Zgurskaya and Nikaido 2000). In contrast to the CueO/PcoA-type multi-copper oxidases that require oxygen for function, the CusCBA copper efflux system can work under anaerobic as well as aerobic conditions. It was recently shown that mutation of the Cus system strongly affected copper sensitivity under anaerobic conditions, while both *cueO* and *cusCFBA* had to be mutated to affect copper sensitivity in aerobic conditions (Outten et al. 2001).

A surprising additional function of CusC (IbeB) from *E. coli* K1 was recently described. *E. coli* K1 is the most common Gram-negative organism causing neonatal meningitis. Invasion of brain microvascular endothelial cells is a prerequisite for penetration into the human central nervous system. It was shown that CusC is an important contributing determinant in the process (Huang et al. 1999; Wang and Kim 2002). However, it is currently not apparent how CusC functions in bacterial penetration.

3.5 Copper Biosensors

A reporter plasmid was constructed by inserting the *E. coli* CopA promoter into a plasmid containing the *lux* gene cluster of *Vibrio fischeri* without a promoter (Rogowsky et al. 1987; Stoyanov et al. 2003). *E. coli* cells containing the biosensor exhibited luminescence in response to externally added $CuSO_4$, $AgNO_3$, or $HAuCl_4$. As expected, the biosensor in wild-type cells and in an *E. coli* strain deleted in the CopA copper-exporting ATPase (Δ*copA*) responded differently to metal ions. In a Δ*copA* strain, an eight-fold lower concentration of silver and a 15-fold lower concentration of copper (1 μM), compared to wild-type, was required to elicit maximal luminescence (Stoyanov et al. 2003). At the same time, overall luminescence intensities were enhanced five to seven fold. The more sensitive and more pronounced response of luminescence to copper and silver in the Δ*copA* strain is supposedly due to increased intracellular levels of these ions. This suggests that the homeostatic controls of both copper and silver is impaired in the Δ*copA* strain and that CopA not only effluxes copper ions, but also silver ions. Since the expression of the CopA ATPase is also regulated by silver via the CueR activator (Stoyanov et al. 2001), this appears to represent a true homeostatic control system for silver. Indeed, it could be shown that wild-type cells were more silver-resistant than Δ*copA* cells (Stoyanov et al. 2003). The wild-type still grew at half the maximal growth rate in 0.2 μM Ag^+, which fully inhibited growth of the Δ*copA* strain. The biosensor also responded to gold, with maximal activation observed in 30–40 μM gold. However, there was no significant difference in sensitivity and intensity of the luminescence response between wild-type and Δ*copA* cells. This argues against a participation of the CopA ATPase in gold homeostasis.

A similar biosensor with the *lux* genes under the control of the *E. hirae cop* promoter and carrying the corresponding CopY repressor and CopZ chaperone genes of *E. hirae* was also introduced into an *E. coli* Δ*copA* strain. In this case, half-maximal luminescence was reached at 5 μM extracellular copper (Portmann et al. 2004). It appears that a Δ*copA* strain is completely devoid of a cytoplasmic copper export mechanism. Thus, copper concentrations in the μM range may well represent physiological, cytoplasmic copper concentrations (see also the chapter by Harms for an extended discussion of biosensors).

3.6 Copper Regulons

Our knowledge of copper homeostasis in *E. coli* is still limited, and experimental evidence suggests that it is mainly maintained by controlling the export of excess copper out of the cell. A systematic search for copper-responsive genes by using the DNA microarray technique was initiated with the aim to better understand the genome-wide response to copper. It was found that 29 genes were specifically and significantly affected by copper. These genes are organized into a hierarchy of the regulation network and form at least four regulons (Yamamoto and Ishihama 2005):

CusR regulon – DNA array experiments confirmed the activation of the *cusCFB* and *cusRS* operons by copper. CusR, the response regulator, binds to the CusR box with the inverted repeat AAAATGACAANNTTGCATTTT. This sequence was only found between the divergent *cusCFBA* and *cusRS* operons. In a CusR-defective mutant, gene expression by copper was absent for only three genes, *cusC*, *cusF*, and *cusB*, which reside in the *cusCFBA* operon. First, this confirms that CusR regulates the transcription of the *cusCFBA* operon, and secondly it shows that no other gene is regulated by this regulator. Similar experiments with DNA microarrays confirmed the strong induction of the *cue* and *cus* systems by copper. Elevated copper concentrations also positively affected the transcription of regulons for superoxide stress response, iron homeostasis, and envelope stress (Kershaw et al. 2005).

CueR regulon – The MerR-like CueR activator binds to the CueR box in the *copA* and *cueO* promoters. An additional 129 CueR box-like sequences in the *E. coli* genome suggest other binding sites for CueR. Among the identified CueR sites, 74 are associated with known genes and the rest are found next to genes with unknown function. The *moa* operon encoding enzymes required for molybdopterin biosynthesis was also shown to be induced by copper. This is not surprising in the light of the recent discovery that molybdenum cofactor synthesis is linked to copper: a copper ion is bound to molybdopterin dithiolate sulfurs as an intermediate in the biosynthetic pathway (Kuper et al. 2004).

CpxR regulon – The CpxR regulon is mainly responsible for the response to cell envelope stress (Raivio and Silhavy 2001; Ruiz et al. 2006). *CpxP* and *spy* are both markedly induced by copper and are under the control of the CpxA-CpxR two-component system. Transcription of 11 genes (*cpxP, spy, JW1832, ybaJ, yccA, ycfS, ydeH, yebE, yecI* and *yqjA*) was also strongly upregulated by copper. In the *cpxAR* null-mutant, these gene products cannot be detected. The promoter from these ten genes all carry the proposed CpxR recognition site GTAAANNNNNGTAAA.

YedW regulon – Microarray experiments revealed that the two-component system *yedVW* is induced considerably by copper. In the CusR-null mutant no activation by copper could be shown, but interestingly, no CusR box-like sequence is located in the promoter region of *yedVW*. YedV is thought to be a sensor kinase and YedW the corresponding response regulator. Recently, it was shown that the YedV kinase transfers phosphate not only to its partner YedW, but also to CusR, and may also be important for the activation of CusR (Yamamoto and Ishihama 2005).

Extracytoplasmic function protein family sigma factor – The sigma factor of the extracytoplasmic function protein family, RpoE, is apparently important for full resistance to Zn(II), Cd(II), and Cu(II). Deletion of *rpoE* leads to a decrease in copper tolerance and promoter gene fusions of copper-resistance genes require lower concentrations of copper for optimum induction in the mutant strain. RpoE is involved in maintaining the integrity of periplasmic and outer membrane proteins (Ruiz et al. 2006). Promoter gene fusion showed that RpoE is induced by metals (Egler et al. 2005).

4
Plasmid-Borne Copper Resistance

A plasmid-borne copper-resistance system, identified in *Pseudomonas syringae* pathovar tomato, was one of the first bacterial copper-resistance systems that was studied in detail. The system was discovered in copper-resistant bacteria isolated from tomato cultures in Southern California that had been sprayed with copper sulfate for disease control (Mellano and Cooksey 1988b). The copper resistance of these bacteria is conferred by the *copABCDRS* operon on plasmid pPT23D. It is homologous to the Pco system on plasmid pRJ1004 of *E. coli*. It was isolated from a strain from the gut flora of pigs fed a copper-enriched diet to promote growth (Tetaz and Luke 1983). Gene clusters similar to *pcoABCDRS* (and to *cop* of *P. syringae*) operons can be identified in the genomes of many bacteria. Often only homologues of PcoA and PcoB are encoded on a genome, suggesting that CopCD is an accessory system required for maximal resistance. The following discussion of the function of the *pco* system of *E. coli* in most parts also applies to the *P. syringae copABCDRS* system and related operons of other bacteria.

The *pco* gene cluster of the pPT23D plasmid of *E. coli* contains, in addition to the *pcoABCDRS* operon present in many organisms, also the *pcoE* gene, which is transcribed from its own promoter (Brown et al. 1995; Cooksey 1994; Rouch and Brown 1997). Radioactive ^{64}Cu transport measurements with whole cells showed that resistant cells containing the *pco* determinants exhibited decreased copper accumulation during logarithmic growth. It was suggested that an energy-dependent copper efflux mechanism is associated with the *pco* copper-resistance system of plasmid pRJ1004 (Brown et al. 1995). However, the observed decrease in copper accumulation could also have been due to diminished uptake. Thus, the molecular mechanism of copper resistance by the *pco* and the related *P. syringae cop* systems is still not fully understood (see also below).

Copper resistance directed by the *E. coli pco* (and *P. syringae cop*) operons have been shown to be copper inducible (Mellano and Cooksey 1988a; Rouch et al. 1985). A two-component regulatory system, PcoRS, induces transcription of the *pcoABCDRS* operon, where PcoR acts as a transcriptional activator that binds to DNA (Mills et al. 1994). PcoS appears to be a periplasmic histidine kinase that senses copper. The *pcoE* gene is transcribed from a separate promoter which is also under the control of PcoRS (Rouch and Brown 1997). Both promoters are also induced by copper in the absence of the plasmid-borne *pcoRS* two-component regulatory system due to the presence of a second, related two-component regulatory system on the chromosome, named cusRS, which is also required for the copper-dependent expression of the cusCFBA periplasmic efflux system (Franke et al. 2003; Munson et al. 2000).

A different type of regulation was identified for the *pco*-like, plasmid-borne copper-resistance system of *Xanthomonas axonopodis* pv. vesicatoria. Instead of the CopRS two-component regulatory system, a novel protein, CopL, was found to regulate copper-inducible expression of the downstream multi-copper oxidase CopA in *X. axonopodis*. CopL encodes a histidine and cysteine-rich 130-kDa protein. Inactivation of *copL* caused a loss of transcriptional activation by copper, which could not be complemented by the *E. coli cusRS* or the *P. syringae pcoRS* systems. This suggests that CopL is a novel transcriptional activator and shows an interesting divergence in the mechanisms of regulation of the copper resistance by Pco-like systems.

Recently, progress has been made in understanding the function of some of the components of *cop* and *pco* proteins involved in conferring copper resistance (Fig. 3). PcoA of *E. coli* is a 66-kDa protein and appears to be a multi-copper oxidase similar to CueO of *E. coli* and CopA of *P. syringae*.

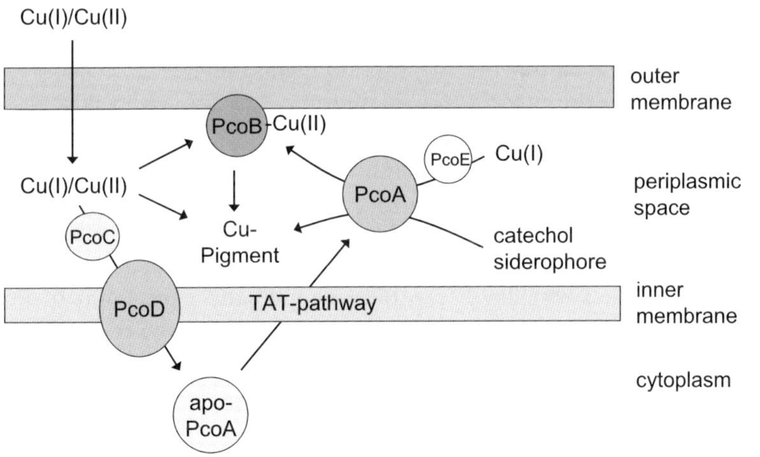

Fig. 3 The Pco/Cop systems of *E. coli* and *P. syringae*. Copper enters the periplasmic space by an unknown pathway. In the periplasm, PcoC shuttles copper to PcoD, which could provide the necessary copper for the PcoA multi-copper oxidase, but the copper may not necessarily transit through the cytoplasm in the process. Periplasmic copper can be detoxified by binding to oxidized catechol siderophores, generated by the PcoA oxidase, or to the outer membrane protein PcoB. PcoE shuttles copper Cu(I) to PcoA for oxidation to less toxic Cu(II)

PcoA exhibits oxidase activity (Huffman et al. 2002) and may, in analogy to CueO, function in copper detoxification by oxidizing Cu(I) to less toxic Cu(II) and/or by oxidizing catechol siderophores, such as enterobactin, which can then sequester copper (Fig. 1). Indeed, the formation of brown colonies by *E. coli* harboring the Pco-plasmid pRJ1004 and grown in the presence of $CuSO_4$ has been reported more than two decades ago (Tetaz and Luke 1983). Interestingly, *P. syringae* harboring the *copABCDRS*-system turn bright blue when grown in the presence of copper. The similarity of the function of PcoA and CueO is underlined by the observation that PcoA could functionally complement *E. coli cueO* deletion strain (Rensing and Grass 2003). Also similar to CueO, PcoA possesses a arginine-arginine motif in its leader sequence and is probably exported through the cytoplasmic membrane by the TAT-pathway.

PcoB is a 33-kDa protein, predicted to be localized in the outer membrane. PcoA and PcoB together confer higher copper resistance than PcoA alone, indicating that they act in concert, but the exact function of PcoB remains elusive. Its most feasible function at the current state of knowledge is that of a copper sequestration protein to buffer periplasmic copper excess.

PcoC is a 12-kDa periplasmic protein (Huffman et al. 2002; Lee et al. 2002). Recent structural studies showed that PcoC can bind copper(I) as well as copper(II) concomitantly (Zhang et al. 2006). A closely related protein from *Bacillus subtilis*, YcnJ, appears to be a fusion of two proteins, one homologous to PcoC and one homologous to PcoD. This observation and the biochemical properties of PcoC from *P. syringae* suggests that PcoC serves as a copper chaperone, delivering copper to PcoD. A PcoC homologue, YobA, is encoded by the *E. coli* chromosome but its function is unknown.

PcoD is a 34-kDa cytoplasmic membrane protein with eight predicted transmembranous helices. A homologue of PcoD (YebZ) with unknown function can be predicted from the *E. coli* genome sequence.

PcoE, which is transcribed from its own promoter, is a small periplasmic protein that is required for full copper resistance. It is related to SilE, which is a silver binding protein of the plasmid-borne *sil* silver-resistance system (Gupta et al. 1999). PcoE can bind copper and expression of PcoE by itself leads to copper accumulation in the periplasm. Likely, PcoE functions as a periplasmic copper chaperone (Rensing and Grass 2003).

5
Copper Homeostasis in *Enterococcus hirae*

Starting with the discovery of the two copper ATPases in *E. hirae* in 1992, copper homeostasis has been systematically elaborated in this Gram-positive bacterium. It possesses a *cop* operon that is required for copper homeostasis and which consists of the four genes *copY*, *copZ*, *copA*, and *copB*. *copY* en-

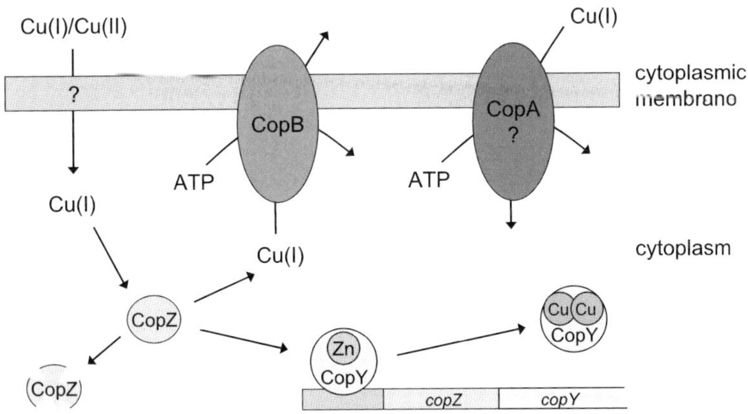

Fig. 4 Copper homeostasis in *E. hirae*. Under high copper conditions, copper enters cells by unknown pathways. Inside the cell, CopZ picks up copper(I) and can deliver it to CopB for expulsion and to the CopY repressor to induce transcription of the *cop* operon. CopY in its zinc form binds to the promoter and represses transcription. Two copper(I) can replace the zinc(II) in CopY, thus abolishing DNA binding and inducing transcription. If intracellular copper is excessive, CopZ is degraded by a copper-activated protease. CopA is a putative copper uptake ATPase under copper-limiting conditions

codes a copper-responsive repressor, *copZ* a copper chaperone, and *copA* and *copB* code for CPx-type copper ATPases. Figure 4 depicts the current working model of copper circulation in *E. hirae*.

copA and *copB* were the first genes encoding copper ATPases to be described (Odermatt et al. 1992). The function of CopB in copper excretion had been shown by direct demonstration of $^{64}Cu^{+}$ as well as $^{m110}Ag^{+}$ transport (Solioz and Odermatt 1995). The evidence for CopA being involved in copper uptake is still indirect. It rests on the following four properties of a *copA* knock-out strain: (i) it grows like wild-type under normal or elevated copper conditions, (ii) it is more silver resistant than the wild-type, presumably because CopA can be a route for silver entry into the cell, (iii) it is not deficient in silver extrusion from the cytoplasm, and (iv) cytoplasmic copper accumulation in high copper is similar to that of wild-type (Odermatt et al. 1993, 1994). However, it cannot be ruled out at the present time that CopA has another function.

5.1 Intracellular Routing of Copper by CopZ

Intracellular routing of copper is accomplished by copper chaperones, specialized proteins which deliver copper intracellularly to copper-utilizing enzymes (Harrison et al. 2000; O'Halloran and Culotta 2000). In *E. hirae*, the 69-amino-acid protein CopZ has been shown to function as a chaperone and to specifically deliver copper to the CopY repressor. In its zinc form, the CopY repressor binds to the *cop* promoter and represses transcription of the four *cop* genes, *copY*, *copZ*, *copA*, and *copB*. When CopZ donates copper to CopY, its bound zinc is displaced by copper and the repressor dissociates from the promoter, allowing expression of the downstream genes (Cobine et al. 1999, 2002a; Lu et al. 2003; Magnani and Solioz 2005) (see also below).

The solution structure of CopZ has been solved by NMR (Wimmer et al. 1999). It exhibits a $\beta\alpha\beta\beta\alpha\beta$ global structure: two α-helices laying on a four-stranded, antiparallel β-sheet, a structure colloquially called an "open-face sandwich". CopZ belongs to a conserved family of copper chaperones, which have been identified in yeast, plants, and higher cells (Huffman and O'Halloran 2001). The same structural element is also found in the N-terminal metal binding domain of heavy metal transporting ATPases in one to six copies, but the function in these enzymes remains unclear (Arnesano et al. 2002). A key element of CopZ and related copper chaperones and metal binding domains is the CxxC metal-binding motif located between the first β-sheet and the first α-helix. It binds copper(I) or silver(I) in a solvent-exposed binding site. The solution structure of CopZ allowed to make predictions about surface residues involved in the interaction with other proteins.

Using surface plasmon resonance analysis (Biacore apparatus), it was shown that CopZ also interacts with the CopA copper ATPase (Multhaup et al. 2001). The binding of CopZ to CopA was modulated by copper. Maximal binding of CopZ to CopA was observed in the presence of roughly stoichiometric amounts of copper(I) (10 μM) corresponding to a molar stoichiometry of 0.9 copper(I) per CopZ. The association rate k_a was $2.4 \times 10^3\ M^{-1}\ s^{-1}$ without added copper and was only slightly affected by copper. The dissociation rate k_d was $14 \times 10^{-3}\ s^{-1}$ and was strongly influenced by copper, with a decrease of up to 15-fold. This results in a 16-fold increase of the affinity constant K_D in the presence of 10 μM copper(I). It was also shown that mutating the CxxC copper-binding motif in the N-terminus of CopA to SxxS abolished the copper-induced decrease in the dissociation rate without significantly affecting the association rate. Thus, the CopZ chaperone can interact with the CopA ATPase in a copper-dependent manner. Preliminary studies showed that CopZ also interacts with the CopB copper ATPase.

5.2 Proteolytic Degradation of CopZ

Under high copper conditions, proteolysis of CopZ was observed. Levels of *copZ* mRNA increased with increasing ambient copper levels and reached a 1000-fold induction at 0.25 mM copper, as assessed by real-time quantitative PCR. However, CopZ protein increased only up to 0.5 mM copper and declined at higher copper concentrations, to become nearly undetectable at 3 mM copper (Lu and Solioz 2001). It was concluded that CopZ overexpression is toxic to cells, because growth of a strain overexpressing CopZ from a plasmid ceased to grow in 1.5 mM copper and was more sensitive to oxidative stress induced by to H_2O_2 or paraquat (Lu and Solioz 2001).

Proteolysis of CopZ could be demonstrated in vitro. When cytosolic extracts were mixed with purified CopZ, it was rapidly degraded. Cu(I)CopZ was more rapidly degraded than *apo*-CopZ, in line with the notion that the exposed copper on CopZ can be toxic. The serine protease inhibitors *p*-phenylmethylsulfonyl fluoride and *p*-aminobenzamidine inhibited the degradation of CopZ, while N-α-tosyl-L-lysine chloromethyl ketone and N-tosyl-L-phenylalanine chloromethyl ketone, which are also serine protease inhibitors, were without effect. The metallo-proteinase inhibitor *o*-phenanthroline also did not inhibit CopZ degradation. It was thus concluded that the protease-degrading CopZ is a serine-type protease. On zymograms, the CopZ degrading activity was tentatively identified as a protein of 58 kDa. This protein displayed the expected properties, namely activation by copper and inhibition by *p*-aminobenzamidine (Magnani and Solioz 2005).

5.3
Regulation of the *cop* Operon

CopY is a copper-responsive repressor that binds as a homodimer to two distinct regions in the *cop* promoter (Strausak and Solioz 1997). It contains one zinc per monomer, which plays an important structural and functional role. CopY has a modular structure with a DNA-binding and a copper-binding domain. The N-terminal part shows approximately 30% sequence identity to bacterial transcription repressors of β-lactamase genes such as MecI of *Staphylococcus aureus*, PenI of *Bacillus licheniformis*, or BlaI from an *S. aureus* transposon (Himeno et al. 1986; Suzuki et al. 1993; Wittman and Wong 1988). These are members of the family of "winged-helix" proteins, which form a winged-helix structure in the N-terminal DNA-binding domain. Winged-helix proteins belong to the superfamily of helix-turn-helix DNA-binding proteins (Gajiwala and Burley 2000; Portmann et al. 2006).

The C-terminal domain contains four cysteine residues in the arrangement Cys-x-Cys-x_4-Cys-x-Cys. The same consensus motif is also present in other copper-responsive transcriptional regulators such as MacI, a transcription factor for the Ctr1 and Ctr3 copper transporters of *Saccharomyces cerevisiae*, or ACE1 and AMT1, which regulate transcription of metallothionein (Dameron et al. 1993; Zhou and Thiele 1991).

When transcription of the *cop* operon is activated, Cu(I)CopZ transfers two copper ions per Zn(II)CopY monomer, displacing the single Zn(II) (Cobine et al. 2002b). This appears to lead to conformational changes and reduced affinity to the operator. The copper transfer reaction is rapid, for there was no stable complex detected by gel filtration (Cobine et al. 1999). Two copper ions are required to remove the zinc(II) ion bound per CopY monomer and CopY has a higher affinity for copper than CopZ. Lysine residues 30, 31, 37, and 38 of CopZ appeared to be important for the interaction with the CopY repressor. MNKr2, the second CopZ-like copper-binding domain of the human Menkes copper ATPase, cannot donate copper to CopY in spite of its similar predicted structure. However, if four extra lysine residues are introduced into MNKr2 at positions corresponding to the lysine residues in CopZ, the resultant mutant molecule MNKr2K4 becomes competent in donating copper to CopY (Cobine et al. 2002a; Magnani and Solioz 2005). This "gain-of-function" mutation of MNKr2 delineates lysine residues as key features in the interaction of the CopZ chaperone with the CopY repressor and supports the concept that protein–protein interaction governs copper transfer from the chaperone to the repressor. Finally, the kinetic parameters of *apo*-CopZ and Cu(I)CopZ interaction with Zn(II)CopY were determined by surface plasmon resonance. The affinity for Cu(I)CopZ was 440-fold higher than that for *apo*-CopZ (Portmann et al. 2004).

There seems to be no cooperativity of CopY binding to the two different binding sites in the operator region, but the half-association concentration of CopY differs at the two sites: 5 nM for the upstream site and 2 nM for the downstream site. A common conserved motif consisting of an inverted repeat of the sequence TACAnnTGTA, the "cop-box", appears to be the binding site for CopY-like copper-responsive repressors. The *E. hirae* CopY has the same binding activity to the operator region of the *Lactococcus lactis* and *Streptococcus mutans cop* operons as to its endemic promoter. The equilibrium dissociation constant for Zn(II)CopY–DNA interaction in absence of copper was 1.7×10^{-10} and the dissociation rate of Zn(II)CopY was increased sevenfold in the presence of copper, showing that copper reduces the affinity of CopY to the repressor site (Portmann et al. 2004).

Previously, it was reported that mutations of ACA to TCA at the positions A-61 and A-30 strongly reduces the affinity of CopY to the operator sequence highlighting the importance of these bases for the interaction of CopY with this operator region. However, only the double mutant completely abolished the interaction of CopY with the operator region and lead to hyperinduction of the *cop* operon by copper stimulation. When only one of the inverted repeats was mutated, the interaction of CopY still was allowed with the other site and the expression was not affected. The *cop* operon seems to be induced not only by higher copper concentrations, but also by low copper. The mechanism of this regulation is not known and seems to involve other components (Strausak and Solioz 1997; Wunderli-Ye and Solioz 1999).

5.4
CopZ-Interacting Protein Gls24

Using the yeast two-hybrid system with truncated CopZ as a bait, Gls24 was identified (Stoyanov and Solioz, unpublished). Gls24 belongs to the family of

Gls24-like stress response regulators (Giard et al. 2000), which may also have a role in pathogenicity (Teng et al. 2005). Gls24 was overexpressed and purified and was shown by surface plasmon resonance analysis to interact with CopZ in vitro. Cu(I)CopZ exhibited a high affinity for Gls24, with a K_D of 7.5×10^{-6} M. This affinity is five-fold higher than the affinity of Cu(I)CopZ for the CopY repressor or the CopA copper ATPase, which are both around 3.5×10^{-5} M. This suggests that the interaction between CopZ and Gls24 is biologically relevant. The expression of Gls24 was induced by glucose starvation and exposure of cells to copper. By NMR and circular dichroism, the protein appeared to contain unfolded domains.

Because of its interaction with Cu(I)CopZ, one could envision that ehGls24 functions as a protective cap. Copper bound to CopZ and other metallochaperones is in a solvent-exposed position at one end of the molecule. This copper could conceivably still exert toxic effects. A "cap" protein could protect the exposed copper in CopZ and prevent it from interacting in transit. While attractive, this concept remains hypothetical and requires experimental verification.

5.5 CopA Copper ATPase

For the in vitro study of CopA, an expression and purification scheme was developed. CopA was expressed in *E. coli* with a cleavable 6xHis tag. Membranes were extracted with dodecyl maltoside and CopA purified by Ni-affinity chromatography. The enzyme was better than 95% pure after a single chromatographic step, appeared fully active, and exhibited essentially the same properties with or without the His-tag (Wunderli-Ye and Solioz 2001).

CopA was reconstituted into soybean phospholipid vesicles by detergent dialysis. In proteoliposomes, ATPase activity was 300 to 400% of the starting activity. This suggested that the reconstituted ATPase was of high functional integrity and suitable to assess enzyme kinetic parameters. ATPase activity was maximal at pH 6.25, but exhibited a broad pH optimum in the range of pH 5.7 to 6.5. From kinetic analysis, a K_m for ATP of 0.2 mM and a V_{max} of 0.15 μmol/min/mg were derived (Wunderli-Ye and Solioz 2001). Turnover of CopA was slow in comparison to non-heavy metal ATPases, such as eukaryotic Ca^{2+} or Na^+/K^+-ATPases (Heyse et al. 1994; Pickart and Jencks 1984). Slow turnover was also observed for purified CopB of *E. hirae* and may be a general property of copper ATPases (Solioz and Camakaris 1997). CopA was fully active without added copper, presumably through the presence of contaminating copper ions (see below). Vanadate inhibited CopA with an I_{50} of 200 μM. The formation of an acyl phosphate intermediate could also be shown for CopA. The slow turnover of CopA is a major obstacle in the logical extension of this work, namely the measurement of copper transport by CopA in proteoliposomes.

5.6 CopB Copper ATPase

For the overexpression and purification of CopB, an engineered strain of *E. hirae* with a deletion of the CopY repressor was used. The enzyme was solubilized from the membranes with dodecyl maltoside and purified by Ni-affinity chromatography (Bissig et al. 2001b). Since CopB has a very histidine-rich N-terminus, it contains an "endogenous" His-tag and avidly binds to Ni-NTA resins.

The *E. hirae* CopB ATPase confers copper resistance by expelling excess copper and is related to the human ATPases *ATP7A* and *ATP7B* which are defective in Menkes and Wilson disease, respectively. Many mutations in these latter genes have been identified in patients (Gitlin 2003; Mercer 2001). Some mutations occur in highly conserved sequence motifs, which has allowed us to test their effect using the homologous CopB enzyme. The Menkes disease mutation C1000R changing the conserved CPC motif was mimicked in CopB as a C396S mutation. This mutant enzyme was unable to restore copper resistance in a CopB knock-out mutant in vivo. The purified mutant ATPase still formed an acyl phosphate intermediate, but had no detectable ATPase activity. The most frequent Wilson disease mutation H1069Q was introduced into CopB as H480Q. This mutant ATPase also failed to confer copper resistance to a CopB knock-out strain. Purified H480Q CopB formed an acyl phosphate intermediate and retained small, but significant ATPase activity. These results show that S396 and H480 of CopB are key residues for ATPase function and suggest similar roles for S1000 and H1069 of Menkes and Wilson ATPase, respectively (Bissig et al. 2001b). This work opened a new approach to test structure-function relationships in eukaryotic copper ATPases, which have so far not been purified.

An enigma had been the lack of activation of purified CopB by copper. We thus had hypothesized that there is sufficient contaminating copper in the buffers to fully activate the enzyme. Common copper chelators such as bicinchoninic acid or bathocuproine disulfonate did inhibit the ATPase activity, but only at concentrations in the high millimolar range. Tetrathiomolybdate avidly interacts with copper and has recently been employed to reduce excess copper in patients with Wilson disease (Brewer et al. 1994, 1996). It was found that tetrathiomolybdate inhibits CopB with an IC_{50} of 34 nM. Dithiomolybdate and trithiomolybdate, which commonly contaminate tetrathiomolybdate, inhibited the copper ATPases with similar potency. Inhibition could be reversed by copper or silver, suggesting inhibition by substrate binding. These findings for the first time allowed an estimate of the high affinity of CopB for copper and silver (Bissig et al. 2001a). Tetrathiomolybdate is thus a new tool for the study of copper ATPases.

6 Chalkophores

For iron acquisition, so-called siderophores are widely used in nature. They are extracellular iron-chelators that are secreted by cells. Their high complexing power for iron help microorganism acquire the scarce iron via uptake of the siderophore-iron complexes by specialized transport systems. Recently, evidence of analogous molecules for copper transport from methane-oxidizing bacteria has emerged. A fluorescent 1.2-kDa chromopeptide secreted by *Methylosinus trichosporium* OB3b was isolated and structurally analyzed (Kim et al. 2004). The molecule, which was called methanobactin, is composed of a tetrapeptide, a tripeptide, and several unusual moieties, including two 4-thionyl-5-hydroxy-imidazole chromophores that coordinate the copper, a pyrrolidine, and an amino-terminal isopropyl ester group. Copper is coordinated by a dual nitrogen- and sulfur-donating system derived from the thionyl imidazolate moieties. In methanotrophs, copper is required for the biosynthesis of methane monooxygenases and these organisms thus have a high demand for copper. They must therefore have an effective mechanism for acquiring copper from the environment. The secretion of a copper "siderophore" or chalkophore would support growth of the organism in a copper-limiting environment (Choi et al. 2006).

7 Conclusion and Outlook

There has been major progress in our understanding of copper homeostasis (and trace metal ion homeostasis in general) in the last decade. However, major questions still need to be tackled. The first unresolved question is that of the intracellular copper concentration and the experimental difficulty of distinguishing free, bioavailable copper from bound copper that is not available for participation in copper-homeostatic processes. Estimates for the concentration of free, intracellular copper range from 10^{-21} M (= less than one free copper ion per cell) based on thermodynamic calibration of the *E. coli* CueR copper response regulator of *E. coli* (Changela et al. 2003), through 3×10^{-8} M based on half-maximal activation of the purified CopB copper ATPase of *E. hirae* (Bissig et al. 2001a) to 10^{-6} M using the copper-responsive repressors CopY of *E. hirae* and CueR of *E. coli* in *lux*-based biosensors in vivo (Stoyanov et al. 2003). New tools such as the membrane-permeable, copper(I)-selective fluorescent indicator CTAP-1 could aid in defining the state of cytoplasmic copper (Yang et al. 2005).

A second unresolved question is that of the bacterial requirement for cytoplasmic copper. Known cuproenzymes like cytochrome *c* oxidase or superoxide dismutase are embedded in the cytoplasmic membrane and face the periplasmic space. Copper loading of these enzymes during biosynthesis could occur with periplasmic copper. Currently, the only know system that suggests a need for cytoplasmic copper is the biosynthesis of the molybdenum cofactor (Kuper et al. 2004). Its synthesis involves the insertion of molybdenum into molybdopterin by the Cnx1 G-domain. The identification of copper bound to the molybdopterin dithiolate sulphurs in Cnx1G, coupled with the observation that copper inhibited Cnx1G activity, suggests a link between molybdenum and copper metabolism, which would require cytoplasmic copper (Schwarz and Mendel 2006).

The third unanswered question is that of how copper enters bacterial cells. It is not yet known what the pathways for entry across the outer and inner membranes are. Porins such as OmpC or OmpF have been proposed as entry pathways for copper and silver into the periplasmic space but isogenic OmpF and/or OmpC deficient mutants of *E. coli* did not differ significantly in copper or silver sensitivity (Egler et al. 2005; Li et al. 1997). It is also not clear how copper enters the cytoplasm. In higher cells, copper is taken up actively by CtrI-like transporters in the cytoplasmic membrane (Puig and Thiele 2002). No similar system has so far been discovered in bacteria. In *E. hirae*, CopA has been proposed to serve in copper uptake, but final proof of this hypothesis is lacking. Conceivably, copper enters the cytoplasm as substrate-copper complexes by substrate transporters such as sugar or amino acid transporters. Alternatively, transporters for other metal ions may be leak pathways for copper. Finally, the recent discovery of copper "siderophores", or chalkophores, in methanogenic bacteria opens entirely new avenues of copper entry into bacterial cells.

Acknowledgements We thank Jivko Stoyanov, Reto Portmann, and Thomas Weber for their contributions to the work on copper homeostasis in *E. hirae* and *L. lactis*. Part of this work was supported by grant 3100A0-109703 from the Swiss National Foundation and by a grant from the International Copper Association.

References

Arnesano F, Banci L, Bertini I, Ciofi-Baffoni S, Molteni E, Huffman DL, O'Halloran TV (2002) Metallochaperones and metal-transporting ATPases: a comparative analysis of sequences and structures. Genome Res 12:255–271

Ascone I, Longo A, Dexpert H, Ciriolo MR, Rotilio G, Desideri A (1993) An X-ray absorption study of the reconstitution process of bovine Cu, Zn superoxide dismutase by Cu(I)-glutathione complex. FEBS Lett 322:165–167

Bell PF, Chen Y, Potts WE, Chaney RL, Angle JS (1991) A reevaluation of the Fe(III), Ca(II), Zn(II), and proton formation constants of 4,7-diphenyl-1,10-phenanthroline-disulfonate. Biol Trace Elem Res 30:125–144

Bissig K-D, Voegelin TC, Solioz M (2001a) Tetrathiomolybdate inhibition of the *Enterococcus hirae* CopB copper ATPase. FEBS Lett 507:367–370

Bissig K-D, Wunderli-Ye H, Duda P, Solioz M (2001b) Structure-function analysis of purified *Enterococcus hirae* CopB copper ATPase: effect of Menkes/Wilson disease mutation homologues. Biochem J 357:217–223

Brewer GJ, Dick RD, Johnson V, Wang Y, Yuzbasiyan-Gurkan V, Kluin K, Fink JK, Aisen A (1994) Treatment of Wilson's disease with ammonium tetrathiomolybdate: I. Initial therapy in 17 neurologically affected patients. Arch Neurol 51:545–554

Brewer GJ, Johnson V, Dick RD, Kluin KJ, Fink JK, Brunberg JA (1996) Treatment of Wilson disease with ammonium tetrathiomolybdate. II. Initial therapy in 33 neurologically affected patients and follow-up with zinc therapy. Arch Neurol 53:1017–1025

Brown NL, Barrett SR, Camakaris J, Lee BTO, Rouch DA (1995) Molecular genetics and transport analysis of the copper-resistance determinant (*pco*) from *Escherichia coli* plasmid pRJ1004. Mol Microbiol 17:1153–1166

Brown NL, Stoyanov JV, Kidd SP, Hobman JL (2003) The MerR family of transcriptional regulators. FEMS Microbiol Rev 27:145–163

Cha JS, Cooksey DA (1991) Copper resistance in *Pseudomonas syringae* mediated by periplasmic and outer membrane proteins. Proc Natl Acad Sci USA 88:8915–8919

Changela A, Chen K, Xue Y, Holschen J, Outten CE, O'Halloran TV, Mondragon A (2003) Molecular basis of metal-ion selectivity and zeptomolar sensitivity by CueR. Science 301:1383–1387

Choi DW, Zea CJ, Do YS, Semrau JD, Antholine WE, Hargrove MS, Pohl NL, Boyd ES, Geesey GG, Hartsel SC, Shafe PH, McEllistrem MT, Kisting CJ, Campbell D, Rao V, de la Mora AM, DiSpirito AA (2006) Spectral, kinetic, and thermodynamic properties of Cu(I) and Cu(II) binding by methanobactin from *Methylosinus trichosporium* OB3b. Biochemistry 45:1442–1453

Ciriolo MR, Desideri A, Paci M, Rotilio G (1990) Reconstitution of Cu, Zn-superoxide dismutase by the Cu(I). glutathione complex. J Biol Chem 265:11030–11034

Cobine P, Wickramasinghe WA, Harrison MD, Weber T, Solioz M, Dameron CT (1999) The *Enterococcus hirae* copper chaperone CopZ delivers copper(I) to the CopY repressor. FEBS Lett 445:27–30

Cobine PA, George GN, Jones CE, Wickramasinghe WA, Solioz M, Dameron CT (2002a) Copper transfer from the Cu(I) chaperone, CopZ, to the repressor, Zn(II)CopY: metal coordination environments and protein interactions. Biochemistry 41:5822–5829

Cobine PA, Jones CE, Dameron CT (2002b) Role for zinc(II) in the copper(I) regulated protein CopY. J Inorg Biochem 88:192–196

Cooksey DA (1994) Molecular mechanisms of copper resistance and accumulation in bacteria. FEMS Microbiol Rev 14:381–386

Crichton RR, Pierre J-L (2001) Old iron, young copper: from Mars to Venus. Biometals 14:99–112

Dameron CT, George GN, Arnold P, Santhanagopalan V, Winge DR (1993) Distinct metal binding configurations in ACE1. Biochemistry 32:7294–7301

Egler M, Grosse C, Grass G, Nies DH (2005) Role of the extracytoplasmic function protein family sigma factor RpoE in metal resistance of *Escherichia coli*. J Bacteriol 187:2297–2307

Fan B, Grass G, Rensing C, Rosen BP (2001) *Escherichia coli* CopA N-terminal $Cys(X)_2Cys$ motifs are not required for copper resistance or transport. Biochem Biophys Res Commun 286:414–418

Ferreira AM, Ciriolo MR, Marcocci L, Rotilio G (1993) Copper(I) transfer into metallothionein mediated by glutathione. Biochem J 292:673–676

Finney LA, O'Halloran TV (2003) Transition metal speciation in the cell: insights from the chemistry of metal ion receptors. Science 300:931–936

Franke S, Grass G, Nies DH (2001) The product of the *ybdE* gene of the *Escherichia coli* chromosome is involved in detoxification of silver ions. Microbiology 147:965–972

Franke S, Grass G, Rensing C, Nies DH (2003) Molecular analysis of the copper-transporting efflux system CusCFBA of *Escherichia coli.* J Bacteriol 185:3804–3812

Freedman JH, Ciriolo MR, Peisach J (1989) The role of glutathione in copper metabolism and toxicity. J Biol Chem 264:5598–5605

Freedman JH, Peisach J (1989a) Intracellular copper transport in cultured hepatoma cells. Biochem Biophys Res Commun 164:134–140

Freedman JH, Peisach J (1989b) Resistance of cultured hepatoma cells to copper toxicity. Purification and characterization of the hepatoma metallothionein. Biochim Biophys Acta 992:145–154

Gajiwala KS, Burley SK (2000) Winged helix proteins. Curr Opin Struct Biol 10:110–116

Giard JC, Rince A, Capiaux H, Auffray Y, Hartke A (2000) Inactivation of the stress- and starvation-inducible *gls24* operon has a pleiotrophic effect on cell morphology, stress sensitivity, and gene expression in *Enterococcus faecalis.* J Bacteriol 182:4512–4520

Gitlin JD (2003) Wilson disease. Gastroenterology 125:1868–1877

Grass G, Rensing C (2001) CueO is a multi-copper oxidase that confers copper tolerance in Escherichia coli. Biochem Biophys Res Commun 286:902–908

Grass G, Thakali K, Klebba PE, Thieme D, Muller A, Wildner GF, Rensing C (2004) Linkage between catecholate siderophores and the multi-copper oxidase CueO in *Escherichia coli.* J Bacteriol 186:5826–5833

Gupta A, Matsui K, Lo JF, Silver S (1999) Molecular basis for resistance to silver cations in *Salmonella.* Nat Med 5:183–188

Harrison MD, Jones CE, Solioz M, Dameron CT (2000) Intracellular copper routing: the role of copper chaperones. Trends Biochem Sci 25:29–32

Heyse S, Wuddel I, Apell HJ, Sturmer W (1994) Partial reactions of the Na, K-ATPase: determination of rate constants. J Gen Physiol 104:197–240

Himeno T, Imanaka T, Aiba S (1986) Nucleotide sequence of the penicillinase repressor gene penI of *Bacillus licheniformis* and regulation of *penP* and *penI* by the repressor. J Bacteriol 168:1128–1132

Huang SH, Chen YH, Fu Q, Stins M, Wang Y, Wass C, Kim KS (1999) Identification and characterization of an *Escherichia coli* invasion gene locus, *ibeB*, required for penetration of brain microvascular endothelial cells. Infect Immun 67:2103–2109

Huffman DL, Huyett J, Outten FW, Doan PE, Finney LA, Hoffman BM, O'Halloran TV (2002) Spectroscopy of Cu(II)-PcoC and the multi-copper oxidase function of PcoA, two essential components of *Escherichia coli pco* copper resistance operon. Biochemistry 41:10046–10055

Huffman DL, O'Halloran TV (2001) Function, structure, and mechanism of intracellular copper trafficking proteins. Annu Rev Biochem 70:677–701

Kaim W, Rall J (1996) Copper-a "modern" bioelement. Angew Chem Int Ed Engl 35:43–60

Karlin KD (1993) Metalloenzymes, structural motifs, and inorganic models. Science 261:701–708

Kershaw CJ, Brown NL, Constantinidou C, Patel MD, Hobman JL (2005) The expression profile of *Escherichia coli* K-12 in response to minimal, optimal and excess copper concentrations. Microbiology 151:1187–1198

Kim HJ, Graham DW, DiSpirito AA, Alterman MA, Galeva N, Larive CK, Asunskis D, Sherwood PM (2004) Methanobactin, a copper-acquisition compound from methane-oxidizing bacteria. Science 305:1612–1615

Koronakis V, Sharff A, Koronakis E, Luisi B, Hughes C (2000) Crystal structure of the bacterial membrane protein TolC central to multidrug efflux and protein export. Nature 405:914–919
Kuper J, Llamas A, Hecht HJ, Mendel RR, Schwarz G (2004) Structure of the molybdopterin-bound Cnx1G domain links molybdenum and copper metabolism. Nature 430:803–806
Lee SM, Grass G, Rensing C, Barrett SR, Yates CJ, Stoyanov JV, Brown NL (2002) The Pco proteins are involved in periplasmic copper handling in *Escherichia coli*. Biochem Biophys Res Commun 295:616–620
Li XZ, Nikaido H, Williams KE (1997) Silver-resistant mutants of *Escherichia coli* display active efflux of Ag^+ and are deficient in porins. J Bacteriol 179:6127–6132
Loftin IR, Franke S, Roberts SA, Weichsel A, Heroux A, Montfort WR, Rensing C, McEvoy MM (2005) A novel copper-binding fold for the periplasmic copper resistance protein CusF. Biochemistry 44:10533–10540
Lu ZH, Dameron CT, Solioz M (2003) The *Enterococcus hirae* paradigm of copper homeostasis: copper chaperone turnover, interactions, and transactions. Biometals 16:137–143
Lu ZH, Solioz M (2001) Copper-induced proteolysis of the CopZ copper chaperone of *Enterococcus hirae*. J Biol Chem 276:47822–47827
Magnani D, Solioz M (2005) Copper chaperone cycling and degradation in the regulation of the *cop* operon of *Enterococcus hirae*. Biometals 18:407–412
McLaggan D, Logan TM, Lynn DG, Epstein W (1990) Involvement of gamma-glutamyl peptides in osmoadaptation of *Escherichia coli*. J Bacteriol 172:3631–3636
McPhail DB, Goodman BA (1984) Tris buffer—a case for caution in its use in copper-containing systems (letter). Biochem J 221:559–560
Mellano MA, Cooksey DA (1988a) Induction of the copper resistance operon from *Pseudomonas syringae*. J Bacteriol 170:4399–4401
Mellano MA, Cooksey DA (1988b) Nucleotide sequence and organization of copper resistance genes from *Pseudomonas syringae* pv. tomato. J Bacteriol 170:2879–2883
Mercer JF (2001) The molecular basis of copper-transport diseases. Trends Mol Med 7:64–69
Mills SD, Lim C-K, Cooksey DA (1994) Purification and characterization of CopR, a transcriptional activator protein that binds to a conserved domain (*cop* box) in copper-inducible promoters of *Pseudomonas syringae*. Mol Gen Genet 244:341–351
Multhaup G, Strausak D, Bissig K-D, Solioz M (2001) Interaction of the CopZ copper chaperone with the CopA copper ATPase of *Enterococcus hirae* assessed by surface plasmon resonance. Biochem Biophys Res Commun 288:172–177
Munson GP, Lam DL, Outten FW, O'Halloran TV (2000) Identification of a copper-responsive two-component system on the chromosome of *Escherichia coli* K-12. J Bacteriol 182:5864–5871
Nies DH (2003) Efflux-mediated heavy metal resistance in prokaryotes. FEMS Microbiol Rev 27:313–339
O'Halloran TV, Culotta VC (2000) Metallochaperones, an intracellular shuttle service for metal ions. J Biol Chem 275:25057–25060
Odermatt A, Krapf R, Solioz M (1994) Induction of the putative copper ATPases, CopA and CopB, of *Enterococcus hirae* by Ag^+ and Cu^{2+}, and Ag^+ extrusion by CopB. Biochem Biophys Res Commun 202:44–48
Odermatt A, Suter H, Krapf R, Solioz M (1992) An ATPase operon involved in copper resistance by *Enterococcus hirae*. Ann NY Acad Sci 671:484–486

Odermatt A, Suter H, Krapf R, Solioz M (1993) Primary structure of two P-type ATPases involved in copper homeostasis in *Enterococcus hirae*. J Biol Chem 268:12775–12779
Outten FW, Huffman DL, Hale JA, O'Halloran TV (2001) The independent *cue* and *cus* systems confer copper tolerance during aerobic and anaerobic growth in *Escherichia coli*. J Biol Chem 276:30670–30677
Outten FW, Outten CE, Hale J, O'Halloran TV (2000) Transcriptional activation of an *Escherichia coli* copper efflux regulon by the chromosomal MerR homologue, CueR. J Biol Chem 275:31024–31029
Petersen C, Moller LB (2000) Control of copper homeostasis in *Escherichia coli* by a P-type ATPase, CopA, and a MerR-like transcriptional activator, CopR. Gene 261:289–298
Pickart CM, Jencks WP (1984) Energetics of the calcium-transporting ATPase. J Biol Chem 259:1629–1643
Portmann R, Magnani D, Stoyanov JV, Schmechel A, Multhaup G, Solioz M (2004) Interaction kinetics of the copper-responsive CopY repressor with the *cop* promoter of *Enterococcus hirae*. J Biol Inorg Chem 9:396–402
Portmann R, Poulsen KR, Wimmer R, Solioz M (2006) CopY-like copper inducible repressors are putative "winged helix" proteins. Biometals 19:61–70
Puig S, Thiele DJ (2002) Molecular mechanisms of copper uptake and distribution. Curr Opin Chem Biol 6:171–180
Rae TD, Schmidt PJ, Pufahl RA, Culotta VC, O'Halloran TV (1999) Undetectable intracellular free copper: the requirement of a copper chaperone for superoxide dismutase. Science 284:805–808
Raivio TL, Silhavy TJ (2001) Periplasmic stress and ECF sigma factors. Annu Rev Microbiol 55:591–624
Rensing C, Fan B, Sharma R, Mitra B, Rosen BP (2000) CopA: an *Escherichia coli* Cu(I)-translocating P-type ATPase. Proc Natl Acad Sci USA 97:652–656
Rensing C, Grass G (2003) *Escherichia coli* mechanisms of copper homeostasis in a changing environment. FEMS Microbiol Rev 27:197–213
Rodriguez-Montelongo L, Volentini SI, Farias RN, Massa EM, Rapisarda VA (2006) The Cu(II)-reductase NADH dehydrogenase-2 of *Escherichia coli* improves the bacterial growth in extreme copper concentrations and increases the resistance to the damage caused by copper and hydroperoxide. Arch Biochem Biophys 451:1–7
Rogowsky PM, Close TJ, Chimera JA, Shaw JJ, Kado CI (1987) Regulation of the *vir* genes of *Agrobacterium tumefaciens* plasmid pTiC58. J Bacteriol 169:5101–5112
Rouch D, Camakaris J, Lee BT, Luke RK (1985) Inducible plasmid-mediated copper resistance in *Escherichia coli*. J Gen Microbiol 131:939–943
Rouch DA, Brown NL (1997) Copper-inducible transcriptional regulation at two promoters in the *Escherichia coli* copper resistance determinant pco. Microbiology 143(Pt 4):1191–1202
Ruiz N, Kahne D, Silhavy TJ (2006) Advances in understanding bacterial outer-membrane biogenesis. Nat Rev Microbiol 4:57–66
Samuni A, Chevion M, Czapski G (1981) Unusual copper-induced sensitization of the biological damage due to superoxide radicals. J Biol Chem 256:12632–12635
Schwarz G, Mendel RR (2006) Molybdenum cofactor biosynthesis and molybdenum enzymes. Annu Rev Plant Biol 57:623–647
Singh SK, Grass G, Rensing C, Montfort WR (2004) Cuprous oxidase activity of CueO from *Escherichia coli*. J Bacteriol 186:7815–7817
Solioz M, Camakaris J (1997) Acylphosphate formation by the Menkes copper ATPase. FEBS Lett 412:165–168

Solioz M, Odermatt A (1995) Copper and silver transport by CopB-ATPase in membrane vesicles of *Enterococcus hirae*. J Biol Chem 270:9217–9221

Solioz M, Vulpe C (1996) CPx-type ATPases: a class of P-type ATPases that pump heavy metals. Trends Biochem Sci 21:237–241

Stoyanov JV, Brown NL (2003) The *Escherichia coli* copper-responsive *copA* promoter is activated by gold. J Biol Chem 278:1407–1410

Stoyanov JV, Hobman JL, Brown NL (2001) CueR (YbbI) of *Escherichia coli* is a MerR family regulator controlling expression of the copper exporter CopA. Mol Microbiol 39:502–512

Stoyanov JV, Magnani D, Solioz M (2003) Measurement of cytoplasmic copper, silver, and gold with a *lux* biosensor shows copper and silver, but not gold, efflux by the CopA ATPase of *Escherichia coli*. FEBS Lett 546:391–394

Strausak D, Solioz M (1997) CopY is a copper-inducible repressor of the *Enterococcus hirae* copper ATPases. J Biol Chem 272:8932–8936

Suzuki E, Kuwahara Arai K, Richardson JF, Hiramatsu K (1993) Distribution of *mec* regulator genes in methicillin-resistant *Staphylococcus clinical* strains. Antimicrob Agents Chemother 37:1219–1226

Teng F, Nannini EC, Murray BE (2005) Importance of gls24 in virulence and stress response of *Enterococcus faecalis* and use of the Gls24 protein as a possible immunotherapy target. J Infect Dis 191:472–480

Tetaz TJ, Luke RK (1983) Plasmid-controlled resistance to copper in *Escherichia coli*. J Bacteriol 154:1263–1268

Tylecote RF (1992) A history of metallurgy. Institute of Materials, London

Wang Y, Kim KS (2002) Role of OmpA and IbeB in *Escherichia coli* K1 invasion of brain microvascular endothelial cells in vitro and in vivo. Pediatr Res 51:559–563

Wimmer R, Herrmann T, Solioz M, Wüthrich K (1999) NMR structure and metal interactions of the CopZ copper chaperone. J Biol Chem 274:22597–22603

Wittman V, Wong HC (1988) Regulation of the penicillinase genes of *Bacillus licheniformis*: interaction of the *pen* repressor with its operators. J Bacteriol 170:3206–3212

Wunderli-Ye H, Solioz M (1999) Effects of promoter mutations on the in vivo regulation of the *cop* operon of *Enterococcus hirae* by copper(I) and copper(II). Biochem Biophys Res Commun 259:443–449

Wunderli-Ye H, Solioz M (2001) Purification and functional analysis of the copper ATPase CopA of *Enterococcus hirae*. Biochem Biophys Res Commun 280:713–719

Yamamoto K, Ishihama A (2005) Transcriptional response of *Escherichia coli* to external copper. Mol Microbiol 56:215–227

Yang L, McRae R, Henary MM, Patel R, Lai B, Vogt S, Fahrni CJ (2005) Imaging of the intracellular topography of copper with a fluorescent sensor and by synchrotron X-ray fluorescence microscopy. Proc Natl Acad Sci USA 102:11179–11184

Yoshida Y, Furuta S, Niki E (1993) Effects of metal chelating agents on the oxidation of lipids induced by copper and iron. Biochim Biophys Acta 1210:81–88

Zgurskaya HI, Nikaido H (2000) Multidrug resistance mechanisms: drug efflux across two membranes. Mol Microbiol 37:219–225

Zhang L, Koay M, Maher MJ, Xiao Z, Wedd AG (2006) Intermolecular transfer of copper ions from the CopC protein of *Pseudomonas syringae*. Crystal structures of fully loaded Cu(I)Cu(II) forms. J Am Chem Soc 128:5834–5850

Zhou PB, Thiele DJ (1991) Isolation of a metal-activated transcription factor gene from *Candida glabrata* by complementation in *Saccharomyces cerevisiae*. Proc Natl Acad Sci USA 88:6112–6116

Microbiol Monogr (6)
D. H. Nies, S. Silver: Molecular Microbiology of Heavy Metals
DOI 10.1007/7171_2006_082/Published online: 3 February 2007

Microbial Physiology of Nickel and Cobalt

Robert P. Hausinger[1] (✉) · Deborah B. Zamble[2]

[1]6193 Biomedical Physical Sciences,
Dept. Microbiology & Molecular Genetics and Dept. Biochemistry & Molecular Biology,
Michigan State University, East Lansing, MI 48824-4320, USA
hausinge@msu.edu

[2]Department of Chemistry, University of Toronto, 80 St. George St.,
Toronto, ON M5S 3H6, Canada

Abstract Nickel and cobalt are essential micronutrients for many microorganisms and serve as enzyme cofactors that catalyze a diverse array of reactions. One complication is that high concentrations of these transition metal ions are toxic to cells, leading some prokaryotes to evolve sophisticated homeostatic mechanisms to regulate their transmembrane uptake or efflux. The biosynthesis of nickel and cobalt metalloenzymes requires the intracellular allocation of the metals to the appropriate apoproteins, often in an intricate process that involves the cooperative activity of accessory proteins. Here, we highlight the molecular physiology of nickel and cobalt cation metabolism in *Escherichia coli* and summarize additional nickel- or cobalt-dependent processes and homeostatic mechanisms found in other microorganisms.

1 Introduction: Microbial Physiology of Nickel and Cobalt

Over 50 years ago a requirement for Co^{2+} or vitamin B_{12} was established in cyanobacteria that fix nitrogen or utilize nitrate (Holm-Hansen et al. 1954). A decade later, Bartha and Ordal noted that Ni^{2+} is needed for the growth of two hydrogen-oxidizing bacteria (Bartha and Ordal 1965). These micronutrients are transported into cells by highly specific uptake systems (Eitinger and Mandrand-Berthelot 2000; Eitinger et al. 2005). By contrast, microbes that grow at elevated concentrations of Ni^{2+} or Co^{2+} can utilize metal-specific efflux systems as a powerful resistance mechanism (Nies 2003). The Ni^{2+} or Co^{2+} uptake and efflux systems must be carefully regulated for optimal cell growth, and certain bacteria have devised ingenious mechanisms to sense the concentrations of these and other cations (see the work by Helman, Soosanga, and Gabriel, 2007, in this volume). Within the cell these metal ions are incorporated into specific metalloenzymes (see Table 1) by a process that often requires additional accessory proteins.

Our goals in this contribution are to illustrate in some detail the physiology of Ni^{2+} and Co^{2+} metabolism in *Escherichia coli*, in some cases extending to related better-studied systems, and to briefly review novel aspects of the physiology of these metal ions in other microorganisms. We cannot be comprehensive in our coverage of the microbial physiology of Ni^{2+} and Co^{2+}, so we refer to many key papers and more extensive reviews of these general topics.

2 Nickel and Cobalt Physiology of *E. coli*

Physiological studies of *E. coli* have been used to address the mechanisms of cell uptake of Ni^{2+}, Co^{2+}, and vitamin B_{12}, to provide evidence for a Ni^{2+} and Co^{2+} efflux transporter, and to examine for Ni^{2+} storage in these cells. Additional investigations in this microorganism have characterized the mechan-

Table 1 Representative nickel- and cobalt-dependent enzymes

Metallocenter	Enzyme	Function
Mononuclear		
[Ni]	Glyoxalase	glutathione + methylglyoxal → *S*-D-lactoylglutathione
[Ni]	Superoxide dismutase	$2O_2^- + 2\,H^+ \rightarrow H_2O_2 + O_2$
[Ni]	Aci-reductone dioxygenase	aci-reductone[a] + O_2 → HCOOH + CO + methylthiopropionate
[Co]	Nitrile hydratase	$R-CN + H_2O \rightarrow R-CO-NH_2$
[Co]	Thiocyanate hydrolase	$HS-CN + H_2O \rightarrow S=C=O + NH_3$
Dinuclear		
[NiNi]	Urease	urea + $2H_2O \rightarrow H_2CO_3 + NH_3$
[NiFe]	Hydrogenase	$H_2 \leftrightarrow 2H^+ + 2e^-$
[CoCo]	Prolidase	Xxx[b]-Pro + H_2O → Xxx + Pro
Complex		
[Ni – 4Fe – 4/5S]	Carbon monoxide dehydrogenase	$CO + H_2O \leftrightarrow CO_2 + 2e^- + 2H^+$
[Ni – Ni – 4Fe – 4S]	Acetyl-CoA[c] synthase/decarbonylase	$CO + CH_3-Co(III)P$[d] $+ CoA-SH \leftrightarrow CH_3-CO-S-CoA + Co(I)P$
Tetrapyrroles		
F_{430} [Ni]	Methyl-coenzyme M reductase	$CH_3-S-CoM$[e] $+ CoB-SH$[f] $\rightarrow CH_4 + CoB-S-S-CoM$
Vitamin B_{12} [Co]	Isomerases	Rearrangement reactions
Corrinoid [Co]	Methyl transferases	Methyl transfers
Corrinoid [Co]	Dehalogenases	Halide elimination reactions

[a] 1,2-dihydroxy-3-keto-5-methylthiopentane
[b] Xxx is a nonpolar amino acid such as Met, Leu, Val, Phe, or Ala
[c] CoA is coenzyme A
[d] P is a corrinoid protein containing a [4Fe – 4S] cluster
[e] CoM – SH is coenzyme M or thioethanesulfonate and $CH_3-S-CoM$ is the methylated species
[f] CoB-SH is coenzyme B or *N*-7-mercaptoheptanoylthreonine phosphate

isms by which these cations participate its transcriptional regulation and have studied its nickel- and cobalt-containing enzymes.

2.1 Ni^{2+}, Co^{2+}, and Vitamin B_{12} Uptake in *E. coli*

The mechanisms for cellular uptake of Ni^{2+} and vitamin B_{12} are particularly well characterized in *E. coli*. Less is known about how Co^{2+} is taken up by these cells.

2.1.1
Ni^{2+} Uptake in *E. coli*

Initial evidence for a Ni^{2+} transporter in *E. coli* was provided by lesions at the *hydC* locus that produced low hydrogenase activity and defective Ni^{2+} uptake (Wu and Mandrand-Berthelot 1986; Wu et al. 1989). Hydrogenase activity, which requires a nickel cofactor, was restored by growing the mutants in μM concentrations of Ni^{2+} thought to enter the bacteria through high-capacity Mg^{2+} transport systems. Cloning of the *hydC* locus (renamed *nik*) revealed five genes (Wu et al. 1991; Navarro et al. 1993) with homology to the ATP-binding cassette (ABC) family of transporters that couple ATP hydrolysis to substrate translocation across a membrane. Members of this family of transporters are fairly substrate selective (Higgins 2001), and it is interesting that the Nik transporter has much higher sequence homology to the oligopeptide and dipeptide transporters than the other metal ion ABC transporters (Tam and Saier 1993). The Nik transporter consists of two cytoplasmic, membrane-associated ATPases (NikD and NikE), two transmembrane components (NikB and NikC), and the periplasmic binding protein NikA that also participates in *E. coli*'s negative chemotaxis response to Ni^{2+} (de Pina et al. 1995).

The soluble NikA protein was characterized by biochemical methods and high-resolution structural analysis. Two different methods were used to estimate Ni^{2+} dissociation constants of $< 0.1\,\mu M$ to $10\,\mu M$ and revealed that NikA is selective for Ni^{2+} over other divalent metal ions by at least an order of magnitude (de Pina et al. 1995; Heddle et al. 2003). Initial X-ray absorption spectroscopy (XAS) suggested that the metal is 6-coordinate with primarily oxygen-based ligands and a 7th, longer-range Met ligand (Allan et al. 1998). The first X-ray crystal structure of NikA revealed a fold characteristic of the periplasmic binding proteins (Heddle et al. 2003), with two globular domains connected by a hinge that closes when the substrate binds between the lobes. This NikA structure is unusual because, although there is clearly some hinge closing in comparison with the apoprotein structure, the metal-binding site remains solvent-accessible (Heddle et al. 2003). Furthermore, the metal does not contact the protein and is surrounded by water molecules, with the binding pocket composed of several aromatics as well as an Arg and Met. Neither the coordination number nor the ligand distances correspond with the spectroscopic data, and it is unclear how such a complex would have strong affinity or any selectivity for Ni^{2+} ions. One possible explanation was suggested by a new structure of NikA (Fig. 1A) with an unexpected bound small molecule that was best fit as $FeEDTA(H_2O)^-$ (Cherrier et al. 2005). Although iron was not added to the protein, the EDTA was present in the purification protocol and the authors speculate that the iron derived from trace amounts in the medium. This complex remained intact throughout the purification so it must be tightly bound, and it raises the possibility

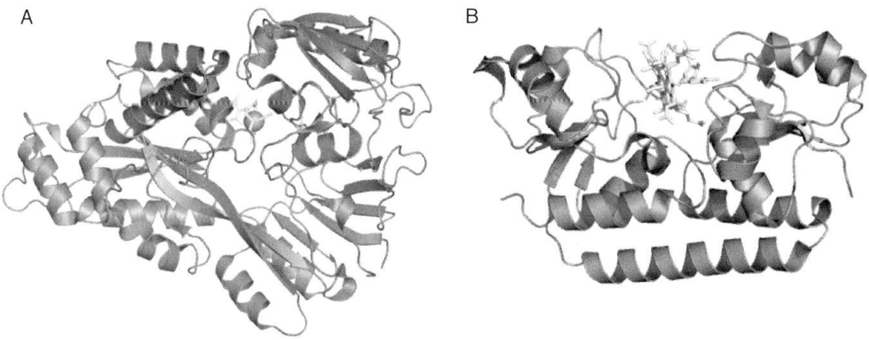

Fig. 1 Structures of NikA and BtuF involved in Ni^{2+} and vitamin B_{12} uptake by *E. coli*. **A** NikA in complex with Fe-EDTA (PDB #1zlq), which is thought to resemble the Ni^{2+}-containing complex recognized by this protein. **B** B_{12}-bound to BtuF (#1n2z). These components are periplasmic binding proteins of the ATP-binding cassette transport systems for these cofactors. All figures were generated with MacPymol

that NikA binds Ni^{2+} ions in a complex with a metallophore of a similar nature as EDTA.

Of added interest, a mutant strain lacking both the Nik system and the CorA Mg^{2+} transporter is still sensitive to Ni^{2+} because of a mechanism that is blocked by Mg^{2+} (Navarro et al. 1993), suggesting that there is at least one additional pathway for Ni^{2+} uptake. This hypothesis was recently supported by a study demonstrating that expression from the Ni^{2+}-responsive *nik* promoter is not influenced by the absence of the Nik proteins (Rowe et al. 2005).

2.1.2
Co^{2+} Uptake in *E. coli*

Although Co^{2+}-specific transporters have been identified in various prokaryotes (Rodionov et al. 2006), how Co^{2+} is brought into *E. coli* remains unclear. Several ion transporters have broad specificity for divalent metals and can mediate influx of Co^{2+}, including the Zn^{2+} transporter ZupT (Grass et al. 2005) and to a lesser extent the Mg^{2+} transporter CorA (Hmiel et al. 1986), the Mn^{2+} transporter MntH (Kehres et al. 2000; Makui et al. 2000), and the Nik transporter (Wu et al. 1994). Affinities for Co^{2+} have been estimated directly or indirectly to be in the μM range, except for ZupT where Co^{2+} affinity is unknown, so the amount of Co^{2+} uptake may be low and depend on competition with other metals. However, Co^{2+} is not detected in aerobically grown *E. coli* (Outten and O'Halloran 2001), so it is unlikely that more than traces amounts of Co^{2+} are needed and this nonspecific "leakiness" may provide a sufficient supply.

2.1.3
Vitamin B_{12} Uptake in *E. coli*

Unlike many bacteria, *E. coli* does not synthesize cobalamin de novo, so it must be imported from the external environment or synthesized from cobinamide, a closely related precursor (Lawrence and Roth 1995). Uptake of vitamin B_{12} requires two sequential transporters because this cofactor is too large to move passively through the outer-membrane porins. The genes for both transporters were identified by using the absolute cobalamin requirement of strains lacking the cobalamin-independent methionine synthase grown on methionine-deficient media; the transporter mutants require much higher concentrations of exogenous B_{12} to survive (Roth et al. 1996). Both transporters from *E. coli* have been characterized by biochemical studies and high-resolution structural analysis, providing an unusually comprehensive picture.

BtuB, the outer membrane transporter, binds cobalamin with nM affinity only if Ca^{2+} is present (Bradbeer et al. 1986). Active transport through BtuB is driven by the TonB inner membrane complex (Reynolds et al. 1980), which is energetically coupled to the proton gradient. These biochemical results are explained by several crystal structures of BtuB, which consists of a 22-stranded β-barrel stoppered with a 4-stranded β-sheet referred to as "the hatch" (Chimento et al. 2003). The structures reveal that two Ca^{2+} ions bind to and order several extracellular protein loops, opening access to the high-affinity B_{12} site. Subsequently, substrate binding induces a significant conformational change in the periplasmic Ton box, a highly conserved seven-residue stretch found in TonB-dependent transporters, which may activate TonB for the energy-dependent passage of the substrate.

The inner-membrane transporter, which belongs to the same superfamily of ABC transporters as the Nik transporter, is encoded in the *btuCED* operon (DeVeaux et al. 1986). BtuC is the transmembrane component and BtuD is the ATPase, but surprisingly BtuE is not the soluble periplasmic component and its requirement for cobalamin transport is unclear. The periplasmic-binding protein BtuF, encoded in an unconnected gene, binds cobalamin with nM affinity (Cadieux et al. 2002). The structure of BtuCD was solved in the absence of BtuF as a C_2D_2 complex (Locher et al. 2002). $BtuC_2$ has 20 transmembrane helices that surround a hydrophobic periplasmic cavity that is large enough to bind B_{12} and is closed to the cytoplasm by several residues that form a putative gate. On the cytoplasmic side the contacts between $BtuC_2$ and $BtuD_2$ encompass a large opening to the cytoplasm, a likely escape route for the substrate. The interprotein contacts include a sequence conserved in ABC transporters that is involved in transmitting the "substrate bound" signal to the ATPases that triggers gate opening and substrate translocation during ATP binding and hydrolysis.

As with NikA, BtuF has the characteristic bilobal structure of periplasmic binding proteins (Borths et al. 2002; Karpowich et al. 2003), but the substrate-binding cleft is much more open in BtuF to accommodate the larger substrate (Fig. 1B). The two lobes are connected by a rigid alpha helix, in contrast to the flexible β-strands in NikA, so there is likely to be less movement of the two lobes with respect to each other. Manual docking between several BtuF Glu and BtuC Arg residues within salt-bridging distances places the BtuF-bound B_{12} over the BtuC cavity (Borths et al. 2002). The apo-BtuF structure appears much more mobile than the B_{12}-bound protein because of elevated B-factors and several unfolded helices distant from the B_{12}-binding site. This increase in mobility and modest changes in conformation upon substrate release could promote BtuF dissociation from the transmembrane components (Karpowich et al. 2003), but biochemical studies suggest that apo-BtuF does not dissociate from the transporter (Borths et al. 2002, 2005).

2.2
Ni^{2+} and Co^{2+} Efflux in *E. coli*

A likely efflux transporter in *E. coli* is RcnA, a membrane protein induced by Ni^{2+} that was identified by a database search for His-rich sequences (Rodrigue et al. 2005). Mutation of *rcnA* resulted in enhanced sensitivity to Ni^{2+} as well as intracellular accumulation of Ni^{2+}, suggesting that this protein serves to promote efflux of excess Ni^{2+}. Furthermore, although the addition of several other divalent metals to the growth media had no effect, Co^{2+} also induced expression of RcnA and blocked Ni^{2+} efflux, suggesting that excess Co^{2+} may be removed through the same transporter as Ni^{2+}. Negative regulation of *rcnA* is controlled, at least in part, by the product of the proximal gene *rcnR* (Iwig et al. 2006), and Fur binding to the *rcnA* promoter regions suggests a connection to iron homeostasis (Koch et al. 2006).

2.3
Ni^{2+} Storage in *E. coli*

As described later, many homologs of the hydrogenase accessory protein HypB have His-rich regions that bind multiple Ni^{2+} ions and may function in the storage of Ni^{2+}. In contrast, *E. coli* HypB does not contain such a sequence. Instead, a Ni^{2+} storage role may be filled by SlyD, a member of the superfamily of peptidyl-prolyl isomerases that has a C-terminal domain rich in His/Cys/Asp/Glu residues and can bind multiple metal ions (Roof et al. 1994; Wülfing et al. 1994; Hottenrott et al. 1997). This protein was recently connected with the hydrogenase biosynthetic pathway because it interacts directly with HypB, and *ΔslyD* strains exhibit a partially hydrogenase-deficient phenotype that is complemented by Ni^{2+} (Zhang et al. 2005). It is feasible that

SlyD could store Ni^{2+} and supply the Ni^{2+} for the hydrogenase, but whether it has a more involved role has yet to be determined.

2.4 Ni^{2+} and B_{12} Regulation in *E. coli*

E. coli serves as the paradigm microorganism for understanding the mechanisms of cellular regulation by Ni^{2+} and vitamin B_{12}.

2.4.1 Ni^{2+}-Dependent Transcriptional Regulation in *E. coli*

Genetic studies of the *nik* operon revealed that it is induced under anaerobic conditions by the oxygen-sensitive regulator Fnr (Wu and Mandrand-Berthelot 1986; Wu et al. 1989), it responds to cellular metabolites that also affect hydrogenase expression, such as nitrate (Rowe et al. 2005), and it is repressed by Ni^{2+} (Wu and Mandrand-Berthelot 1986; Wu et al. 1989; de Pina et al. 1999). This latter activity is due to the protein encoded by *nikR*, which is just downstream from the *nik* operon and expressed under the control of its own constitutive promoter in addition to partial regulation by the *nik* promoter (de Pina et al. 1999). NikR has an N-terminal ribbon-helix-helix DNA-binding domain and a C-terminal metal-binding domain (Chivers and Sauer 1999; Schreiter et al. 2003), and it functions as a dimer of dimers with two intertwined DNA-binding domains flanking a core tetramer of metal-binding domains (Fig. 2) (Schreiter et al. 2003; Bloom and Zamble 2004). Four Ni^{2+} ions are bound per tetramer with pM affinity in square-planar sites at the dimer interface (Chivers and Sauer 2002; Carrington et al. 2003; Schreiter et al. 2003; Bloom and Zamble 2004). Ni^{2+} activates binding to a DNA recognition sequence that overlaps the transcription start site of the *nik* operon (Chivers and Sauer 2000), so it is likely that NikR regulates transcription by blocking RNA polymerase. Biochemical studies indicate that Ni^{2+} induces a conformational change in the protein that is not observed with other metals (Wang et al. 2004; Dias and Zamble 2005). The crystal structures of holo-NikR in the absence and presence of DNA (Schreiter et al. 2006), in comparison with the apo structure (Schreiter et al. 2003), reveal that the Ni^{2+} organizes several pieces of secondary structure that interact with the DNA, but does not reorient the DNA-binding domains to the optimal structure for DNA binding. Instead, it appears that the connection between the two domains is flexible, and that Ni^{2+} activates nonspecific DNA binding, allowing the DNA-binding domains to then swing down and contact the recognition sequence in a specific manner. In addition, it is possible that Ni^{2+} affects the quaternary structure of NikR and stabilizes the active tetramer conformation (Fauquant et al. 2006).

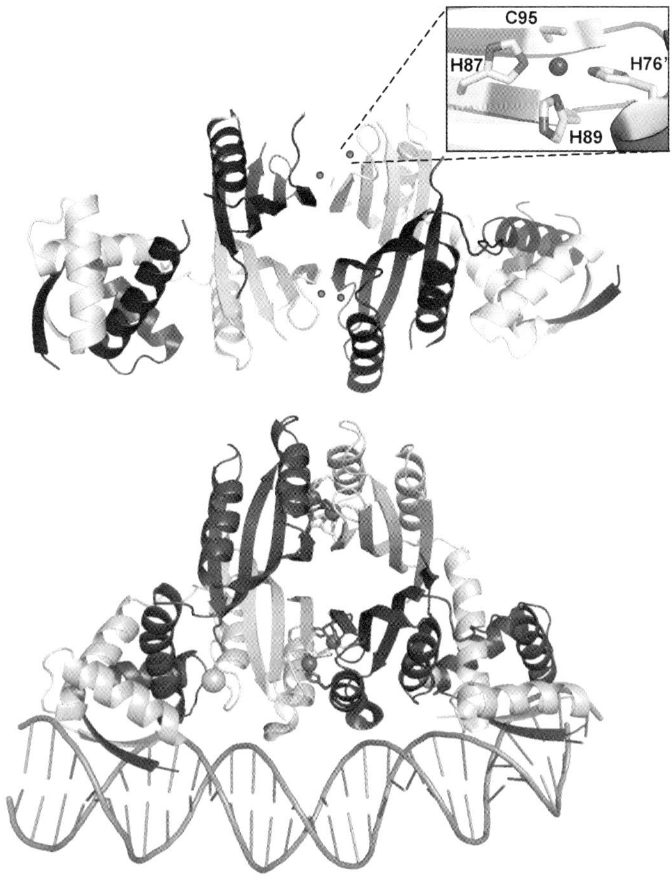

Fig. 2 The Ni^{2+}-sensing transcriptional repressor NikR of *E. coli*. *Top*: The tetrameric apoprotein is shown with four Ni^{2+} ions modeled into the high-affinity sites (from a combination of #1q5v and #1q5y). The square-planar Ni^{2+}-coordination site of the holoprotein is shown in the *inset*, H76′ is from the subunit opposing the one containing the other three ligands. Binding of Ni^{2+} organizes pieces of the secondary structure that are not observed in the apo structure, but does not dramatically alter the tertiary structure of the protein. *Bottom*: Holo-NikR in a complex with DNA (#2hzy). The Ni^{2+} ions are shown in *dark grey* and the K^+ ions as larger, *lighter spheres*

Solution studies demonstrate that stoichiometric amounts of multiple divalent metals activate DNA binding by NikR (Bloom and Zamble 2004) so Ni^{2+} selectivity would only be achieved by limited availability of other metals in the *E. coli* cytoplasm. An additional metal-binding site was suggested upon observing that the affinity for DNA increases several orders of magnitude with excess metal ions (Chivers and Sauer 2002; Bloom and Zamble 2004). The tighter DNA complex is selective for Ni^{2+} in both the high-affinity site and the supplementary site (Bloom and Zamble 2004). The location of the additional metal site(s) is not clear, but the structures of NikR orthologs from *Pyrococcus horikoshii* (PhNikR) and *Helicobacter pylori* (HpNikR) revealed several bound Ni^{2+} (Chivers and Tahirov 2005; Dian et al. 2006). One of these sites bridges the metal- and DNA-binding domains, a reasonable location to enhance DNA binding, but it is occupied by K^+ in the *E. coli* NikR DNA-bound structure (Schreiter et al. 2006) and the octahedral coordination site is composed of carboxyl groups and backbone carbonyls, not a likely site for Ni^{2+} selectivity. The affinity of this second site, indirectly estimated in the mid-nM to μM range (Chivers and Sauer 2002; Bloom and Zamble 2004), as well as its metal selectivity, suggest it is physiologically relevant. In support of this hypothesis, a comparison of expression from the *nik* promoter with mutant NikR versus wild type suggested there are two stages of regulation (Rowe et al. 2005).

2.4.2 Regulation of B_{12} Uptake in *E. coli*

Even before the BtuB protein was isolated it was clear that B_{12} regulates its own uptake (Kadner 1978). In contrast to other regulatory systems, B_{12}-dependent control of BtuB expression occurs primarily at the translational level, not transcription (Lundrigan et al. 1991). BtuB repression disappears upon mutating *btuR*, the gene for cobalamin adenosyltransferase that converts B_{12} into adenosylcobalamin (AdoCbl), demonstrating that the active repressor is AdoCbl (Lundrigan and Kadner 1989). AdoCbl specifically binds to the 5′-untranslated region of *btuB* mRNA and alters its structure with a $K_D \approx 300\,\text{nM}$ (Nahvi et al. 2002), the same concentration needed to inhibit ribosome binding to the *btuB* mRNA (Nou and Kadner 2000). These experiments suggested that AdoCbl directly controls translation initiation without the activity of a protein regulator, probably by stabilizing a hairpin structure that sequesters the Shine-Dalgarno sequence (Nou and Kadner 1998). This system was one of the first examples of metabolite control of RNA regulatory structures called riboswitches (Vitreschak et al. 2004), and these B_{12}-elements are found in most of the sequenced bacterial genomes (Rodionov et al. 2003). The conservation of this regulatory sequence allows it to be used as a tool to identify genes for cobalt transporters, cobalamin biosynthesis, and B_{12}-independent enzymes in organisms that have B_{12}-dependent isoenzymes (Rodionov et al. 2003, 2006).

2.5 Nickel- and Cobalt-Enzymes of *E. coli*

This section describes the known nickel- and cobalt-containing enzymes of *E. coli* and provides cautionary comments with regard to assigning such metal ion dependence.

2.5.1
[NiFe]-Hydrogenases in *E. coli*

Nickel is required in *E. coli* as a catalytic cofactor of several [NiFe]-hydrogenase isoenzymes (Vignais et al. 2001) that are components of certain types of anaerobic metabolism (Böck and Sawers 1996). Hydrogenases 1 and 2 are primarily involved in consuming H_2 as a source of reducing energy whereas hydrogenase 3, as part of a formate hydrogenlyase fermentation complex, produces H_2 to disperse excess electrons and protons. The genes for a fourth isoenzyme have been noted in the *E. coli* genome (Andrews et al. 1997) but it is unclear if this enzyme is active under normal growth conditions (Bagramyan et al. 2002; Skibinski et al. 2002; Self et al. 2004). The [NiFe]-hydrogenase enzymes are heterodimers with the [NiFe] active site (Fig. 3A) coordinated to the large subunit near the dimer interface (Volbeda and Fontecilla-Camps 2003). The nickel is bound to four Cys thiolates, two of which bridge to the iron center that also has three CN/CO ligands.

As with many metalloenzymes (Kuchar and Hausinger 2004), the biosynthesis of the [NiFe] cluster is achieved through the coordinated activity of accessory proteins. In *E. coli*, most of these proteins are encoded by genes given the *hyp* designation (Lutz et al. 1991; Jacobi et al. 1992; Maier et al. 1996) because mutations affect the production of all three *hy*drogenases *p*leiotropically (Lee et al. 1985; Waugh and Boxer 1986). The exceptions are HypC and HypA, which are involved with hydrogenase 3 biosynthesis and are replaced by HybG and HybF for the biosynthesis of hydrogenases 1 and 2 (Blokesch et al. 2001; Hube et al. 2002). Several additional accessory proteins not found in *E. coli* have been noted in other organisms, including HupK, a protein of unknown function that resembles a hydrogenase large subunit

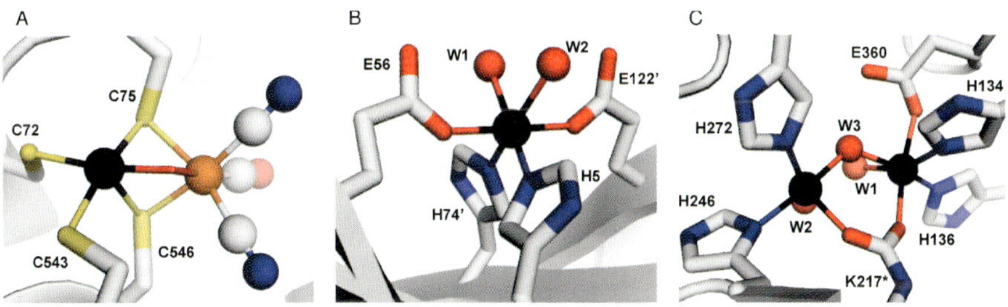

Fig. 3 The active sites of nickel-enzymes found in *E. coli*. **A** The dinuclear center of the [NiFe] hydrogenase, based on the structure of the protein from *Desulfovibrio fructosovorans* (#1yrq). **B** The mononuclear center of glyoxalase (#1f9z) with two bound solvent molecules. **C** The dinuclear nickel center of urease bridged by a carbamylated Lys residue and a water molecule with an additional water coordinated to each metal, based on the structure of the *K. aerogenes* protein (#1fwj). In all structures the Ni^{2+} ions are in *dark grey*. Apostrophes on the residue numbers indicate amino acids from different polypeptide chains

(Kovács et al. 2005), and HypX, a protein found in aerobic organisms that is thought to provide oxygen tolerance to the hydrogenases by delivering an additional CN ligand to the metallocenter (Bleijlevens et al. 2004).

Iron insertion occurs first (Winter et al. 2005) and HypCDEF are responsible for the formation and incorporation of the iron into hydrogenase 3 precursor protein HycE (for recent reviews see Casalot and Rousset 2001; Blokesch et al. 2002; Mulrooney and Hausinger 2003; Kuchar and Hausinger 2004). HypC remains associated with HycE until the metal center is complete, and this interaction is mediated by one of the nickel-ligating Cys of HycE (Magalan and Böck 2000), suggesting that HypC may act as a chaperone that holds the precursor protein in a conformation appropriate for metal insertion. In addition, the promiscuous chaperone pair GroEL/GroES is required for optimal production of both hydrogenase 1 and 3 (Rodrigue et al. 1996).

Although lesions in all of the *hyp* genes block biosynthesis of hydrogenase 3 (Lutz et al. 1991; Jacobi et al. 1992; Maier et al. 1996), bacterial growth in media with excess Ni^{2+} partially restores activity in the $\Delta hypB$ and $\Delta hypA$ strains (Lutz et al. 1991; Jacobi et al. 1992; Maier et al. 1995; Hube et al. 2002). In vitro experiments demonstrated that each protein binds Ni^{2+} and that they interact with each other (Atanassova and Zamble 2005; Leach et al. 2005), suggesting that HypA-HypB may cooperate as a Ni^{2+} insertion complex. In *H. pylori*, the HypA-HypB complex is also implicated in the production of nickel-containing urease (Olson et al. 2001).

HypB has weak GTPase activity required for Ni^{2+} incorporation (Maier et al. 1993, 1995), and it binds stoichiometric Ni^{2+} with sub-pM affinity in an oxygen-sensitive site that includes the cysteines in the N-terminal NH_2-CXXCGC sequence (Leach et al. 2005). Given that this motif is not conserved in all HypB homologs (Robson 2001) it may only be required in certain growth environments, such as Ni^{2+} limiting. In contrast, a second metal ion-binding site of HypB located in the GTPase domain includes several highly-conserved residues and binds Ni^{2+} with μM affinity (Leach et al. 2005), as do HypA and HybF (Blokesch et al. 2004; Atanassova and Zamble 2005). A recent structure of a HypB ortholog lacking the high-affinity site from *Methanocaldococcus jannaschii* revealed a dimer bridged by a nonhydrolyzable GTP analog as well as an asymmetric dinuclear zinc cluster (Gasper et al. 2006), although how this dimer functions in hydrogenase biosynthesis is not yet clear. In addition to Ni^{2+}, HypA/HybF bind Zn^{2+} (Blokesch et al. 2004; Atanassova and Zamble 2005) with the tetrathiolate coordination commonly found in zinc proteins that mediate biomolecular interactions (Berg and Shi 1996; Auld 2001). This structure may be responsible for complex formation with HypB, and HybF/HypA may direct the nickel insertion complex to the distinctive sequences of the large subunits of hydrogenases 1/2 and hydrogenase 3 (Vignais et al. 2001; Blokesch et al. 2004), but such an architectural role has not yet been proven experimentally.

The final step in hydrogenase maturation is proteolytic cleavage of a C-terminal fragment by an isoenzyme-specific endopeptidase that recognizes complete assembly of the metal center (reviewed in Theodoratou and Böck 2005). The crystal structure of the hydrogenase 2-specific protease HybD revealed topological similarities to the family of metzincins and possessed a Cd^{2+} ion supplied by the crystallization buffer (Fritsche et al. 1999), although there is no evidence for metal ion-binding to the protease in solution. Mutation of the residues corresponding to the Cd^{2+} ligands in the homologous peptidase for hydrogenase 3, HycI, disrupted proteolysis (Theodoratou and Böck 2005) and these residues may take part in sensing that the Ni^{2+} is loaded in the hydrogenase precursor protein, thus preventing premature cleavage.

2.5.2 Nickel-Glyoxalase in *E. coli*

A second *E. coli* enzyme with a nickel-containing active site is glyoxalase I (GlxI), the first enzyme in the ubiquitous pathway for degrading methylglyoxal and other toxic α-ketoaldehydes (Thornalley 2003). The human and yeast homologs are Zn^{2+} enzymes (Aronsson et al. 1978), so it was a surprise that the *E. coli* protein was inactive with Zn^{2+} and expressed maximal activity with Ni^{2+} (Clugston et al. 1998) as well as stoichiometric amounts of other divalent metals ($Ni^{2+} > Co^{2+} > Mn^{2+} > Fe^{2+} > Cd^{2+}$) (Clugston et al. 2004). Both X-ray crystallography and XAS analysis revealed that the active metal ion is bound in an octahedral geometry to two His, two carboxylates and two water molecules (Fig. 3B), while the Zn^{2+} center has one less water bound to produce a trigonal bipyramidal center (Davidson et al. 2000; He et al. 2000). Why the Zn^{2+} in the *E. coli* protein does not form a six-coordinate site is not clear. Changing a residue in the *E. coli* site to more closely resemble the human protein yields some activation with Zn^{2+} as well as the other metals (Clugston et al. 2004), but the affinity of the mutant protein for metal ions is noticeably decreased and there are GlxI Zn^{2+} proteins that have the same ligand sphere as the *E. coli* protein, so there must be other critical factors involved. In addition, the crystal structures revealed two metal ion-binding sites in the protein dimer (He et al. 2000), but the maximum enzyme activity only requires one metal ion equivalent per dimer (Clugston et al. 1998), so the function of the second site is not known.

2.5.3 Nickel-Urease in *E. coli*

Most laboratory strains of *E. coli* do not contain the urease gene cluster, however these genes are widely found in enterohemorrhagic strains including the notorious strain O157:H7 (Nakano et al. 2001). Curiously, most of these pathogenic *E. coli* strains do not express urease activity because of a specific one-base substitution in *ureD* resulting in a premature stop codon (Nakano et al. 2004). *E. coli* urease has not been well studied, but the sequences of its *ure* genes are closely related to those of the best-characterized system from *Klebsiella aerogenes* (Hausinger et al. 2001); thus, the following paragraphs summarize our understanding of urease derived from this alternate enteric bacterium as expressed in recombinant form in *E. coli*.

The urease gene cluster of *K. aerogenes* and most urealytic *E. coli* strains is comprised of seven genes (*ureDABCEFG*) that encode a heterotrimeric urease and four accessory proteins needed for urease activity (Lee et al. 1992). The $(UreABC)_3$ enzyme contains three deeply buried dinuclear active sites, where each pair of Ni^{2+} is bridged by a carbamylated Lys residue (Fig. 3C) (Jabri et al. 1995). Cellular synthesis of the urease metallocenter is a complex process that utilizes all four accessory proteins (UreDEFG) and requires Ni^{2+}, CO_2, and MgGTP. When synthesized alone, UreABC is primarily an apoprotein with only trace levels of urease activity (Kim et al. 2005); however, $\sim$ 15% of this protein is activated in vitro by addition of high concentrations of Ni^{2+} and bicarbonate (the latter supplies CO_2 for Lys carbamylation) (Park and Hausinger 1995b). UreD, UreF, and the GTPase UreG bind sequentially to the urease apoprotein to form a UreGFD-UreABC complex and each accessory protein incrementally enhances enzyme activation (Park et al. 1994; Park and Hausinger 1995a, 1996; Moncrief and Hausinger 1996; Soriano and Hausinger 1999). Upon further addition of UreE to the activation mixture, fully active urease is generated from this apoprotein complex by using near-physiological concentrations of Ni^{2+} and bicarbonate (Soriano et al. 2000).

The urease accessory proteins serve two distinct roles. $UreE_2$ is proposed to be a metallochaperone that delivers Ni^{2+} to the apoprotein complex (Mulrooney et al. 2005). The structure of this Ni^{2+}-binding protein (Song et al. 2001) combined with results from site-directed mutagenesis studies (Colpas and Hausinger 2000) reveals a critical Ni^{2+}-binding site at the interface of the homodimer, with additional but nonessential metal ion-binding sites elsewhere in the protein. In contrast, the UreDFG complex is thought to function as a GTP-dependent molecular chaperone that alters the conformation of UreABC to facilitate metallocenter assembly, a hypothesis supported by chemical cross-linking of various complexes followed by proteolysis and mass spectrometric analyses (Chang et al. 2004). The central role of UreD in the urease activation machinery explains why most *E. coli* strains that encode a truncated version of this protein lack urease activity (Nakano et al. 2004).

2.5.4
Cobalt-Enzymes in *E. coli*

E. coli has multiple corrinoid-containing enzymes, such as methionine synthase, methylmalonyl-CoA mutase, and ethanolamine lyase (Roth et al. 1996).

In contrast, there is no conclusive evidence yet for a noncorrinoid cobalt enzyme in *E. coli*.

2.5.5 Cautionary Comments about Assigning Ni^{2+} and Co^{2+} Dependence

Although many enzymes from *E. coli* and other microorganisms are active in vitro with Ni^{2+} or Co^{2+} as a cofactor, the metal employed in vivo depends on many factors including availability, so the assignment of the physiological metal ion can be difficult. An illustrative example is the case of peptide deformylase. The protein sequence contains a potential metal-binding HEXXH motif, the activity is inhibited by metal chelators, and the enzyme purified from *E. coli* contains Zn^{2+} (Meinnel and Blanquet 1995); however, subsequent studies showed that Zn^{2+} was unlikely to be the in vivo cofactor while Ni^{2+} provided full activity and the Ni^{2+}-bound enzyme was extensively characterized, including structural characterization by NMR and crystallographic methods (Becker et al. 1998; Dardel et al. 1998). Decisive studies established that the in vivo form of the enzyme actually contains Fe^{2+} (Rajagopalan et al. 1997). In a similar manner, methionine aminopeptidase was initially labeled a Co^{2+} enzyme because the purified protein was more active with Co^{2+} than other metals (Ben-Bassat et al. 1987). Careful anaerobic experiments revealed that Fe^{2+} was a comparable cofactor to Co^{2+} and metal analysis of cell extracts demonstrated that the level of iron increased upon overexpression of the enzyme whereas the amount of cobalt remained undetectable (D'souza and Holz 1999). Other enzymes activated by Co^{2+} include more aminopeptidases (Zheng et al. 2005), several nucleotidases (Neu 1967; Proudfoot et al. 2004), dehydroquinate synthase (Bender et al. 1989), and a carbohydrate reductoisomerase (Takahashi et al. 1998), but again these enzymes can function with other divalent metal ions.

3 Novel Aspects of Nickel and Cobalt Physiology in Other Microorganisms

The mechanisms used by *E. coli* for uptake of Ni^{2+}, Co^{2+}, and vitamin B_{12}, for efflux of Ni^{2+} and Co^{2+}, and for storage of Ni^{2+} also are found in many other microorganisms, in some cases with interesting variations of these themes. Similarly, the regulatory aspects of Ni^{2+} and Co^{2+} found in *E. coli* are reflected in and expanded upon by other microbes. Finally, a wide range of microorganisms contains the same metalloenzymes already described for *E. coli*, but several additional nickel- and cobalt-dependent enzymes are known. This section focuses on the novel aspects of nickel and cobalt physiology.

3.1
Ni^{2+} and Co^{2+} Uptake

Both Ni^{2+} and Co^{2+} are indiscriminately taken up by cells using various Mg^{2+} transport systems (Moncrief and Maguire 1999). In addition, several microorganisms possess Ni^{2+}- or Co^{2+}-specific ABC transporters similar to the archetype *nik* system in *E. coli* (Rodionov et al. 2006). There is also a separate family of high-affinity permeases specific to Ni^{2+} and/or Co^{2+} and widely distributed in microorganisms (Eitinger et al. 2005).

The Ni^{2+}/Co^{2+} permeases contain eight transmembrane helices, with domain II containing the signature metal ion-binding motif HX_4DH. The first representative discovered is encoded by *hoxN* of *Cuprividus necator* (formerly *Alcaligenes eutrophus* or *Ralstonia eutropha*) and mutations lead to reduced activities of hydrogenase and urease (Eberz et al. 1989; Eitinger and Friedrich 1991). Several orthologues of HoxN transport Ni^{2+}, including NixA of *H. pylori* (Mobley et al. 1995), HupN of *Bradyrhizobium japonicum* (Fu et al. 1994), UreH of a thermophilic *Bacillus* species (Maeda et al. 1994), Nic1p of the filamentous fungus *Schizosaccharomyces pombe* (Eitinger et al. 2000), and many other examples as recently reviewed (Eitinger et al. 2005; Rodionov et al. 2006). Alternatively, some homologues such as NhlF of *Rhodococcus rhodochrous* preferentially transport Co^{2+} (Komeda et al. 1997). Notably, many of the genes encoding these permeases are located adjacent to genes encoding nickel-dependent enzymes (e.g., hydrogenase, urease, or superoxide dismutase) or cobalt-containing proteins (e.g., nitrile hydratase).

3.2
Ni^{2+} and Co^{2+} Efflux

Bacteria utilize three major mechanisms to achieve resistance to metal ions, all related to efflux of the toxic species from the cells (Nies 2003). Representatives of the HME (heavy metal efflux) family of RND (resistance-nodulation-cell division) proteins extend from the cell interior through the cytoplasmic membrane, periplasm, and outer membrane, and use a proton antiport mechanism to pump metal ions from the periplasm to the outside of the cell. The cation diffusion facilitator family (Haney et al. 2005) also uses an antiport mechanism, but in this case they simply pump the metal ions across the cytoplasmic membrane. Finally, the P-type ATPases utilize the energy of ATP hydrolysis to catalyze metal ion efflux across the cytoplasmic membrane. Bacteria appear to utilize primarily the first two mechanisms for dealing with Ni^{2+} and Co^{2+} (Nies 1999, 2003; Haney et al. 2005). Efflux systems are described in more detail in a contribution by Nies in this volume.

3.3
Nickel Storage

HypB and UreE sequences in some microorganisms contain His-rich motifs that bind, and potentially store, large amounts of Ni^{2+}. For example, HypB of *Bradyrhizobium japonicum* has 24 histidines in the first 39 residues and binds 18 Ni^{2+}/dimer (Fu et al. 1995). Deletion of this region of the protein has no effect on its ability to activate the hydrogenase in this microorganism when grown with 50 μM Ni^{2+} in the medium, but results in less activity at lower Ni^{2+} concentrations and less Ni^{2+} within the cell (Olson et al. 1997). Similarly, UreE of *K. aerogenes* contains 10 His among the 15 C-terminal residues and binds approximately six Ni^{2+}/dimer (Lee et al. 1993). Truncation of the protein at the first of these His residues results in H144*UreE that is fully capable of activating urease, but binds only two Ni^{2+}/dimer (Brayman and Hausinger 1996). Two additional potential Ni storage proteins are proposed in *H. pylori*: the GroES homologue HspA and the small polypeptide Hpn, both of which have His-dense sequences and bind multiple Ni^{2+} ions (Kansau et al. 1996; Ge et al. 2006).

3.4
Ni^{2+} and Co^{2+} Regulation

Homologs of *E. coli* NikR have been identified in multiple organisms including *H. pylori* (Chivers and Sauer 1999), a pathogen that uses nickel as a cofactor for [NiFe]-hydrogenase and urease—both required for efficient colonization (Olson and Maier 2002; Mulrooney and Hausinger 2003). HpNikR has broader responsibilities than the *E. coli* protein (Dosanjh and Michel 2006) and regulates the transcription of multiple genes including the urease operon, the Ni^{2+} uptake transporter, and its own gene (van Vliet et al. 2004). Direct binding to several promoters in response to Ni^{2+} has been observed (Abraham et al. 2005; Delany et al. 2005; Ernst et al. 2005), and although the sequences vary, there is a correlation between the location of the HpNikR site with respect to the transcription-start site and whether HpNikR acts as a positive or negative regulator. In addition to Ni^{2+} pathways, HpNikR is also linked to the responses of *H. pylori* to pH and iron (van Vliet et al. 2004; Delany et al. 2005).

Additional Ni^{2+}- and/or Co^{2+}-sensing regulatory proteins include representatives of the SmtB/ArsR family of repressors (Busenlehner et al. 2003; Tottey et al. 2007, in this volume) and Nur, a Fur homologue specific to Ni^{2+} (Ahn et al. 2006) (Helman et al. 2007, in this volume). Two examples of the first class of regulators include NmtR and SrnQ. The Ni^{2+}- and Co^{2+}-sensing NmtR was identified in *Mycobacterium tuberculosis* and shown by mutagenesis studies to bind the hexacoordinate metal ions at its αC5 sites (Cavet et al. 2002). The distantly related SrnQ of *Streptomyces griseus* forms a heterodimeric repressor complex with SrnR and functions to down-regulate *sodF*, encoding an iron-dependent superoxide dismutase, while allowing expression of the nickel-dependent form of the enzyme (Kim et al. 2003). Repression of *sodF* also occurs in *Streptomyces coelicolor* with Ni^{2+}, but in this case the repressor is encoded by one of the four Fur homologues in the cell (Ahn et al. 2006). Fur repressors were originally identified in relationship to iron regulation, but homologues have since been shown to have regulatory roles involving Zn^{2+}, Mn^{2+}, and now Ni^{2+} (Ahn et al. 2006).

3.5 Nickel Enzymes

The nickel-dependent enzymes found in *E. coli* are widely distributed in prokaryotes. For example, [NiFe]-hydrogenases have been identified and characterized from a large range of bacteria and archaea (Vignais et al. 2001; Vignais and Colbeau 2004), and a complex array of accessory proteins is uniformly required for their activation (Casalot and Rousset 2001). Nickel-dependent glyoxalases have been demonstrated in *Yersinia pestis*, *Pseudomonas aeruginosa*, and *Neisseria meningitidis* (Sukdeo et al. 2004), and are likely to be found in other microbes. Finally, every urease examined has been shown to contain nickel (Hausinger and Karplus 2001) and, with the possible exception of *Bacillus subtilis* (Kim et al. 2005), multiple accessory proteins are required for metallocenter assembly. This section summarizes other nickel-containing enzymes (Hausinger 1987; Mulrooney and Hausinger 2003) not represented in *E. coli*.

3.5.1 Superoxide Dismutase

Copper/zinc-, iron-, and manganese-containing superoxide dismutases have long been known to exist, but in 1996 a nickel-containing form of this detoxifying enzyme was discovered in a species of *Streptomyces* (Youn et al. 1996). Subsequently, all tested clinical and soil isolates of this genus were shown to possess the cytoplasmic, nickel-containing enzyme (Leclere et al. 1999) and the range of microorganisms with this protein, encoded by *sodN*, was expanded to include several cyanobacteria (Eitinger 2004; Wuerges et al. 2004). The crystal structures of the homohexameric nickel-enzymes from *Streptomyces seoulensis* and *S. coelicolor* reveal a mononuclear Ni(III) coordinated by the amino group and imidazolate of His-1, the amide group and thiolate of Cys-2, and the thiolate of Cys-6, where the His-1 axial ligand is lost upon reduction of the metal (Barondeau et al. 2004; Wuerges et al. 2004). Synthesis of active enzyme requires cleavage of a 14-residue amino terminal extension to free the resulting His-1 amine to act as a ligand. Genes encoding putative serine proteases are located near *sodN* in several microorganisms, and evidence was obtained using *Prochlorococcus marinus* MIT9313 that this gene, denoted *sodX*, assists in superoxide dismutase maturation (Eitinger 2004).

3.5.2 Carbon Monoxide Dehydrogenase

In 1976, an anaerobic phototrophic bacterium was shown to grow on CO as the sole carbon and energy source (Uffen 1976). Subsequent studies with the related microorganism *Rhodospirillum rubrum* resulted in purification of an enzyme that reversibly oxidizes CO to CO_2 and the demonstration that it contains nickel, zinc, iron, and inorganic sulfur (Bonam and Ludden 1987). Many additional efforts culminated in the structural characterization of the homodimeric enzymes from *R. rubrum* (Drennan et al. 2001) and *Carboxythermus hydrogenoformans* (Dobbek et al. 2001). Both structures reveal a bridging [4Fe – 4S] cluster accessible to external electron transfer components, two additional [4Fe – 4S] clusters likely to function as internal electron transfer sites, and the [Ni – 4Fe – 4S] (Drennan et al. 2001) or [Ni – 4Fe – 5S] (Dobbek et al. 2001) active site cluster. Genetic and physiological studies in *R. rubrum* reveal that *cooC* and *cooJ*, located downstream of the *cooS* structural gene, function in enzyme activation (Kerby et al. 1997). CooC is related in sequence to the HypB and UreG GTPases (facilitating synthesis of hydrogenase and urease), and the purified CooC protein exhibits both ATPase and GTPase activity (Jeon et al. 2001). CooJ was purified (Watt and Ludden 1998) and shown to be a Ni^{2+}-binding protein (4 Ni/monomer) with 16 His in the C-terminal 34 residues, reminiscent of the His-rich sequences involved in hydrogenase and urease activation. The precise roles of these accessory proteins in CO dehydrogenase synthesis are unknown.

3.5.3 Acetyl-CoA Synthase/Decarbonylase

Nickel-containing CO dehydrogenases also were identified in acetogenic and selected other anaerobic bacteria as well as in methanogenic archaea, but this activity proved to be a side-reaction of a much more intricate and fascinating enzyme: acetyl-CoA synthase/decarbonylase (ACS/D) (Lindahl 2002; Drennan et al. 2004). ACS/D uses two distinct active sites to synthesize or decompose acetyl-CoA, depending on the microorganism. The enzyme of acetogens reduces CO_2 to CO at one active site then combines this intermediate with a methyl group (donated by a corrinoid iron-sulfur protein) plus coenzyme A to form acetyl-CoA at a separate active site. By contrast, the enzyme of homoacetogenic methanogens decomposes acetyl-CoA to form a methylated species (eventually yielding methyl-coenzyme M that gives rise to methane, see below) plus CoA and CO, and oxidizes CO to CO_2 at a second active site. Remarkably, both active sites of this enzyme contain nickel. The crystal structure of ACS/D from *Moorella thermoacetica* (formerly *Clostridium thermoaceticum*) reveals a β_2 core resembling CO dehydrogenase containing two [Ni – Fe – S] clusters and three additional [4Fe – 4S] clusters, along with flanking α subunits containing what are now known to be [Ni – Ni – 4Fe – 4S] clusters (Doukov et al. 2002; Darnault et al. 2003). The latter site has a typical [4Fe – 4S] cluster bridged by one of its coordinating Cys thiolates to a labile nickel site (partially substituted by copper or zinc in the published structures); two additional Cys in a Cys-Gly-Cys motif bind both the proximal nickel and a second, distal nickel that additionally interacts with two peptide amides. The same [Ni – Ni – 4Fe – 4S] active site was observed in the crystal structure of monomeric acetyl-CoA synthase from *C. hydrogenoformans* that also exists in complex with the CO dehydrogenase (Svetlitchnyi et al. 2004). In complexes with both activities, a molecular tunnel connects the two nickel-containing active sites allowing for the internal passage of CO (Maynard and Lindahl 1999; Seravalli and Ragsdale 2000). Although the gene clusters encoding ACS/D appear to contain accessory genes, including a CooC-like ATPase termed AcsF (Loke and Lindahl 2003), heterologous expression of only the structural genes from *M. thermoacetica* in *E. coli* resulted in active enzyme (Loke et al. 2000) consistent with self-assembly of the unique clusters or with hijacking of the *E. coli* machinery.

3.5.4 Methyl-Coenzyme M Reductase

Methanogenic archaea, known to require Ni^{2+} for growth since 1979 (Schönheit et al. 1979), typically possess three nickel-containing enzymes: hydrogenase, ACS/D, and, of greatest importance, methyl-coenzyme M (2-thioethanesulfonate, CoM) reductase. The latter enzyme catalyzes the terminal step of methane production in which methyl-CoM and *N*-7-mercaptoheptanoylthreonine phosphate (coenzyme B or CoB) react to release methane and the CoM-CoB mixed disulfide (Thauer 1998). The nickel of methyl-CoM reductase is present as a Ni(I)-tetrapyrrole, termed F_{430} for its characteristic yellow color when in the inactive Ni(II) state. Crystal structures of methyl-CoM reductase (Ermler et al. 1997; Grabarse et al. 2000, 2001) demonstrate that the cofactor is deeply buried in the protein, but connected by a 30 Å channel to the surface. The biosynthesis of F_{430} (Thauer and Bonacker 1994) is an offshoot of the pathways used to form siroheme and corrinoids, with 5-aminolevulinic acid, uroporphyrinogen III, and dihydrosirohydrochlorin being common intermediates. The conversion of dihydrosirohydrochlorin to F_{430} requires amidation of the acetate groups on two rings, reduction of two double bonds, cyclization of an acetamide to form a five-membered ring, cyclization of a propionic acid to form a six-membered ring, and insertion of Ni^{2+}; however, the order of these steps and the mechanism underlying Ni^{2+} insertion and F_{430} incorporation into the protein remain unknown.

Methanotrophic archaea that catalyze the anaerobic oxidation of methane possess an enzyme closely related to methyl-CoM reductase that functions in the reverse direction (Shima and Thauer 2005). This enzyme possesses an F_{430}-like molecule, but of 46 Da increased mass (Krüger et al. 2003). These recently discovered microorganisms were first found in microbial mats associated with anoxic methane seeps in the Black Sea, but have since been shown to be widely distributed in the environment (Hallam et al. 2004).

3.5.5 Aci-Reductone Dioxygenase

A key step in the methionine salvage pathway involves the conversion of 1,2-dihydroxy-3-keto-5-methylthiopentane (aci-reductone) to formate and the ketoacid precursor of methionine by an iron-dependent enzyme. In *Klebsiella pneumoniae*, this intermediate was shown to undergo an alternative, non-productive, dioxygenase reaction to produce formate, carbon monoxide, and methylthiobutyric acid. Remarkably, the latter reaction is carried out by the same protein, but with Ni^{2+} rather than Fe^{2+} bound to its active site (Dai et al. 1999). The solution structure of the Ni^{2+}-containing enzyme was determined by NMR methods (Pochapsky et al. 2002), and XAS studies confirmed the metal binding to three His and three other nitrogen or oxygen ligands (Al-Mjeni et al. 2002). The cellular relevance of the Ni^{2+}-bound form of the protein is unclear and the mechanism of Ni^{2+} insertion is unknown.

3.6 Cobalt Enzymes

The vitamin B_{12}-dependent enzymes found in *E. coli* are also present in many other microorganisms. Furthermore, a wealth of additional cobalamin-dependent enzymes are found in selected bacteria that carry out methyl-transfer, isomerization, or dehalogenase reactions (Banerjee and Ragsdale 2003). Rather than attempting to cover this field, the following comments focus on selected examples of noncorrinoid cobalt-dependent enzymes (Kobayashi and Shimizu 1999).

3.6.1 Nitrile Hydratase and Thiocyanate Hydrolase

Nitrile hydratases catalyze the hydration of nitriles to form amides (Kobayashi and Shimizu 1998) while thiocyanate hydrolase converts the substrate to carbonyl sulfide and ammonia (Katayama et al. 2006). Mononuclear cobalt-containing forms of both of these enzymes have been identified. The nitrile hydratase of *R. rhodochrous* J1 is best characterized, including determination of its three-dimensional structure (Miyanaga et al. 2001). Three Cys side chains (108, 111, and 113) along with the backbone amides of residues 112, and 113 coordinate the active site cobalt. Of interest, Cys-111 and Cys-113 are oxidized to the sulfinic and sulfenic states and participate in a hydrogen-bonding network. Mass spectrometric studies reveal that a Cys-sulfinic acid also exists in thiocyanate hydrolase (Kobayashi and Shimizu 1998).

3.6.2 Prolidase

Proline dipeptidase, or prolidase, of *Pyrococcus furiosus* is a Co^{2+}-containing enzyme when isolated directly from the hyperthermophilic archaeon or when obtained from recombinant *E. coli* cells (Ghosh et al. 1998). The enzyme contains two Co^{2+}/dimer, but is greatly stimulated by additional Co^{2+} (and less so by Mn^{2+}) suggesting a second metal-binding site. Crystallographic analyses reveal that the enzyme possesses a dinuclear metallocenter, albeit the crystalline sample is inactive and possesses Zn^{2+} rather than Co^{2+} (Maher et al. 2004). Although many microorganisms possess the ability to hydrolyze Xxx-Pro dipeptides (where Xxx is a hydrophobic residue such as Met, Leu, Val, Phe, or Ala), it remains unclear how widely the requirement for Co^{2+} is retained or whether this metal is uniquely found in the protein from hyperthermophilic archaea.

3.6.3 Transcarboxylase

Methylmalonyl-CoA carboxytransferase (transcarboxylase) from *Propionibacterium shermanii* transfers a carboxyl group from methylmalonyl-CoA to pyruvate, forming propionyl-CoA and oxaloacetate. In pioneering studies carried out nearly 40 years ago, the 5S subunit of this enzyme was shown to contain zinc and cobalt (Northrop and Wood 1969). The cobalt site was verified by crystallographic studies (Hall et al. 2004) that reveals its coordination by two His, a bidentate carbamylated Lys (converted to monodentate when substrate binds), Asp, and a solvent molecule.

3.6.4 Other Cobalt Enzymes

Although several other enzymes have been reported to contain cobalt (Kobayashi and Shimizu 1999), caution must be exercised in assigning metal dependence as described in Sect. 2.5.5. For example, cobalt initially was stated to be essential for glucose isomerase activity of selected *Streptomyces* strains (Tsumura et al. 1967; Hemker et al. 1987). This enzyme often is activated by Co^{2+} and stabilized to high temperatures by this metal ion (Epting et al. 2005), but Mn^{2+} and Mg^{2+} also typically activate the enzyme and may be used in vivo by most microorganisms. Many types of haloperoxidases have been described, including one in *Pseudomonas putida* IF-3 that is supposedly activated specifically by Co^{2+} (Itoh et al. 1994). More recent studies suggest this enzyme is both a bromoperoxidase and a perhydrolase that decomposes peracetic acid (Itoh et al. 2001; Kawanami et al. 2002). Other reported cobalt-containing enzymes are even less well characterized (Kobayashi and Shimizu 1999).

Acknowledgements This work was supported by grants from the Canadian Research Chairs Program (to D.B.Z.) as well as the National Institutes of Health (DK45686 to R.P.H.).

References

Abraham LO, Li Y, Zamble DB (2005) Metal-binding and DNA-binding activities of *Helicobacter pylori* NikR. J Inorg Biochem 100:1005–1014

Ahn B-E, Cha J, Lee E-J, Han A-R, Thompson CJ, Roe J-H (2006) Nur, a nickel-responsive regulator of the Fur family, regulates superoxide dismutases and nickel transport in *Streptomyces coelicolor*. Molec Microbiol 59:1848–1858

Allan CB, Wu L-F, Gu Z, Choudhury SB, Al-Mjeni F, Sharma ML, Mandrand-Berthelot M-A, Maroney MJ (1998) An X-ray absorption spectroscopic structural investigation of the nickel site in *Escherichia coli* NikA protein. Inorg Chem 37:5952–5955

Al-Mjeni F, Ju T, Pochapsky TC, Maroney MJ (2002) XAS investigation of the structure and function of Ni in acireductone dioxygenase. Biochemistry 41:6761–6769

Andrews SC, Berks BC, McClay J, Ambler A, Quail MA, Golby P, Guest JR (1997) A 12-cistron *Escherichia coli* operon (*hyf*) encoding a putative proton-translocating formate hydrogenlyase system. Microbiol 143:3633–3647

Aronsson A-C, Marmstål E, Mannervik B (1978) Glyoxalase I, a zinc metalloenzyme of mammals and yeast. Biochem Biophys Res Commun 81:1235–1240

Atanassova A, Zamble DB (2005) *Escherichia coli* HypA is a zinc metalloprotein with a weak affinity for nickel. J Bacteriol 187:4689–4697

Auld DS (2001) Zinc coordination sphere in biochemical zinc sites. BioMetals 14:271–313

Bagramyan K, Mnatsakanyan N, Poladian A, Vassilian A, Trchounian A (2002) The roles of hydrogenases 3 and 4, and the F_0F_1-ATPase, in H_2 production by *Escherichia coli* at alkaline and acidic pH. FEBS Lett 516:172–178

Banerjee R, Ragsdale SW (2003) The many faces of vitamin B_{12}: catalysis by cobalamin-dependent enzymes. Annu Rev Biochem 72:209–247

Barondeau DP, Kassman CJ, Bruns CK, Tainer JA, Getzoff ED (2004) Nickel superoxide dismutase structure and mechanism. Biochemistry 43:8038–8047

Bartha R, Ordal EJ (1965) Nickel-dependent chemolithotrophic growth of two *Hydrogenomonas* strains. J Bacteriol 89:1015–1019

Becker A, Schlichtling I, Kabsch W, Schultz S, Wagner AFV (1998) Structure of peptide deformylase and identification of the substrate binding site. J Biol Chem 273:11413–11416

Ben-Bassat A, Bauer K, Chang S-Y, Myambo K, Boosman A, Chang S (1987) Processing of the initiation methionine from proteins: properties of the *Escherichia coli* methionine aminopeptidase and its gene structure. J Bacteriol 169:751–757

Bender SL, Mehdi S, Knowles JR (1989) Dehydroquinate synthase: the role of divalent metal cations and of nicotinamide adenine dinucleotide in catalysis. Biochemistry 28:7555–7560

Berg JM, Shi Y (1996) The galvanization of biology: a growing appreciation for the roles of zinc. Science 271:1081–1085

Bleijlevens B, Buhrke T, van der Linden E, Friedrich B, Albracht SPJ (2004) The auxiliary protein HypX provides oxygen tolerance to the soluble [NiFe]-hydrogenase of *Ralstonia eutropha* H16 by way of a cyanide ligand to nickel. J Biol Chem 279:46686–46691

Blokesch M, Magalon A, Böck A (2001) Interplay between the specific chaperone-like proteins HybG and HypC in maturation of hydrogenases 1, 2, and 3 from *Escherichia coli*. J Bacteriol 183:2817–2822

Blokesch M, Paschos A, Theodoratou E, Bauer A, Hube M, Huth S, Böck A (2002) Metal insertion into NiFe-hydrogenases. Biochem Soc Trans 30:674–680

Blokesch M, Rohrmoser M, Rode S, Böck A (2004) HybF, a zinc-containing protein involved in NiFe hydrogenase biosynthesis. J Bacteriol 186:2603–2611

Bloom SB, Zamble DB (2004) The metal-selective DNA-binding response of *Escherichia coli* NikR. Biochemistry 43:10029–10038

Böck A, Sawers G (1996) Fermentation. In: Neidhardt FC (ed) *Escherichia coli* and *Salmonella*, Cellular and Molecular Biology, 2nd edn. ASM Press, Washington, DC

Bonam D, Ludden PW (1987) Purification and characterization of carbon monoxide dehydrogenase, a nickel, zinc, iron-sulfur protein, from *Rhodospirillum rubrum*. J Biol Chem 262:2980–2987

Borths EL, Locher KP, Lee AT, Rees DC (2002) The structure of *Escherichia coli* BtuF and binding to its cognate ATP binding cassette transporter. Proc Natl Acad Sci USA 99:16642–16647

Borths EL, Poolman B, Hvorup RN, Locher KP, Rees DC (2005) In vitro functional characterization of BtuCD-F, the *Escherichia coli* ABC transporter for vitamin B_{12} uptake. Biochemistry 44:16301–16309

Bradbeer C, Reynolds PR, Bauler GM, Fernandez MT (1986) A requirement for calcium in the transport of cobalamin across the outer membrane of *Escherichia coli*. J Biol Chem 261:2520–2523

Brayman TG, Hausinger RP (1996) Purification, characterization, and functional analysis of a truncated *Klebsiella aerogenes* UreE urease accessory protein lacking the histidine-rich carboxyl terminus. J Bacteriol 178:5410–5416

Busenlehner LS, Pennella MA, Giedroc DP (2003) The SmtB/ArsR family of metalloregulatory transcriptional repressors: structural insights into prokaryotic metal resistance. FEMS Microbiol Rev 27:131–143

Cadieux N, Bradbeer C, Reeger-Schneider E, Köster W, Mohanty AK, Wiener MC, Kadner RJ (2002) Identification of the periplasmic cobalamin-binding protein BtuF of *Escherichia coli*. J Bacteriol 184:706–717

Carrington PE, Chivers PT, Al-Mjeni F, Sauer RT, Maroney MJ (2003) Nickel coordination is regulated by the DNA-bound state of NikR. Nat Struct Biol 10:126–130

Casalot L, Rousset M (2001) Maturation of [NiFe] hydrogenases. Trends Microbiol 9:228–237

Cavet JS, Meng W, Pennella MA, Appelhoff RJ, Giedroc DP, Robinson NJ (2002) A nickel-cobalt sensing ArsR-SmtB family repressor: contributions of cytosol and effector binding sites to metal selectivity. J Biol Chem 277:38441–38448

Chang Z, Kuchar J, Hausinger RP (2004) Chemical crosslinking and mass spectrometric identification of sites of interaction for UreD, UreF, and urease. J Biol Chem 279:15305–15313

Cherrier MV, Martin L, Cavazza C, Jacquamet L, Lemaire D, Gaillard J, Fontecilla-Camps JC (2005) Crystallographic and spectroscopic evidence for high affinity binding of $FeEDTA(H_2O)^-$ to the periplasmic nickel transporter NikA. J Am Chem Soc 127:10075–10082

Chimento DP, Mohanty AK, Kadner RJ, Wiener MC (2003) Substrate-induced transmembrane signaling in the cobalamin transporter BtuB. Nat Struct Biol 10:394–401

Chivers PT, Sauer RT (1999) NikR is a ribbon-helix-helix DNA-binding protein. Prot Sci 8:2494–2500

Chivers PT, Sauer RT (2000) Regulation of high affinity nickel uptake in bacteria. Ni^{2+}-dependent interaction of NikR with wild-type and mutant operator sites. J Biol Chem 275:19735–19741

Chivers PT, Sauer RT (2002) NikR repressor: high-affinity nickel binding to the C-terminal domain regulates binding to operator DNA. Chem Biol 9:1141–1148

Chivers PT, Tahirov TH (2005) Structure of *Pyrococcus horikoshi* NikR: nickel sensing and implications for the regulation of DNA recognition. J Mol Biol 348:597–607

Clugston SL, Barnard JFJ, Kinach R, Miedema D, Ruman R, Daub E, Honek JF (1998) Overproduction and characterization of a dimeric non-zinc glyoxylase I from *Escherichia coli*: evidence for optimal activation by nickel ions. Biochemistry 37:8754–8763

Clugston SL, Yajima R, Honek JF (2004) Investigation of metal binding and activation of *Escherichia coli* glyoxalase I: kinetic, thermodynamic and mutagenesis studies. Biochem J 377:309–316

Colpas GJ, Hausinger RP (2000) In vivo and in vitro kinetics of metal transfer by the *Klebsiella aerogenes* urease nickel metallochaperone, UreE. J Biol Chem 275:10731–10737

Dai Y, Wensink PC, Abeles RH (1999) One protein, two enzymes. J Biol Chem 274:1193–1195

Dardel F, Ragusa S, Lazennec C, Blanquet S, Meinnel T (1998) Solution structure of nickel-peptide deformylase. J Mol Biol 280:501–513

Darnault C, Volbeda A, Kim EJ, Legrand P, Vernède X, Lindahl PA, Fontecilla-Camps JC (2003) $NiZn[Fe_4S_4]$ and $NiNi[Fe_4S_4]$ clusters in closed and open α subunits of acetyl-CoA synthase/carbon monoxide dehydrogenase. Nat Struct Biol 10:271–279

Davidson G, Clugston SL, Honek JF, Maroney MJ (2000) XAS investigation of the nickel active site structure in *Escherichia coli* glyoxalase I. Inorg Chem 39:2962–2963

de Pina K, Desjardin V, Mandrand-Berthelot M-A, Giordano G, Wu L-F (1999) Isolation and characterization of the *nikR* gene encoding a nickel-responsive regulator in *Escherichia coli*. J Bacteriol 181:670–674

de Pina K, Navarro C, McWalter L, Boxer DH, Price NC, Kelly SM, Mandrand-Berthelot M-A, Wu L-F (1995) Purification and characterization of the periplasmic nickel-binding protein NikA of *Escherichia coli* K12. Eur J Biochem 227:857–865

Delany I, Ieva R, Soragni A, Hilleringmann M, Rappuoli R, Scarlato V (2005) In vitro analysis of protein-operator interactions of the NikR and Fur metal-responsive regulators of coregulated genes in *Helicobacter pylori*. J Bacteriol 187:7703–7715

DeVeaux LC, Clevenson DS, Bradbeer C, Kadner RJ (1986) Identification of the BtuCED polypeptides and evidence for their role in vitamin B12 transport in *Escherichia coli*. J Bacteriol 167:920–927

Dian C, Schauer K, Kapp U, McSweeney SM, Labigne A, Terradot L (2006) Structural basis of the nickel response in Helicobacter pylori: Crystal structures of HpNikR in apo and nickel-bound states. J Mol Biol 361:715–730

Dias AV, Zamble DB (2005) Protease digestion analysis of *E. coli* NikR: Evidence for conformational stabilization with Ni(II). J Biol Inorg Chem 10:605–612

Dobbek H, Svetlitchnyi V, Gremer L, Huber R, Meyer O (2001) Crystal structure of a carbon monoxide dehydrogenase reveals a [Ni – 4Fe – 5S] cluster. Science 293:1281–1285

Dosanjh NS, Michel SLJ (2006) Microbial nickel metalloregulation: NikRs for nickel ions. Curr Opin Chem Biol 10:123–130

Doukov TI, Iverson TM, Seravalli J, Ragsdale SW, Drennan CL (2002) A Ni – Fe – Cu center in a bifunctional carbon monoxide dehydrogenase/acetyl-CoA synthase. Science 298:567–572

Drennan CL, Doukov TI, Ragsdale SW (2004) The metalloclusters of carbon monoxide dehydrogenase/acetyl-CoA synthase: a story in pictures. J Biol Inorg Chem 9:511–515

Drennan CL, Heo J, Sintchak MD, Schreiter E, Ludden PW (2001) Life on carbon monoxide: X-ray structure of *Rhodospirillum rubrum* Ni – Fe – S carbon monoxide dehydrogenase. Proc Natl Acad Sci USA 98:11973–11978

D'souza VM, Holz RC (1999) The methionyl aminopeptidase from *Escherichia coli* can function as an iron(II) enzyme. Biochemistry 38:11079–11085

Eberz G, Eitinger T, Friedrich B (1989) Genetic determinants of a nickel-specific transport system are part of the plasmid-encoded hydrogenase gene cluster in *Alcaligenes eutrophus*. J Bacteriol 171:1340–1345

Eitinger T (2004) In vivo production of active nickel superoxide dismutase from *Prochlorococcus marinus* MIT9313 is dependent on its cognate peptidase. J Bacteriol 186:7812–7825

Eitinger T, Degen O, Böhnke U, Müller M (2000) Nic1p, a relative of bacterial transition metal permeases in *Schizosaccharomyces pombe*, provides nickel ion for urease biosynthesis. J Biol Chem 275:18029–18033

Eitinger T, Friedrich B (1991) Cloning, nucleotide sequence, and heterologous expression of the high-affinity nickel transport gene from *Alcaligenes eutrophus*. J Biol Chem 266:3222–3227

Eitinger T, Mandrand-Berthelot M-A (2000) Nickel transport systems in microorganisms. Arch Microbiol 173:1–9

Eitinger T, Suhr J, Moore L, Smith JAC (2005) Secondary transporters for nickel and cobalt ions: theme and variations. BioMetals 18:399–405

Epting KL, Vieille C, Zeikus JG, Kelly RM (2005) Influence of divalent cations on the structural stability and thermal inactivation kinetics of class II xylose isomerases. FEBS J 272:1454–1464

Ermler U, Grabarse W, Shima S, Goubeaud M, Thauer RK (1997) Crystal structure of methyl-coenzyme M reductase: the key enzyme of biological methane formation. Science 278:1457–1462

Ernst FD, Kuipers EJ, Heijens A, Sarwari R, Stoof J, Penn CW, Kusters JG, van Vliet AH (2005) The nickel-responsive regulator NikR controls activation and repression of gene transcription in *Helicobacter pylori*. Infect Immun 73:7252–7258

Fauquant C, Diederix RE, Rodrigue A, Dian C, Kapp U, Terradot L, Mandrand-Berthelot MA, Michaud-Soret I (2006) pH dependent Ni(II) binding and aggregation of *Escherichia coli* and *Helicobacter pylori* NikR. Biochimie 88:1693–1705

Fritsche E, Paschos A, Beisel H-G, Böck A, Huber R (1999) Crystal structure of the hydrogenase maturating endopeptidase HydD from *Escherichia coli*. J Mol Biol 288:989–998

Fu C, Javedan S, Moshiri F, Maier RJ (1994) Bacterial genes involved in incorporation of nickel into a hydrogenase enzyme. Proc Natl Acad Sci USA 91:5099–5103

Fu C, Olson JW, Maier RJ (1995) HypB protein of *Bradyrhizobium japonicum* is a metal-binding GTPase capable of binding 18 divalent nickel ions per dimer. Proc Natl Acad Sci USA 92:2333–2337

Gasper R, Scrima A, Wittinghofer A (2006) Structural insights into HypB, a GTP-binding protein that regulates metal binding. J Biol Chem 281:27492–27502

Ge R, Watt RM, Sun X, Tanner JA, He Q-Y, Huang J-D, Sun H (2006) Expression and characterization of a histidine-rich protein, Hpn: potential for Ni^{2+} storage in *Helicobacter pylori*. Biochem J 393:285–293

Ghosh M, Grunden AM, Dunn DM, Weiss R, Adams MW (1998) Characterization of native and recombinant forms of an unusual cobalt-dependent proline dipeptidase (prolidase) from the hyperthermophilic archaeon *Pyrococcus furiosus*. J Bacteriol 180:4781–4789

Grabarse W, Mahlert F, Duin EC, Goubeaud M, Shima S, Thauer RK, Lamzin V, Ermler U (2001) On the mechanism of biological methane formation: structural evidence for conformational changes in methyl-coenzyme M reductase upon substrate binding. J Mol Biol 309:315–330

Grabarse W, Mahlert F, Shima S, Thauer RK, Ermler U (2000) Comparison of three methyl-coenzyme M reductases from phylogenetically distant microorganisms: unusual amino acid modification, conservation, and adaptation. J Mol Biol 303:329–344

Grass G, Franke S, Taudte N, Nies DH, Kucharski LM, Maguire ME, Rensing C (2005) The metal permease ZupT from *Escherichia coli* is a transporter with a broad substrate spectrum. J Bacteriol 187:1604–1611

Hall PR, Zheng R, Anthony L, Pusztai-Carey M, Carey PR, Yee VC (2004) Transcarboxylase 5S structures: assembly and catalytic mechanism of a multienzyme complex subunit. EMBO J 23:3621–3631

Hallam SJ, Putnam N, Preston CM, Detter JC, Rokhsar D, Richardson PM, DeLong EF (2004) Reverse methanogenesis: testing the hypothesis with environmental genomics. Science 305:1457–1462

Haney CJ, Grass G, Frake S, Rensing C (2005) New developments in the understanding of the cation diffusion facilitator family. J Indust Microbiol Biotechnol 32:215–226

Hausinger RP (1987) Nickel utilization by microorganisms. Microbiol Rev 51:22–42

Hausinger RP, Colpas GJ, Soriano A (2001) Urease: a paradigm for protein-assisted metallocenter assembly. ASM News 67:78–84

Hausinger RP, Karplus PA (2001) Urease. In: Wieghardt K, Huber R, Poulos TL, Messerschmidt A (eds) Handbook of Metalloproteins. Wiley, Chichester, West Sussex, UK, p 867–879

He MM, Clugston SL, Honek JF, Matthews BW (2000) Determination of the structure of *Escherichia coli* glyoxylase I suggests a structural basis for differential metal activation. Biochemistry 39:8719–8727

Heddle J, Scott DJ, Unzai S, Park S-Y, Tame JRH (2003) Crystal structures of the liganded and unliganded nickel-binding protein NikA from *Escherichia coli*. J Biol Chem 278:50322–50329

Hemker J, Kleinschmitdt L, Witzel H (1987) Glucoseisomerase, obviously the first native cobalt enzyme. Reueil des Traveaux Chemiques des Pays-Bas 106:350

Higgins CF (2001) ABC transporters: physiology, structure and mechanism - an overview. Res Microbiol 152:205–210

Hmiel SP, Snavely MD, Miller CG, Maguire ME (1986) Magnesium transport in *Salmonella typhimurium*: characterization of magnesium influx and cloning of a transport gene. J Bacteriol 168:1444–1450

Holm-Hansen O, Gerloff G, Skoog F (1954) Cobalt as an essential element for blue-green algae. Physiol Plant 7:665–675

Hottenrott S, Schumann T, Pluckthun A, Fischer G, Rahfeld JU (1997) The *Escherichia coli* SlyD is a metal ion-regulated peptidyl-prolyl cis/trans-isomerase. J Biol Chem 272:15697–15701

Hube M, Blokesch M, Böck A (2002) Network of hydrogenase maturation in Escherichia coli: role of accessory proteins HypA and HybF. J Bacteriol 184:3879–3885

Itoh N, Kawanami T, Liu J-Q, Dairi T, Miyakoshi M, Nitta C, Kimoto Y (2001) Cloning and biochemical characterization of Co^{2+}-activated bromoperoxidase-esterase (perhydrolase) from *Pseudomonas putida* IF-3 strain. Biochim Biophys Acta 1545:53–66

Itoh N, Morinaga N, Kouzai T (1994) Purification and characterization of a novel metal-containing nonheme bromoperoxidase from *Pseudomonas putida*. Biochim Biophys Acta 1207:208–216

Iwig JS, Rowe JL, Chivers PT (2006) Nickel homeostasis in *Escherichia coli* – the *rcnR-rcnA* efflux pathway and its linkage to NikR function. Molec Microbiol 62:252–262

Jabri E, Carr MB, Hausinger RP, Karplus PA (1995) The crystal structure of urease from *Klebsiella aerogenes*. Science 268:998–1004

Jacobi A, Rossman R, Böck A (1992) The *hyp* operon gene products are required for maturation of catalytically active hydrogenase isoenzymes in *Escherichia coli*. Arch Microbiol 158:444–451

Jeon WB, Cheng J, Ludden PW (2001) Purification and characterization of membrane-associated CooC protein and its functional role in the insertion of nickel into carbon monoxide dehydrogenase from *Rhodospirillum rubrum*. J Biol Chem 276:38602–38609

Kadner RJ (1978) Repression of synthesis of the vitamin B12 receptor in *Escherichia coli*. J Bacteriol 136:1050–1057

Kansau I, Guillain F, Thiberge J-M, Labigne A (1996) Nickel binding and immunological properties of the C-terminal domain of the *Helicobacter pylori* GroES homologue (HspA). Molec Microbiol 22:1013–1023

Karpowich NK, Huang HH, Smith PC, Hunt JF (2003) Crystal structures of the BtuF periplasmic-binding protein for vitamin B12 suggest a functionally important reduction in protein mobility upon ligand binding. J Biol Chem 278:8429–8434

Katayama Y, Hoshimoto K, Nakayama H, Mino H, Nojiri M, Ono TA, Nyunoya H, Yohda M, Takio K, Odaka M (2006) Thiocyanate hydrolase is a cobalt-containing metalloenzyme with a cysteine-sulfinic acid ligand. J Am Chem Soc 128:728–729

Kawanami T, Miyakoshi M, Dairi T, Itoh N (2002) Reaction mechanism of the Co^{2+}-activated multifunctional bromoperoxidase-esterase from *Pseudomonas putida* IF-3. Arch Biochem Biophys 398:94–100

Kehres DG, Zaharik ML, Finlay BB, Maguire ME (2000) The Nramp proteins of *Salmonella typhimurium* and *Escherichia coli* are selective manganese transporters involved in the response to reactive oxygen. J Bacteriol 36:1085–1100

Kerby RL, Ludden PW, Roberts GP (1997) In vivo nickel insertion into carbon monoxide dehydrogenase of *Rhodosprillum rubrum*: molecular and physiological characterization of *cooCTJ*. J Bacteriol 179:2259–2266

Kim JK, Mulrooney SB, Hausinger RP (2005) Biosynthesis of active *Bacillus subtilis* urease in the absence of known urease accessory proteins. J Bacteriol 187:7150–7154

Kim J-S, Kang S-O, Lee J-K (2003) The protein complex composed of nickel-binding SrnQ and DNA binding motif-bearing SrnR of *Streptomyces griseus* represses sodF trnascription in the presence of nickel ions. J Biol Chem 278:18455–18463

Kobayashi M, Shimizu S (1998) Metalloenzyme nitrile hydratase: structure, regulation, and application to biotechnology. Nat Biotechnol 16:733–736

Kobayashi M, Shimizu S (1999) Cobalt proteins. Eur J Biochem 261:1–9

Koch D, Nies DH, Grass G (2007) The RcnA (YohLM) system of *Escherichia coli*: a connection between nickel, cobalt and iron homeostasis. BioMetals (in press)

Komeda H, Kobayashi M, Shimizu S (1997) A novel transporter involved in cobalt transport. Proc Natl Acad Sci USA 94:36–41

Kovács KL, Kovács AT, Maróti G, Mészáros LS, Balogh J, Latinovics D, Fülöp A, Dávid R, Dorogházi E, Rákhely G (2005) The hydrogenases of *Thiocapsa roseopersicina*. Biochem Soc Trans 33:61–63

Krüger M, Meyerdierks A, Glöckner FO, Amann R, Widdel F, Kube M, Reinhardt R, Kahnt J, Böcher R, Thauer RK, Shima S (2003) A conspicuous nickel protein in microbial mats that oxidize methane anaerobically. Nature 426:878–881

Kuchar J, Hausinger RP (2004) Biosynthesis of metal sites. Chem Rev 104:509–526

Lawrence JG, Roth JR (1995) The cobalamin (coenzyme B12) biosynthetic genes of *Escherichia coli*. J Bacteriol 177:6371–6380

Leach MR, Sandal S, Sun H, Zamble DB (2005) Metal binding activity of the *Escherichia coli* hydrogenase maturation factor HypB. Biochemistry 44:12229–12238

Leclere V, Boiron P, Blondeau R (1999) Diversity of superoxide-dismutases among clinical and soil isolates of *Streptomyces* species. Curr Microbiol 39:365–368

Lee JH, Patel P, Sankar P, Shanmugam KT (1985) Isolation and characterization of mutant strains of *Escherichia coli* altered in H_2 metabolism. J Bacteriol 162:344–352

Lee MH, Mulrooney SB, Renner MJ, Markowicz Y, Hausinger RP (1992) *Klebsiella aerogenes* urease gene cluster: sequence of *ureD* and demonstration that four accessory genes (*ureD*, *ureE*, *ureF*, and *ureG*) are involved in nickel metallocenter biosynthesis. J Bacteriol 174:4324–4330

Lee MH, Pankratz HS, Wang S, Scott RA, Finnegan MG, Johnson MK, Ippolito JA, Christianson DW, Hausinger RP (1993) Purification and characterization of *Klebsiella aerogenes* UreE protein: a nickel-binding protein that functions in urease metallocenter assembly. Prot Sci 2:1042–1052

Lindahl PA (2002) The Ni-containing carbon monoxide dehydrogenase family: light at the end of the tunnel? Biochemistry 41:2097–2105

Locher KP, Lee AT, Rees DC (2002) The *E. coli* BtuCD structure: a framework for ABC transporter architecture and mechanism. Science 296:1091–1098

Loke H-K, Bennett GN, Lindahl PA (2000) Active acetyl-CoA synthase from *Clostridium thermoaceticum* obtained by cloning and heterologous expression of *acsAB* in *Escherichia coli*. Proc Natl Acad Sci USA 97:12530–12535

Loke H-K, Lindahl PA (2003) Identification and preliminary characterization of AcsF, a putative Ni-insertase used in the biosynthesis of acetyl-CoA synthase from *Clostridium thermoaceticum*. J Inorg Biochem 93:33–40

Lundrigan MD, Kadner RJ (1989) Altered cobalamin metabolism in *Escherichia coli btuR* mutants affects *btuB* gene regulation. J Bacteriol 171:154–161

Lundrigan MD, Köster W, Kadner RJ (1991) Transcribed sequences of the *Escherichia coli butB* gene control its expression and regulation by vitamin B12. Proc Natl Acad Sci USA 88:1479–1483

Lutz A, Jacobi A, Schlensog V, Böhm R, Sawers G, Böck A (1991) Molecular characterization of an operon (*hyp*) necessary for the activity of the three hydrogenase isoenzymes in *Escherichia coli*. Molec Microbiol 5:123–135

Maeda M, Hidaka M, Nakamura A, Masaki H, Uozumi T (1994) Cloning, sequencing, and expression of thermophilic *Bacillus* strain TB-90 urease gene complex in *Escherichia coli*. J Bacteriol 176:432–442

Magalan A, Böck A (2000) Dissection of the maturation reactions of the [NiFe] hydrogenase 3 from *Escherichia coli* taking place after nickel incorporation. FEBS Lett 473:254–258

Maher MJ, Ghosh M, Grunden AM, Menon AO, Adams MW, Freeman HC, Guss JM (2004) Structure of the prolidase from *Pyrococcus furiosus*. Biochemistry 43:2771–2783

Maier T, Binder U, Böck A (1996) Analysis of the *hydA* locus of *Escherichia coli*: two genes (*hydN* and *hypF*) involved in formate and hydrogen metabolism. Arch Microbiol 165:333–341

Maier T, Jacobi A, Sauter M, Böck A (1993) The product of the *hypB* gene, which is required for nickel incorporation into hydrogenases, is a novel guanine nucleotide-binding protein. J Bacteriol 175:630–635

Maier T, Lottspeich F, Böck A (1995) GTP hydrolysis by HypB is essential for nickel insertion into hydrogenases of *Escherichia coli*. Eur J Biochem 230:133–138

Makui H, Roig E, Cole ST, Helmann JD, Gros P, Cellier MFM (2000) Identification of the *Escherichia coli* K-12 Nramp orthologue (MntH) as a selective divalent metal ion transporter. Molec Microbiol 35:1065–1078

Maynard EL, Lindahl PA (1999) Evidence of a molecular tunnel connecting the active sites for CO_2 reduction and acetyl-CoA synthesis in acetyl-CoA synthase from *Clostridium thermoaceticum*. J Am Chem Soc 121:9221–9222

Meinnel T, Blanquet S (1995) Enzymatic properties of *Escherichia coli* peptide deformylase. J Bacteriol 177:7387–7390

Miyanaga A, Fushinobu S, Ito K, Wakagi T (2001) Crystal structure of cobalt-containing nitrile hydratase. Biochem Biophys Res Commun 288:1169–1174

Mobley HLT, Garner RM, Bauerfeind P (1995) *Helicobacter pylori* nickel-transport gene *nixA*: synthesis of catalytically active urease in *Escherichia coli* independent of growth conditions. Molec Microbiol 16:97–109

Moncrief MBC, Hausinger RP (1996) Purification and activation properties of UreD-UreF-urease apoprotein complexes. J Bacteriol 178:5417–5421

Moncrief MBC, Maguire ME (1999) Magnesium transport in prokaryotes. J Biol Inorg Chem 4:523–527

Mulrooney SB, Hausinger RP (2003) Nickel uptake and utilization by microorganisms. FEMS Microbiol Rev 27:239–261

Mulrooney SB, Ward SK, Hausinger RP (2005) Purification and properties of the *Klebsiella aerogenes* UreE metal-binding domain, a functional metallochaperone of urease. J Bacteriol 187:3581–3585

Nahvi A, Sudarsan N, Ebert MS, Zou X, Brown KL, Breaker RR (2002) Genetic control by a metabolite binding mRNA. Chem Biol 9:1043–1049

Nakano M, Iida T, Honda T (2004) Urease activity of enterohaemorrhagic *Escherichia coli* depends on a specific one-base substitution in *ureD*. Microbiol 150:3483–3489

Nakano M, Iida T, Ohnishi M, Kurokawa K, Takahashi A, Tsukamoto T, Yasunaga T, Hayashi T, Honda T (2001) Association of the urease gene with enterohemorrhagic *Escherichia coli* strains irrespective of their serogroups. J Clin Microbiol 39:4541–4543

Navarro C, Wu L-F, Mandrand-Berthelot M-A (1993) The *nik* operon of *Escherichia coli* encodes a periplasmic binding-protein-dependent transport system for nickel. Molec Microbiol 9:1181–1191

Neu HC (1967) The 5′-nucleotidase of *Escherichia coli*. J Biol Chem 242:3896–3904

Nies DH (1999) Microbial heavy-metal resistance. Appl Microbiol Biotechnol 51:730–750

Nies DH (2003) Efflux-mediated heavy metal resistance in prokaryotes. FEMS Microbiol Rev 27:313–339

Northrop DB, Wood HG (1969) Transcarboxylase. V. The presence of bound zinc and cobalt. J Biol Chem 244:5801–5807

Nou X, Kadner RJ (1998) Coupled changes in translation and transcription during cobalamin-dependent regulation of *btuB* expression in *Escherichia coli*. J Bacteriol 180:6719–6728

Nou X, Kadner RJ (2000) Adenosylcobalamin inhibits ribosome binding to *btuB* RNA. Proc Natl Acad Sci USA 97:7190–7195

Olson JW, Fu C, Maier RJ (1997) The HypB protein from *Bradyrhizobium japonicum* can store nickel and is required for the nickel-dependent transcriptional regulation of hydrogenase. Molec Microbiol 24:119–128

Olson JW, Maier RJ (2002) Molecular hydrogen as an energy source for *Helicobacter pylori*. Science 298:1788–1790

Olson JW, Mehta NS, Maier RJ (2001) Requirement of nickel metabolism proteins HypA and HypB for full activity of both hydrogenase and urease in *Helicobacter pylori*. Molec Microbiol 39:176

Outten CE, O'Halloran TV (2001) Femtomolar sensitivity of metalloregulatory proteins controlling zinc homeostasis. Science 292:2488–2492

Park I-S, Carr MB, Hausinger RP (1994) In vitro activation of urease apoprotein and role of UreD as a chaperone required for nickel metallocenter assembly. Proc Natl Acad Sci USA 91:3233–3237

Park I-S, Hausinger RP (1995a) Evidence for the presence of urease apoprotein complexes containing UreD, UreF, and UreG in cells that are competent for in vivo enzyme activation. J Bacteriol 177:1947–1951

Park I-S, Hausinger RP (1995b) Requirement of carbon dioxide for in vitro assembly of the urease nickel metallocenter. Science 267:1156–1158

Park I-S, Hausinger RP (1996) Metal ion interactions with urease and UreD-urease apoproteins. Biochemistry 35:5345–5352

Pochapsky TC, Pochapsky SS, Ju T, Mo H, Al-Mjeni F, Maroney MJ (2002) Modeling and experiment yields the structure of acireductone dioxygenase from *Klebsiella pneumoniae*. Nat Struct Biol 9:966–972

Proudfoot M, Kuznetsova K, Brown G, Rao NN, Kitagawa M, Mori H, Savchenko A, Yakunin AF (2004) General enzymatic screens identify three new nucleotidases in *Escherichia coli*. J Biol Chem 279:54687–54694

Rajagopalan PTR, Yu XC, Pei D (1997) Peptide deformylase: a new type of mononuclear iron protein. J Am Chem Soc 119:12418–12419

Reynolds PR, Mottur GP, Bradbeer C (1980) Transport of vitamin B12 in *Escherichia coli*. Some observations on the roles of the gene products of *BtuC* and *TonB*. J Biol Chem 255:4313–4319

Robson R (2001) The assembly line. In: Robson R (ed) Hydrogen as a fuel: Learning from nature. Taylor and Francis, New York, pp 57–72

Rodionov DA, Hebbeln P, Gelfand MS, Eitinger T (2006) Comparative and functional genomic analysis of prokaryotic nickel and cobalt uptake transporters: evidence for a novel group of ATP-binding cassette transporters. J Bacteriol 188:317–327

Rodionov DA, Vitreschak AG, Mironov AA, Gelfand MS (2003) Comparative genomics of the vitamin B12 metabolism and regulation in prokaryotes. J Biol Chem 278:41148–41159

Rodrigue A, Batia N, Müller M, Fayet O, Böhm R, Mandrand-Berthelot M-A, Wu L-F (1996) Involvement of the GroE chaperonins in the nickel-dependent anaerobic biosynthesis of NiFe-hydrogenases of *Escherichia coli*. J Bacteriol 178:4453–4460

Rodrigue A, Effantin G, Mandrand-Bethelot MA (2005) Identification of *rcnA* (*yohM*), a nickel and cobalt resistance gene in *Escherichia coli*. J Bacteriol 187:2912–2916

Roof WD, Horne SM, Young KD, Young R (1994) *slyD*, a host gene required for PhiX174 lysis, is related to the FK506-binding protein family of peptidyl-prolyl *cis-trans*-isomerases. J Biol Chem 269:2902–2910

Roth JR, Lawrence JG, Bobik TA (1996) Cobalamin (coenzyme B12): synthesis and biological significance. Annu Rev Microbiol 50:137–181

Rowe JL, Starnes GL, Chivers PT (2005) Complex transcriptional control links NikABCDE-dependent nickel transport with hydrogenase expression in *Escherichia coli*. J Bacteriol 187:6317–6323

Schönheit P, Moll J, Thauer RK (1979) Nickel, cobalt, and molybdenum requirement for growth of *Methanobacterium thermoautotrophicum*. Arch Microbiol 123:105–107

Schreiter ER, Sintchak MD, Guo Y, Chivers PT, Sauer RT, Drennan CL (2003) Crystal structure of the nickel-reponsive transcriptional factor NikR. Nat Struct Biol 10:794–799

Schreiter ER, Wang SC, Zamble DB, Drennan CL (2006) NikR-operator complex structure and the mechanism of repressor activation by metal ions. Proc Natl Acad Sci USA 103:13676–13681

Self WT, Hasona A, Shanmugam KT (2004) Expression and regulation of a silent operon, *hyf*, coding for hydrogenase 4 isoenzyme in *Escherichia coli*. J Bacteriol 186:580–587

Seravalli J, Ragsdale SW (2000) Channeling of carbon monoxide during anaerobic carbon dioxide fixation. Biochemistry 39:1274–1277

Shima S, Thauer RK (2005) Methyl-coenzyme M reductase and anaerobic oxidation of methane in methanotrophic archaea. Curr Opin Microbiol 8:643–648

Skibinski DAG, Golby P, Chang Y-S, Sargent F, Hoffman R, Harper R, Guest JR, Attwood MM, Berks BC, Andrews SC (2002) Regulation of the hydrogenase-4 operon of *Escherichia coli* by the σ^{54}-dependent transcriptional activators FhlA and HyfR. J Bacteriol 184:6642–6653

Song HK, Mulrooney SB, Huber R, Hausinger RP (2001) Crystal structure of *Klebsiella aerogenes* UreE, a nickel-binding metallochaperone for urease activation. J Biol Chem 276:49359–49364

Soriano A, Colpas GJ, Hausinger RP (2000) UreE stimulation of GTP-dependent urease activation in the UreD-UreF-UreG-urease apoprotein complex. Biochemistry 39:12435–12440

Soriano A, Hausinger RP (1999) GTP-dependent activation of urease apoprotein in complex with the UreD, UreF, and UreG accessory proteins. Proc Natl Acad Sci USA 96:11140–11144

Sukdeo N, Clugston SL, Daub E, Honek JF (2004) Distinct classes of glyoxylase I: Metal specificity of the *Yersinia pestis*, *Pseudomonas aeruginosa* and *Neisseria meningitidis* enzymes. Biochem J 384:111–117

Svetlitchnyi V, Dobbek H, Meyer-Klaucke W, Meins T, Thiele B, Römer P, Huber R, Meyer O (2004) A functional Ni – Ni – [4Fe – 4S] cluster in the monomeric acetyl-CoA synthase from *Carboxydothermus hydrogenoformans*. Proc Natl Acad Sci USA 101:446–451

Takahashi S, Kuzuyama T, Watanabe H, Seto H (1998) A 1-deoxy-D-xylulose 5-phosphate reductoisomerase catalyzing the formation of 2-C-methyl-D-erythritol 4-phosphate in an alternative nonmevalonate pathway for terpenoid biosynthesis. Proc Natl Acad Sci USA 95:9879–9884

Tam R, Saier MHJ (1993) Structural, functional, and evolutionary relationships among extracellular solute-binding receptors of bacteria. Microbiol Rev 57:320–346

Thauer RK (1998) Biochemistry of methanogenesis: a tribute to Marjory Stephenson. Microbiol 144:2377–2406

Thauer RK, Bonacker LG (1994) Biosynthesis of coenzyme F430, a nickel porphinoid involved in methanogenesis. Ciba Found Symp 180:210–227

Theodoratou E, Böck A (2005) [NiFe]-hydrogenase maturation endopeptidase: structure and function. Biochem Soc Trans 33:108–111

Thornalley PJ (2003) Glyoxalase I—structure, function and a critical role in the enzymatic defense against glycation. Biochem Soc Trans 31:1343–1348

Tsumura N, Hagi M, Sato T (1967) Enzymatic conversion of D-glucose to D-fructose. III. Propagation of *Streptomyces phaeochromogenes* in the presence of cobaltous ion. Agric Biol Chem 31:902–907

Uffen RL (1976) Anaerobic growth of *Rhodopseudomonas* species in the dark with carbon monoxide as sole carbon and energy substrate. Proc Natl Acad Sci USA 73:3298–3302

van Vliet AHM, Ernst FD, Kusters JG (2004) NikR-mediated regulation of *Helicobacter pylori* acid adaptation. Trends Microbiol 12:489–494

Vignais PM, Billoud B, Meyer J (2001) Classification and phylogeny of hydrogenases. FEMS Microbiol Rev 25:455–501

Vignais PM, Colbeau A (2004) Molecular biology of microbial hydrogenases. Curr Issues Molec Biol 6:159–188

Vitreschak AG, Rodionov DA, Mironov AA, Gelfand MS (2004) Riboswitches: the oldest mechanism for the regulation of gene expression? Trends Genet 20:44–50

Volbeda A, Fontecilla-Camps JC (2003) The active site and catalytic mechanism of NiFe hydrogenases. Dalton Trans: 4030–4038

Wang SC, Dias AV, Bloom SL, Zamble DB (2004) The selectivity of metal binding and the metal-induced stability of *Escherichia coli* NikR. Biochemistry 43:10018–10028

Watt RK, Ludden PW (1998) The identification, purification and characterization of CooJ. A nickel-binding protein that is co-regulated with the Ni-containing CO dehydrogenase from *Rhodospirillum rubrum*. J Biol Chem 273:10019–10025

Waugh R, Boxer DH (1986) Pleiotropic hydrogenase mutants of *Escherichia coli* K-12: growth in the presence of nickel can restore hydrogenase activity. Biochimie 68:157–166

Winter G, Buhrke T, Lenz O, Jones AK, Forqber M, Friedrich B (2005) A model system for [NiFe] hydrogenase maturation studies: Purification of an active site-containing hydrogenase large subunit with small subunit. FEBS Lett 579:4292–4296

Wu L-F, Mandrand-Berthelot M-A (1986) Genetic and physiological characterization of new *Escherichia coli* mutants impaired in hydrogenase activity. Biochimie 68:167–179

Wu L-F, Mandrand-Berthelot M-A, Waugh R, Edmonds CJ, Holt SE, Boxer DH (1989) Nickel deficiency gives rise to the defective hydrogenase phenotype of *hydC* and *fnr* mutants in *Escherichia coli*. Molec Microbiol 3:1709–1718

Wu L-F, Navarro C, de Pina K, Quénard M, Mandrand-Berthelot M-A (1994) Antagonistic effect of nickel on the fermentative growth of *Escherichia coli* K-12 and comparison of nickel and cobalt toxicity on the aerobic and anerobic growth. Environ Health Persp 102:297–300

Wu L-F, Navarro C, Mandrand-Berthelot M-A (1991) The *hydC* region contains a multicistronic operon (*nik*) involved in nickel transport in *Escherichia coli*. Gene 107:37–42

Wuerges J, Lee J-W, Yim Y-I, Kang SO, Carugo KD (2004) Crystal structure of nickel-containing superoxide dismutase reveals another type of active site. Proc Natl Acad Sci USA 101:8569–8574

Wülfing C, Lombardero J, Plückthun A (1994) An *Escherichia coli* protein consisting of a domain homologous to FK506-binding proteins (FKBP) and a new metal binding motif. J Biol Chem 269:2895–2901
Youn H-D, Kim E-J, Roe J-H, Hah YC, Kang S-O (1996) A novel nickel-containing superoxide dismutase from *Streptomyces* spp. Biochem J 318:889–896
Zhang JW, Butland G, Greenblatt JF, Emili A, Zamble DB (2005) A role for SlyD in the *Escherichia coli* hydrogenase biosynthetic pathway. J Biol Chem 280:4360–4366
Zheng Y, Roberts RJ, Kasif S, Guan C (2005) Characterization of two new aminopeptidases in *Escherichia coli*. J Bacteriol 187:3671–3677

Microbiol Monogr (6)
D. H. Nies, S. Silver: Molecular Microbiology of Heavy Metals
DOI 10.1007/7171_2006_083/Published online: 24 February 2007

Zinc, Cadmium, and Lead Resistance and Homeostasis

Christopher Rensing[1] (✉) · Bharati Mitra[2]

[1]Department of Soil, Water, and Environmental Science, University of Arizona, Shantz Bldg #38, Rm 429, P.O. Box 210038, Tucson, AZ 85721, USA
rensingc@ag.arizona.edu

[2]Department of Biochemistry and Molecular Biology, Wayne State University School of Medicine, 540 E. Canfield Avenue, Detroit, MI 48201, USA

Abstract Metals such as zinc are required for life but can be toxic in excess. Other metals such as cadmium and lead have almost no known biological function, and can lead to cell damage and death even at low concentrations. Interestingly, all three metals are often recognized by the same gene regulators and membrane transporters. Therefore, an examination of the inherent chemical properties of these three metal ions is essential in understanding the basis of metal specificity displayed by target proteins responsible for metal homeostasis and resistance. The relationship between the chemical properties of these metals and similarities in structural responses they may elicit are discussed. The core elements regulating uptake, efflux, and sequestration of these metals are described and interpreted both biologically and chemically. Additional mechanisms aiding cell survival, such as precipitation of metal salts on the cell surface, are also mentioned.

1 Introduction

Divalent cation "micronutrients" include cobalt, copper, zinc, manganese, nickel, and the macronutrient iron. Zinc is an essential metal for all organisms and has many biological roles. In contrast, a biological function for cadmium has only been suggested in a few instances, such as carbonic anhydrase from a marine diatom (Lane et al. 2005). Though lead can substitute zinc in carboxypeptidase resulting in a still functioning enzyme, it does not occur naturally as a required element in biology. Thus in most organisms, cadmium, lead, and mercury have no known beneficial role, making resistance mechanisms necessary. The mercury resistance mechanisms are described in the chapter by Silver and Hobman (this volume). Zinc levels have to be carefully regulated to ensure sufficient supply of zinc while avoiding toxicity. This process is often referred to as metal homeostasis as opposed to resistance, which only applies to metals with no biological role in a specific organism. However, in the course of this chapter it will become evident that often the transporters responsible for zinc uptake and efflux are also responsible for transport of other metal ions, such as Cd(II), Pb(II), and Hg(II).

All organisms need to maintain homeostasis of different essential metals, such as zinc, iron, manganese, and copper. In order to do this, mechanisms had to evolve to differentiate between closely related metals. The initial expectation was that regulatory proteins or transporters would have domains that allow them to bind only specific metals with high affinity. The basis for how metal specificity is achieved is only now beginning to be understood. In recent work, it has become clear that metal specificity is not just dependent on differences in metal binding affinity. Other important factors include the chemical properties of different metals, such as their coordination geometry preference and softness according to the Lewis acid–base concept. As we will later see, these distinct properties can lead to different conformational changes that are transmitted through a protein domain and affect function. The softness of a metal ion determines its affinity for thiolate ligands and is another factor that determines specificity (Lippard and Berg 1994). Size and charge of the metal ion are other important factors that play a role in determining specificity. Therefore, in order to understand how zinc homeostasis and resistance to cadmium and lead is achieved without negatively affecting the concentration of other essential metals, we have to review some basic inorganic principles and coordination preferences of these metals.

2
Chemical Properties of Zinc, Cadmium, and Lead Critical for Conferring Metal Specificity

Zinc (Zn), cadmium (Cd), and mercury (Hg) are classified as group 12 elements. They all contain an $(n\text{-}1)d$ shell plus two ns electrons. The chemistry of Zn and Cd is closely related whereas that of Hg is not. All members of this group have II as the primary oxidation state; for Zn and Cd, this is the only observed oxidation state (Cotton et al. 1999). Hg can also exist in the I oxidation state as the Hg_2^{2+} complex. As regards coordination geometry there is a clear trend, with Zn(II) preferring four to six ligands to Hg(II), which prefers two. Lead (Pb) is a group 14 element with no known biological function. Lead with an outer electronic configuration of $4f^{14}5d^{10}6s^{2}6p^{2}$ is also usually found in the II oxidation state, although the higher oxidation state, IV, is possible (Cotton et al. 1999). Pb(II) prefers trigonal coordination in thiolate-rich environments, in contrast to zinc and cadmium which prefer four-coordinated tetrahedral geometry (Magyar et al. 2005). Pb(II) is the largest ion with a radius of 1.33Å, followed by Cd(II) at 1.09Å and Zn(II) at 0.88Å. In terms of hard–soft classification, Zn(II) is considered borderline between hard and soft, while Cd(II) and Pb(II) ions are considered to be soft cations (Lippard and Berg 1994). Thus, Zn(II) preferentially coordinates with both hard and soft ligands, whereas Cd(II) and Pb(II) prefer soft ligands.

Studies aimed at understanding factors that confer metal specificity have been performed mostly with the MerR and ArsR/SmtB regulatory families and the cation translocating P-type ATPases (Harvie et al. 2006; Banci et al. 2006). Three factors influence metal specificity in vivo: (1) differences in metal affinity, (2) differences in intracellular concentration of metals (for example, presence of certain uptake or efflux systems), and (3) preferences in coordination geometries of different metals and ability to trigger desired conformational change (Harvie et al. 2006; Rensing 2005; Liu et al. 2006; Dutta et al. 2006).

3
Regulating Metal Flow and Homeostasis

Cellular homeostasis involves regulation of uptake, sequestration, and efflux of metals. In order to maintain optimal levels while expending the least amount of energy, cells need regulatory proteins to orchestrate an adequate response to metal deficiency or excess. Well-studied examples of zinc-, cadmium-, and lead-responsive regulatory proteins include: Zur, a homologue of the iron-responsive repressor Fur; CadC, a member of the ArsR/SmtB family of repressors; and ZntR, a member of the MerR-type activator family.

These regulators induce or repress specific operons or genes. The respective gene products are involved in uptake, storage, or efflux of these metals.

3.1 The Zur Regulon

In times of zinc deficiency many bacteria up- or downregulate an array of operons that are all under the control of the Zur regulatory protein (Bsat et al. 1998; Hantke 2005; Gaballa and Helmann 1998; Patzer and Hantke 2000). Zur is a member of the Fur family of bacterial metal-responsive regulators and in many bacteria controls not only expression of genes involved in zinc uptake, such as *znuABC* and *zinT*, but also expression of up to four ribosomal proteins without a zinc binding site to replace paralogues that contain a zinc ribbon motif and bind zinc (Akanuma et al. 2006; Panina et al. 2003).

Both Zur and Fur contain two very different metal binding sites. In both proteins site A is a four-coordinate site occupied by zinc and is largely conserved between the two proteins. In Zur, Zn(II) is bound at site A with a coordination environment of $S_3(N/O)$ (Outten et al. 2001). Since both Zur and Fur contain zinc at this site, this site is probably not a Zn(II)- or Fe(II)-responsive sensor. In contrast, site B in Zur is specific for Zn(II), with a probable $S(N/O)_3$ binding environment. Again this is a tetrahedral environment favored by zinc but not too thiolate-rich to prevent softer metals such as Cd(II) replacing Zn(II). As discussed below, zinc is a d^{10} metal ion that is not subject to ligand field stabilizations and can therefore bind in a tetrahedral site without loss in ligand field stabilization energy. Fe(II), in contrast, is a d^6 metal ion and therefore tetrahedral coordination is unfavorable in relation to octahedral or pentacoordination. This is also reflected by the presence of five to six N/O ligands in the site B Fe(II) binding site of Fur (Outten et al. 2001).

3.2 Repressors

Homodimeric proteins that bind to DNA operator regulatory regions in the absence of substrate and that change conformation (allostery) so that they dissociate from the operator DNA in the presence of the small activator substrate are called repressors, as the process of transcriptional regulation is basically negative and "turning on" of genes is essentially relief from repression. For proteins responding to elevated levels of Zn(II), Cd(II), and Pb(II), mostly transporters but also metallothionines, repressors include CadC, for Cd/Pb/Zn resistance in Gram-positive bacteria (Silver and Phung 1996), CzrA, involved in *Streptococcus aureus* chromosomal zinc efflux (Singh et al. 1999; Kuroda et al. 1999), ZiaR from *Synechocystis*, SmtB in *Synechococcus*, and AztR from the cyanobacterium *Anabaena* PCC 7120 (Liu et al. 2005a).

These repressors belong to the large ArsR/SmtB family of metalloregulatory proteins that all contain a conserved helix-turn-helix domain involved in DNA recognition. Members of the ArsR/SmtB family bind as homodimers to their operator–promoter (O/P) sequence, thereby repressing transcription of their respective operon (Rensing 2005; Ye et al. 2005). Metal binding then induces a conformational change or steric hindrance decreasing the affinity of the repressor to the O/P sequence and subsequently leading to a release of the metal-bound repressor. It is important to note that it is not affinity of a specific metal to the metal binding sites of the repressor that confer metal specificity, but rather the correct conformational change induced by the metals. The conformational changes are dependent on the coordination preferences of particular metals and metalloids.

There are two well-characterized metal binding sites present in members of the ArsR/SmtB family. However, some members such as ArsR only have a type 1 metal binding site, whereas others such as SmtB only contain a type 2 metal binding site and still others such as CadC have both. The type 1 metal binding site, or $\alpha3$ in ArsR and $\alpha4N'$ in CadC (Giedroc terminology), is thiolate-rich with two or more Cys residues and preferentially binds soft metals/metalloids. In contrast, the type 2 site, or $\alpha5$ in SmtB and $\alpha6$ in CadC, is mostly involved in regulation of borderline soft/hard metals such as Zn(II), Co(II), and Ni(II) and usually contains histidines and acidic residues. In the cases examined, Zn(II) and Cd(II) were coordinated in four-coordinate geometry whereas Pb(II) was trigonal. For example, SmtB, the repressor for the metallothionein SmtA, binds Zn(II) as a homodimer in four-coordinate tetrahedral geometry at the type 2 binding site. Cd(II) and Zn(II) bound to CadC adopted a four-coordinate S_4 structure. On the other hand, Pb(II) has a trigonal S_3 structure when bound to CadC (Rensing 2005; Ye et al. 2005). AztR from the cyanobacterium *Anabaena* responds to the same metals, as does CadC but with a higher affinity for zinc. This is presumably due to a tetrahedral $S_3(N/O)$ type 1 metal site, not S_4 as in CadC. Pb(II) in CadC forms a trigonal pyramidial $Pb-S_3$ complex (Liu et al. 2005a; Harvie et al. 2006))

3.3 Activators

As the name implies, the MerR family of transcriptional activators was first discovered and characterized in mercury resistance operons. Regulation of mercury resistance was studied in both Gram-negative and Gram-positive bacteria (Brown et al. 2003; Barkay et al. 2003). However, it became subsequently clear that the MerR family regulates not only Hg(II) resistance but also homeostasis of a number of other divalent cations. In *Escherichia coli*, ZntR regulates the Zn(II)/Cd(II)/Pb(II)-transporting P-type ATPase ZntA (Brocklehurst et al. 1998; Outten et al. 1999), and CueR upregulates the Cu(I)/Ag(I)-transporting P-type ATPase CopA and the Cu(I)-oxidizing multicopper oxidase CueO (Outten et al. 2000; Grass and Rensing 2001; Singh et al. 2004). In *Synechocystis*, the transcriptonal activator CoaR regulates the P-type ATPase, CoaT, in response to elevated levels of Co(II) (Rutherford et al. 1999).

The operator–promoter region regulated by members of the MerR family has a suboptimal spacing of 19 bp between the –35 and –10 sites. In most prokaryotic promoters 16–18 bp spacing can be found. This suboptimal spacing interferes with proper binding and open complex formation by RNA polymerase (RNAP). The MerR homodimer (or other member of the MerR family) binds to the merO/P region and further bends the DNA making binding by RNAP even more difficult (Ansari et al. 1995). Upon binding of the appropriate metal to MerR or other paralogues, a conformational change takes place resulting in a relaxation of the DNA bend and an allosteric unwinding of the promoter DNA by approximately 33°. This results in an alignment of the –35 and –10 sequences and allows RNAP to initiate transcription (Hobman et al. 2005).

In MerR itself, three cysteines in the homodimer are involved in forming a stable tricoordinate $HgSR_3$ complex. Two of these cysteines are from one monomer, one from the other (Utschig et al. 1995). Although MerR will bind several other metals, it is quite specific for Hg(II) as an inducer in vivo. This variation on a theme is again related to Hg(II) being able to induce the proper conformational change in response to binding of Hg(II) but less so for other metal ions. In ZntR two Zn(II) ions are bound by the metal-binding loop. In *E. coli* ZntR the first Zn(II) binds to Cys114 and Cys125 from one monomer and to Cys79 from the other monomer; the second Zn(II) is coordinated by Cys115 and His119 from one monomer and Cys79 from the other monomer. A bridging phosphate ion completes the four-coordinate binding geometry for zinc (Changela et al. 2003). Interestingly, four cysteine residues and a histidine appear to be essential in binding Zn(II) to ZntR, whereas only three cysteines appear to be sufficient for Pb(II) and Cd(II) (Hobman et al. 2005). PbrR contains six cysteines so possibly there are multiple Pb(II) binding sites in PbrR, as Pb(II) prefers trigonal coordination geometries in a thiolate-rich environment (Magyar et al. 2005).

3.4 Two-Component Kinases and Responders

The best-studied examples of sensor kinase/DNA-binding responder "RS" systems for transcriptional regulation of divalent cation transport, such as zinc and cadmium in bacteria, are *czcRS* in *Cupriavidus metallidurans* CH34 and *zraRS* in *E. coli*. The CzcRS system of *C. metallidurans* regulates expression of the *czc* operon encoding an efflux transport system governing Cd(II), Zn(II), and Co(II) in exchange for protons (see below). ZraRS regulate expression of the periplasmic zinc-binding protein ZraP. A similar system, CzrRS from *Pseudomonas aeruginosa*, was shown to regulate *czrCBA* (Hassan et al. 1999). In all these examples, how the metal specificity of these systems is conferred has not been studied.

4
Biology and Chemistry of Zinc

Zinc with an outer configuration of $3d^{10}4s^2$ combines both partial "soft" metal character and a high charge-to-radius ratio. This combination, together with its redox-inert properties and high natural abundance, helps us to understand why microorganisms recruited zinc early on in the history of life to perform a variety of functions (Auld 2001). In fact, zinc is the only metal having representatives in all six recognized classes of enzymes. Zinc uses nitrogen and oxygen as ligands as readily as sulfur. In contrast, cadmium, lead, and mercury ions favor sulfur since they are much "softer" cations (i.e., they are more easily polarized). An examination of zinc-containing protein structures indicates four primary zinc sites: catalytic, structural, cocatalytic, and regulatory (Auld 2001; Rensing 2005). The most common amino acid ligands in these proteins are Cys, His, Asp, and Glu.

The coordination geometry of the catalytic zinc site is usually distorted tetrahedral, sometimes trigonal bipyramidal, with coordination numbers of four and five, respectively. One of the ligands in the catalytic zinc site is always water. The zinc-bound water can be displaced by substrate resulting in Lewis acid catalysis by the catalytic zinc atom, as has been described for alcohol dehydrogenase (Esposito et al. 2002; Lippard and Berg 1994).

The zinc-bound water can also be activated by ionization or polarization leading to the presence of hydroxide ion at neutral pH. Prominent examples of these types of catalytic zinc enzymes are the carboxypeptidase family of exopeptidases and the thermolysin family of endopeptidases. Although unrelated these two enzymes show striking similarities at their active site. Zinc is coordinated by two His, an acidic residue contributing a bidentate carboxylate group, and water in the unligated form. This results in a five-coordinated zinc at the active site. In the presence of substrate the Glu or Asp carboxylate group becomes coordinated in a unidentate fashion allowing maintenance of a constant coordination number. This process has been termed the carboxylate shift (Lippard and Berg 1994).

The first described zinc enzyme was carbonic anhydrase (CA) catalyzing the reaction $CO_2(aq) + OH^- \rightarrow HCO_3^-$ at pH 7. CAs are examples of zinc enzymes with an ionization-activated zinc-bound water mechanism. There are four distinct families of CAs based on the metal site:

1. Tetrahedral coordination, ligands usually three His and one water
2. Two Cys, one His, one water

3. Three His and one water but the structure is not tetrahedral
4. Two to three Cys but Cd instead of Zn

A fourth protein ligand has only been observed in a few catalytic zinc sites, resulting in trigonal bipyramidal coordination geometry. Examples include Adh from *Chlostridium beijerincki* (Korkhin et al. 1998) and *E. coli* fuculose aldolase (Dreyer and Schulz 1993).

Structural zinc sites have four protein ligands and no bound water in the tetrahedral configuration. The preferred ligands are Cys and His (Vallee et al. 1991; Auld 2001). Examples include the *E. coli* tRNA synthetase family and the DNA polymerase III d subunit. Regulatory zinc sites have already been described in the sections on activators and repressors.

Finally, in cocatalytic zinc sites the preferred amino acid ligands are Asp and His and almost never Cys. The novel feature of these sites is the bridging of two of the metal sites by a side chain of an amino acid or a water molecule. Prominent examples include Cu,Zn superoxide dismutase (SOD) and phosphatases, the best-studied being alkaline phosphatase. In human Cu,Zn SOD, Zn(II) is coordinated to three His residues and one Asp residue resulting in a distorted tetrahedral geometry, while Cu(II) coordination is accomplished by a shared His residue (His63 in human SOD), three more His residues, and a water molecule forming a distorted square pyramidal geometry (Strange et al. 2003).

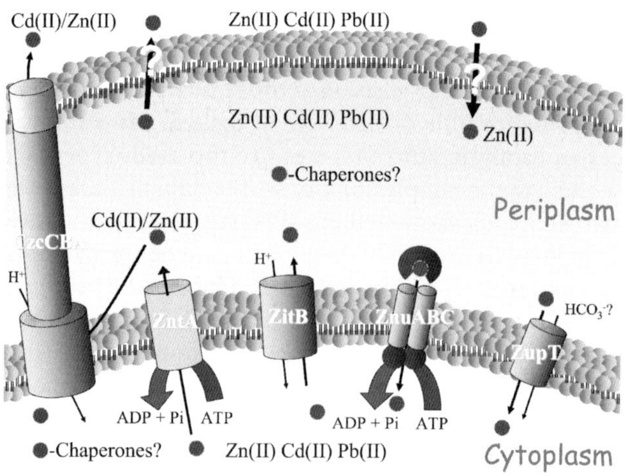

Fig. 1 Zinc homeostasis mechanisms in Gram-negative bacteria. The individual components are discussed in the text. ZntA is a Zn(II)/Cd(II)/Pb(II)-translocating P-type ATPase, ZitB a PMF-dependent efflux pump of the CDF family, ZnuABC a multicomponent, ATP-dependent zinc uptake system, ZupT a broad-spectrum uptake system of the ZIP family, and CzcCBA a three-component Cd(II)/Zn(II)/Co(II) efflux pump. Specific zinc chaperones have not been identified

This section is not intended as an extensive overview on the manifold roles of zinc in biology, but to make clear that the very properties that make zinc attractive as a catalyst in biology are also used in recognition and transport.

5 Transport Systems for Zinc

Since zinc is essential for all organisms, low concentrations of zinc in the environment have to be overcome by specific zinc uptake systems (Fig. 1). In *E. coli* the *znuABC* operon is specifically induced under zinc limiting conditions by Zur (Hantke 2005; Blencowe and Morby 2003). Under conditions of zinc sufficiency, expression of the pump is repressed by the Fur homologue Zur, which presumably binds to the bidirectional promoter region of *znuA* and *znuBC*. The ZnuABC complex catalyzes ATP-dependent uptake of zinc (Patzer and Hantke 1998). ZnuA is a periplasmic binding protein, ZnuB is the membrane permease of the pump, and ZnuC is the ATPase catalytic subunit. Genes encoding ZnuABC homologues have also been found in *Haemophilus influenza*, *Bacillus subtilis*, and many others.

The crystal structure of several related periplasmic Zn- or Mn-binding proteins has been determined (Banerjee et al. 2003; Lawrence et al. 1998; Lee et al. 2002). One unresolved question centers on the elements in these transporters that are responsible for metal specificity. Again the coordination geometry and the amino acid residues involved appear to play a crucial role. In ZnuA from *Synechocystis* 6803, Banerjee et al. (2003) have suggested that zinc uptake can be favored over that of manganese by a decreased coordination number of only three His and a water molecule in Zn(II) uptake proteins, whereas binding of Mn(II) requires three His and a bidentate Asp, resulting in a five-coordinate trigonal pyramid arrangement. In addition, many zinc-specific periplasmic binding proteins possess a flexible loop that is rich in acidic and histidine residues. This loop, located in proximity to the metal binding site, could serve as zinc chaperone (Banerjee et al. 2003). It is also likely that binding affinity for Mn(II) and Zn(II) in different uptake systems plays a role in these cases, given that Mn(II) concentrations are much higher than those of Zn(II) in the periplasm.

5.1 Broad Spectrum Uptake Systems

Under conditions of sufficient zinc, ATP does not have to be expended for the uptake of zinc. Transporters with a broad substrate spectrum are then sufficient for the zinc needs of the cell. The advantage of relying on transporters such as ZupT, belonging to the ZIP family of proteins, is that they use far less energy than ATPases such as ZnuABC (Fig. 1). However, the energy source driving uptake of zinc and other metals by ZupT is currently still unknown, although carbonate has been suggested in other ZIP transporters. ZupT contains eight transmembrane helixes (TMHs) and was shown to be involved in uptake of Zn(II), Fe(II), Co(II), Cd(II), and possibly Mn(II) (Grass et al. 2005; Franke, unpublished results).

Under toxic conditions Zn(II) and Pb(II) enter the cells by unknown pathways. Several studies have yielded preliminary results. The phosphate uptake system seems to be implicated in uptake of Zn(II), possibly as a metal phosphate (Beard et al. 2000). Cd(II) and Pb(II) compete for transport with other physiologically important cations; Pb(II) may be taken up by Ca(II) uptake systems. Transport inhibitor studies demonstrate that in Gram-positive bacteria, such as *S. aureus*, *B. subtilis*, and *Lactobacillus plantarum* ATCC14917, Cd(II) is taken up by the manganese transporter MntH, an NRAMP orthologue (Archibald and Duong 1984; Hao et al. 1999; Que and Helmann 2000). Interestingly, in *B. subtilis* and perhaps other bacteria MntH is repressed by MntR, not only under conditions of Mn(II) sufficiency but also by the presence of Cd(II) (Moore et al. 2005). Recent results in *E. coli* and *Salmonella typhimurium* suggest that Cd(II) is also taken up by MntH (Makui et al. 2000; Kehres et al. 2000). The physiological function of MntH seems to be uptake of Mn(II) possibly as protection from hydrogen peroxide.

5.2
ATP-Driven Efflux Pumps and their Regulation

Growth of *E. coli* in high concentrations of Zn(II), Cd(II), and Pb(II) results in induction of ZntA, a Zn(II)-, Cd(II)-, and Pb(II)-transporting P-type ATPase by the MerR homologue ZntR (Brocklehurst et al. 1998; Outten et al. 1999) (Fig. 1). Disruption of *zntA* results in sensitivity to Zn(II), Cd(II), and Pb(II), and to a lesser extent, sensitivity to Ni(II) and Co(II) (Muralidharan and Mitra, unpublished observations). The Δ*zntA* strain did not exhibit any sensitivity to copper and silver, suggesting that this P-type ATPase is a relatively specific divalent metal ion pump that is selective for Zn(II), Cd(II), and Pb(II). ZntA has been shown to catalyze ATP-coupled accumulation of ^{65}Zn(II) and ^{109}Cd(II) in everted (inside out) vesicles of *E. coli*, where accumulation in everted membrane vesicles is equivalent to efflux from intact cells (Rensing et al. 1997a, 1998a; Beard et al. 1997). ATPase activity of purified ZntA is stimulated by Zn(II), Cd(II), Pb(II), and Hg(II) (Sharma et al. 2000) and weakly by Cu(II), Ni(II), and Co(II) (Dutta et al. 2006b). ZntA homologues are widespread in bacteria, archaea, and plants (Rensing et al. 1998a, 1999).

Another well-studied homologue of ZntA is CadA from *S. aureus* plasmid pI258. The *cad* operon of pI258 has two genes: *cadC*, which encodes a transcriptional repressor; and *cadA*, which encodes CadA, a P-type ATPase that is about 30% identical to ZntA. CadA has been shown to catalyze ATP-dependent ^{109}Cd(II) and ^{65}Zn(II) uptake in everted membrane vesicles from *B. subtilis* or *E. coli* (Tsai et al. 1992; Rensing et al. 1998b). CadA homologues have been identified in plasmids and chromosomes from a number of bacteria and archaea. Recently, the CadA homologue from *Synechocystis* sp. Strain PCC 6803 has been named ZiaA and shown to confer Zn(II) and Cd(II) tolerance due to reduced accumulation (Thelwell et al. 1998). The pI258 *cadA* gene also functionally substitutes the Zn(II)-, Cd(II)-, and Pb(II)-sensitive phenotypes of an *E. coli zntA*-disrupted strain. While the *E. coli zntA* gene is positively regulated by ZntR, *cadA* and *ziaA* are negatively regulated by the CadC and ZiaR repressors, respectively, both of which are members of the ArsR family of metalloregulatory proteins. CadC responds to Zn(II), Cd(II), and Pb(II), the same metal ion specificity as the pumps themselves.

Metal-transporting P_{1B}-type (or CPx) ATPases can be grouped into several subfamilies based on sequence alignment; these subfamilies appear to have different substrate specificity. The most obvious distinction is between P-type ATPases transporting monovalent cations and those transporting divalent cations. Most contain one or more independent high-affinity metal binding sites in the hydrophilic N-terminal domain (Liu et al. 2005b). In addition, all contain one independent and essential high-affinity metal binding site in the transmembrane region (Liu et al. 2006). The N-terminal region is not necessary for transport but for achieving maximum velocity; this kinetic advantage is important in vivo (Mitra and Sharma 2001; Liu et al. 2005b).

The apo- and metal-containing structures of the N-terminus of a number of P_{1B}-type ATPases have been elucidated using a variety of techniques including NMR, X-ray spectroscopy, and extended X-ray absorption fine structure (EXAFS). Cu(I) and Ag(I) appear to bind to the N-terminal CXXC motif in a bidentate two-coordinate fashion. In divalent cation transporters, such as CadA from *Listeria monocytogenes* and ZntA, Cd(II) and Zn(II) are bound with tetrahedral geometry at the N-terminus by two Cys and one or two oxygen atoms of a conserved acidic residue (Glu in CadA and Asp in ZntA) (Banci et al. 2002, 2006). The acidic residue appears to play a key role in the coordination of Cd(II) and Zn(II) as opposed to monovalent cations like Cu(I) and Ag(I). However, other divalent cations such as Cu(II) and Ni(II) also bind with high affinity to the N-terminal metal site (Liu et al. 2005b). These observations again point to a variation on a common theme: metal selectivity is not determined by the affinity of certain metals to this site, but rather by the preferred coordination geometries of specific metals and the conformational change elicited as a result of this coordination.

Binding of Pb(II) to the N-terminus of ZntA requires more than the two cysteines of the CXXC motif; additional cysteines from an upstream cysteine-rich motif play a role (Liu et al. 2005b). It is possible that, due to its large radius and pronounced soft ion character, Pb(II) binds with the trigonal configuration $PbCys_3$ with two Cys from the conserved CXXC motif and one additional cysteine from the upstream CCX(D,E)XXC motif supplying ligands.

The transmembrane metal binding site in P_{1B}-type ATPases is composed at least in part of the strictly conserved (C,S,T)P(C,H) motif in the sixth TMH. There are distinct patterns of conserved residues in TMHs 6, 7, and 8 that are likely to define selectivity (Arguello 2003; Dutta et al. 2006). For example, in the P_{1B} subfamily containing ZntA, CadA, and homologues, that is, pumps transporting Zn(II), Cd(II), and Pb(II), there is an invariant Asp in TMH 8 in addition to the cysteines of the CPC motif in TMH 6. All three residues, Asp714, Cys392, and Cys394, are essential for activity and have been shown to be ligands for the metal ion in ZntA (Dutta et al. 2006, 2007). Comparison with the N-terminal metal binding site suggests a four-coordinate geometry for the binding of Zn(II) and Cd(II) at the transmembrane site as well (Fig. 2). While binding of Pb(II) at the transmembrane site also involves these three residues, it appears to be slightly different from that of Zn(II) and Cd(II) in that histidine and serine do not act as ligands for Pb(II) (Dutta et al. 2006, 2007). Interestingly, different cations, including Cu(II), Ni(II), and Co(II), can bind to the transmembrane site with high affinity. Even more interestingly, conservative mutations of serine and histidine at Cys392 and Cys394 and glutamate at Asp714 in ZntA result in no decrease in the binding affinity for Zn(II) and Cd(II), but do result in greatly diminished ATPase and transport activities. This again clearly demonstrates that the conserved residues in the

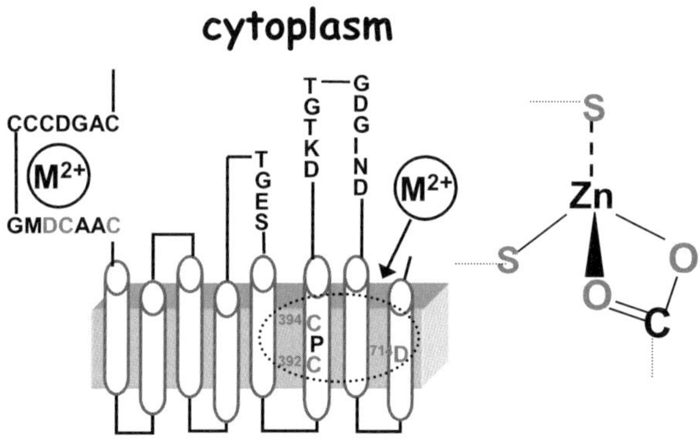

Fig. 2 Schematic representation of ZntA, a P_{1B}-type ATPase, showing the location of the two metal binding sites in the N-terminal and transmembrane domains. The positions of the zinc-binding residues in the N-terminal metal site, as well as those in the transmembrane site, are indicated in *red*. NMR structural data show that zinc is bound at the N-terminal site in tetrahedral coordination, to the two cysteine and the aspartate residues of the GMD-CAAC motif. Mutagenesis and metal binding studies have established that the two cysteines of the CPC motif in TMH 6 and the conserved aspartate residue in TMH 8 are ligands to zinc at the transmembrane site. A possible tetrahedral coordination of zinc at this site is proposed based on the N-terminal site. Cadmium binding is also through tetrahedral coordination at the N-terminal site; however, lead binding appears to be different

transmembrane domain of different subfamilies of P_{1B}-type ATPases give rise to both cation specificity and selectivity through specific metal coordination geometry that lead to the correct structural response necessary for activity. Binding affinity is not important in determining metal specificity among different divalent metal ions.

Metal binding for monovalent cation transporting ATPases like the Cu(I)/Ag(I) pumps is through linear two-coordinate geometry at the N-terminal site. It is possible that the transmembrane site also binds monovalent cations in a linear geometry through the two cysteines of the CPC motif; however, there is no evidence for this yet.

5.3 CDF Transporters

The CDF family has been found in all three domains of life: bacteria, eukaryotes, and archeae. Representative bacterial examples include CzcD from *C. metallidurans*, ZitB and FieF from *E. coli*, and CzcD from *S. aureus* and *B. subtilis* (Grass et al. 2005; Anton et al. 2004; Moore et al. 2005) (Fig. 1). The CDF proteins encompass six predicted transmembrane domains and variable potential metal binding sites. The CDF family can be subdivided into two main branches. One subbranch of CDF proteins transports zinc in representative examples studied, but some members of this subbranch are able to transport additional cations such as cobalt, mercury, and cadmium (Anton et al. 1999; Haney et al., unpublished results). The other subbranch appears to be involved in iron export, as the only two characterized members FieF or MamB of *Magnetospirillum gryphiswaldense* suggest (Grass et al. 2005; Grünberg et al. 2001). CDF proteins couple the transport of cations against their concentration gradient with the energy provided by the proton motive force.

How metal specificity is conferred in members of the CDF family is not yet known. Sequence alignments and site-directed mutagenesis of conserved residues suggests that several His and acidic residues are required for function (Anton et al. 2004). Some of these essential acidic residues are conserved in all CDF proteins regardless of the metals transported, suggesting they might form the proton channel. Other residues, such as His_{53}, His_{159}, and Asp_{163} (ZitB numbering), are conserved in zinc transporters such as ZitB from *E. coli* and CzcD from *C. metallidurans* CH34 and could be involved in metal binding and transport (Haney et al. 2005). These highly conserved residues in TM5, ZitB Asp_{163}, and the homologous Asp_{158} in CzcD from *C. metallidurans* were altered to Glu and Ala. $ZitB_{D163E}$, $ZitB_{D163A}$, $CzcD_{D158E}$, and $CzcD_{D158A}$ were no longer able to confer zinc resistance and are probably required for function and metal binding (Anton et al. 2004). Potentially these three residues could form a tetrahedral coordination geometry favored by zinc (two binding sites provided by Glu). However, at this stage this is mere speculation. This hypothesis was recently strengthened by examining the conserved Asp157 in FieF of *E. coli*. This residue appears to be a metal-coordination residue for Cd(II) and Zn(II) binding but not for other metals and is required for transport (Wei and Fu 2005).

5.4 Periplasmic Vacuum Cleaners and Binding Proteins

The previous sections have primarily dealt with transport of metal ions across the cytoplasmic membrane and intracellular regulation. However, microorganisms also have mechanisms at their disposal to handle periplasmic metals. Many Gram-negative bacteria possess multicomponent systems to pump metals from the periplasm across the outer membrane. The three-component CzcCBA from *C. metallidurans* CH34 is a well-characterized example (Dong and Mergeay 1994; Rensing et al. 1997b) (Fig. 1).

CzcA encompasses 12 transmembrane domains and two large periplasmic loops (Goldberg et al. 1999). These loops contain potential metal binding sites and CzcA may prevent metal cations from entering the cell by pumping them across the outer membrane while they are at or near the cytoplasmic membrane. CzcB belongs to the membrane fusion protein (MFP) family (Dinh et al. 1994), which spans the periplasm and brings the outer and inner membranes in close apposition (Saier et al. 1995). CzcC belongs to the outer membrane channel-tunnel efflux-duct family of proteins (Koronakis et al. 2000) with TolC being the best-studied member; frequently this family is also named the outer membrane factor (OMF) family of proteins (Paulsen et al. 1997). It could be shown that CzcC possesses a periplasmic leader peptide, which is cleaved after 22 amino acids (Grass, unpublished results). The crystal structures of the related outer membrane efflux duct TolC (Koronakis et al. 2000), the antiporter AcrB (Murakami et al. 2002), and the membrane fusion protein AcrA (Mikolosko et al. 2006) of *E. coli* have recently been determined. In addition, the structure of the related MFP MexA from *P. aeruginosa* has been determined (Akama et al. 2004). It is therefore very likely that the overall structure of the CzcCBA complex is similar. AcrB (homologous to CzcA) and TolC (homologous to CzcC) form trimers creating a channel that can be modeled as connecting the cytoplasm to the extracellular medium. In contrast, MexA (homologous to CzcB) appears to assemble as a tridecamer. However, in the models presented of the CBA efflux complex, a ring formed by nine or 12 molecules of MexA is proposed.

It is likely that these systems also transport substrates from the periplasm to the extracellular space since other transporters of the resistance nodulation division (RND) superfamily, such as AcrB and MexB, can efflux substrates that do not cross the cytoplasmic membrane. It was therefore suggested that the binding of substrate might occur on the periplasmic side of the transporter (Yu et al. 2003). The structure of the antiporter AcrB shows a channel leading from the periplasmic space to the central cavity that is a possible mode of entry of substrates (Murakami et al. 2002). Furthermore, increasing evidence points to the periplasmic domains of RND proteins as conferring substrate specificity (Yu et al. 2005).

Several lines of evidence in CzcCBA and related systems corroborate these findings. Expression of CzcA and CnrA alone confers only very weak resistance (Rensing et al. 1997b; Grass, unpublished results). However, it could recently be shown that a *cnrC*-deletion strain could be complemented *in trans* by *czcC* or by the related *nccC* gene from the highly nickel resistant bacterium *C. metallidurans* 31A (formerly *Alcaligenes xylosoxydans* 31A) (Schmidt and Schlegel 1994; Grass 2000; Grass and Nies, unpublished results). This indicates that only CnrBA is specific for the transported metals and CnrC is interchangeable with other homologous proteins. There is increasing evidence that these transport systems pump specific metals directly from the periplasm across the outer membrane (Grass and Rensing 2001; Franke et al. 2003; Legatzki et al. 2003; Munkelt et al. 2004). It was therefore concluded that the initial metal binding, and hence metal specificity, should occur in periplasmic domains of CzcA.

Probably *czc* possesses six regulators, with their genes arranged up- and downstream of *czcCBA*, respectively. It could be shown that the response regulator CzcR binds upstream of the *czcN* gene (Grosse et al. 1999). However, deletion of *czcR* did not lead to a change of β-galactosidase activity in a Φ(*czcC-lacZ-czcBA*) reporter strain (Anton 2001). Since all *czc* regulators (except CzcR) are either membrane bound (CzcN, CzcS), periplasmic (CzcI, CzcE) (Anton 2001), or metal transporters (CzcD) (Anton et al. 1999), it seems possible that the activity of CzcCBA may be modulated posttranslationally.

6 Intracellular Sequestration: Metallothioneins and Ribosomal Binding

In addition to controlling the transport of Zn, Cd, and Pb, cells can also control intracellular binding and release of these metals. Metallothionein proteins can sequester metals such as Zn, Cd, and Pb intracellularly. These cysteine-rich proteins bind Zn(II) and Cd(II) in four-coordinate geometry as $Zn-S_4$ and $Cd-S_4$. In contrast, Pb(II) appears to prefer a three-coordinate $Pb-S_3$ trigonal geometry in a thiolate-rich environment (Magyar et al. 2005; Lippard and Berg 1994). Bacteria that have so far been shown to possess metallothioneins include cyanobacteria and pseudomonads (Morby et al. 1993; Blindauer et al. 2002). Metallothioneins can sequester excess intracellular metal, thereby preventing toxicity.

Under conditions of zinc scarcity, many bacteria are able to release zinc bound to ribosomal proteins. This was recently discovered during an *in silico* examination of genes presumed to be controlled by Zur (Panina et al. 2003). The authors discovered that several paralogues of ribosomal proteins were predicted to be repressed by zinc. These paralogues differed from the ribosomal proteins usually present by not having a predicted Zn-ribbon motif. The authors correctly predicted that under zinc depletion the ribosomal proteins without a zinc binding motif replaced the original proteins with bound zinc. Essential zinc requiring proteins can then use the zinc liberated from these ribosomal proteins. This model was later validated in *B. subtilis* (Akanuma et al. 2006).

7 Extracellular Binding and Precipitation: The Lead Resistance Operon in *Cupriavidus Metallidurans* Strain CH34

C. metallidurans CH34 harbors a unique lead resistance operon *pbr* that contains components that have already been discussed, such as the P_{1B}-ATPase PbrA and a repressor of the ArsR/SmtB family, PbrR. In addition, there are components where the function has not been elucidated (Borremans et al. 2001). PbrT is thought to be involved in Pb(II) uptake since expression of PbrT makes cells hypersensitive to lead. This is reminiscent of the role of Hg transporters encoded on *mer* operons. These transporters are thought to protect the periplasm and respiratory chain from mercury-induced damage by delivering Hg(II) directly from the periplasm to the cytoplasmic mercuric reductase MerA. In analogy, PbrT could directly deliver Pb(II) to the putative chaperone PbrD or the P-type ATPase PbrA. However, without subsequent transport of Pb(II) across the outer membrane this would be a futile cycle. Therefore the fusion protein PbrBC is thought to be involved in transporting Pb(II) across the outer membrane. At this point there is no evidence supporting this hypothesis. In addition, Pb(II) could be immobilized on the surface of the outer membrane because, as the pH changes to about 9, there is significant lead carbonate precipitation.

Acknowledgements This work was supported by NSF grant 421860 and USPHS grant GM61689.

References

Akama H, Matsuura T, Kashiwagi S, Yoneyama H, Narita S, Tsukihara T, Nakagawa A, Nakae T (2004) Crystal structure of the membrane fusion protein, MexA, of the multidrug transporter in *Pseudomonas aeruginosa*. J Biol Chem 279:25939–25942

Akanuma G, Nanamiya H, Natori Y, Nomura N, Kawamura F (2006) Liberation of zinc-containing L31 (RpmE) from ribosomes by its paralogous gene product, YtiA, in *Bacillus subtilis*. J Bacteriol 188:2715–2720

Ansari AZ, Bradner JE, O'Halloran TV (1995) DNA-bend modulation in a repressor-to-activator switching mechanism. Nature 374:371–375

Anton A (2001) Genetische und biochemische Charakterisierung von CzcD und anderen Regulatoren der *czc*-vermittelten Schwermetallresistanz in *Ralstonia metallidurans.* Dissertation, Martin-Luther-Universität, Halle-Wittenberg

Anton A, Grosse C, Reissmann J, Pribyl T, Nies DH (1999) CzcD is a heavy metal ion transporter involved in regulation of heavy metal resistance in *Ralstonia* sp. strain CH34. J Bacteriol 181:6876–6881

Anton A, Weltrowski A, Haney CJ, Franke S, Grass G, Rensing C, Nies DH (2004) Characteristics of zinc transport by two bacterial cation diffusion facilitators from *Ralstonia metallidurans* and *Escherichia coli.* J Bacteriol 186:7499–7507

Archibald FS, Duong MN (1984) Manganese acquisition by *Lactobacillus plantarum.* J Bacteriol 158:1–8

Arguello JM (2003) Identification of ion-selectivity determinants in heavy-metal transport P_{1B}-type ATPases. J Membr Biol 195:93–108

Auld DS (2001) Zinc coordination sphere in biochemical zinc sites. BioMetals 14:271–313

Banci L, Bertini I, Ciofi-Baffoni S, Finney LA, Outten CE, O'Halloran TV (2002) A new zinc–protein coordination site in intracellular metal trafficking: solution structure of the apo and Zn(II) forms of ZntA(46-118). J Mol Biol 323:883–897

Banci L, Bertini I, Ciofi-Baffoni S, Su XC, Miras R, Bal N, Mintz E, Catty P, Shokes JE, Scott RA (2006) Structural basis for metal binding specificity: the N-terminal cadmium binding domain of the P1-type ATPase CadA. J Mol Biol 356:638–650

Banerjee S, Wei B, Bhattacharyya-Pakresi M, Pakrasi HB, Smith TJ (2003) Structural determinants of metal specificity in the zinc transport protein ZnuA from *Synechocystis* 6803. J Mol Biol 333:1061–1069

Barkay T, Miller SM, Summers AO (2003) Bacterial mercury resistance from atoms to ecosystems. FEMS Microbiol Rev 27:355–384

Beard SJ, Hashim R, Membrillo-Hernandez J, Hughes MN, Poole RK (1997) Zinc(II) tolerance in *Escherichia coli* K12; evidence that the *zntA* gene (*o*732) encodes a cation transport ATPase. Mol Microbiol 25:883–891

Beard SJ, Hashim R, Wu G, Binet MRB, Hughes MN, Poole RK (2000) Evidence for the transport of zinc(II) ions via the Pit inorganic phosphate transport system in *Escherichia coli.* FEMS Microbiol Lett 184:231–235

Bhattacharyya P (1975) Active transport of manganese in isolated membrane vesicles of *Bacillus subtilis.* J Bacteriol 123:123–127

Blencowe DK, Morby AP (2003) Zn(II) metabolism in prokaryotes. FEMS Microbiol Rev 27:291–311

Blindauer CA, Harrison MD, Robinson AK, Parkinson JA, Bowness PW, Sadler PJ, Robinson NJ (2002) Multiple bacteria encode metallothioneins and SmtA-like zinc fingers. Mol Microbiol 45:1421–1432

Borremans B, Hobman JL, Provoost A, Brown NL, van der Lelie D (2001) Cloning and functional analysis of the *pbr* lead resistance determinant of *Ralstonia metallidurans* CH34. J Bacteriol 183:551–568

Brocklehurst KR, Hobman JL, Lawley B, Blank L, Marshall SJ, Brown NL, Morby AP (1999) ZntR is a Zn(II)-responsive MerR-like transcriptional regulator of *zntA* in *Escherichia coli.* Mol Microbiol 31:893–902

Brown NL, Stoyanov JV, Kidd SP, Hobman JL (2003) The MerR family of transcriptional regulators. FEMS Microbiol Rev 27:145–163

Bsat N, Herbig A, Casillas-Martinez L, Setlow P, Helmann JD (1998) *Bacillus subtilis* contains multiple Fur homologues: identification of the iron uptake (Fur) and peroxide regulon (PerR) repressors. Mol Microbiol 29:189–198

Cotton FA, Wilkinson G, Murillo CA, Bochmann M (1999) Advanced inorganic chemistry, 6th edn. Wiley, New York

Dinh T, Paulsen IT, Saier MH Jr (1994) A family of extracytoplasmic proteins that allow transport of large molecules across the outer membranes of gram-negative bacteria. J Bacteriol 176:3825–3831

Dong Q, Mergeay M (1994) Czc/Cnr efflux: a three-component chemiosmotic antiport pathway with a 12-transmembrane-helix protein. Mol Microbiol 14:185–187

Dreyer MK, Schulz GE (1993) The spatial structure of the class II L-fuculose-1-phosphate aldolase from *Escherichia coli.* J Mol Biol 231:549–553

Dutta SJ, Liu J, Mitra B (2005) Kinetics of metal binding to the amino-terminal domain of ZntA by monitoring metal-thiolate charge-transfer complexes. Biochemistry 44:14268–14274

Dutta SJ, Liu J, Hou Z, Mitra B (2006) Conserved Asp714 in transmembrane segment 8 of the ZntA subgroup of P_{1B}-type ATPases is a metal-binding residue. Biochemistry 45:5923–5931

Dutta SJ, Liu J, Stemmler AJ, Mitra B (2007) Conservative and non-conservative mutation of the cysteine residues of the transmembrane CPC motif in ZntA: effect on metal binding and activity. Biochemistry 46 (in press)

Esposito L, Sica F, Raia CA, Giordano A, Rossi M, Mazzarella L, Zagari A (2002) Crystal structure of the alcohol dehydrogenase from the hyperthermophile archaeon *Sulfolobus solfataricus* at 1.85 Å resolution. J Mol Biol 318:463–477

Gaballa A, Helmann JD (1998) Identification of a zinc-specific metalloregulatory protein, Zur, controlling zinc transport operons in *Bacillus subtilis.* J Bacteriol 180:5815–5821

Grass G (2000) Molekulargenetische und biochemische Charakterisierung der *cnr* Cobalt/Nickel-Resistenz-Determinante aus *Ralstonia metallidurans* CH34. Dissertation, Martin-Luther-Universität, Halle-Wittenberg

Grass G, Rensing C (2001) CueO is a multi-copper oxidase that confers copper tolerance in *Escherichia coli.* Biochem Biophys Res Commun 286:902–908

Grass G, Otto M, Fricke B, Haney CJ, Rensing C, Nies DH, Munkelt D (2005) FieF (YiiP) from *Escherichia coli* mediates decreased cellular accumulation of iron and relieves iron stress. Arch Microbiol 183:9–18

Grünberg K, Wawer C, Tebo BM, Schüler D (2001) A large gene cluster encoding several magnetosome proteins is conserved in different species of magnetotactic bacteria. Appl Environ Microbiol 67:4573–4582

Haney CJ, Grass G, Franke S, Rensing C (2005) New developments in functional understanding of the cation-diffusion-facilitator family. J Ind Microbiol Biotechnol 32:215–226

Hantke K (2005) Bacterial zinc uptake and regulators. Curr Opin Microbiol 8:196–202

Harvie DR, Andreini C, Cavallaro G, Meng W, Connolly BA, Yoshida KI, Fujita Y, Harwood CR, Radford DS, Tottey S, Cavet JS, Robinson NJ (2006) Predicting metals sensed by ArsR-SmtB repressors: allosteric interference by a non-effector metal. Mol Microbiol 59:1341–1356

Hassan MT, van der Lelie D, Springael D, Romling U, Ahmed N, Mergeay M (1999) Identification of a gene cluster, *czr*, involved in cadmium and zinc resistance in *Pseudomonas aeruginosa.* Gene 238:417–425

Hobman JL, Wilkie J, Brown NL (2005) A design for life: prokaryotic metal-binding MerR family regulators. BioMetals 18:429–436

Kehres DG, Lawyer CH, Maguire ME (1998) The CorA magnesium transporter gene family. Microb Comp Genomics 3:151–169

Kehres DG, Zaharik ML, Finlay BB, Maguire ME (2000) The NRAMP proteins of *Salmonella typhimurium* and *Escherichia coli* are selective manganese transporters involved in the response to reactive oxygen. Mol Microbiol 36:1085–1100

Korkhin Y, Kalb AJ, Peretz M, Bogin O, Burstein Y, Frolow F (1998) NADP-dependent bacterial alcohol dehydrogenase: crystal structure, cofactor-binding and cofactor specificity of the ADHs of *Chlostridium beijerinckii* and *Thermoanaerobacter brockii*. J Mol Biol 278:967–981

Koronakis V, Sharff A, Koronakis E, Luisi B, Hughes C (2000) Crystal structure of the bacterial membrane protein TolC central to multidrug efflux and protein export. Nature 405:914–919

Kuroda M, Hayashi H, Ohta T (1999) Chromosome-determined zinc-responsible operon *czr* in *Staphylococcus aureus* strain 912. Microbiol Immunol 43:115–125

Lane TW, Saito MA, George GN, Pickering IJ, Prince RC, Morel FM (2005) Biochemistry: a cadmium enzyme from a marine diatom. Nature 435:42

Lawrence MC, Pilling PA, Epa VC, Berry AM, Ogunniyi AD, Paton JC (1998) The crystal structure of pneumococcal surface antigen PsaA reveals a metal-binding site and a novel structure for a putative ABC-type binding protein. Structure 6:1553–1561

Lee YH, Dorwart MR, Hazlett KR, Deka RK, Norgard MV, Radolf JD, Hasemann CA (2002) The crystal structure of Zn(II)-free *Treponema pallidum* TroA, a periplasmic metal-binding protein, reveals a closed conformation. J Bacteriol 184:2300–2304

Legatzki A, Grass G, Anton A, Rensing C, Nies DH (2003) Interplay of the Czc system and two P-type ATPases in conferring metal resistance in *Ralstonia metallidurans*. J Bacteriol 185:4354–4361

Lippard SJ, Berg JM (1994) Principles of bioinorganic chemistry. University Science Books, Mill Valley, CA

Liu T, Golden JW, Giedroc DP (2005a) A zinc(II)/lead(II)/cadmium(II)-inducible operon from the cyanobacterium *Anabaena* is regulated by AztR, an a3N ArsR/SmtB metalloregulator. Biochemistry 44:8673–8683

Liu J, Stemmler AJ, Fatima J, Mitra B (2005b) Metal-binding characteristics of the amino-terminal domain of ZntA: lead binds to different ligands compared to cadmium and zinc. Biochemistry 44:5159–5167

Liu J, Dutta SJ, Stemmler AJ, Mitra B (2006) Metal-binding affinity of the transmembrane site in ZntA: implications for metal selectivity. Biochemistry 45:763–772

Magyar JS, Weng TC, Stern CM, Dye DF, Rous BW, Payne JC, Bridgewater BM, Mijovilovich A, Parkin G, Zaleski JM, Penner-Hahn JE, Godwin HA (2005) Reexamination of lead(II) coordination preferences in sulfur-rich sites: implications for a critical mechanism of lead poisoning. J Am Chem Soc 127:9495–9505

Makui H, Roig E, Cole ST, Helmann JD, Gros P, Cellier MF (2000) Identification of the *Escherichia coli* K-12 Nramp orthologue (MntH) as a selective divalent metal ion transporter. Mol Microbiol 35:1065–1078

Mikolosko J, Bobyk K, Zgurskaya HI, Ghosh P (2006) Conformational flexibility in the multidrug efflux system protein AcrA. Structure 14:577–587

Mitra B, Sharma R (2001) The cysteine-rich amino-terminal domain of ZntA, a Pb(II)/Cd(II)/Zn(II)-translocating ATPase from *Escherichia coli*, is not essential for its function. Biochemistry 40:7694–7699

Moore CM, Gaballa A, Hui M, Ye RW, Helmann JD (2005) Genetic and physiological responses of *Bacillus subtilis* to metal ion stress. Mol Microbiol 57:27–40

Morby AP, Turner JS, Huckle JW, Robinson NJ (1993) SmtB is a metal-dependent repressor of the cyanobacterial metallothionein gene SmtA: identification of a Zn-inhibited DNA-protein complex. Nucleic Acids Res 21:921–925

Munkelt D, Grass G, Nies DH (2004) The chromosomally encoded cation diffusion facilitator proteins DmeF and FieF from *Wautersia metallidurans* CH34 are transporters of broad metal specificity. J Bacteriol 186:8036–8043

Murakami S, Nakashima R, Yamashita E, Yamaguchi A (2002) Crystal structure of bacterial multidrug efflux transporter AcrB. Nature 419:587–593

Nies DH (1995) The cobalt, zinc and cadmium efflux system CzcABC from *Alcaligenes eutrophus* functions as a cation–proton antiporter in *Escherichia coli.* J Bacteriol 177:2707–2712

Outten CE, Outten FW, O'Halloran TV (1999) DNA distortion mechanism for transcriptional activation by ZntR, a Zn(II)-responsive MerR homologue in *Escherichia coli.* J Biol Chem 274:37517–37524

Outten CE, Tobin DA, Penner-Hahn JE, O'Halloran TV (2001) Characterization of the metal receptor sites in *Escherichia coli* Zur, an ultrasensitive zinc(II) metalloregulatory protein. Biochemistry 40:10417–10423

Outten FW, Outten CE, Hale J, O'Halloran TV (2000) Transcriptional activation of an *Escherichia coli* copper efflux regulon by the chromosomal MerR homologue, CueR. J Biol Chem 275:31024–31029

Panina EM, Mironov AA, Gelfand MS (2003) Comparative genomics of bacterial zinc regulons: enhanced ion transport, pathogenesis, and rearrangement of ribosomal proteins. Proc Natl Acad Sci USA 100:9912–9917

Patzer SI, Hantke K (1998) The ZnuABC high-affinity zinc uptake system and its regulator Zur in *Escherichia coli.* Mol Microbiol 28:1199–1210

Patzer SI, Hantke K (2000) The zinc-responsive regulator Zur and its control of the *znu* gene cluster encoding the ZnuABC zinc uptake system in *Escherichia coli.* J Biol Chem 275:24321–24332

Paulsen IT, Saier MJ (1997) A novel family of ubiquitous heavy metal ion transport proteins. J Membr Biol 156:99–103

Perry RD, Silver S (1982) Cadmium and manganese transport in *Staphylococcus aureus* membrane vesicles. J Bacteriol 150:973–976

Que Q, Helmann JD (2000) Manganese homeostasis in *Bacillus subtilis* is regulated by MntR, a bifunctional regulator related to the diphtheria toxin repressor family of proteins. Mol Microbiol 35:1454–1468.

Rensing C (2005) Form and function in metal-dependent transcriptional regulation: dawn of the enlightenment. J Bacteriol 187:3909–3912

Rensing C, Mitra B, Rosen BP (1997a) The *zntA* gene of *Escherichia coli* encodes a Zn(II)-translocating P-type ATPase. Proc Natl Acad Sci USA 94:14326–14331

Rensing C, Pribyl T, Nies DH (1997b) New functions for the three subunits of the CzcCBA cation–proton antiporter. J Bacteriol 179:6871:6879

Rensing C, Mitra B, Rosen BP (1998a) A Zn(II)-translocating P-type ATPase from *Proteus mirabilis.* Biochem Cell Biol 76:787–790

Rensing C, Sun Y, Mitra B, Rosen BP (1998b) Pb(II)-translocating P-type ATPases. J Biol Chem 273:32614–32617.

Rensing C, Ghosh M, Rosen BP (1999) Families of soft-metal-ion-transporting ATPases. J Bacteriol 181:5891–5897

Rutherford JC, Cavet JS, Robinson NJ (1999) Cobalt-dependent transcriptional switching by a dual-effector MerR-like protein regulates a cobalt-exporting variant CPx-type ATPase. J Biol Chem 274:25827–25832

Sharma R, Rensing C, Rosen BP, Mitra B (2000) The ATP hydrolytic activity of purified ZntA, a Pb(II)/Cd(II)/Zn(II)-translocating ATPase from *Escherichia coli.* J Biol Chem 275:3873–3878

Silver S (1998) Genes for all metals: a bacterial view of the Periodic Table. J Ind Microbiol Biotechnol 20:1–12

Silver S, Hobman JL (2007) Mercury Microbiology: Resistance Systems, Environmental Aspects, Methylation and Human Health. Springer, Heidelberg (in this volume)

Silver S, Phung LT (1996) Bacterial heavy metal resistance: new surprises. Annu Rev Microbiol 50:753–789

Singh SK, Grass G, Rensing C, Montfort WR (2004) Cuprous oxidase activity of CueO from *Escherichia coli*. J Bacteriol 186:7815–7817

Singh VK, Xiong A, Usgaard TR, Chakrabarti S, Deora R, Misra TK, Jayaswal RK (1999) ZntR is an autoregulatory protein and negatively regulates the chromosomal zinc resistance operon *znt* of *Staphylococcus aureus*. Mol Microbiol 33:200–207

Solioz M, Vulpe C (1996) CPx-type ATPases: a class of P-type ATPases that pump heavy metals. Trends Biochem Sci 21:237–241

Strange RW, Antonyuk S, Hough MA, Doucette PA, Rodriguez JA, Hart J, Hayward LJ, Valentine JS, Hasnain SS (2003) The structure of holo and metal-deficient wild-type human Cu,Zn superoxide dismutase and its relevance to familial amyotrophic lateral sclerosis. J Mol Biol 328:877–891

Thelwell C, Robinson NJ, Turner-Cavet JS (1998) An SmtB-like repressor from *Synechocystis* PCC 6803 regulates a zinc exporter. Proc Natl Acad Sci USA 95:10728–10733

Tsai KJ, Yoon KP, Lynn AR (1992) ATP-dependent cadmium transport by the *cadA* cadmium resistance determinant in everted membrane vesicles of *Bacillus subtilis*. J Bacteriol 174:116–121

Utschig LM, Bryson JW, O'Halloran TV (1995) Mercury-199 NMR of the metal receptor site in MerR and its protein–DNA complex. Science 268:380–385

Vallee BL, Coleman JE, Auld DS (1991) Zinc fingers, zinc clusters, and zinc twists in DNA-binding protein domains. Proc Natl Acad Sci USA 88:999–1003

Wei Y, Fu D (2005) Selective metal binding to a membrane-embedded aspartate in the *Escherichia coli* transporter YiiP (FieF). J Biol Chem 280:33716–33724

Xiong A, Jayaswal RK (1998) Molecular characterization of a chromosomal determinant conferring resistance to zinc and cobalt ions in *Staphylococcus aureus*. J Bacteriol 180:4024–4029

Ye J, Kandegedara A, Martin P, Rosen BP (2005) Crystal structure of the *Staphylococcus aureus* pI258 CadC Cd(II)/Pb(II)/Zn(II)-responsive repressor. J Bacteriol 187:4214–4221

Yu EW, McDermott G, Zgurskaya HI, Nikaido H, Koshland DE Jr (2003) Structural basis of multiple drug-binding capacity of the AcrB multidrug efflux pump. Science 300:976–980

Yu EW, Aires JR, McDermott G, Nikaido H (2005) A periplasmic drug-binding site of the AcrB multidrug efflux pump: a crystallographic and site-directed mutagenesis study. J Bacteriol 187:6804–6815

Microbiol Monogr (6)
D. H. Nies, S. Silver: Molecular Microbiology of Heavy Metals
DOI 10.1007/7171_2006_084/Published online: 2 February 2007

Microbiology of the Toxic Noble Metal Silver

Sylvia Franke

Biology Department, Skidmore College, 815 North Broadway,
Saratoga Springs, NY 12866, USA
sfranke@skidmore.edu

Abstract One of the transition metals without a function in biological systems is silver. Silver interferes with the normal protein function of the organism and is extremely toxic because of its ability to bind to the metal binding sites in proteins. The use of silver as a biocide has long been known and today there is an increasing number of applications for silver, not only in hospitals but also in everyday life. In addition, silver resistant bacteria have been isolated from hospitals, silver mines, and silver-contaminated areas. There are different resistance mechanisms, as strains of *Pseudomonas* appear to precipitate silver to remove it from the medium. In contrast, silver resistance in *Salmonella typhimurium* is plasmid-encoded and based on silver binding and export. Genome sequencing projects increasingly reveal the presence of this *sil* determinant, indicating a more widespread occurrence than previously expected.

1 Introduction

The transition element silver (Ag) is not relevant for living organisms as a trace element. Naturally occurring silver is composed of two stable isotopes, ^{107}Ag and ^{109}Ag with 51.9 and 48.1% natural abundance, respectively, and 28 radioisotopes have been characterized. Silver combined with arsenic, sulfur, antimony, or chlorine is mainly found in copper, lead, and lead–zinc ores as elemental silver or silver compounds, e.g. horn silver (AgCl) and argentite (Ag_2S).

Silver has a long history. Silver mining in Mesopotamia (now Iraq, east Syria, and southeast Turkey) goes back as far as 2500 B.C. In ancient Egypt the bones of gods were thought to be of silver, and there it was first described as "white metal". Since it was very rare, silver was considered more valuable than gold (http://touregypt.net/featurestories/silver.html). Silver mining then spread to Europe, where first Greece and later Spain were the mining centers but also regions of Saxony in Germany. The "discovery" of the "New World" opened a new resource of the highly treasured metal (http://www.silverinstitute.org/facts/). For thousands of years, silver was used for trade, ornaments, and utensils.

Currently silver is mined in more than 50 countries (19 700 t in 2004) with Peru being the world's leading producer followed by Mexico, China, and Australia. Because of its physical properties, silver is an excellent electrical and thermal conductor. It has a wide range of industrial applications, such as in batteries, fuses, electrical switches, or heated automobile windows. The light sensitivity of silver is used in photography. And still today's best-known use of silver is in jewelry and silverware (Brooks 2004).

2 Use of Silver as a Biocide

The antibiotic effect of silver has long been known and goes back to ancient Greece and Rome, where silver coins were used to disinfect water even before understanding of infectious diseases (Brett 2006). Today silver compounds are still used to control the formation of bacterial biofilms in water piping (Rogers et al. 1995).

Medical use of silver nitrate goes back to the Middle Ages when it was probably first used to treat chronic wounds and ulcers, and was eventually used as an alternative method to cauterizing iron in the treatment of fistulas (Klasen 2000a). In the nineteenth century silver was also used in burn wound treatment, one of today's best-known uses of silver in medicine. Early publications refer to concentrations of 3.4 to 8.4% silver nitrate (Klasen 2000a), whereas today silver nitrate in wound treatment is largely replaced by silver sulfadiazine (Klasen 2000b; Graham 2005) (usually 1% silver sulfadiazine plus 0.2% chlorhexidine in a hydrophilic cream, market name Silvazine) and silver ion releasing dressings (Cooper 2004), which are shown to be effective to generally prevent the growth of a variety of burn wound pathogens (Bowler et al. 2004). Besides silver salts, at the beginning of the twentieth century silver foil was used in wound treatment (Klasen 2000a). Silver-containing bandages are now also commercially available from different companies for in-home wound treatment (Silver et al. 2006).

Another medical application of silver ions is the use of a 2% $AgNO_3$ solution in the eyes of newborn children to prevent gonorrhea infections, introduced in 1880. The so-called silver filling amalgam in dental restoration, which contains 50% mercury and 35% silver, is still common in dental restoration although it is being replaced by other materials (Klasen 2000a). Silver-coated catheters are used to prevent the formation of biofilms and resulting infections (Spencer 1999; Tobin and Bambauer 2003). Furthermore, many people use colloidal silver as a "miracle weapon" against all kinds of diseases (http://www.soul-guidance.com/health/colloidalsilver.html, http://www.all-natural.com/silver-1.html).

The use of nanoparticulate silver as an alternative antimicrobial agent in bone cement to replace the currently used antibiotics, in order to solve the problem of infections with antibiotic resistant bacteria, is relatively recent. Preliminary results using 1% nanoparticulate silver show a complete inhibition of the bacteria tested, and give no indication of cytotoxic effects as known from several cases where probably silver salts were used.

Today silver ions, which are constantly released from silver-metal-coated surfaces, are being used in a wide range of industrial and medical applications as well as consumer products, such as household appliances (clothes and dish washers), food processing, and water treatment (e.g. AgION Technologies, Inc.: http://www.agion-tech.com).

The antibacterial effect of silver has long been studied (Yudkins 1937). Silver has a high affinity to sulfhydryl groups in proteins, which leads to changes in conformation and inactivation of enzymes. In *Vibrio cholerae* low concentrations of silver ions induce a leakage of protons, which results in deenergization and cell death (Dibrov et al. 2002). Silver ions also block the respiratory chain reversibly (low silver concentrations) or irreversibly (high silver concentrations) (Bragg and Rainnie 1974). Binding to DNA increases its stability and inhibits proliferation. However, to deal with these toxic effects of silver, a number of microorganisms have acquired mechanisms of silver resistance.

3
Silver Resistant Bacteria

Reports of silver resistant bacteria and viruses date back to the early 1950s (Lobo and Vasconcelos 1950; Lacorte et al. 1955). Silver as silver nitrate was commonly used in a 0.5% solution in hospitals, mainly for burn wound treatment. It was in this environment that silver resistant microorganisms were first identified. In 1968 the first silver resistant enterobacterium *Escherichia coli* was isolated from a silver nitrate treated burn wound and characterized (Jelenko 1969). Whereas normally a 0.5% solution of aqueous silver inhibits *E. coli* completely, the isolate tolerated exposure to a more than 4% solution. Since then, several cases of silver resistant *Enterobacteriaceae* isolated from burn wounds treated with silver dressing have been reported (McHugh et al. 1975; Gupta et al. 2001; Pirnay et al. 2003). Also, silver resistant strains of *Pseudomonas aeruginosa* were isolated from burn wounds (Bridges et al. 1979; Pirnay et al. 2003). Nevertheless, silver compounds are still the best antiseptic treatment against burn wound infection (George et al. 1997; Wright et al. 1998; Brady et al. 2003; Bowler et al. 2004; Cooper 2004). With the widespread use of silver compounds in various dental applications, silver resistant bacteria can also be isolated from teeth, i.e. *Enterobacter cloacae* (Davis et al. 2005).

Yet, resistance against silver not only arises when silver compounds are used as biocides. Silver can be found in the environment, where it either occurs naturally, for example in silver ores, or is introduced through industrial waste water. From such habitats silver resistant bacteria can also be isolated. Strains of *Klebsiella pneumoniae* able to grow in the presence of silver were isolated from shrimp of coastal waters (Choudhury and Kumar 1998). *Acinetobacter* was used to remove silver from photographic waste water (Shakibaie et al. 1999), and a consortium of a strain of *Citrobacter* and several strains of *Pseudomonas* was used for biodetoxification of silver cyanide (Patil and Paknikar 2000). A silver resistant strain of *P. stutzeri* was isolated from a silver mine (Haefeli et al. 1984,) and *Thiobacillus ferrooxidans* and *T. thiooxidans* were used in silver leaching (Pooley 1982). The question arises: how are these microorganisms able to protect themselves from silver mediated damage and cell death?

3.1 The *Sil* Determinant of *Salmonella* pMG101 and Homologous Systems

To date, the best-characterized silver resistance system is Sil, encoded on pMG101 of *Salmonella typhimurium*. This particular strain of *S. typhimurium* was isolated from patients of the burns unit of Massachusetts General Hospital (McHugh et al. 1975). It was noted that the silver resistance was transferable by cell to cell conjugation. This was the first reported case of silver resistance encoded on a conjugative plasmid, pMG101.

The molecular basis for this resistance was not understood until Gupta et al. (1999) published the sequence of the *sil* determinant from pMG101 (GenBank accession # AF067954). The *sil* determinant consists of nine open reading frames (ORFs) organized in three operons: *silCFBA(ORF105aa)P*, *silRS*, and *silE*. Proteins encoded by *silRS* show homology to a two-component regulatory system with a transmembrane histidine sensor kinase SilS and the response regulator SilR, with the highest homologies to two-component regulatory systems involved in regulation of other metal cation resistance determinants.

Downstream of *silRS* and under control of its own promoter is the gene *silE*. Expression of this gene is strongly induced by the presence of silver, and the encoded protein is located in the periplasm. Initial biochemical characterization showed its ability to bind five Ag^+ ions per protein molecule. The ten histidine residues of SilE are considered to be the metal cation ligands. Proton NMR spectroscopy was used to show the involvement of the histidine imidazole N atoms in Ag^+ binding. Whereas SilE without bound metal ions is lacking any secondary structure, addition of Ag^+ resulted in a mostly α-helical protein (Silver 2003). However, the structure of the SilE protein has not yet been solved. Also still not understood is how SilE is involved in silver resistance. One hypothesis is that SilE is the first defense of *Salmonella* against silver, by binding Ag^+ before it can enter the cytoplasm. This theory is supported by the localization of the protein and the high expression rate of *silE* in the presence of Ag^+. SilE is encoded as part of additional hypothetical silver resistance determinants on virulence plasmids from various enterobacteria. In addition, the encoded copper resistance system of *E. coli*, *pco*, encodes a protein PcoE which is 46.5% identical in sequence to SilE and similarly possesses ten histidine residues (Brown et al. 1995; see chapter by Magnani and Solioz, in this volume). Again, however, the role of the periplasmic protein is not yet experimentally tested.

The third transcript is *silCFBA(ORF105)P*, encoding a CPx-type ATPase, SilP; a CBA transport complex, SilCFBA(ORF96aa) was recently renamed to SilF (Silver et al. 2006). This gene cluster also contains an uncharacterized open reading frame, *ORF105aa* (Gupta et al. 1999). CPx-type ATPases are a family of ATPases involved in the transport of heavy metal cations, such as Zn^{2+}, Cd^{2+}, Cu^+, and Ag^+ (Solioz and Vulpe 1996; see chapters by Nies; Magnani and Solioz; Mitra and Rensing, this volume). SilP contains the key features of this P-type ATPase protein family: the CPx motif in transmembrane helix six, an ATP binding domain with a conserved motif including the aspartate residue which is initially phosphorylated and a conserved HisPro between the ATP binding and phosphorylation domains. In contrast to other members of the CPx-type ATPases, SilP lacks N-terminal metal binding domains with the GMXCX2C motif. However, SilP contains N-terminal CX2C and CX3C as well as several histidine residues as potential metal ligands (Gupta et al. 1999; Rensing et al. 1999; Silver 2003). The involvement of SilP in silver resistance was shown in a partial *silP* deletion which becomes more silver sensitive. In the model of silver resistance in *Salmonella* (Fig. 1) SilP transports Ag^+ across the cytoplasmic membrane into the periplasm, from where the other proteins encoded by this *silCFBA(ORF105aa)P* operon are responsible for the further transport into the surrounding medium.

SilP was found as part of the silver resistance determinants identified during sequencing projects of several enterobacterial virulence plasmids, like pLVPK (Chen et al. 2004), pK2044 (Tsai et al. 2005) of *K. pneumoniae*, pAPEC-O2-R from *E. coli* (Johnson et al. 2005), and R478 of *Serratia marcescens* (Gilmour et al. 2004). Analysis of plasmids from several clinical isolates shows the presence of *silP* genes in six out of 19 tested plasmids (Gupta et al. 2001). Two of them were R478 and pMG101. The other four plasmids were originally from three different isolates of *S. typhimurium* and *S. marcescens* (Gupta et al. 2001). None of these proteins has been characterized so far. However, because of their close identity to SilP from pMG101 (more than 90%) it is

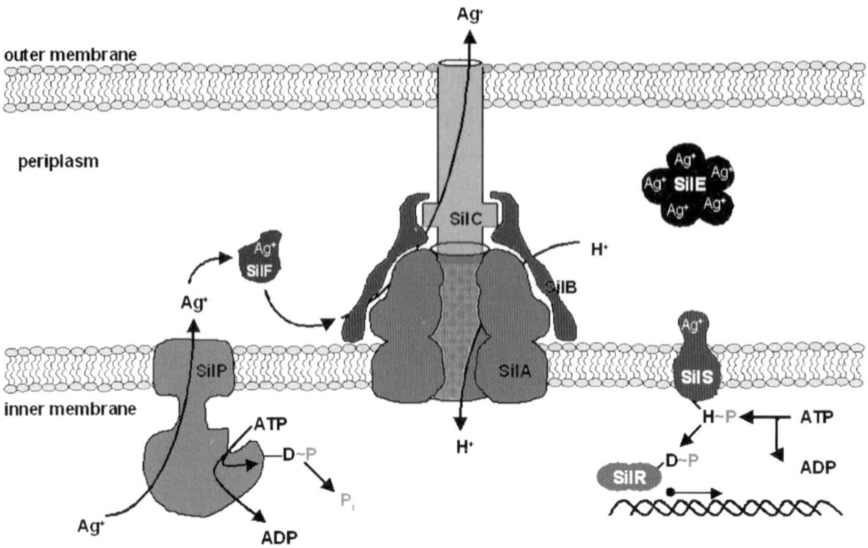

Fig. 1 Model of silver resistance conferred by the Sil system of plasmid pMG101 (modified from Silver 2003, 2006). Ag^+ is exported from the cytoplasm into the periplasm via the CPx-type ATPase SilP, forming a phosphorylated (D~P) intermediate. In the periplasm it is bound to the metal chaperone SilF and delivered to the SilCBA efflux complex, which, in a proton dependent process, transports Ag^+ into the surrounding medium. Excess silver in the periplasm can be bound to SilE to prevent cell damage. Expression of the system is regulated by the two-component regulatory system SilRS, forming phosphorylated histidine (SilS: H~P) and aspartate (SilR: D~P) intermediates

most likely that these proteins are involved in silver resistance. Other CPx-type ATPases which are not part of *sil* determinants have also been found to be able to transport Ag^+. Solioz and Odermatt (1995) demonstrated the ability of the copper transporting ATPase CopB, from *E. hirae*, to also accumulate silver ions in inverted membrane vesicles of this organism (see chapter by Magnani and Solioz, this volume). However, transport of silver ions by CopB in vesicles does not seem to be biologically relevant, since growth of *E. hirae* is inhibited by 5 μM Ag^+. Another example of a copper transporting ATPase able to accept Ag^+ as substrate is CopA of *E. coli*. Expression of *copA* is not only induced by copper, but also by Ag^+ and Au^+ (Rensing et al. 2000; Outten et al. 2001; Stoyanov et al. 2003). Fan and Rosen (2002) showed the Ag^+ dependent phosphorylation of the conserved aspartate residue, forming a covalent acylphosphate intermediate which is a signature property of all P-type ATPases. Initial characterization of a *copA* knockout strain of *E. coli* in the presence of silver ions did not indicate a physiological role of CopA in silver tolerance. Recent studies show a reduced growth rate at concentrations below 0.2 μM Ag^+ and then complete inhibition, whereas the growth rate of the *E. coli* wild type was still about 50% of the maximum in the presence of 0.2 μM Ag^+ compared to the control (Stoyanov et al. 2003). Other examples of CPx-ATPases involved in transport of Cu^+ and Ag^+ are CopA of *Archaeoglobus fulgidus* (Arguello et al. 2003), CRD1 of *Candida albicans* (Riggle and Kumamoto 2000), and Bxa1 of *Oscillatoria brevis* (Tong et al. 2002). These proteins were first described as copper transporters but are now shown also to be able to transport silver ions. To what extent SilP and its close homologues are also able to transport Cu^+ in addition to Ag^+ has not yet been tested.

Between *silCFBA* and *silP* is a small uncharacterized open reading frame, *ORF105*. Homologues of the hypothetical gene product have been found in sequencing projects, where they are also part of annotated but not yet characterized *sil* determinants.

The first four genes of the *silCFBA*(*ORF105*)*P* transcript encode proteins to form a CBA efflux complex such as the ones involved in export of a wide variety of metal cations, xenobiotics, and drugs in Gram-negative bacteria (Paulsen et al. 1996; Nies 2003; see chapter by Nies, this volume). These efflux complexes consist of: a transport protein in the cytoplasmic membrane of the resistance nodulation cell division (RND) family (Saier et al. 1994; Tseng et al. 1999), SilA in this case; an outer membrane factor (OMF) (Paulsen et al. 1997), SilC; and a membrane fusion protein (MFP) (Saier et al. 1994), SilB. Together SilCBA are thought to form a complex spanning across both membranes. Silver ions are exported into the periplasm by SilP, from where they are transported into the surrounding medium via the SilCBA efflux complex (Fig. 1). In the current model, SilF is a periplasmic chaperon, transporting silver ions from SilP to the SilCBA complex (Fig. 1). Homologous systems are found on some other plasmids from enterobacteria. Like genes encoding SilP, the ones from sequencing projects, encoded on plasmids pLVPK and pK2044 of *K. pneumoniae*, pAPEC-O2-R from *E. coli*, and R478 of *S. marcescens*, are part of identified *sil* determinants which consist of all nine genes, such as *sil* from pMG101. Other *silA* homologues were identified via DNA–DNA hybridization and are on plasmids where the genes encoding SilP, SilE, and SilS are also located. Because of the operon structure of the silver resistance determinant, it seems likely that complete resistance systems are encoded on R476b, MIP233, pWR23, and MIP235 as well (Gupta et al. 2001).

However, analysis of clinical isolates of enterobacteria indicates that the genes encoding the two-component regulatory system and the CBA efflux complex are always present together, whereas *silP* and *silE* are occasionally missing (Silver 2003). If this is an indication, whether determinants are encoded chromosomally or on plasmids still has to be determined.

A quite different aspect of bacterial silver resistance was studied by Li et al. (1997) who described a proton dependent silver efflux in *E. coli* with no plasmids present. Analysis of the genomic sequence of *E. coli* strains identified a chromosomal determinant with homology to *sil* called *cus*. The proteins encoded by *cusRS* and *cusCFBA* are mainly involved in copper homeostasis (see chapter by Magnani and Solioz, this volume), but it was shown that *cus* confers a certain degree of silver resistance as well (Franke et al. 2001; Gupta et al. 2001).

3.2
Other "Strategies" of Silver Resistance

The best-understood resistance mechanism against silver ions is based on efflux, like the previously described silver resistance system of *S. typhimurium* pMG101. This strategy seems to be common in enterobacteria, since homologous systems are present on various clinical isolates of enterobacteria (Silver 2003). The silver resistance mechanism in Gram-positive bacteria (Solioz and Odermatt 1995), Archaea (Arguello et al. 2003), yeast (Riggle and Kumamoto 2000), and cyanobacteria (Tong et al. 2002) is also achieved by efflux. In contrast, some Gram-negative bacteria may have alternative strategies of handling high concentrations of silver ions.

Li et al. (1997) characterized an *E. coli* strain able to tolerate higher silver concentrations compared to the wild type. They found that deficiency in porins, either OmpF or OmpF plus OmpC, leads to a lower silver sensitivity of mutant strains. However, the lower permeability of the outer membrane alone does not explain the observed silver resistance of the mutants.

Changes in the outer membrane resulting in its lower permeability seem to be a general response to stepwise exposure to higher Ag^+ concentrations, as commonly used to isolate or create silver resistant bacteria. Using this method, Gupta et al. (1992) created a silver resistant mutant of *K. pneumoniae* showing changes in cell morphology and capsular composition, leading to higher silver tolerance (tenfold) but reduced virulence. Earlier studies also indicate a change in outer membrane composition as the reason for silver resistance of an experimentally derived silver resistant *K. pneumoniae* (Kaur and Vadehra 1986). In *P. aeruginosa*, silver resistance was the result of changes in capsular composition. However, at the same time it might result in higher pathogenicity (Vasishta et al. 1991).

The silver resistance mechanisms described so far are based on reduced uptake and/or active efflux of Ag^+. Another strategy for dealing with high concentrations of silver ions is the removal of silver ions from the medium. This observation was first reported by Charley and Bull (1979) with a multispecies community of bacteria. Increasing concentrations of Ag^+ during enrichment culture enabled the bacteria to tolerate up to 100 mM Ag^+, whereas a nonadapted culture was silver sensitive. Community analysis showed that about 50% of the bacteria were *Stenotrophomonas* (former name *Pseudomonas*) *maltophilia*, independent of growth in the absence or presence of silver ions. The other organisms of the culture were *Staphylococcus aureus* and a coryneform organism. Their proportion changes depending on the presence of silver, with *S. aureus* being highly silver sensitive. The Ag^+

adapted culture is able to accumulate silver and reduce the silver concentration in the medium. *S. maltophilia* is responsible for the removal of silver from the culture. However, bioaccumulation of silver by the multispecies community is higher than by *S. maltophilia* alone (Charley and Bull 1979). Since that time, a number of silver accumulating Gram-negative bacteria, e.g. *Pseudomonas*, *Thiobacillus*, *Cupridavidus*, and *Acinetobacter*, have been reported (Pooley 1982; Richards et al. 1984; Starodub and Trevors 1990; Deshpande and Chopade 1994; Klaus et al. 1999; Shakibaie et al. 1999; Patil and Paknikar 2000; Ibrahim et al. 2001; Shakibaie et al. 2003). Silver crystals were found in the bacterial envelope as elemental silver, as silver sulfide, or in crystals which also contain carbon, oxygen, phosphate, and chloride (Pooley 1982; Klaus et al. 1999; Ledrich et al. 2005). The mechanism of silver accumulation is not understood. However, several examples of a plasmid encoded mechanism are known. In *Cupridavidus metallidurans* CH34 the genetic information for silver resistance by accumulation is on the large plasmid pMOL30 (Ledrich et al. 2005). Plasmid pMOL30 contains a *sil* determinant similar to plasmid encoded silver resistance determinants of enterobacteria, which efflux the silver ions (Mergeay et al. 2003). To what extent these genes are involved in silver resistance of this organism is not known. In *A. baumannii* BL88 the presence of plasmid pUPI199 leads to silver accumulation and storage, whereas *E. coli* containing this plasmid releases silver (Deshpande and Chopade 1994). *A. baumannii* BL54 containing plasmid pUPI276 also accumulates silver. Here a cysteine-rich, silver-binding, 22-kDa protein was isolated, which was only present when the cells harboring pUPI276 were grown in the presence of Ag^+ (Shakibaie et al. 1999, 2003). In *P. diminuta* the presence of Ag^+ leads to the synthesis of two types of proteins with a probable role in silver binding: low molecular weight proteins released into the medium and cytoplasmic high molecular weight proteins (Ibrahim et al. 2001). The specific nature of these proteins needs to be analyzed.

3.3 Bacterial Silver Resistance: Where to Go from Here

The abundance of silver in the earth's crust is very low at 10^{-5}%. In some habitats, however, silver is present in much higher concentrations, due to either naturally occurring high local abundance or human introduction through industry. For more than 50 years bacteria that are able to tolerate "normally" toxic concentrations of silver ions have been known. Silver resistance by bacteria is a mixed blessing. On the one hand, the antibacterial character of silver ions has long been used in medicine to prevent wound infection. Yet, the occurrence of silver resistant pathogens, especially in a time of more and more multidrug resistant bacteria, is a disadvantage. On the other hand, some of these organisms can be used in bioremediation of silver-contaminated industrial waste water and silver recycling, as well as in silver mining. It remains the responsibility of science, medicine, and industry to seek better understanding of the effects of silver on microorganisms, human health, and the environment as well as to weigh the pros and cons of using silver in each specific application.

References

Arguello JM, Mandal AK, Mana-Capelli S (2003) Heavy metal transport CPx-ATPases from the thermophile *Archaeoglobus fulgidus*. Ann NY Acad Sci 986:212–218

Bowler PG, Jones SA, Walker M, Parsons D (2004) Microbicidal properties of silver-containing Hydrofiber(R) dressing against a variety of burn wound pathogens. J Burn Care Rehabil 25:192–196

Brady MJ, Lisay CM, Yurkovetskiy AV, Sawan SP (2003) Persistent silver disinfectant for the environmental control of pathogenic bacteria. Am J Infect Control 31:208–214

Bragg PD, Rainnie DJ (1974) The effect of silver ions on the respiratory chain of *Escherichia coli*. Can J Microbiol 20:883–889

Brett DW (2006) A discussion of silver as an antimicrobial agent: alleviating the confusion. Ostomy Wound Manage 52:34–41

Bridges K, Kidson A, Lowbury EJ, Wilkins MD (1979) Gentamicin- and silver-resistant *Pseudomonas* in a burns unit. Br Med J 1:446–449

Brooks WE (2004) In: Silver S (ed) Minerals yearbook, Vol I: Metals and minerals. US Government Printing Office, pp 68.1–68.3

Brown NL, Barrett SR, Camakaris J, Lee BT, Rouch DA (1995) Molecular genetics and transport analysis of the copper-resistance determinant (*pco*) from *Escherichia coli* plasmid pRJ1004. Mol Microbiol 17:1153–1166

Charley RC, Bull AT (1979) Bioaccumulation of silver by a multispecies community of bacteria. Arch Microbiol 123:239–244

Chen YT, Chang HY, Lai YC, Pan CC, Tsai SF, Peng HL (2004) Sequencing and analysis of the large virulence plasmid pLVPK of *Klebsiella pneumoniae* CG43. Gene 337:189–198

Choudhury P, Kumar R (1998) Multidrug- and metal-resistant strains of *Klebsiella pneumoniae* isolated from *Penaeus monodon* of the coastal waters of deltaic Sandarban. Can J Microbiol 44:186–189

Cooper R (2004) A review of the evidence for the use of topical antimicrobial agents in wound care. http://www.worldwidewounds.com/2004/february/Cooper/Topical-Antimicrobial-Agents.html

Davis IJ, Richards H, Mullany P (2005) Isolation of silver- and antibiotic-resistant *Enterobacter cloacae* from teeth. Oral Microbiol Immunol 20:191–194

Deshpande LM, Chopade BA (1994) Plasmid mediated silver resistance in *Acinetobacter baumannii*. BioMetals 7:49–56

Dibrov P, Dzioba J, Gosink KK, Hase CC (2002) Chemiosmotic mechanism of antimicrobial activity of Ag^+ in *Vibrio cholerae*. Antimicrob Agents Chemother 46:2668–2670

Fan B, Rosen BP (2002) Biochemical characterization of CopA, the *Escherichia coli* Cu(I)-translocating P-type ATPase. J Biol Chem 277:46987–46992

Franke S, Grass G, Nies DH (2001) The product of the *ybdE* gene of the *Escherichia coli* chromosome is involved in detoxification of silver ions. Microbiology 147:965–972

George N, Faoagali J, Muller M (1997) Silvazine(TM) (silver sulfadiazine and chlorhexidine) activity against 200 clinical isolates. Burns 23:493–495

Gilmour MW, Thomson NR, Sanders M, Parkhill J, Taylor DE (2004) The complete nucleotide sequence of the resistance plasmid R478: defining the backbone components of incompatibility group H conjugative plasmids through comparative genomics. Plasmid 52:182–202

Graham C (2005) The role of silver in wound healing. Br J Nurs 14:S22–S26

Gupta A, Matsui K, Lo J-F, Silver S (1999) Molecular basis for resistance to silver cations in *Salmonella*. Nat Med 5:183–188

Gupta A, Phung LT, Taylor DE, Silver S (2001) Diversity of silver resistance genes in IncH incompatibility group plasmids. Microbiology 147:3393–3402

Gupta LK, Jindal R, Beri HK, Chhibber S (1992) Virulence of silver-resistant mutant of *Klebsiella pneumoniae* in burn wound model. Folia Microbiol (Praha) 37:245–248

Haefeli C, Franklin C, Hardy K (1984) Plasmid-determined silver resistance in *Pseudomonas stutzeri* isolated from a silver mine. J Bacteriol 158:389–392

Ibrahim Z, Ahmad WA, Baba AB (2001) Bioaccumulation of silver and the isolation of metal-binding protein from *P. diminuta*. Braz Arch Biol Technol 44:223–225

Jelenko C 3rd (1969) Silver nitrate resistant *E. coli*: report of case. Ann Surg 170:296–299

Johnson TJ, Siek KE, Johnson SJ, Nolan LK (2005) DNA sequence and comparative genomics of pAPEC-O2-R, an avian pathogenic *Escherichia coli* transmissible R plasmid. Antimicrob Agents Chemother 49:4681–4688

Kaur P, Vadehra DV (1986) Mechanism of resistance to silver ions in *Klebsiella pneumoniae*. Antimicrob Agents Chemother 29:165–167

Klasen HJ (2000a) Historical review of the use of silver in the treatment of burns. I. Early uses. Burns 26:117–130

Klasen HJ (2000b) Historical review of the use of silver in the treatment of burns. II. Renewed interest for silver. Burns 26:131–138

Klaus T, Joerger R, Olsson E, Granqvist C-G (1999) Silver-based crystalline nanoparticles, microbially fabricated. Proc Natl Acad Sci USA 96:13611–13614

Lacorte JG, Loures JC, Monteiro E (1955) Resistance of influenza virus to the oligodynamic action of silver. Mem Inst Oswaldo Cruz 53:537–544

Ledrich M-L, Stemmler S, Laval-Gilly P, Foucaud L, Falla J (2005) Precipitation of silver thiosulfate complex and immobilization of silver by *Cupriavidus metallidurans* CH34. BioMetals 18:643–650

Li X, Nikaido H, Williams K (1997) Silver-resistant mutants of *Escherichia coli* display active efflux of Ag^+ and are deficient in porins. J Bacteriol 179:6127–6132

Lobo MB, Vasconcelos JV (1950) Resistance of *Mycobacterium tuberculosis* to oligodynamic action of silver. Rev Bras Tuberc Doencas Torac 18:647–654

Magnani D, Solioz M (2007) How Bacteria Handle Copper, in this volume. Springer, Heidelberg

McHugh GL, Moellering R, Hopkins C, Swartz M (1975) *Salmonella typhimurium* resistant to silver nitrate, chloramphenicol and ampicillin: a new threat in burn units? Lancet 305:235–240

Mergeay M, Monchy S, Vallaeys T, Auquier V, Benotmane A, Bertin P, Taghavi S, Dunn J, van der Lelie D, Wattiez R (2003) *Ralstonia metallidurans*, a bacterium specifically adapted to toxic metals: towards a catalogue of metal-responsive genes. FEMS Microbiol Rev 27:385–410

Mitra B, Rensing C (2007) Zinc, Cadmium and Lead Resistance and Homeostasis, in this volume. Springer, Heidelberg

Nies DH (2003) Efflux-mediated heavy metal resistance in prokaryotes. FEMS Microbiol Rev 27:313–339

Nies DH (2007) Bacterial Transition Metal Homeostasis, in this volume. Springer, Heidelberg

Outten FW, Huffman DL, Hale JA, O'Halloran TV (2001) The independent *cue* and *cus* systems confer copper tolerance during aerobic and anaerobic growth in *Escherichia coli*. J Biol Chem 276:30670–30677

Patil YB, Paknikar KM (2000) Biodetoxification of silver cyanide from electroplating industry wastewater. Lett Appl Microbiol 30:33–37

Paulsen IT, Brown MH, Skurray RA (1996) Proton-dependent multidrug efflux systems. Microbiol Rev 60:575–608

Paulsen IT, Park JH, Choi PS, Saier MH Jr (1997) A family of Gram-negative bacterial outer membrane factors that function in the export of proteins, carbohydrates, drugs and heavy metals from Gram-negative bacteria. FEMS Microbiol Lett 156:1–8

Pirnay J-P, De Vos D, Cochez C, Bilocq F, Pirson J, Struelens M, Duinslaeger L, Cornelis P, Zizi M, Vanderkelen A (2003) Molecular epidemiology of *Pseudomonas aeruginosa* colonization in a burn unit: persistence of a multidrug-resistant clone and a silver sulfadiazine-resistant clone. J Clin Microbiol 41:1192–1202

Pooley FD (1982) Bacteria accumulate silver during leaching of sulphide ore minerals. Nature 296:642–643

Rensing C, Ghosh M, Rosen BP (1999) Families of soft-metal-ion-transporting ATPases. J Bacteriol 181:5891–5897

Rensing C, Fan B, Sharma R, Mitra B, Rosen BP (2000) CopA: an *Escherichia coli* Cu(I)-translocating P-type ATPase. Proc Natl Acad Sci USA 97:652–656

Richards RM, Odelola HA, Anderson B (1984) Effect of silver on whole cells and spheroplasts of a silver resistant *Pseudomonas aeruginosa*. Microbios 39:151–157

Riggle PJ, Kumamoto CA (2000) Role of a *Candida albicans* P1-type ATPase in resistance to copper and silver ion toxicity. J Bacteriol 182:4899–4905

Rogers J, Dowsett AB, Keevil CW (1995) A paint incorporating silver to control mixed biofilms containing *Legionella pneumophila*. J Ind Microbiol 15:377–383

Saier MH Jr, Tam R, Reizer A, Reizer J (1994) Two novel families of bacterial membrane proteins concerned with nodulation, cell division and transport. Mol Microbiol 11:841–847

Shakibaie MR, Kapadnis BP, Dhakephalker P, Chopade BA (1999) Removal of silver from photographic wastewater effluent using *Acinetobacter baumannii* BL54. Can J Microbiol 45:995–1000

Shakibaie MR, Dhakephalker BA, Kapadnis BP, Chopade BA (2003) Silver resistance in *Acinetobacter baumannii* BL54 occurs through binding to a Ag-binding protein. Ind J Biotechnol 1:41–46

Silver S (2003) Bacterial silver resistance: molecular biology and uses and misuses of silver compounds. FEMS Microbiol Rev 27:341–353

Silver S, Phung L, Silver G (2006) Silver as biocides in burn and wound dressings and bacterial resistance to silver compounds. J Ind Microbiol Biotechnol 33:627–634

Solioz M, Odermatt A (1995) Copper and silver transport by CopB-ATPase in membrane vesicles of *Enterococcus hirae*. J Biol Chem 270:9217–9221

Solioz M, Vulpe C (1996) CPx-type ATPases: a class of P-type ATPases that pump heavy metals. Trends Biochem Sci 21:237–241

Spencer RC (1999) Novel methods for the prevention of infection of intravascular devices. J Hosp Infect 43:S127-S135

Starodub ME, Trevors JT (1990) Silver accumulation and resistance in *Escherichia coli* R1. J Inorg Biochem 39:317–325

Stoyanov JV, Magnani D, Solioz M (2003) Measurement of cytoplasmic copper, silver, and gold with a *lux* biosensor shows copper and silver, but not gold, efflux by the CopA ATPase of *Escherichia coli*. FEBS Lett 546:391–394

Tobin EJ, Bambauer R (2003) Silver coating of dialysis catheters to reduce bacterial colonization and infection. Ther Apher Dial 7:504–509

Tong L, Nakashima S, Shibasaka M, Katsuhara M, Kasamo K (2002) A novel histidine-rich CPx-ATPase from the filamentous cyanobacterium *Oscillatoria brevis* related to multiple-heavy-metal cotolerance. J Bacteriol 184:5027–5035

Tsai HC, Wu KM, Liao TL, Liu YM, Chen HJ, Chang YC, Chang CH, Kirby R, Chen C, Chen CWS, Chang HY, Fung CP, Wang JT, Tsai SF (2005) *Klebsiella pneumoniae* plasmid pK2044, complete sequence. Direct submission: http://www.ncbi.nlm.nih.gov/entrez/viewerfcgi?val=NC_006625

Tseng TT, Gratwick KS, Kollman J, Park D, Nies DH, Goffeau A, Saier MH Jr (1999) The RND permease superfamily: an ancient, ubiquitous and diverse family that includes human disease and development proteins. J Mol Microbiol Biotechnol 1:107–125

Vasishta R, Saxena M, Chhibber S (1991) Contribution of silver ion resistance to the pathogenicity of *Pseudomonas aeruginosa* with special reference to burn wound sepsis. Folia Microbiol (Praha) 36:498–501

Wright JB, Lam K, Burrell RE (1998) Wound management in an era of increasing bacterial antibiotic resistance: a role for topical silver treatment. Am J Infect Control 26:572–577

Yudkins J (1937) The effect of silver ions on some enzymes of *Bacterium coli*. Enzymologia 2:161–170

Microbiol Monogr (6)
D. H. Nies, S. Silver: Molecular Microbiology of Heavy Metals
DOI 10.1007/7171_2006_085/Published online: 17 February 2007

Mercury Microbiology: Resistance Systems, Environmental Aspects, Methylation, and Human Health

Simon Silver[1] (✉) · Jon L. Hobman[2] (✉)

[1]Dept. Microbiology and Immunology, University of Illinois at Chicago, M/C 790, Room E-704, 835 S. Wolcott Avenue, Chicago, IL 60612, USA
simon@uic.edu

[2]School of Biosciences, The University of Birmingham, Edgbaston, Birmingham B15 2TT, UK
J.L.Hobman@bham.ac.uk

Abstract Mercury has no beneficial biological role, and is highly toxic to all forms of life. Bacteria are involved in the global environmental cycling of mercury, both by reducing Hg^{2+} to metallic Hg^0, which is less soluble in aqueous systems and therefore less bioavailable, and by oxidizing and methylating Hg species, and in the process making Hg more bioavailable and more highly toxic. The most thoroughly studied bacterial biotransformation of mercury is reduction by the widely distributed *mer* resistance operons found on plasmids and transposons in Gram-negative and -positive bacteria. The products of these resistance operons transport ionic Hg^{2+} from outside the cell to the cellular cytoplasm, where mercuric reductase reduces divalent Hg^{2+} to Hg^0, which is less toxic than Hg^{2+}. Metallic mercury vapor, Hg^0, is volatile under aerobic conditions, leaves the cell by passive diffusion, and is volatilized from the growth environment. Sometimes, additional gene(s) determine organomercurial lyase, the enzyme that cleaves organomercurial compounds to inorganic Hg^{2+}, which is then reduced to Hg^0. Two types of *mer* operons ("narrow spectrum" with inorganic Hg^{2+} resistance and "broad spectrum" with both organomercurial and inorganic mercury resistances) confer high levels of resistance on host bacteria. The expression of *mer* resistance genes is primarily regulated by the MerR protein, which is the prototype of an increasing family of metal and other effector-responsive transcriptional activators. Methylation of inorganic Hg^{2+} to CH_3Hg^+ is thought to occur nonenzymatically (perhaps even extracellularly) with microbially synthesized *S*-adenosylmethionine as methyl donor.

1 Introduction

Resistance to mercury and organomercurials was the first studied and is still the best understood of toxic metal resistance systems. Other than that for arsenic, it might be the most widely found toxic inorganic ion resistance system and occurs in all bacterial divisions where it has been sought. Mercury resistance genes are frequently found in new microbial total genome sequences. The current best overview of bacterial mercury resistance is found in Barkay et al. (2003). Mercury resistance occurs widely in Gram-negative and -positive bacteria and in environmental, clinical, and industrial isolates, and mercury resistance genes are frequently found on plasmids and encoded by transposons. The mercury resistance transposon Tn*21* occupies about 8 kb of the 94-kb plasmid R100 (NCBI accessions NC_002134; gi: 9507549), the first multidrug resistance plasmid found in Japan 50 years ago, together with several antibiotic resistance determinants. However, chromosomal mercury (organomercurial) resistance is also common, for example in *Bacillus* isolates (Narita et al. 2003).

The mechanism of Hg^{2+} resistance is specific and unlike those for other toxic metal ions. In the *mer* resistance mechanism, Hg^{2+} is bound in the periplasm of Gram-negative bacteria or outer surface of Gram-positive bacteria by MerP, imported into the cytoplasm of the cell by specific membrane-bound transporters (three types have been found: MerT, MerC, and MerF), and then reduced from Hg^{2+} to Hg^0 by the enzyme mercuric reductase, MerA.

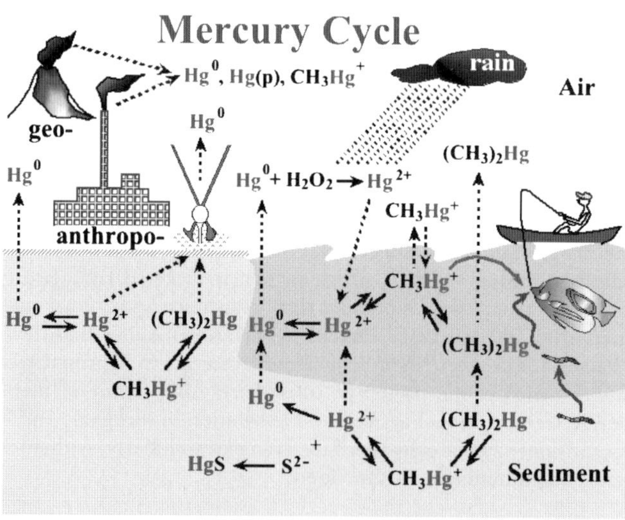

Fig. 1 The global mercury geocycle (modified from Barkay et al. 2003)

Along with abiotic and other microbial factors, the *mer* resistance mechanism contributes in major ways to the global Hg cycle. Hg^0 is oxidized to Hg^{2+} abiotically in the atmosphere or by the enzyme catalase (by bacteria and in red blood cells in animal lungs) and converted to methylmercury (CH_3Hg^+) by methyl cobalamin, perhaps predominantly in the natural environment by anaerobic sulfate-reducing bacteria. Thus, a global Hg geocycle (Fig. 1) functions primarily by microbial processes.

2
Mercury and Organomercurial Resistance Systems

The protein gene products and their functions for the range of genes found in mercury resistance *mer* operons in Gram-negative bacteria are shown in Fig. 2. Starting with Hg^{2+} (or organomercurials) outside the cell surface, the problem for the bacteria is to bring the toxic divalent mercury cations to the cytoplasmic enzyme (where they are detoxified), without allowing the toxic mercury to be free and cause damage. This is different from most microbial metal cation resistance systems that use inner-membrane efflux pump proteins to remove metals from the cytoplasm.

2.1
Mercury Uptake Systems

Counterintuitively, the *mer* resistance system has a Hg^{2+}-import system that binds extracellular Hg^{2+} and moves it along in a cascade of cysteine pairs on different proteins (called a thiol "bucket brigade") to the active site cysteines on mercuric reductase, located in the cytoplasm. No outer membrane protein has been associated with this transport process (Fig. 2A). However, *mer* resistance operons (Fig. 2B) encode the small MerP protein that binds Hg^{2+} in the periplasm (of Gram-negative bacteria) or cell surface (of Gram-positive bacteria). This protein (approximately 72 amino acids in length after processing of a leader peptide) has had its structure solved, and is made up of four antiparallel beta sheets plus two short alpha-helical regions and a "loop" containing the GlyMetAspCys33-AlaAla-Cys36Pro motif that binds Hg^{2+} in a linear S-Hg-S manner (Steele and Opella 1997; Miller 1999). (The Gram-positive bacterial MerP has a similarly located GlyMetAspCys60Cys61Pro sequence in an otherwise nonhomologous sequence.) MerP is thought to pass Hg^{2+} in rapid exchange with two of the four cysteines in the inner membrane protein MerT (Fig. 2A).

MerT is modeled as passing across the membrane three times as alpha-helical regions (Barkay et al. 2003; Wilson et al. 2000), with a cysteine pair in the first membrane helix thought to accept Hg^{2+} initially from MerP. Then a cytoplasmic cysteine pair may receive Hg^{2+} from the first pair (Wilson et al.

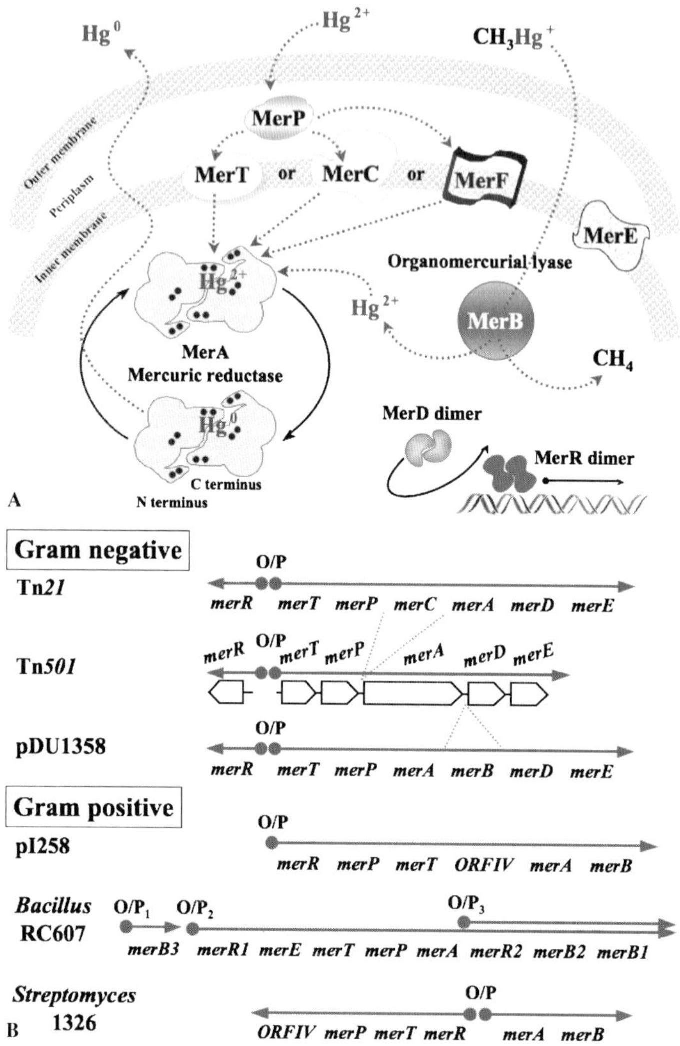

Fig. 2 Resistance by the *mer* operon. **A** The gene products. MerR is the transcriptional regulatory protein; MerA mercuric reductase enzyme; MerB organomercurial lyase enzyme; MerC and MerT alternative membrane uptake proteins; MerD a corepressor or chaperone involved in transcriptional regulation; MerE a hypothesized membrane protein that has not been studied; MerF an additional alternative (to MerT and MerC) membrane uptake protein. Transport across both the inner and outer membranes without indicted proteins are either by diffusion through the lipid bilayer or via yet unidentified proteins. **B** Gene organization of representative *mer* operons

2000). Since there is no ATP-binding motif, Hg^{2+} uptake across the inner cell membrane MerT protein may involve cysteine-to-cysteine transfer without overall energy coupling. While MerT is found most often in *mer* operon products, alternative membrane transport proteins MerC or MerF are also known. MerC crosses the membrane four times and MerF just twice, but all three of these proteins appear to function in similar transport processes (Wilson et al. 2000). A single cell (or operon) can contain more than one membrane transport protein (determinant). Finally, MerE is an additional membrane protein of currently unknown function (Fig. 2A).

2.2 Mercury Reductase

Once at the inner surface of the inner membrane, Hg^{2+} is thought to be transferred directly from the membrane MerT protein to mercuric reductase, rather than from small intercellular thiols such as glutathione (or mercaptoethanol, which is used for cell-free enzyme assays). This is thought to occur as yet another cysteine pair-to-cysteine pair exchange to the N-terminal cysteine pair of the large homodimeric enzyme mercuric reductase, MerA (Fig. 2). All recognized MerA sequences (with the single exception of that from *Streptomyces*) have this N-terminal domain once (or twice in *Bacillus*), which is closely homologous in sequence and structure (Ledwidge et al. 2005) and thought to function similarly to MerP. The N-terminal Hg^{2+} binding domain of mercuric reductase is often lost in proteolytic activity during expression or purification of the enzyme. It also is missing in the available crystal structure of mercuric reductase of *Bacillus* (Pai et al. 1991), suggesting that the domain lacked a fixed position within the crystal, but is present in the structure of the Gram-negative MerA from Tn*501* (Ledwidge et al. 2005). The *Bacillus* enzyme that was purified and solved for crystal structure has two fused back-to-back MerP-like sequences at its N-terminus. Yet the processed MerA lacking the N-terminal domain functions enzymatically in vitro, and the "donor" function of the N-terminus can only be seen in careful kinetic experiments (Engst and Miller 1998). It is hypothesized that the Hg^{2+} is next transferred by still another cysteine pair-to-cysteine pair exchange to the C-terminal cysteine pair (C558–C559 in Tn*21*-determined MerA numbering; Fig. 3A) of the MerA subunit. These cysteines are needed for cellular enzyme activity (Barkay et al. 2003), but can be bypassed with some Hg^{2+} adduct substrates (Engst and Miller 1999).

Hg^{2+} bound to mercuric reductase at the carboxyl-terminal Cys557–Cys558 of one subunit is then transferred by rapid thiol–thiol exchange to the Cys135–Cys140 thiol pair of the other monomer (Fig. 3A) (Barkay et al. 2003). Finally, the Hg^{2+} on to the active site cysteine pair Cys135–Cys140, is reduced by electron transfer from an FAD cofactor. The four Cys residues are close together in the protein structure (Pai et al. 1991) and additionally a Tyr residue is involved in the binding of Hg^{2+}. The enzymatic mechanism of mercuric reductase has been studied by Engst and Miller (1999). Mercuric reductase is an FAD-containing yellow flavoprotein, functionally related to glutathione re-

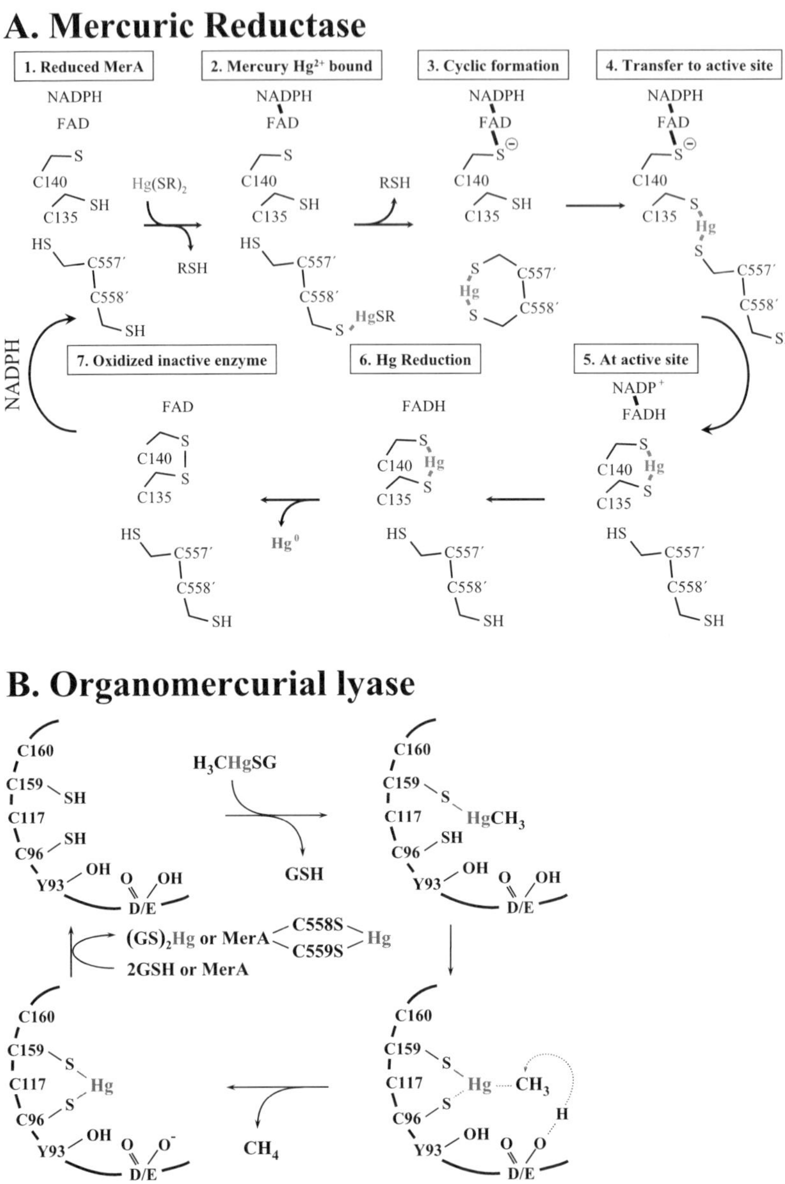

Fig. 3 Properties of the enzymes for mercury resistance. **A** Mercuric reductase. **B** Organomercurial lyase

ductase and lipoamide dehydrogenase (of wide occurrence in prokaryote and eukaryote cells).

2.3
Organomercurial Lyase

Organomercurial lyase is the small monomeric enzyme characteristic of "broad spectrum" mercury resistances that cleaves the Hg – C covalent bond releasing Hg^{2+} (the substrate of mercuric reductase) and reduced organic compounds, such as methane from methyl mercury or benzene from phenyl mercury (Figs. 2B and 3B) (Silver and Phung 1996; Barkay et al. 2003). Organomercurial lyase is encoded by the *merB* gene found only in a few *mer* operon-containing plasmids in *Escherichia coli*, but perhaps 50% in *Pseudomonas*, and in all *Staphylococcus aureus* "penicillinase" plasmids (Barkay et al. 2003; Silver and Phung 1996). The *mer* operons with the *merB* gene for organomercurial lyase are also found on the chromosome of methicillin-resistant *S. aureus* (MRSA) and on the chromosomes of *Bacillus* isolates from as different sources as North American polluted estuaries, Russian mine soil, and the sediment from Minamata Bay, Japan. Some Gram-negative bacteria have two versions of the *merB* gene and the bacilli frequently possess three *merB* genes in close chromosomal proximity. The MerB primary sequence is unusual, having no paralogue enzymes with related sequences but different substrates (Silver and Phung 1996; Pitts and Summers 2002). A wide range of MerB sequences have been identified (accessible in GenBank) and a wide range of organomercurial substrates are known, although we do not know in any case how sequence here determines substrate specificity. The enzymatic reaction for organomercurial lyase was shown to be a concerted proton attack on the Hg – C bond by an SE2 reaction mechanism (Begley and Ealick 2004; Barkay et al. 2003), after the organomercurial is initially bound to a cysteine thiol (Fig. 3B). It is unclear whether a dicarboxylic acid residue (Begley and Ealick 2004), a tyrosine, or a second cysteine thiol is the source of a proton that adds, for example, to methylmercury forming methane or to phenylmercury forming benzene. Recent mutagenesis and structural studies of organomercurial lyase (Pitts and Summers 2002; Benison et al. 2004) have led to the suggestion that a cysteine thiol might be the source of the proton that attacks the Hg – C bond. Thus, Fig. 3B is more explicit than in previous models (Begley and Ealick 2004; Barkay et al. 2003), but still tentative.

Three cysteines (Cys96, Cys117, and Cys159 in the MerB from plasmid R831 numbering) are conserved among organomercurial lyase sequences. Cys96 and Cys159 (but not Cys117) are required for full in vitro activity (Pitts and Summers 2002), leading to the proposal that Cys96 and Cys159 play essential roles in the reaction mechanism (Fig. 3B) (Pitts and Summers 2002; Benison et al. 2004) but that Cys117 (although completely conserved) may play a noncatalytic role (Benison et al. 2004; Barkay et al. 2003). The nonessential Cys160 is found in organomercurial lyases of Gram-negative bacteria (forming a vicinal cysteine pair with Cys159) but not Gram-positive bacteria, and the Cys215–Cys216 (*Bacillus* RC607 MerB1 numbering) form a C-terminal vicinal cysteine pair in organomercurial lyases of Gram-positive bacteria. Four steps are pictured (Fig. 3B), with the organomercurial (methylmercury as an example, although experimental work is generally done with less toxic mercurials) initially forming a thiol–mercury covalent bond with the invariant Cys159. A second cysteine thiol, from Cys96, possibly forms a bond with the Hg and the proton from Cys96 or from Tyr93, or from a conserved aspartate or glutamate (Asp98 is conserved in organomercurial lyases) sequence attacking the Hg – C bond, releasing the methane or other alkyl or aryl organic product (Benison et al. 2004; Barkay et al. 2003). The Hg is then bound to the organomercurial lyase (step 3) by the two thiols and the Hg^{2+} released in cell-free assays generally to added small thiols, such as glutathione, and in intact cells either to soluble thiols or perhaps directly transferred to the C-terminal vicinal cysteines of mercuric reductase (Fig. 3B) (Benison et al. 2004). Surprisingly, with the wide variety of available MerB sequences and genes, there has been little effort to determine differences in organomercurial substrate specificity among the wide range of aliphatic and aromatic organomercurials known to be substrates for the enzyme.

2.4 Regulatory Genes, *merR*, and Others

MerR, the regulator of mercuric cation resistance operons, was the first described and investigated member of the eponymously named family of metal responsive transcriptional activators, which now includes regulators for zinc, copper, cadmium, and lead (Permina et al. 2006), as well as regulators for nonmetal stress responses such as oxidative stress and drug responses (Hobman et al. 2005; Brown et al. 2003; chapters by Helmann et al. and by Tottey et al., in this volume). MerR regulators from both Gram-positive and -negative mercuric ion resistances are activators of approximately 140 amino acids (with the single exception of MerR from *Streptomyces lividans*, which is called MerR but is a sequence-unrelated transcriptional repressor). Unfortunately there is no crystal structure published for MerR, but structures of MerR homologues, especially CueR and ZntR, the regulators of *E. coli* copper and zinc efflux (Changela et al. 2003), have provided a model for a homodimeric protein. This protein has N-terminal winged helix DNA binding domains held together by long N-terminal coiled-coil alpha helices with three cysteine Hg^{2+} binding sites at the two ends of the intertwined alpha helices (Song et al. 2004; Barkay et al. 2003).

Additional experimental evidence from studies of MerR–promoter interactions, MerR site-directed mutants, and studies where small fragments of MerR have been characterized is consistent with the protein structures of MerR family regulators CueR and ZntR, and has led to the development of a model for how MerR regulates *mer* gene expression in response to the presence of divalent Hg^{2+} in the cytoplasm. Based on experimental data from the *Bacillus* RC607 MerR regulator and the MerR regulators from Tn*21* and Tn*501*, the current model for how MerR works is that a MerR homodimer binds to a dyad symmetrical DNA sequence between the -10 and -35 elements of the *mer* structural gene promoter, and recruits RNA polymerase to the promoter. The apo-MerR/RNA polymerase/promoter ternary complex slightly represses transcription from the promoter, by bending the promoter DNA. The *mer* promoter is suboptimal for gene expression because the spacing between the -10 and -35 sequences is unusually long at 19 bp, compared to the standard canonical promoter which has a spacing of 16–18 bp. The long spacer makes it nonideal for RNA polymerase open complex formation and transcription activation. Divalent mercuric ions bind to one cysteine (Cys82) in one of the MerR monomers and two cysteines (Cys117 and Cys126) in the other monomer of the *mer* promoter-bound MerR homodimer in an unusual trigonal coordination. The current model of MerR activation of *mer* gene expression, based on metal cation-containing structures of the related metal sensing MerR family regulators ZntR and CueR, and studies on the *mer* promoter, is that binding of the divalent Hg^{2+} to the MerR homodimer causes DNA distortion and an under-winding of the DNA in the region of the *mer* promoter that the MerR homodimer binds to, aligning the -10 and -35 sequences optimally for initiation of transcription from the *mer* promoter. This mechanism of repression of expression in the absence of Hg^{2+} and activation of expression in the presence of Hg^{2+} leads to a highly sensitive induction of transcription in response to increasing levels of mercury.

DNA sequence analysis of *merD* predicts that the MerD protein its N-terminal DNA-binding domain has some similarities to MerR. MerD was shown to coregulate *mer* promoter expression in earlier work. Recent work shows that MerD dissociates the Hg^{2+}-MerR/*merO*/*P* complex, allowing for the synthesis of new apo-MerR which can then bind to the *mer* promoter (Champier et al. 2004).

3
Environmental and Microbial Methylation of Mercury

Mercury is naturally released from the Earth's crust by erosion or volcanic activity. Human activities such as mining, manufacture, and burning of fossil fuels release arguably the majority of mercury that is newly entering the global mercury cycle, where it can undergo a large number of transformations.

Mercury has been used for over two millennia by humans, because it is relatively easy to extract from ores, such as mercuric sulfide (cinnabar), simply by crushing the ore and heating it to release Hg^0 vapor. Amongst the properties that have made mercury attractive for human use are that it readily makes amalgams with other metals, has antimicrobial properties, and has properties that have made it attractive to the electrical industry. Mercury has been used as a catalyst in vinyl chloride production, and very widely used in the chloralkali industry as a liquid electrode during brine electrolysis. Human activities contribute to the global mercury cycle by the release of mercury through mining, industrial or manufacturing use, or by the burning of fossil fuels. Mercury then enters the global mercury cycle as shown in Fig. 1, where it can become more or less biologically available, through both biotic and abiotic processes.

Although all forms of mercury are to a greater or lesser extent toxic (the exception to this has been thought to be metallic mercury, Hg^0, but ingestion of this over long time periods has adverse health effects (Yoong 2006)), the alkyl organic mercury compounds methylmercury and dimethylmercury are generally considered to be the most toxic. This is because they are readily absorbed by the gut, can cross the blood/brain barrier, bioaccumulate, and are not easily eliminated from the body (Gochfeld 2003). Organomercurials cause serious neurological disorders and their effect on neonates is more severe than on adults. The organomercurials also bioaccumalate and biomagnify so that top predators tend to carry much higher levels of organomercurials than organisms lower down the food chain.

Many fish contain organomercurial compounds at sufficiently high levels to be a health hazard to humans consuming them. The process by which methylmercury is accumulated in aquatic environments by fish is shown in Fig. 1, and involves the conversion of inorganic mercury into organic mercury which is accumulated up through the food chain. The process of environmental mercury methylation involves prokaryote microbes, with the first evidence that methanogenic Archaea were involved being shown by Wood and coworkers (Wood et al. 1968). Subsequently, sulfate-reducing bacteria in anaerobic sediments have also been shown to methylate mercury (Compeau and Bartha 1985), with vitamin B12 (cobalamin) acting as the methyl carrier in the methylation reaction in *Desulfovibrio desulfuricans* (Choi and Bartha 1993).

Alkyl (methyl and ethyl) mercury compounds were first synthesized in the laboratory in 1865, and killed two of the laboratory workers involved in the synthesis, after what became recognized as a characteristic delayed onset of symptoms associated with poisoning by these compounds (Nierenberg et al. 1998; Clarkson et al. 2003). The acute toxicity of alkyl mercurials and lack of effective treatments have led to several fatalities amongst laboratory workers since then. In one case of dimethylmercury poisoning that occurred as a single exposure incident in 1996 and resulted in the death of the laboratory worker the following year, the dimethylmercury permeated the protective gloves of the individual (Nierenberg et al. 1998; Blayney 2001). Despite the known acute toxicity of the alkyl mercurials, they, and in particular methylmercury, were widely used during the middle of the twentieth century in agriculture, paper making, leather production, and paint formulations because they inhibit the growth of bacteria, fungi, and algae. Alkyl mercurials were found to be highly effective seed-dressing treatments for inhibiting fungal growth on grain, both in storage and after planting. Unfortunately, the widespread use of organomercurial seed dressings led to both ecological problems and some of the worst cases of mass poisoning of humans by mercury compounds. Predatory birds were found to be suffering from organomercurial poisoning in Sweden in the 1950s, probably as a result of their ingestion of small mammals that had eaten treated grain (Clarkson et al. 2003). The list of alkyl mercury mass poisoning incidents and the numbers of affected people make grim reading, and have a common factor running through them. In each case the mass poisoning occurred because organomercurial treated wheat or barley seed, which was only intended for planting as a crop, was used directly for human consumption. During the 1950s and 1960s cases of accidental consumption of alkyl mercury treated cereal seed occurred in Iraq, Pakistan, and Guatemala. The worst reported case occurred in rural Iraq (1971–1972), where over 6000 people were affected and 500 died because they ate food containing flour produced from wheat or barley grains treated with methyl- and ethylmercury compounds (Bakir et al. 1973; Lenihan 1988), with subsequent reports suggesting that up to 40 000 people may have been affected (Clarkson et al. 2003). The highly toxic nature of organic mercury compounds is now well recognized, and their use has been severely restricted in recent years.

Perhaps the most notorious case of mass alkyl mercury poisoning is that centered around Minamata in Japan in the late 1950s/early 1960s, and the lesser-known case that occurred around the Agano River in Niigata, Japan, in 1965. In both cases exposure to methylmercury was again associated with food consumption by humans, but in this case the methylmercury was ingested through contaminated fish that had bioaccumulated methylmercury. The source of the mercury pollution in both cases was industrial effluent from acetaldehyde/vinyl chloride manufacture, in which Hg was used as a catalyst in the production of acetaldehyde from acetylene and was discharged into water bodies. Organomercurials may also have been released in effluents with the inorganic mercury, but as inorganic mercury is methylated in anaerobic sediments this is to some extent irrelevant. What is clear is that methylmercury bioaccumulated in the aquatic food chain and alkyl mercury contaminated seafood was consumed by humans and animals, which led to severe methylmercury poisoning with over 980 deaths recorded as a consequence in the Minamata incident.

4 Mercury and Medicine

Mercury metal or inorganic mercury compounds have been used for thousands of years in medicine for treating a wide variety of diseases and conditions, often with questionable efficacy and spectacularly serious side effects (Lenihan 1988). The use of liquid mercury metal as a treatment for constipation was no doubt effective, but the use of calomel (mercurous chloride Hg_2Cl_2) for treating syphilis was at best pointless ("A night with Venus and a lifetime with Mercury"), whilst more aggressive treatment using corrosive sublimate (mercuric chloride $HgCl_2$) was liable to severely or fatally poison the patient. Despite the introduction of the effective antisyphilitic organic arsenic compound Salvarsan at the beginning of the twentieth century, inorganic mercury treatments were still widely used on syphilitic patients until the introduction of penicillin in the 1940s. The toxic effects of mercury and its compounds on humans have been known for many hundreds of years, so the use of mercury compounds as pharmaceuticals, vaccine preservatives, disinfectants, and in dental amalgams has been carefully scrutinized, and when still in use will continue to be carefully scrutinized. An example of the association of a mercury-containing medicine with a medical condition and the withdrawal of that medicine was seen with pink disease (acrodynia), which was a sometimes fatal condition that was first recognized in the early twentieth century. It usually affected young children, and had a key diagnostic feature of painful, swollen, bright pink hands and feet (Black 1999; Weinstein and Bernstein 2003). In Anglo-Saxon countries this disease was recognized as being associated with the use of calomel (Hg_2Cl_2)-containing teething powders, whilst in continental Europe, particularly Switzerland, the disease was associated with the use of calomel preparations such as "Wurmschokolade" to treat ascariasis (Black 1999). Discontinuation of the use of calomel-containing nonprescription preparations has led to the virtual elimination of the disease, with sporadic cases being associated with the use of calomel-containing preparations from Asian countries (Weinstein and Bernstein 2003).

The study of microbial interactions with mercury, and the consequences of the use of mercury as an antimicrobial, have impacted on our understanding of antimicrobial resistance systems, bacterial genetics and horizontal gene transfer, biochemistry, environmental pollution, human health, and toxicology. Despite all of our advances in understanding, we do still wonder whether the biggest lesson of all about toxic metals has been learnt by the human race: when it comes to toxic metals "what goes around, comes around."

Acknowledgements This brief review summarizes efforts over decades. We dedicate this effort to K. Tonomura (Kyoto, Japan) and K. Izaki (Sendai, Japan), who were the first to find that microbes could volatilize mercury, and to John Wood (then at the University of Illinois) who initially advanced understanding of microbial mercury methylation. We and others followed. In particular, Nigel Brown (Bristol, then Birmingham, UK) and Anne O. Summers (St. Louis, later Athens, Georgia, USA) have been long-time major contributors, as have others listed in the references.

References

Bakir F, Damluji SF, Amin-Zaki L, Murtadha M, Khalidi A, Al-Rawi NY, Tikriti S, Dhahir HI, Clarkson TW, Smith JC, Doherty RA (1973) Methylmercury poisoning in Iraq. Science 181:230–241

Barkay T, Miller SM, Summers AO (2003) Bacterial mercury resistance from atoms to ecosystems. FEMS Microbiol Rev 27:355–384

Begley TP, Ealick SE (2004) Enzymatic reactions involving novel mechanisms of carbanion stabilization. Curr Opin Chem Biol 8:508–515

Benison GC, Di Lello P, Shokes JE, Cosper NJ, Scott RA, Legault P, Omichinski JG (2004) A stable mercury-containing complex of the organomercurial lyase MerB: catalysis, product release, and direct transfer to MerA. Biochemistry 43:8333–8345

Black J (1999) The puzzle of pink disease. J R Soc Med 92:478–481

Blayney MB (2001) The need for empirically derived permeation data for personal protective equipment: the death of Dr Karen E. Wetterhahn. Appl Occup Environ Hyg 16:233–236

Brown NL, Stoyanov JV, Kidd SP, Hobman JL (2003) The MerR family of transcriptional regulators. FEMS Microbiol Rev 27:145–163

Champier L, Duarte V, Michaud-Soret I, Coves J (2004) Characterization of the MerD protein from *Ralstonia metallidurans* CH34: a possible role in bacterial mercury resistance by switching off the induction of the *mer* operon. Mol Microbiol 52:1475–1485

Changela A, Chen K, Xue Y, Holschen J, Outten CE, O'Halloran TV, Mondragon A (2003) Molecular basis of metal-ion selectivity and zeptomolar sensitivity by CueR. Science 301:1383–1387

Choi SC, Bartha R (1993) Cobalamin-mediated mercury methylation by *Desulfovibrio desulfuricans* LS. Appl Environ Microbiol 59:290–295

Clarkson TW, Magos L, Myers GJ (2003) Human exposure to mercury: the three modern dilemmas. J Trace Elem Exp Med 16:321–343

Compeau GC, Bartha R (1985) Sulfate-reducing bacteria: principal methylators of mercury in anoxic estuarine sediment. Appl Environ Microbiol 50:498–502

Engst S, Miller SM (1998) Rapid reduction of Hg(II) by mercuric ion reductase does not require the conserved C-terminal cysteine pair using HgBr2 as the substrate. Biochemistry 37:11496–11507

Engst S, Miller SM (1999) Alternative routes for entry of HgX2 into the active site of mercuric ion reductase depend on the nature of the X ligands. Biochemistry 38:3519–3529

Gochfeld M (2003) Cases of mercury exposure, bioavailability, and absorption. Ecotoxicol Environ Saf 56:174–179

Helmann JD, Soonsanga S, Gabriel S (2007) Metalloregulators: Arbiters of Metal Sufficiency. Microbiol Monogr, Springer, Heidelberg, (in this volume)

Hobman JL, Wilkie J, Brown NL (2005) A design for life: prokaryotic metal-binding MerR family regulators. Biometals 18:429–436

Ledwidge R, Patel B, Dong A, Fiedler D, Falkowski M, Zelikova J, Summers AO, Pai EF, Miller SM (2005) NmerA, the metal binding domain of mercuric ion reductase, removes Hg^{2+} from proteins, delivers it to the catalytic core, and protects cells under glutathione-depleted conditions. Biochemistry 44:11402–11416

Lenihan J (1988) Mercury and Venus. In: The crumbs of creation. Taylor and Francis, Oxford (ISBN: 0852743904, p 388)

Lenihan J (1988) Mercury and Venus. In: The crumbs of creation. Taylor and Francis, Oxford (ISBN: 0852743904, pp 76–94)

Miller SM (1999) Bacterial detoxification of Hg(II) and organomercurials. Essays Biochem 34:17–30
Narita M, Chiba K, Nishizawa H, Ishii H, Huang CC, Kawabata Z, Silver S, Endo G (2003) Diversity of mercury resistance determinants among *Bacillus* strains isolated from sediment of Minamata Bay. FEMS Microbiol Lett 223:73–82
Nierenberg DW, Nordgren RE, Chang MB, Siegler RW, Blayney MB, Hochberg F, Toribara TY, Cernichiari E, Clarkson T (1998) Delayed cerebellar disease and death after accidental exposure to dimethylmercury. N Engl J Med 338:1672–1676
Permina EA, Kazakov AE, Kalinina OV, Gelfand MS (2006) Comparative genomics of regulation of heavy metal resistance in eubacteria. BMC Microbiol 6:49
Pitts KE, Summers AO (2002) The roles of thiols in the bacterial organomercurial lyase (MerB). Biochemistry 41:10287–10296
Schiering N, Kabsch W, Moore MJ, Distefano MD, Walsh CT, Pai EF (1991) Structure of the detoxification catalyst mercuric ion reductase from *Bacillus* sp. strain RC607. Nature 352:168–172
Silver S, Phung LT (1996) Bacterial heavy metal resistance: new surprises. Annu Rev Microbiol 50:753–789
Smith T, Pitts K, McGarvey JA, Summers AO (1998) Bacterial oxidation of mercury metal vapor, Hg(0). J Bacteriol 64:1328–1332
Song L, Caguiat J, Li Z, Shokes J, Scott RA, Olliff L, Summers AO (2004) Engineered single-chain, antiparallel, coiled coil mimics the MerR metal binding site. J Bacteriol 186:1861–1868
Steele RA, Opella SJ (1997) Structures of the reduced and mercury-bound forms of MerP, the periplasmic protein from the bacterial mercury detoxification system. Biochemistry 36:6885–6895
Tottey S, Harvie DR, Robinson NJ (2007) Understanding How Cells Allocate Metals. Microbiol Monogr, Springer, Heidelberg (in this volume)
Weinstein M, Bernstein S (2003) Pink ladies: mercury poisoning in twin girls. Can Med Assoc J 168:201
Wilson JR, Leang C, Morby AP, Hobman JL, Brown NL (2000) MerF is a mercury transport protein: different structures but a common mechanism for mercuric ion transporters? FEBS Lett 472:78–82
Wood JM, Scott-Kennedy F, Rosen CG (1968) Synthesis of methylmercury compounds by extracts of a methanogenic bacterium. Nature 220:173–174
Yoong KPC (2006) Heavy metal-meals of mercury. N Engl J Med 354:e3

Microbiol Monogr (6)
D. H. Nies, S. Silver: Molecular Microbiology of Heavy Metals
DOI 10.1007/7171_2006_086/Published online: 17 February 2007

Arsenic Metabolism in Prokaryotic and Eukaryotic Microbes

Hiranmoy Bhattacharjee · Barry P. Rosen (✉)

Department of Biochemistry and Molecular Biology,
Wayne State University School of Medicine,
540 E. Canfield Avenue, Detroit, MI 48201, USA
brosen@med.wayne.edu

Abstract This chapter will focus on recent progress on the mechanisms of metalloid uptake, metabolism, and detoxification in bacteria, archaea, and eukaryotic microbes. One of the initial challenges of the earliest cells would have been the ability to detoxify heavy metal ions, transition metal ions, and metalloids, including arsenic and antimony. The presence of arsenic resistance (*ars*) genes in the genome of by far most living organisms sequenced to date illustrates firstly that *ars* genes must be ancient and secondly that arsenic is still ubiquitous in the environment, providing the selective pressure that maintains these genes in present-day organisms. Some early cells also probably could use arsenite as an electron acceptor, giving selective pressure for the evolution of respiratory arsenate reductase. As atmospheric O_2 levels increased, arsenite was oxidized to arsenate abiotically. This provided an advantage for the evolution of arsenate reductases, some for arsenate respiration and energy production, and others for arsenate detoxification. Present-day selective pressure for metalloid resistance also comes from sources such as natural release of arsenic from volcanic activities, mining activities, the burning of coal, and other human activities. In addition is the use of arsenicals and antimonials as chemotherapeutic drugs for the treatment of parasitic diseases and cancer. Resistance to these drugs is becoming a major dilemma. Thus, an understanding of the molecular details of metalloid transport systems and detoxification enzymes is essential for the rational design of new drugs, and for treating drug-resistant microorganisms and tumor cells. Finally, this chapter will summarize recent identification of novel enzymes for arsenic reduction, oxidation, and methylation that expand the possibilities for metalloid metabolism and transformations.

1 Introduction: The Arsenic Geocycle

Arsenic is widely distributed in the Earth's crust and occurs primarily in four oxidation states: arsenate [As(V)], arsenite [As(III)], elemental arsenic [As(0)], and arsenide [As(–III)]. Volcanic eruptions are a source of human exposure to arsenic. Mining, copper smelting, coal burning, and other combustion processes also bring arsenic into our environment. Anthropogenic sources of arsenic include both inorganic and organic forms. Arsenic serves as an active ingredient in various commonly used herbicides, insecticides, rodenticides, wood preservatives, animal feeds, paints, dyes, and semiconductors.

Microbes play an important role in cycling arsenic between its various oxidation states (Fig. 1) (Mukhopadhyay et al. 2002). Inorganic arsenate entering the microbial cytosol through the phosphate transport system is reduced to arsenite, which is then extruded out of the cell, either through channels or secondary transporters (Rosen 2002). Arsenite is also generated by certain microbes that use arsenate as the terminal electron acceptor in anaerobic respiration (Oremland and Stolz 2003). These arsenate-respiring microbes can release arsenite from arsenate-rich sediments, leading to arsenic contamination of ground water (Oremland and Stolz 2005). Arsenite-oxidizing microbes utilize the reducing power from As(III) oxidation to gain energy for cell growth (Stolz et al. 2006). Microbes can also convert inorganic arsenic into gaseous methylated arsenide (Bentley and Chasteen 2002; Qin et al. 2006). However, whether microbes can metabolize arsenic salts to elemental arsenic remains to be determined. Marine microorganisms can convert inorganic arsenicals to various water- or lipid-soluble organic arsenic species. These include generation of di- and trimethylated arsenic derivatives (DMA, TMA), arsenocholine, arsenobetaine, arsenosugars, and arsenolipids. Arsenobetaine can be degraded to inorganic arsenic by microbial metabolism, completing the arsenic cycle in marine ecosystems (Dembitsky and Levitsky 2004).

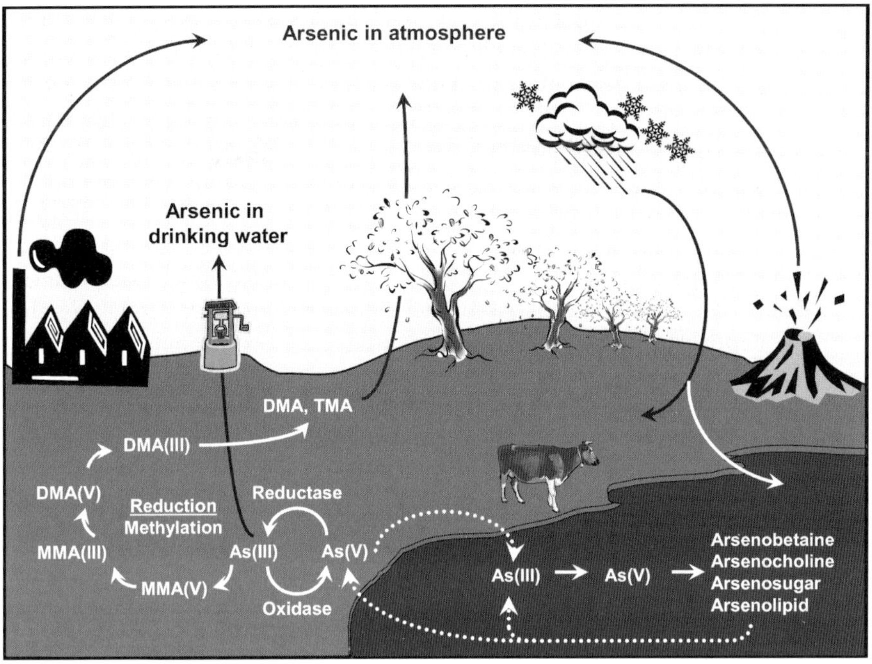

Fig. 1 The arsenic geocycle (see text for details)

2
Uptake Systems for Metalloid Oxyanions and Oxyacids: Phosphate Permeases, Aquaglyceroporin Channels, and Glucose Permeases

Inorganic arsenic has two biologically important oxidation states: pentavalent (As(V)) and trivalent (As(III)). In solution the pentavalent form, H_3AsO_4, exists as the oxyanion arsenate. As a solid, the unhydrated trivalent form is arsenic trioxide (As_2O_3). Reflecting a pK_a of 9.2, in solution arsenic trioxide is the undissociated acid, $As(OH)_3$. Even though it is not an oxyanion in solution, As(III) is frequently called arsenite, and so this term will be used interchangeably with arsenic trioxide in this chapter. Inorganic arsenic in both the pentavalent and trivalent oxidation states uses uptake systems that bring these toxic compounds into cells adventitiously.

Arsenate is an analogue of phosphate, and most organisms take up arsenate via phosphate transporters. In *Escherichia coli* there are two phosphate transporters, Pit and Pst (Rosenberg et al. 1977), both of which catalyze uptake of arsenate. Of the two, the Pit system is the major arsenate uptake system (Willsky and Malamy 1980a,b). In eukaryotic microbes arsenate is also taken up by phosphate transporters, although these are unrelated to the bacterial systems. For example, in *Saccharomyces cerevisiae*, several phosphate transporters catalyze arsenate uptake (Bun-ya et al. 1996).

Two pathways for cellular uptake of trivalent metalloids As(III) and Sb(III) have been identified (Fig. 2). In *E. coli*, GlpF was the first uptake system for As(III) and Sb(III) to be identified (Sanders et al. 1997; Meng et al. 2004). GlpF is the glycerol facilitator of *E. coli* (Heller et al. 1980). This gene was identified using a genetic screen mutant resistant to Sb(III) (Sanders et al. 1997). The assumption was that mutants unable to take up metalloids would become resistant, and the *glpF* mutant is highly resistant to Sb(III). Although As(III) transport is reduced by approximately 80% in the mutant (Meng et al. 2004), enough As(III) gets into the cells by as yet unidentified transporter(s) to keep the cells sensitive to As(III). GlpF was the first identified member of the aquaporin superfamily. The superfamily has two branches, the aquaporin channels that have small pores sufficient only to allow water conduction (Agre et al. 2002), and the aquaglyceroporins, which have larger pores of sufficient diameter to conduct neutral organic solutes such as glycerol and

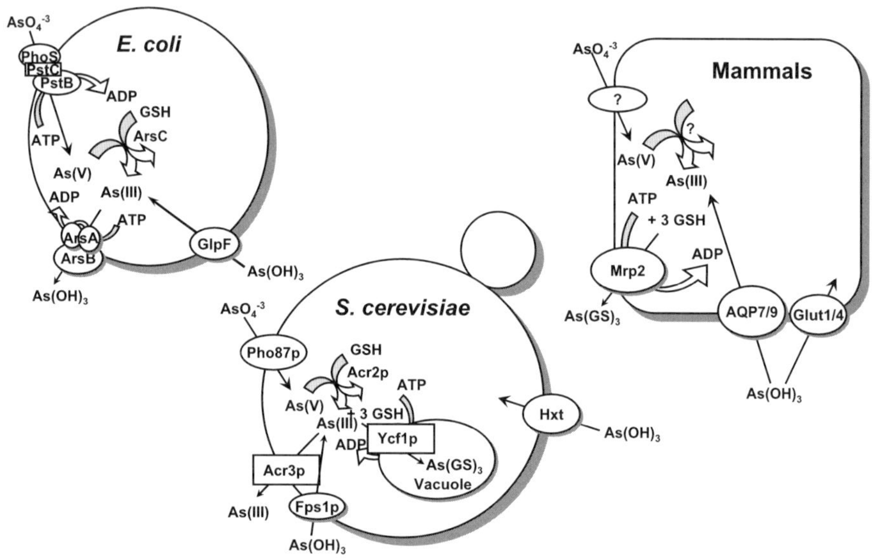

Fig. 2 Pathways of arsenical uptake and detoxification in prokaryotes and eukaryotes. Arsenate (As(V)) is taken up by phosphate transporters, while As(III) is taken up by aquaglyceroporins (GlpF in *E. coli*, Fps1p in yeast, and AQP7 and AQP9 in mammals) and hexose permeases (HXT1, HXT3, HXT4, HXT5, HXT7, or HXT9 in yeast, and GLUT1 and GLUT4 in mammals). In both *E. coli* and *S. cerevisiae*, arsenate is reduced to arsenite by the bacterial ArsC or yeast Acr2p enzymes. In both organisms, glutathione and glutaredoxin serve as the source of reducing potential. The proteins responsible for arsenate uptake and reduction in mammals have not yet been identified. In *E. coli*, arsenite is extruded from the cells by ArsB alone or by the ArsAB ATPase. In yeast, Acr3p is a plasma membrane arsenite efflux protein, and Ycf1p, which is a member of the MRP family of the ABC superfamily of drug-resistance pumps, transports $As(GS)_3$ into the vacuole. In mammals, Mrp isoforms, such as Mrp2, pump $As(GS)_3$ out of cells

urea (Borgnia et al. 1999). Most bacteria have GlpF homologues that render them sensitive to arsenite. Again, this is chiefly adventitious uptake of a toxic compound by a transporter with a physiological role in normal metabolism. However, in some organisms the gene for an aquaglyceroporin homologue, *aqpS*, is found in *ars* operons and appears to have evolved to confer arsenate resistance (Yang et al. 2005), as described in more detail below. As mentioned, arsenate is an analogue of phosphate, so it is easy to understand why it should be able to get into cells through phosphate transporters. It is not as obvious why arsenite should be taken up by channels for organic solutes such as glycerol. From the extended X-ray absorption fine structure (EXAFS) spectrum at neutral pH, the primary solution form of arsenic trioxide is $As(OH)_3$ (Ramirez-Solis et al. 2004) (Fig. 3). This polyhydroxylated arsenical is an inorganic analogue of glycerol, which indicates how it is recognized by GlpF.

In eukaryotic microbes aquaglyceroporin channels also conduct uptake of As(III) and Sb(III). Fps1p, the yeast homologue of GlpF, conducts arsenite uptake in *S. cerevisiae* (Wysocki et al. 2001). By functional complementation in yeast, we showed that mammalian aquaglyceroporins AQP7 and AQP9 catalyze uptake of trivalent metalloids (Liu et al. 2002). The capped RNA for the four human aquaglyceroporins was expressed in frog oocytes, and the results showed that they conduct As(III) in order of effectiveness as AQP9 $>$ AQP7 $\gg$ AQP3 $\gg$ AQP10 (Liu et al. 2004b). This has importance for human health and disease. First, arsenic contamination of drinking water is a health problem in countries such as India and Bangladesh, and considerable individual variability in sensitivity to arsenic is found (Acharyya et al. 1999). One reason for this variability could be differential expression of AQP9, the liver isoform, in those individuals. Also, there appears to be a relationship between arsenic sensitivity and nutrition, and expression of AQP9 has been shown to be elevated by nutritional restriction (Carbrey et al. 2003). We propose that these two facts are linked: inhabitants of villages in West Bengal and Bangladesh often have poor diets, which leads to elevation in AQP9 in the liver and hence higher rates of uptake of arsenite. Thus, these individuals are more sensitive to arsenite than inhabitants of cities who have better nutrition. Arsenic trioxide (Trisenox) is used clinically as a chemotherapeutic agent for the treatment of acute promyelocytic leukemia (Soignet et al. 1998; Soignet 2001). Expression of AQP9 in leukemia cells sensitizes the cells to Trisenox, leading to the proposal that differential expression of AQP9 in leukemia patients could lead to variable effectiveness of Trisenox (Bhattacharjee et al. 2004). The discovery of pharmacological agents that selectively increase expression of AQP9 in leukemia cells could make the drug more effective and selective.

Another eukaryotic microbe, the human pathogen *Leishmania major*, also takes up As(III) and Sb(III) by an aquaglyceroporin, LmAQP1 (Gourbal et al. 2004). This is also clinically relevant because the drug of choice for treatment

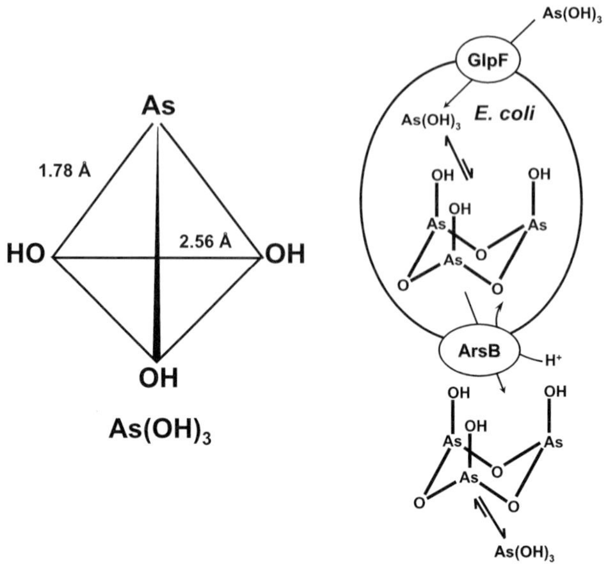

Fig. 3 The substrates of GlpF and ArsB. *Left*: Solution structure of arsenic trioxide. In solution at physiological pH, arsenic trioxide (As_2O_3) dissolves to form $As(OH)_3$, a trigonal pyramidal structure with three As – O bonds of 1.78 Å. *Right*: Postulated structures of the metalloid substrates of GlpF and ArsB. In *E. coli*, $As(OH)_3$ (or $Sb(OH)_3$) uptake is facilitated by the GlpF channel. Polymerization of three $As(OH)_3$ units is predicted to form a six-membered, oxo-bridged metalloid ring, with molecular similarity to a hexose. The ArsB antiporter exchanges this metalloid ring with positively charged H^+, thereby coupling efflux to the electrochemical proton gradient

of leishmaniasis is the pentavalent antimonial Pentostam. At least a portion of the drug appears to be reduced to Sb(III), the active form of the drug, in macrophages. The *Leishmania* amastigote, which resides in the phagolysosome of the infected macrophage, takes up Sb(III) by LmAQP1. Thus, it appears that arsenite gets into cells of most, if not all, organisms by aquaglyceroporins.

A second family of membrane transporters that catalyze arsenite uptake is the family of glucose permeases (Liu et al. 2004a). In the presence of glucose, most $As(OH)_3$ is taken up by yeast by the aquaglyceroporin Fps1p. However, in the absence of glucose, the Δ*fps1* strain exhibits only a 25% reduction in $^{73}As(OH)_3$ uptake compared to its parent, suggesting that the majority of $As(OH)_3$ accumulation in *S. cerevisiae* is via hexose permeases. The family of hexose permeases in *S. cerevisiae* is quite large, including 18 hexose transporters, Hxt1p to Hxt17p and Gal2p, and two glucose sensors, Snf3p and Rgt2p (Boles and Hollenberg 1997). These membrane proteins are members of the major facilitator superfamily (Marger and Saier 1993). A strain lacking *FPS1* and the genes for all 18 hexose permeases exhibits only residual uptake of arsenic trioxide, and expression of *HXT1, HXT3, HXT4, HXT5, HXT7*, or *HXT9* restored transport, which indicates that many of the hexose permeases can transport As(III). These results clearly demonstrate that glucose carrier proteins catalyze transport of trivalent arsenic. More recently mammalian GLUT1 and GLUT4, which are homologues of the yeast HXT hexose permeases, have also been shown to transport As(III) (Liu et al. 2006). Thus, hexose transporters are responsible for a portion of the uptake and most likely the toxicity of inorganic trivalent arsenic trioxide in mammals, including humans.

How do hexose permeases transport As(III)? Even though the majority species in solution is $As(OH)_3$, other forms may be present in lower amounts. One form may be a six-membered ring composed of a trimer of $As(OH)_3$ that has molecular similarity to hexoses (Meng et al. 2004) (Fig. 3). The Cambridge Structural Database contains 109 oxo-bridged As – O – As compounds, including ten with six-membered $(As-O)_3$ rings. In addition, the crystal structure of arsenious oxide, As_4O_6, is also a six-membered $(As-O)_3$ ring with the fourth As(III) coordinated to the three axial oxygens (Hamson and Stosick 1938). Thus, it is reasonable to consider that $As(OH)_3$ forms a cyclic trimer that is an inorganic analogue of glucose.

3 Arsenic Respiration: Arsenate Reductases and Arsenite Oxidases

Arsenate may serve as electron acceptor for anaerobic respiration processes, while arsenite may donate electrons to chemolithotrophic bacteria. The next two sections describe the enzymes responsible for the respective chemical reactions.

3.1 Arsenate Reductases

Anaerobic cell growth in several bacterial species is linked to energy generation during the reduction of arsenate to arsenite. For example, the bacterium *Chrysiogenes arsenatis* respires anaerobically using arsenate as the terminal electron acceptor and acetate as the respiratory electron donor (Krafft and Macy 1998). During growth, arsenate is reduced to arsenite, the reduction being catalyzed by an arsenate respiratory reductase (Arr). It is believed that energy is conserved via electron-transport-linked phosphorylation, with the arsenate reductase functioning as a terminal reductase, which is coupled to an electron-transport chain in the membrane. The soluble *C. arsenatis* arsenate reductase is a periplasmic, heterodimeric protein that consists of two subunits with molecular masses of 87 kDa (ArrA) and 29 kDa (ArrB). Sequence analysis indicates that ArrA is a Mo/Fe protein that shows sequence homology to the large dimethyl sulfoxide (DMSO) reductase class of proteins. ArrB is believed to contain a [4Fe – 4S] cluster that may be involved in the transfer of electrons to the molybdenum cofactor of the ArrA subunit. The K_m for arsenate is 0.3 μM, and the V_{max} is 7 millimoles arsenate reduced per minute per milligram of protein. Fumarate, nitrate, sulfate, and selenate could not serve as alternative electron acceptors for the arsenate reductase. The molecular details and mechanism of arsenate reduction and energy coupling remain to be determined.

Arsenate respiratory reductases have also been purified and characterized from *Shewanella* sp. strain ANA-3 (Saltikov and Newman 2003) and *Bacillus selenitireducens* (Afkar et al. 2003), although their structural details are still unknown. *Shewanella* ArrA is predicted to be a Mo/Fe protein with a molecular mass of 95.2 kDa, which is homologous to the DMSO reductase family, and ArrB is predicted to encode a 25.7-kDa iron–sulfur protein. A gene-encoded Tat (twin arginine translocation) motif at the N-terminus of ArrA suggests that the arsenate reductase complex is exported to the periplasm (Saltikov and Newman 2003). The arsenate respiratory reductase from *B. selenitireducens* strain MLS10 is a heterodimer of 150 kDa and is composed of ArrA (110 kDa) and ArrB (34 kDa), again with a putative Tat signal in the gene for ArrA (Afkar et al. 2003), similar to what has been found in *Shewanella* ArrA (Saltikov and Newman 2003), indicating that the enzyme may have an extracytoplasmic location. These arsenate respiratory reductases show significant sequence homology with each other.

While the periplasmic arsenate respiratory reductases are linked to energy metabolism, they are quite different from the cytosolic ArsC arsenate reductases, which play an important role in arsenic detoxification. Both *arr* and *ars* systems are expressed in *Shewanella* sp. strain ANA-3 under different growth conditions (Saltikov et al. 2003). While *arr* is only expressed anaerobically, *ars* is expressed under both aerobic and anaerobic conditions. The *arr* system was activated by a 1000-fold lower concentration of As(III) than that required for the *arsC* system. Under anaerobic conditions, *arr* transcription is activated at nanomolar concentrations of either As(V) or As(III). During the early phases of growth, *Shewanella* preferentially couples the reduction of As(V) to growth, instead of detoxification by the *ars* system. Over the course of growth, as As(III) accumulates to toxic levels, *Shewanella* switches on the *ars* operon to facilitate arsenic resistance.

The arsenate respiratory reductases play an important role in arsenic geochemistry and may lead to arsenic contamination of drinking water supplies. As(V) can bind to minerals commonly found in the environment, such as ferrihydrite and alumina, which can limit its mobility and bioavailability. However, microbial reduction of As(V) to As(III) can facilitate the release and transport of this toxic metal into aquatic environments and into drinking water (Harvey et al. 2002; Islam et al. 2004).

3.2
Arsenite Oxidases

Arsenite oxidase activity has been reported in heterotrophic as well as in chemoautotrophic microorganisms. The best-characterized arsenite oxidase is from the soil pseudomonad *Alcaligenes faecalis* (Anderson et al. 1992; Ellis et al. 2001). The *A. faecalis* arsenite oxidase is a 100-kDa heterodimer of a large and a small subunit (Anderson et al. 1992; Ellis et al. 2001). The large catalytic subunit of 825 residues harbors the molybdenum (Mo) center, the Mo atom being bound to two pterin cofactors and a [3Fe – 4S] cluster, and is structurally related to the DMSO reductase family of molybdenum enzymes and to the Arr arsenate reductases. The small subunit of approximately 134 residues contains a Rieske-type [2Fe – 2S] cluster and is homologous to the Rieske protein domains of cytb complexes and dioxygenases. The protein is likely transported across the cytoplasmic membrane via the Tat protein export pathway, and most likely remains attached to the outer surface of the inner membrane by the N-terminal transmembrane helix of the small Rieske subunit (Lebrun et al. 2003). The catalytically active form of the *A. faecalis* arsenite oxidase has an oxidized Mo(VI) center which is solvent accessible via a hydrophilic channel. Several residues, including His195, Glu203, Arg419, and His423, line the base of the channel and form the arsenite binding site. The minimal mechanism of arsenite oxidation by the *A. faecalis* arsenite oxidase consists of the following steps (Ellis et al. 2001). After arsenite binds at the substrate binding site, it executes a nucleophilic attack on the Mo = O group of the oxidized Mo(VI) center. This reaction yields reduced Mo(IV) coordinated with arsenate. Subsequently, arsenate is released upon its dissociation from the reduced enzyme. Finally, following the addition and deprotonation of water, the reduced enzyme is transformed back to the active form by regenerating the Mo(VI)-oxo group. The reduced electrons generated during the course of the reaction are channeled through the [3Fe – 4S] and [2Fe – 2S] clusters into the periplasmic electron acceptors, perhaps azurin and cytochrome *c*.

Arsenite oxidase genes have also been cloned from the β-proteobacterial strain ULPAs1 (Muller et al. 2003), the heterotrophic *Hydrogenophaga* sp. strain NT-14 (van den Hoven and Santini 2004), and the chemolithoautotrophic bacterium NT-26 (Santini and van den Hoven 2004). Each of these proteins shows significant sequence homology to the *A. faecalis* enzyme. Contrary to ULPAs1 (Muller et al. 2003) and *A. faecalis* arsenite oxidase that are involved in arsenic detoxification (Anderson et al. 1992; Ellis et al. 2001), either of the NT-14 and NT-26 enzymes is implicated in metabolism, as these strains utilize the reducing power from As(III) oxidation to gain energy for cell growth (Santini et al. 2000; Santini and van den Hoven 2004; van den Hoven and Santini 2004).

4 Metalloid Resistances

Arsenate, arsenite, and antimonite are detoxified by an interplay of redox, transport, sequestration, and covalent modification reactions.

4.1 Regulation of Metalloid Detoxification by the ArsR/SmtB Family of Metalloregulatory Proteins

Expression of the genes for resistance to arsenic as well as toxic metals and metalloids is nearly always subject to transcriptional regulation by the toxic ions themselves. One large group is the ArsR/SmtB family of metal(loid)-responsive transcriptional repressors. The 117-residue As(III)/Sb(III)-responsive ArsR repressor encoded by the *arsRDABC* operon of *E. coli* plasmid R773 (Wu and Rosen 1991, 1993; Shi et al. 1994) and the 122-residue Zn(II)-responsive SmtB repressor (Huckle et al. 1993; Morby et al. 1993) were the first identified members of the ArsR/SmtB family of small metalloregulatory proteins. Since the initial reports, the number of identified members of this family has grown to 198, with 192 homologues in Gram-positive and -negative bacteria, and six homologues in archaea. These include proteins that respond to As(III)/Sb(III) (ArsR) (Wu and Rosen 1991), Pb(II)/Cd(II)/Zn(II) (CadC) (Endo and Silver 1995), Cd(II)/Pb(II) (CmtR) (Cavet et al. 2003), Zn(II) (SmtB and ZiaR) (Morby et al. 1993), and Co(II)/Ni(II) (NmtR) (Cavet et al. 2002). ArsR, CadC, and SmtB (and, by extrapolation, all members of the family) are homodimers that repress transcription by binding to DNA in the absence of inducing metal ion. They dissociate from the DNA when metal is bound, resulting in expression of metal ion resistances.

What is the basis of metal ion binding and specificity among members of the ArsR/SmtB family? Selectivity cannot be inferred from genome sequence and annotation, so how can As(III)-responsive regulators be distinguished from those that respond to other metals? We made the assumption that any homologue within or adjacent to putative *ars* operons or arsenic resistance genes responds to As(III), and chose three for further analysis. The first is the well-characterized R773 ArsR, while the other two lack the As(III) binding site of the R773 protein, and yet have been shown to be As(III) responsive. In each subunit of the R773 ArsR repressor there is a metal binding domain consisting of Cys32 and Cys34, which are ligands for As(III) and Sb(III) (Shi et al. 1994). Cys37 is a third ligand, although it is neither required for induction nor conserved in some homologues. This illustrates that a vicinal cysteine pair is sufficient for strong binding of As(III). Consistent with this chemistry, the best inducer of the operon is phenylarsine oxide, which can accommodate only two protein ligands because of the phenyl ring. Note that other types of ligands for As(III) bind much more poorly, so it is unlikely that ligands such as carboxylates, serine hydroxyls, or histidine imidazole nitrogens, which form only weak As(III) binding sites, could participate. Thus, in As(III)-responsive repressors, it is reasonable to expect that a pair of cysteine residues will form the metalloid binding site.

Although the structure of R773 ArsR has not been determined, the structure of the CadC homologue has been solved (Ye et al. 2005), and this structure was used to model other members of the ArsR/SmtB family. In ArsR, the inducer binding site is located in the $\alpha 3$ helix, which is the first helix of the helix–loop–helix DNA binding domain. We hypothesize that distortion of the DNA binding domain upon inducer binding results in dissociation from the operator/promoter site and transcription of the resistance genes (Fig. 4).

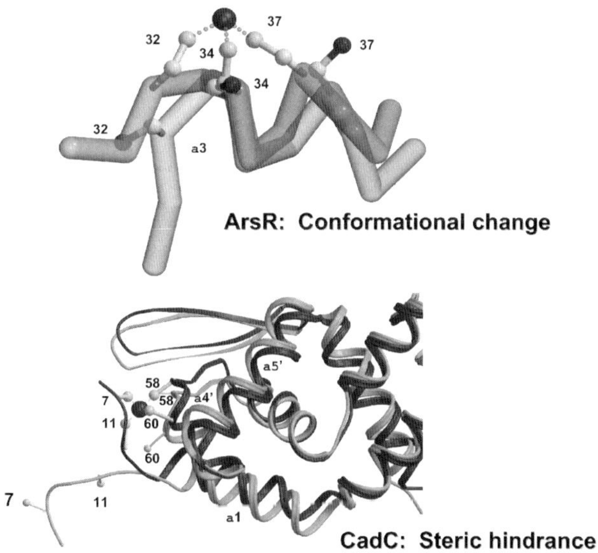

Fig. 4 Models for metalloregulation by the ArsR and CadC repressors. *Top* (ArsR): The apo-ArsR homodimer binds to the operator/promoter region, through a helix–turn–helix domain in each monomer, repressing transcription of the *ars* operon. The sulfur atoms of the three cysteines, Cys32, Cys34, and Cys37, are linearly arrayed along the first helix of the helix–loop–helix DNA binding domain, with more than 10 Å separating Cys32 from Cys37. From EXAFS and biochemical data, As(III) has been shown to bind to Cys32, Cys34, and Cys37 through metal–sulfur bonds, bringing the three sulfur atoms to 2.25 Å from the bound As(III). This is predicted to produce a substantial conformational change in that helix compared with the apo-repressor, resulting in dissociation from the operator/promoter site and transcription of the resistance genes. *Bottom* (CadC): Metalloregulation by CadC is predicted to follow a different mechanism. In this case, binding of Cd(II) to the sulfur atoms of Cys58 and Cys60 from one subunit of CadC, and Cys7 and Cys11 from the other subunit, brings the N-terminus of CadC into close contact with $\alpha 4$, the first part of the helix–loop–helix DNA binding motif, making it inaccessible to the DNA. Thus, binding of Cd(II), Pb(II), or Zn(II) to CadC could sterically block interaction with the *cad* operator/promoter

This idea is based on several lines of evidence. First, from EXAFS analysis, the distance between As(III) and each of the three cysteine thiolates was shown to be 2.25 Å (Shi et al. 1996). From this and the known structures of small-molecule As(III)-thiol compounds, the sulfur-to-sulfur distances would be predicted to be 3.5 Å. In the model of the ArsR aporepressor, the sulfur atoms of the three cysteines are linearly arrayed along the first helix of the DNA binding domain, with more than 10 Å from Cys32 to Cys37 (Fig. 5). To be consistent with the EXAFS results, binding of As(III) must bring the cysteines into proximity, producing a large conformational change in this helix (Fig. 4).

Why does R773 ArsR not bind divalent soft metals such as Cd(II)? From a combination of results from molecular genetics (Sun et al. 2001), biochemistry (Wong et al. 2002), and EXAFS (Busenlehner et al. 2001), the inducer binding domain in CadC was shown to be an S_4 site composed of four cysteine residues, Cys7, Cys11, Cys58, and Cys60, each 2.53 Å from a bound Cd(II) (Fig. 4). Like Cys37 of ArsR, Cys11 is not required for induction. Thus, the location of the inducer binding sites in ArsR and CadC is conserved,

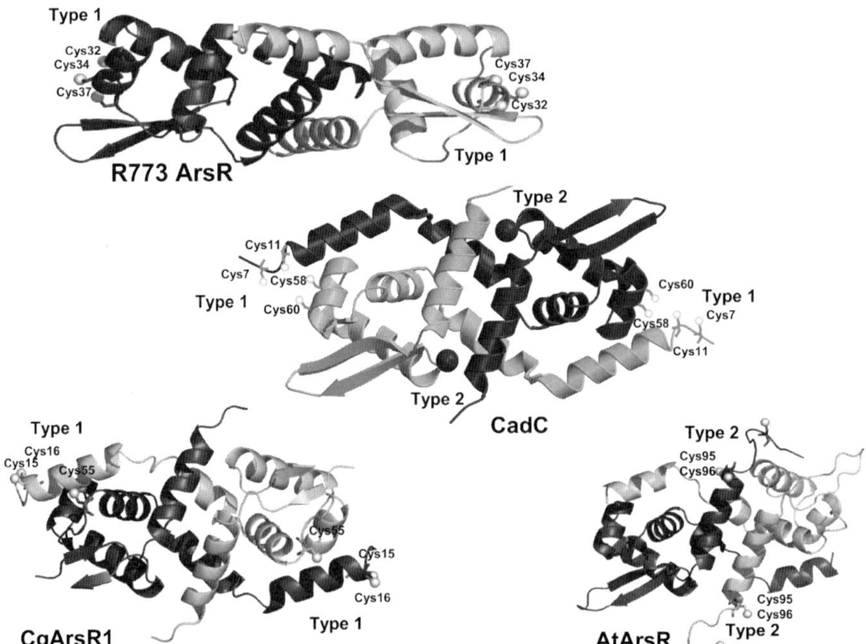

Fig. 5 Model of ArsR aporepressors based on the structure of *Staphylococcus aureus* pI258 CadC. R773 ArsR, CgArsR1, and AtArsR have been modeled based on the structure of pI258 CadC (*center*). The CadC dimer is shown as a ribbon diagram with secondary structural units N-α1-α2-α3-β1-α4-α5-β2-β3-α6-C. Each of the type 1 and type 2 metal binding sites in CadC is composed of two residues from each of the monomers

even though CadC does not respond to As(III) or Sb(III) (Endo and Silver 1995; Tauriainen et al. 1998; Sun et al. 2001) (although it does respond to Bi(III) (Busenlehner et al. 2002)), and ArsR does not respond to Cd(II), Pb(II), or Zn(II) (Wu and Rosen 1993). From modeling the inducer binding sites of ArsR and CadC with bound metals, the mechanism of derepression can be predicted to differ between the two (Fig. 4). While binding of As(III) is proposed to distort the α3 helix of ArsR, causing loss of DNA binding, the binding of Cd(II) to both the α1 and α4 helices of CadC results in the α1 helix sterically hindering the DNA binding site so that the repressor is no longer able to bind to its operator/promoter site.

Both the number and spatial location of available protein ligands are critical for selectivity. ArsR has an S_3 binding contributed by three cysteine residues of a single subunit (Shi et al. 1996), while the site in CadC is S_4, with two of the cysteine thiolates contributed by one subunit and the other two contributed by the other subunit (Busenlehner et al. 2001). CadC has an N-terminal extension that is absent in ArsR. This sequence contains Cys7 and Cys11. The results of cross-linking experiments demonstrated that each of the two binding sites is composed of Cys7 and Cys11 from one subunit and Cys58 and Cys60 from the other (Wong et al. 2002). While this N-terminal sequence is not visible in the SmtB structure, the 1.9-Å structure of CadC extends to residue 10 and clearly shows that the N-terminus of one subunit is located adjacent to α1 of the DNA binding domain of the other subunit (Ye et al. 2005) (Fig. 5). The inducer binding sites located in the DNA binding region have been called type 1 sites, to contrast them with the type 2 sites found in SmtB, which is in a totally different location. These sites have also been called α3N and α5 (VanZile et al. 2000, 2002b), based on a crystal structure of SmtB in which the putative α1 helix is not visible (Cook et al. 1998). If SmtB has an α1 helix, the metal sites would actually be α4N and α6.

The metal binding properties of SmtB have been analyzed in detail by a combination of UV/visible and X-ray absorption spectroscopy (VanZile et al. 2000, 2002a,b). The two metal binding sites in SmtB are at the dimer interface and not in the DNA binding domain, although it has an apparent vestigial type 1 site, with one cysteine corresponding to Cys32 of R773 and another in the other subunit similar to Cys11 of CadC. The type 2 site in SmtB has four ligands, two from one subunit (Asp104 and His106) and two from the other subunit (His117 and Glu120). While there is no trace of a metal binding site at the dimer interface of ArsR, in the CadC structure there are two Zn(II) bound at the dimer interface in sites composed of Asp101 and His103 from one subunit and His114 and Glu117 from the other (Fig. 5). Thus, CadC has the four residues corresponding to the SmtB type 2 site, but this type 2 site is not a physiological inducer binding site (Wong et al. 2002). SmtB has only a type 2 site, while ArsR has only a type 1 site. CadC has both type 1 and type 2 sites. In addition, the CadC type 1 site is more complicated than the type 1 site of R773 ArsR. The two CadC type 1 sites are each formed between the two subunits of the dimer, while the two ArsR type 1 sites are composed of three residues from a single subunit. It is reasonable to consider that the simple intrasubunit site in the common ancestor of these repressors arose first, and a more complicated intersubunit site arose later. Thus, the ArsR type 1 site is closer to the ancestral site than the CadC type 1 site. ArsR has no type 2 site, CadC has a type 2 site that is not used for metalloregulation, and SmtB has a regulatory type 2 site, suggesting that type 2 sites evolved subsequently to type 1 sites. The presence of a vestigial type 1 site in SmtB is consistent with this concept of sequential evolution of two types of metal sites.

However, the situation appears still more complicated. Two other As(III)-responsive ArsR repressors lack cysteine residues corresponding to Cys32, Cys34, or Cys37 of R773 ArsR, indicating that they do not have a simple type 1 As(III) binding site. The first atypical ArsR controls the *ars* operon of *Acidothiobacillus ferrooxidans* (Butcher and Rawlings 2002). Phylogenetically, this AtArsR is more closely related to CadC and SmtB than to the R773 ArsR and lacks any hint of a type 1 site, with no cysteine residues in the DNA binding domain. It has two cysteine residues, Cys95 and Cys96, at the C-terminal end of each subunit and binds one As(III) per subunit (Qin and Rosen, unpublished results). Mutation of Cys95 and/or Cys96 eliminates metal binding. From EXAFS analysis, As(III) is bound in a mixed S and O/N environment. From this result, we propose that As(III) is bound to the thiolates of Cys95 and Cys96, with the third ligand as a hydroxyl. Modeling AtArsR on CadC, Cys95 and Cys96 are congruent with the type 2 metal binding site of CadC (Fig. 5). However, neither corresponds to the exact residues found in the type 2 site of CadC, indicating that this site is not derived from a CadC-like site simply by two substitutions. More importantly, this type 2 site is composed of residues from a single subunit, and is a simpler type 2 site than the intrasubunit type 2 site of CadC.

In *Corynebacterium glutamicum* ATCC 13032 two *ars* operons are controlled by two closely related ArsRs (Ordonez et al. 2005). Both CgArsRs respond to As(III) (Thiyagarajan and Rosen, unpublished results), but they are quite divergent from the R773 ArsR in primary sequence. They have the N-terminal extension characteristic of CadC-like repressors but do not have cysteine residues corresponding to either the ArsR or CadC type 1 sites. Instead they have two cysteine residues in the first α helix and one cysteine just N-terminal to the putative DNA binding domain. While there are no experimental structure–function analyses yet on the role of these residues in As(III) binding, a model of CgArsR1 built on the CadC structure is quite instructive: the two α1 cysteines from one subunit are juxtaposed to the cysteine in the DNA binding domain of the other subunit (Fig. 5). This intrasubunit type 1 site is exactly what would be predicted for an S_3 As(III) binding site. It may resemble the complex intrasubunit type 1 site of CadC, but the locations of the three cysteines in CgArsR1 are different from those of the four cysteines of the S_4 type 1 site of CadC.

What can we conclude from this comparative analysis of ArsR repressors? First, the inducer binding site can be located either near the DNA binding domain or the dimer interface. Second, it is formed by placement of pairs or triads of cysteine residues. Third, it can be formed within a single subunit or between subunits. Fourth, and very interestingly, the As(III) binding sites of R773 ArsR, AtArsR, and CgArsR appear to be the result of three independent and relatively recent evolutionary events, building on the same backbone repressor protein. By placement of four cysteines, four coordinate sites such as found in CadC can be formed. Introduction of harder ligands, such as the oxygen and nitrogen sites of the type 2 sites of CadC and SmtB, allows binding sites for harder metals such as Zn(II) to evolve.

4.2 Families of Arsenite Efflux Carriers: ArsBs and Acr3s

The most common mechanism of arsenite resistance is efflux from cells catalyzed by members of two different and unrelated families of permeases, ArsB and Acr3. The first identified arsenite efflux protein is ArsB encoded by the *arsRDABC* operon of the conjugative R-factor R773, which confers resistance to inorganic As(III) and Sb(III) in *E coli* (Chen et al. 1986; Tisa and Rosen 1990). By itself, ArsB is a secondary efflux protein coupled to the proton-motive force and confers a moderate level of arsenite resistance (Kuroda et al. 1997). ArsB associates with the ArsA ATPase to form a pump that confers high-level resistance (Dey et al. 1994), as described in more detail below. Thus, the Ars efflux system is unique in that it exhibits a dual mode of energy coupling depending on the subunit composition (Dey and Rosen 1995).

ArsB is widespread in bacteria and archaea. It has 12 membrane spanning segments, which is similar to many carrier proteins (Wu et al. 1992). It transports As(III) but has higher affinity for Sb(III). ArsB is an antiporter that catalyzes the exchange of trivalent metalloid for protons, coupling arsenite efflux to the electrochemical proton gradient (Meng et al. 2004). Curiously, As(III) inhibits ArsB-mediated Sb(III) transport, yet Sb(III) stimulates ArsB-mediated As(III) transport. One explanation for these results is that the true substrate of ArsB is a polymer of As(III) or Sb(III) or a copolymer of As(III) and Sb(III) (Fig. 3). Complexes with arsenic are transported with lower affinity than the Sb(III) polymer, so trivalent arsenic appears to inhibit trivalent antimony transport. Reciprocally, complexes with antimony are transported with higher affinity than the As(III) polymer, so trivalent antimony appears to stimulate trivalent arsenic transport. Again, the nature of the polymer is not known, but we have proposed cyclic six-membered oxo-bridged rings (Meng et al. 2004).

The Acr3 family includes members found in bacteria, archaea, and fungi. Unfortunately, the literature is confused by the fact that many members of the Acr3 family have been given the name ArsB, even though they exhibit almost no sequence similarity to ArsB. The first identified member of this family is encoded by the *ars* operon of the skin (*sigK* intervening) element in the chromosome of *B. subtilis* (Sato and Kobayashi 1998). Fungal members of this family include the *S. cerevisiae* Acr3p metalloid efflux protein (Bobrowicz et al. 1997; Ghosh et al. 1999). Interestingly, yeast Acr3p appears to be selective for As(III) over Sb(III). This is unusual because of the similarity in chemical properties between the two metalloids. Recently *C. glutamicum* ATCC 13032 has been shown to have three genes for Acr3 homologues, each of which contributes to the high level of arsenite resistance in this organism (Ordonez et al. 2005). Expression of one of the *acr3* genes in an arsenite-hypersensitive strain of *E. coli* in which all *ars* genes were deleted confers resistance to As(III) (Meng and Rosen, in preparation). This heterologously expressed Acr3 catalyzes efflux of arsenite from *E. coli*. It exhibits significant differences from R773 ArsB. While ArsB exchanges As(III) with protons, Acr3 does not, but how it is coupled to the proton-motive force is unknown at this point. Like the yeast Acr3p, the *C. glutamicum* Acr3 is also more specific for As(III) than Sb(III), in contrast to ArsB, which has higher affinity for Sb(III) than As(III).

4.3 Efflux Pumps: ArsAB ATPases and Multidrug Resistance-Associated Proteins (MRPs)

The most efficient efflux pumps for metalloids are driven by ATP hydrolysis. These are the ArsAB ATPases, mostly in bacteria, and MRPs, members of the ABC family, in eukaryotic organisms.

4.3.1 ArsAB ATPases

The *ars* operon of R-factor R773 encodes an arsenite extrusion pump that confers resistance to the metalloids As(III) and Sb(III) in *E. coli*. This efflux pump has two subunits: ArsA ATPase is the catalytic subunit of the pump and ArsB functions as a membrane anchor for ArsA and also contains the translocation pathway. While ArsB alone can catalyze the extrusion of oxyanions across the membrane by functioning as a secondary transporter (Dey and Rosen 1995), the ArsA ATPase forms the ArsAB pump with the capacity to hydrolyze ATP, and drive active transport of As(III)/Sb(III) against a chemical gradient (Kuroda et al. 1997).

ArsA is normally bound to ArsB in vivo, but it can be overexpressed and purified as a soluble protein from the cytosol (Rosen et al. 1988). The soluble ArsA exhibits ATPase activity which is stimulated by either As(III) or Sb(III). The 583-residue ArsA ATPase has two homologous halves, A1 and A2, connected by a short linker. Each half has a consensus nucleotide binding domain (NBD) and both NBDs are required for ATPase activity and oxyanion transport (Karkaria et al. 1990; Kaur and Rosen 1992). The crystal structure of the MgADP bound form of ArsA has been determined (Fig. 6) at 2.3 Å resolution (Zhou et al. 2000). The A1 and A2 halves of the enzyme are related by a pseudo-twofold axis of symmetry. The two NBDs are in close proximity to each other and are located at the A1/A2 interface. In each NBD, Mg^{2+} is octahedrally coordinated with the β-phosphate of ADP, a threonine hydroxyl group, and several water molecules. In the solved structure, the two NBDs appear to be in different conformations despite extensive topological similarities, with NBD1 almost completely closed and the NBD2 fully open (Zhou et al. 2000, 2001). To determine the role of each NBD in substrate binding and catalysis, a thrombin site was introduced into the linker region that connects the A1 and A2 halves. Following covalent radiolabeling with 8-azidoATP and thrombin cleavage, the A1 and A2 halves migrated with different mobilities on sodium dodecyl sulfate polyacrylamide gel electrophoresis. In the absence of metalloid, both NBDs bind and hydrolyze ATP. However, in the presence of metalloid, although both NBDs hydrolyze ATP, hydrolysis in NBD1 is stimulated to a much greater extent than in NBD2. These experiments suggest that the two homologous halves of ArsA are functionally nonequivalent.

A novel metalloid binding domain (MBD) is also located at the A1/A2 interface and is over 20 Å distant from the NBDs. The MBD contains three metalloid atoms: one As(III) or Sb(III) is coordinated to His148 (A1) and Ser420 (A2), a second to Cys113 (A1) and Cys422 (A2), and the third to Cys172 (A1) and His453 (A2). Kinetic studies indicate that the three metalloid atoms bind with different affinities (Walmsley et al. 2001a), suggesting

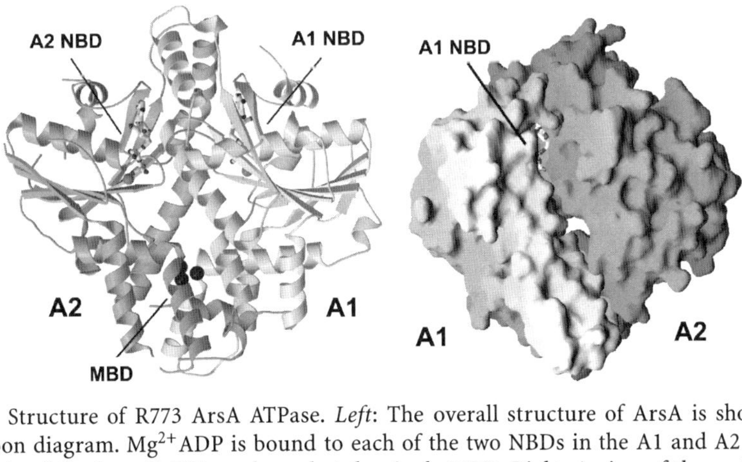

Fig. 6 Structure of R773 ArsA ATPase. *Left*: The overall structure of ArsA is shown as a ribbon diagram. Mg^{2+}ADP is bound to each of the two NBDs in the A1 and A2 halves of ArsA, while three Sb(III) are bound at the single MBD. *Right*: A view of the molecular surface of ArsA showing the relative positions of the A1 and A2 halves and details of ADP bound in the NBD1

that the three metalloid binding sites may allow the pump to operate in different ranges of ion concentration. Each of the amino acids lining the MBD has been altered by site-directed mutagenesis. Alteration of Cys113, Cys172, and Cys422 resulted in substantial loss of metalloid-stimulated ATPase activity (Bhattacharjee et al. 1995; Ruan et al. 2006), while mutation of His148 and His453 showed a modest decrease in allosteric activation of the enzyme (Bhattacharjee and Rosen 2000). Cys113 and Cys422 form the high-affinity metalloid binding site (Ruan et al. 2006). Although, metalloid stimulation of ArsA ATPase activity enhances the ability of the pump to reduce the intracellular concentration of oxyanions, high-affinity binding of metalloid by ArsA is not obligatory for transport or resistance. It appears that binding of metalloid to MBD provides an evolutionary advantage to survive in low levels of arsenite ubiquitously present in the environment (Ruan et al. 2006).

The transfer of information of metal occupancy at MBD to the two NBDs is performed by two signal transduction domains (STDs). Two stretches of residues, $D^{142}TAPTGH^{148}$ in A1 STD and $D^{447}TAPTGH^{453}$ in A2 STD, physically connect the MBD to NBD1 and NBD2, respectively. By introducing tryptophan residues in proximity to the $D^{142/447}TAPTGH^{148/453}$ sequence, conformational changes have been observed in response to binding of nucleotide and metalloid at the NBDs and MBD, respectively (Zhou et al. 1995, 2002; Zhou and Rosen 1997).

The mechanism of allosteric activation of ArsA ATPase has been elucidated at the molecular level from a combination of biochemical (Bhattacharjee et al. 2000), structural (Zhou et al. 2000, 2001), and kinetic experiments (Walmsley et al. 1999, 2001a,b). To summarize, these studies indicated that ArsA exists in at least two different conformational forms, where bound ATP favors one form and bound metalloids the other. In the absence of activator, the enzyme undergoes a slow conformational change between these forms, leading to a lag in attaining maximal steady-state activity. Binding of oxyanions at MBD acts as molecular glue to bring the A1 and A2 halves of ArsA together. This enhances the steady-state ATPase activity by inducing rapid product release and allowing the protein to adopt a conformation that can bind MgATP for the next catalytic cycle. In the presence of activator, ArsA avoids the rate-limiting isomerization at the end of the ATPase reaction and now ATP hydrolysis becomes rate limiting for the reaction. When ArsA is bound to ArsB, the complex probably functions as a reciprocating engine (Zhou et al. 2001), where cytosolic As(III)/Sb(III) accesses the high-affinity metalloid binding site of ArsA. ATP hydrolysis is linked with stepwise transfer of the metalloid to the low-affinity sites, followed by vectorial transport of metalloid into ArsB.

Homologues of the bacterial ArsA ATPase are widespread in nature and are present in most members of all three domains: bacteria, archaea, and eukaryotes (Bhattacharjee et al. 2001). Similar to *E. coli* R773 ArsA, many bacterial ArsAs also have an A1–A2 arrangement, and are part of *ars* operons that reside either on chromosomes or plasmids. On the other hand, many bacteria encode for ArsA homologues with only a single "A" structure. The role of the bacterial single A proteins have not been determined. Archaea have either an A1–A2 structure or just a single A arrangement. For example, halobacteria have an A1–A2 organization while methanogenic archaea have a single A structure. All eukaryotic ArsA homologues known to date are single A domain proteins.

To determine the role of eukaryotic ArsA homologues, the yeast homologue (*ARR4*) was disrupted in *S. cerevisiae.* The null mutant shows no growth defects either on synthetic complete media or on rich media (Zuniga et al. 1999). The disrupted strain, however, does not exhibit an arsenic-related phenotype but shows increased sensitivity to heavy metals and temperature, indicating that Arr4p may be part of a complex that is involved in stress tolerance but not arsenic resistance (Shen et al. 2003). In support of this idea, the ATPase activity of purified Arr4p was not stimulated by As(III) or Sb(III). It has been recently shown that, in the presence of Cu(II), Arr4p binds with the yeast intracellular CLC chloride-transport protein, Gef1p (Metz et al. 2006). The Arr4p–Gef1p complex has been implicated to be involved in yeast copper metabolism (Metz et al. 2006). On the other hand, Schuldiner et al. reported that Arr4p is a subunit of the GET complex that is required for the retrieval of HDEL proteins from the Golgi to the endoplasmic reticulum in an ERD2-dependent fashion (Schuldiner et al. 2005).

The mouse homologue of the bacterial ArsA ATPase also shows a single A structure. To determine the physiological role of the mouse homologue, heterozygous Asna1 knockout mice were generated by homologous recombination. The heterozygous Asna1 knockout mice displayed no obvious phenotype. However, early embryonic lethality was observed in homozygous Asna1 knockout embryos, indicating that Asna1 plays a crucial role during early embryonic development (Mukhopadhyay et al. 2006).

Leipe et al. (2002) suggested that ArsA homologues are most likely of archaeal or archaeoeukaryotic origin and spread to bacteria later to provide selective environmental advantage. Although these homologues have common structural motifs, phylogenetic analysis of ArsA homologues suggests separate evolutionary lines for these proteins, resulting in distinct biochemical activities (Gihring et al. 2003).

4.3.2 MRPs

The yeast protein Ycf1p (yeast cadmium factor), a close homologue of the human MRP1 (multidrug resistance-associated protein), catalyzes the vacuolar sequestration of As(III) (Ghosh et al. 1999). MRP1 is an ABC transporter that confers multidrug resistance in human small cell lung carcinoma (Cole et al. 1994). Both MRP1 and Ycf1p are ATPases that pump glutathione S-conjugated drugs out of the cytosol. MRP1 transports arsenic as a triglutathione conjugate out of the cell (Leslie et al. 2004), while Ycf1p catalyzes the ATP-driven uptake of As(III)-glutathione conjugate into the yeast vacuole (Ghosh et al. 1999).

4.4 ArsD: An As(III) Chaperone

Both prokaryotes and eukaryotes have metallochaperones that sequester metals in the cytoplasm, buffering their concentration, and deliver them to protein targets, such as transporters for extrusion (Rosenzweig 2002). For example, Atx1, the yeast homologue of Atox1, delivers copper to the transport-ATPase Ccc2p in the *trans*-Golgi network for incorporation into the multicopper oxidase Fet3p (Lin et al. 1997).

Although most sequenced bacterial genomes contain an *ars* operon, only a few are five-gene operons such as the *arsRDABC* operon of plasmid R773. The product of the *arsD* gene, a 120-residue polypeptide that is a functional homodimer, is a weak As(III)-responsive transcriptional repressor (Chen and Rosen 1997). Since its regulatory properties are so poor, it is questionable whether that is the primary function of ArsD. Recently, ArsD has been shown to be a metalloid chaperone that delivers As(III) to the ArsAB As(III)-translocating ATPase (Lin et al. 2006). The initial motivation for considering an alternate function for ArsD was analysis of all *ars* operons that contain an *arsD* gene. The *arsD* gene is found in only a small number of operons, 14 to date. Remarkably, in each operon the *arsD* gene precedes an *arsA* gene. This observation suggested that ArsD and ArsA coevolved for a common and related function. A number of lines of evidence support this proposition. (1) When a mixed culture of cells expressing either *arsDAB* or *arsAB* were grown in the presence of a subtoxic concentration of As(III), the cells expressing all three genes took over the culture within 1 week, showing that cells with *arsDAB* have increased fitness in low As(III) compared with cells with only *arsAB*. (2) Expression of ArsD increases the ability of the ArsAB pump to extrude As(III). (3) ArsA and ArsD were shown to interact in vivo by yeast two-hybrid analysis. (4) ArsD and ArsA can be chemically cross-linked in a 1 : 1 complex through the cysteine residues of their metalloid binding sites. (5) The rate of dissociation of metalloid from ArsD is enhanced by four orders of magnitude by interaction with ArsA, consistent with transfer of As(III) from the chaperone to the ATPase. (6) ArsD increases ArsA's affinity for As(III) without altering the V_{max}. This makes the ArsAB pump more effective at low—and environmentally relevant—concentrations of metalloid, a property expected for a metallochaperone. At this time the molecular mechanism of chaperone activity is not known, but we propose that ArsD and ArsB bind to the same site on ArsA sequentially in a cycle of metal transfer from ArsD to ArsA to ArsB concomitant with ATP binding and hydrolysis.

4.5
Arsenate Reductases: ArsCs and Acr2s

In contrast to the membrane-bound respiratory arsenate reductases, enzymes catalyzing arsenate reduction for the purpose of detoxification are soluble proteins.

4.5.1
General Function

Cytosolic arsenate reductases catalyze the two-electron reduction of inorganic As(V) to As(III). These enzymes arose independently at least three times by convergent evolution. One family of arsenate reductases that includes the *E. coli* plasmid R773 ArsC uses glutaredoxin (Grx) and glutathione (GSH) as reductants (Mukhopadhyay and Rosen 2002). A second family, represented by the *Staphylococcus aureus* plasmid pI258 ArsC (Ji and Silver 1992) and the *B. subtilis* chromosomal ArsC, uses thioredoxin (Trx) as a reductant. This bacterial family of arsenate reductases is distantly related to mammalian low molecular weight (LMW) protein tyrosine phosphatases (PTPases) (Messens et al. 1999; Bennett et al. 2001). A third family of eukaryotic arsenate reductases includes Acr2p (Bobrowicz et al. 1997; Mukhopadhyay and Rosen 1998) from the baker's yeast *S. cerevisiae*, and LmAcr2p (Zhou et al. 2004) from the parasitic protozoa *Leishmania major*. The eukaryotic reductases are related to the catalytic domain of the Cdc25 cell-cycle PTPase (Fauman et al. 1998).

4.5.2
R773 Arsenate Reductase

The 141 amino acid R773 ArsC is a monomeric enzyme that catalyzes the reduction of arsenate to arsenite using GSH and Grx as electron donors (Gladysheva et al. 1994). The active site Cys12 is essential for ArsC catalysis (Fig. 7) (Liu et al. 1995) and is surrounded by an arginine triad composed of Arg60, Arg94, and Arg107 (Shi et al. 2003). The arginines are proposed to be involved in arsenate binding and transition-state stabilization (Shi et al. 2003; DeMel et al. 2004). *E. coli* has three glutaredoxins, Grx1, Grx2, and Grx3, each with a classical -Cys-Pro-Tyr-Cys- active-site motif, and Grx2 was found to be the most effective hydrogen donor for the reduction of arsenate by ArsC (Åslund et al. 1994; Shi et al. 1999). Analysis of single and double cysteine-to-serine substitutions in the active site of the three glutaredoxins demonstrated that only the N-terminal cysteine residue is essential for activity. Based on biochemical studies and X-ray crystallographic structures (Martin et al. 2001; DeMel et al. 2004) of the free and ligand-bound forms of ArsC, a mechanism for ArsC-catalyzed arsenate reductase activity has been

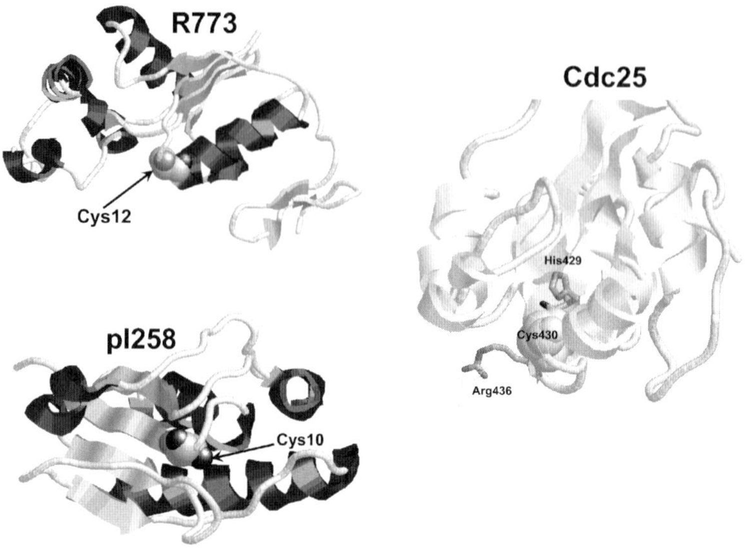

Fig. 7 Structure of arsenate reductase. The structures of R773 and pI258 ArsC are shown, identifying their secondary structural elements. The catalytic cysteines at the active sites of either enzyme complexed with As(V) are also indicated. The yeast arsenate reductase, Acr2p, is predicted to have a similar active site to the human Cdc25a. The structure of the catalytic core of human Cdc25a is shown, identifying the active site Cys430 that charge-pairs with His429, and Arg436, which forms part of the phosphate binding loop

proposed. The reaction starts by a nucleophilic attack by Cys12 on an arsenate that is noncovalently bound at the active site. This leads to the formation of a thioarsenate binary adduct and release of OH^-. In step 2, a nucleophilic attack on the arsenate adduct by GSH results in the formation of an {ArsC-Cys12}S–As(V)–S{glutathione} tertiary complex (Liu and Rosen 1997), and release of water. Step 3 involves binding of Grx, with reduction of arsenate to the dihydroxy monothiol As(III) intermediate, along with release of OH^- and a mixed disulfide complex of glutathione and glutaredoxin. The penultimate step is the formation of a monohydroxy, positively charged As(III), with release of OH^-. In the final step, addition of OH^- releases free arsenite [$As(OH)_3$] and regenerates the free enzyme. Except for the glutathionylated intermediate in step 2, the structures for all other intermediates in the above reaction pathway have been determined (Martin et al. 2001; DeMel et al. 2004).

4.5.3
S. aureus and *B. subtilis* Arsenate Reductases

The ArsC arsenate reductase encoded by *S. aureus* plasmid pI258 and the closely related *B. subtilis* enzyme are 14.8-kDa monomeric proteins that are unrelated to the R773 ArsC. In contrast to R773 ArsC, which requires GSH and Grx for arsenate reductase activity, the pI258 enzyme is coupled to thioredoxin, thioredoxin reductase, and nicotinamide adenine dinucleotide phosphate (NADPH) (Ji et al. 1994). Mutation of Cys10, Cys82, and Cys89 of pI258 ArsC led to inactive enzymes (Messens et al. 1999), and these cysteines were shown to form a unique intramolecular disulfide cascade, essential for the reduction of arsenate to arsenite (Messens et al. 2002). The enzyme has a LMW PTPase anion binding motif that forms a phosphate binding P-loop and has phosphatase activity, catalyzing hydrolysis of *p*-nitrophenyl phosphate (Zegers et al. 2001), so the single protein has both arsenate reductase and PTPase activity.

On the basis of the structures of the reduced and oxidized forms of pI258 ArsC, along with data from NMR and kinetic studies, a multistep reaction mechanism has been proposed (Zegers et al. 2001; Roos et al. 2006). The first step of pI258 ArsC catalyzed arsenate reduction starts with a nucleophilic displacement reaction by Cys10 (Fig. 7) on arsenate, leading to the formation of a covalent Cys10–arseno adduct (Zegers et al. 2001), which is the equivalent of the Cys12–arseno adduct of R773 ArsC (Martin et al. 2001). In the second step, arsenite is released following a nucleophilic attack by Cys82 on the covalent Cys10–arseno adduct. An oxidized Cys10–Cys82 intermediate is formed (Messens et al. 2002). In the third reaction step, Cys89 attacks the Cys10–Cys82 disulfide, resulting in the formation of oxidized Cys82-Cys89 disulfide and the reduction of Cys10 (Zegers et al. 2001; Messens et al. 2002). Finally, ArsC is regenerated by thioredoxin that reduces the Cys82–Cys89 disulfide.

The X-ray crystal structure for another thioredoxin-linked arsenate reductase has been reported (Bennett et al. 2001) for the enzyme encoded by the *B. subtilis* skin element (Sato and Kobayashi 1998). The *B. subtilis* arsenate reductase contains 139 amino acid residues and is a monomer in solution. Like the pI258 ArsC, the *B. subtilis* ArsC also uses thioredoxin and exhibits PTPase activity, although this rate is many-fold lower than that of the *p*-nitrophenyl phosphate hydrolysis by true PTPase. The crystal (Bennett et al. 2001) and solution structures (Guo et al. 2005) of *B. subtilis* ArsC suggest that the catalytic mechanism for arsenate reduction involves the same triple cysteine redox relay system as the pI258 enzyme (Zegers et al. 2001; Messens et al. 2002).

An interesting variant of arsenate reductase has been described from the cyanobacteria *Synechocystis* sp. strain PCC 6803 (Li et al. 2003). The *Synechocystis* ArsC is homologous to pI258 ArsC and shows both phosphatase and arsenate reductase activities. However, its catalytic activity exhibits a unique combination of both R773 and pI258 ArsC mechanisms. Like the R773 ArsC, *Synechocystis* ArsC requires both GSH and Grx as the source of reducing equivalents for arsenate reduction, rather than thioredoxin, as does the pI258 enzyme. While the R773 enzyme requires only one cysteine for catalysis, the *Synechocystis* ArsC has three essential cysteine residues, like the pI258 *S. au-*

reus enzyme. *Synechocystis* and *S. aureus* enzymes may be the products of two independent evolutionary pathways of LMW PTPase into arsenate reductases (Li et al. 2003).

4.5.4
Eukaryotic Arsenate Reductases

Eukaryotic arsenate reductases that reduce arsenate to arsenite have been identified from *S. cerevisiae* (Bobrowicz et al. 1997; Mukhopadhyay and Rosen 1998), *S. douglasii* (Maciaszczyk et al. 2004), fern (*Pteris vittata*) (Ellis et al. 2006), rice (*Oryza sativa*) (Duan, Rosen, and Zhu 2007) and *L. major* (Zhou et al. 2004). The best-characterized eukaryotic arsenate reductases are Acr2p from *S. cerevisiae* (Mukhopadhyay and Rosen 1998; Mukhopadhyay et al. 2000) and LmACR2 from *L. major* (Zhou et al. 2004).

The 130 amino acid Acr2p exists as a homodimer and is unrelated in sequence to bacterial arsenate reductases. Acr2p is a member of the superfamily of PTPases such as the human cell-cycle dual-specific phosphatase Cdc25a (Fig. 7). Members of this family have a consensus active site $HC(X)_5R$ motif (Denu and Dixon 1995; Fauman et al. 1998). Alteration of either Cys76 or Arg82 at the consensus $C^{76}(X)_5R^{82}$ motif of Acr2p resulted in loss of As(V) resistance in vivo and As(V) reduction in vitro (Mukhopadhyay and Rosen 2001). These results suggest that Cys76 is the equivalent of Cys12 in ArsC and may form As(V) and As(III) intermediates. Similar to the *E. coli* ArsC, Acr2p activity required both GSH and glutaredoxin, but not thioredoxin, to support arsenate reduction. Glutaredoxins from both *S. cerevisiae* and *E. coli* were able to serve as reductants. Analysis of *grx* mutants lacking one or both cysteine residues in the -Cys-Pro-Tyr-Cys- active site demonstrated that only the N-terminal cysteine residue is essential for arsenate reductase activity. This suggests that the catalytic mechanism of Acr2p may follow a similar pathway as *E. coli* ArsC (Mukhopadhyay et al. 2000).

Although Acr2p is a member of the superfamily of PTPases, it does not exhibit phosphatase activity. It has the $C^{76}(X)_5R^{82}$ phosphatase motif but not the glycine-rich phosphate binding motif (GXGXXG) that is found in PTPases. Introduction of glycines at positions 79, 81, and 84 of the Acr2p primary sequence resulted in a gain of PTPase activity and a loss of arsenate reductase activity (Mukhopadhyay et al. 2003). The relative ease with which an arsenate reductase can be converted into a protein–tyrosine phosphatase indicates that selective environmental pressure led to the evolution of an arsenate reductase from a protein–tyrosine phosphatase (or, possibly, the reverse).

An arsenate reductase from *L. major*, LmACR2, has been identified and characterized (Zhou et al. 2004). LmACR2 exhibits 28% sequence identity and 44% similarity with *S. cerevisiae* Acr2p. LmACR2 can functionally complement the arsenate-sensitive phenotypes of the *acr2* deletion strain of *S. cerevisiae* and also the Δ*arsC* strain of *E. coli*. In contrast to yeast Acr2p which exists as a dimer, the LmACR2 exists as a monomeric species. LmACR2 also requires GSH and Grx as the source of reducing equivalents to catalyze the reduction of arsenate (Zhou et al. 2004). Additionally, LmACR2 can also reduce pentavalent antimony [Sb(V)] compounds (Zhou et al. 2004). This is the first example of an arsenate reductase that also has antimonate reductase activity. Macrophage-infected amastigotes derived from LmACR2-transfected *L. infantum* promastigotes exhibited increased sensitivity to the pentavalent antimony drug, Pentostam. This is because LmACR2 can reduce the pentavalent antimony in Pentostam to the active trivalent form of the drug. Thus, in contrast to the bacterial and yeast arsenate reductases that are components of metalloid detoxification pathways, expression of LmACR2 in *Leishmania* leads to metalloid sensitivity. Why should *Leishmania* evolve an enzyme solely for drug activation as it is detrimental to the pathogen? We have recently shown that LmACR2 also exhibits PTPase activity at similar levels as other known PTPases (Zhou et al. 2006). Most likely, the physiological function of LmACR2 is to dephosphorylate phosphotyrosine residues in leishmanial proteins, and not detoxification of metalloids.

4.6 Arsenite Methylases

The ability to methylate arsenic methylation is widespread: many bacteria, archaea, fungi, plants, and animals do so. The history of arsenic methylation has been described entertainingly (Bentley and Chasteen 2002) and will be only briefly summarized here. Arsenic metabolism produces a garlicky odor. For example, individuals treated with cacodylic acid (dimethylarsinic acid or DMA(V)) have been known to exhale the odor of garlic. Copper–arsenic compounds such as the pigment Paris green were used in the production of wallpaper. Fungi can convert this to a volatile, garlicky, and toxic arsenical, which has been called "Gosio gas" after the Italian physician Bartolomeo Gosio, who described this process in the 1890s. In 1945 Frederick Challenger demonstrated that Gosio gas is trimethylarsine, $(CH_3)_3As$ or TMA(III). Challenger proposed a pathway in which arsenate can be converted to TMA(III) (Fig. 8) by a series of reductions of pentavalent to trivalent arsenicals coupled to oxidative methylations to mono-, di-, and trimethyl species.

Some mammals, including humans and rat, methylate inorganic arsenic and excrete methylated species such as DMA(V) and, to a lesser extent, MMA(V) in the urine. This has led to the proposal that methylation is a detoxification process. However, other species such as chimpanzees do not methylate inorganic arsenic, and do not seem to be any less arsenic sensitive than humans or rat. Additionally, it has been pointed out that the trivalent intermediates in the Challenger pathway, MMA(III), DMA(III), and TMA(III), are considerably more toxic than inorganic arsenate or arsenite (Petrick et al.

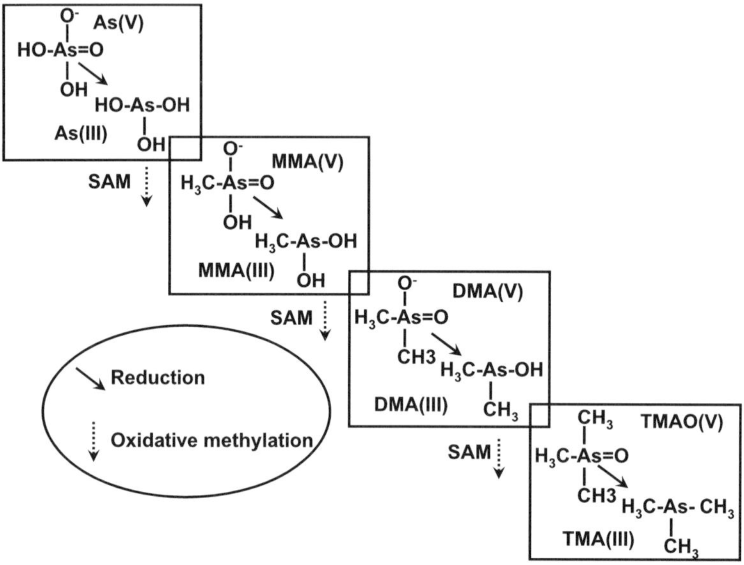

Fig. 8 The Challenger pathway of arsenic methylation. In each step a pentavalent arsenical is reduced to a trivalent arsenical, which is then oxidatively methylated with *S*-adenosylmethionine to form the pentavalent form. The overall scheme involves four reductive steps and three methylations to form the gas TMA(III) from inorganic As(V)

2000; Drobna et al. 2005). In vivo, the toxicity of inorganic and organic arsenicals is DMA(III), MMA(III) > As(III) > As(V) > DMA(V), MMA(V) > TMAO (Akter et al. 2005). The major pentavalent products, DMA(V) and TMAO, are approximately 100-fold and 1000-fold less toxic than As(III), respectively (Hirano et al. 2004). This has led to the proposal that methylation may activate inorganic arsenic to more toxic metabolites. This question remains unresolved to date.

In 2002 Thomas and coworkers cloned the gene for a rat enzyme that methylates As(V) and As(III) using *S*-adenosylmethionine (SAM) as the methyl donor (Lin et al. 2002). The enzyme has been termed Cyt19 or As3MT. It is not known whether arsenic methylation is the primary physiological function of As3MT. However, it does seem likely that this enzyme is the major mammalian arsenic methylase. The enzyme is not made in chimpanzees due to a frame-shift mutation, leading to an aborted protein product, and consequently primates do not methylate inorganic arsenic.

Bacteria and fungi are known to produce volatile and toxic arsines (Bentley and Chasteen 2002), but the physiological roles of arsenic methylation in microorganisms are likewise unclear, as is its biochemical basis. A large number of genes for bacterial and archaeal homologues of As3MT have been identified. To date, 200 homologues have been identified in genome databases, 7 in archaea, 13 in fungi, 17 in metazoa, and 163 in bacteria. A subset of these genes has been termed *arsM* and their protein product ArsM (arsenite S-adenosyl methyltransferase). What sets these homologues apart from As3MT is that they are downstream of an *arsR* gene, encoding the archetypal arsenic-responsive transcriptional repressor that controls expression of *ars* operons (Xu and Rosen 1999), suggesting that these ArsMs evolved to confer arsenic resistance. The gene for the 283-residue ArsM (29 656 Da) (accession number NP_948900.1) was cloned from the soil bacterium *Rhodopseudomonas palustris* and expressed in an arsenic-hypersensitive strain of *E. coli* (Qin et al. 2006). This heterologous expression of *arsM* conferred As(III) resistance to *E. coli* in the absence of any other *ars* genes, demonstrating that methylation is sufficient to detoxify arsenic (Qin et al. 2006). ArsM converted As(III) into a number of methylated intermediates of the Challenger pathway, with volatile trimethylarsine as the end product. MMA(III) and DMA(III) may be more toxic than As(III), but they do not accumulate in cells expressing *arsM*, and, while TMA(III) is more toxic than arsenite, the fact that it is a gas means that it does not accumulate in cells expressing ArsM. TMA(III) volatilization results in loss of arsenic, from both cells and medium. Since ArsM homologues are widespread in nature, this microbial-mediated transformation is proposed to have an important impact on the global arsenic cycle. ArsM was purified and shown to catalyze the transfer of methyl groups from S-adenosylmethionine to arsenite, forming di- and trimethylated species, with TMA(III) gas as the final product (Qin et al. 2006). This is the first demonstration that methylation of environmental arsenic by conversion to soluble and gaseous methylated species is a detoxifying process that may contribute to global cycling of arsenic.

4.7
Other Novel Mechanisms of Arsenic Resistance: AqpS and ArsH

In addition to *arsRBC* and *arsRDABC* operons, several bacterial *ars* operons encode for proteins that participate in novel ways to arsenic resistance. For example, the chromosomal *ars* operon of the legume symbiont *Sinorhizobium meliloti* Rm1021 displays a cluster of four genes: *arsR*, *aqpS*, *arsC*, and *arsH*. Although, *S. meliloti* ArsR shows sequence similarity to the ArsR subfamily of helix–turn–helix bacterial transcription regulatory proteins, it does not contain the N-terminal CXCXXC motif, which is required for binding of the inducer in *E. coli* and R773 ArsR (Shi et al. 1996; Xu et al. 1996). Instead, *S. meliloti* ArsR is more closely related to the *A. ferrooxidans* ArsR discussed above, and has two vicinal cysteine pairs located near the C-terminal end of the protein. However, the involvement of these thiols in metalloid binding remains to be determined. The second gene *aqpS* encodes for a membrane protein that belongs to the major intrinsic protein or aquaporin superfamily (King et al. 2004) and shows sequence homology with the bacterial glycerol facilitator (GlpF), yeast aquaglyceroporin Fps1p, and mammalian aquaglyceroporin AQP9. *S. meliloti* ArsC is homologous to *E. coli* ArsC and both show similar levels of arsenate reductase activity. The fourth gene encodes a novel protein, ArsH, which has conserved domains related to the NADPH-dependent flavin mononucleotide reductase class of proteins.

The presence of *aqpS* in lieu of *arsB* in the *S. meliloti ars* operon is interesting, since aquaglyceroporins have been shown to be responsible for the uptake of arsenite (Meng et al. 2004) and antimonite (Sanders et al. 1997), rendering them sensitive to arsenite. To understand the role of *aqpS* in arsenic resistance, *S. meliloti aqpS* and *arsC* were disrupted individually. Disruption of *aqpS* resulted in increased tolerance to arsenite but not arsenate, while cells with an *arsC* disruption showed selective sensitivity to arsenate. Additionally, transport experiments with intact cells showed that AqpS facilitates transport of arsenite. Coexpression of *S. meliloti aqpS* and *arsC* in an *E. coli* Δ*ars* strain complemented arsenate but not arsenite sensitivity. These results indicated that, when *S. meliloti* is exposed to environmental arsenate, arsenate enters the cell through the phosphate transport system and is reduced to arsenite by ArsC. Internally generated arsenite is extruded out of the cell by downhill movement through AqpS. Therefore, AqpS and ArsC together form a novel pathway of As(V) detoxification in *S. meliloti*. This is the only example of an aquaglyceroporin with a physiological role in arsenic resistance (Yang et al. 2005). Several other organisms, for example, *Mesorhizobium loti*, *Caulobacter crescentus*, and *Ralstonia solanacearum*, also have chromosomal *ars* operons that contain the unique combination of *aqpS* and *arsC* genes. This pathway may be widespread in organisms that are exposed primarily to As(V).

The role of ArsH in arsenic resistance is unclear. ArsH is found to be widely distributed in bacteria but sparsely in fungi, plants, and archaea. The *arsH* gene product in *Yersinia enterocolitica* confers resistance to both arsenite and arsenate (Neyt et al. 1997). Similarly, deletion of the *arsH* gene from the IncH12 plasmid R478 resulted in a total loss of resistance to arsenic compounds (Ryan and Colleran 2002). However, *A. ferrooxidans* and *Synechocystis* ArsHs are not required for arsenic resistance (Butcher et al. 2000; Lopez-Maury et al. 2003). We find that inactivation of the *S. meliloti arsH* gene results in increased As(III) sensitivity. Overexpression of ArsH in Δ*ars* strain of *S. meliloti* shows high-level resistance to trivalent arsenicals. This indicates that ArsH can provide resistance to As(III) by a unique mechanism that does not require AqpS and ArsC activities. ArsH has been crystallized (Jun et al., in preparation), and its catalytic mechanism is currently being investigated.

Acknowledgements This work was supported by EHS Center Grant P30 ES06639 and Michigan Life Sciences Corridor Grant GR-180 085P1000684 (to H.B.) and United States Public Health Service Grants AI43428, GM52216, and GM55425 (to B.P.R.).

References

Acharyya SK, Chakraborty P, Lahiri S, Raymahashay BC, Guha S, Bhowmik A (1999) Arsenic poisoning in the Ganges delta. Nature 401:545; discussion 546–547

Afkar E, Lisak J, Saltikov C, Basu P, Oremland RS, Stolz JF (2003) The respiratory arsenate reductase from *Bacillus selenitireducens* strain MLS10. FEMS Microbiol Lett 226:107–112

Agre P, King LS, Yasui M, Guggino WB, Ottersen OP, Fujiyoshi Y, Engel A, Nielsen S (2002) Aquaporin water channels—from atomic structure to clinical medicine. J Physiol 542:3–16

Akter KF, Owens G, Davey DE, Naidu R (2005) Arsenic speciation and toxicity in biological systems. Rev Environ Contam Toxicol 184:97–149

Anderson GL, Williams J, Hille R (1992) The purification and characterization of arsenite oxidase from *Alcaligenes faecalis*, a molybdenum-containing hydroxylase. J Biol Chem 267:23674–23682

Åslund F, Ehn B, Miranda-Vizuete A, Pueyo C, Holmgren A (1994) Two additional glutaredoxins exist in *Escherichia coli*: glutaredoxin 3 is a hydrogen donor for ribonucleotide reductase in a thioredoxin/glutaredoxin 1 double mutant. Proc Natl Acad Sci USA 91:9813–9817

Bennett MS, Guan Z, Laurberg M, Su XD (2001) *Bacillus subtilis* arsenate reductase is structurally and functionally similar to low molecular weight protein tyrosine phosphatases. Proc Natl Acad Sci USA 98:13577–13582

Bentley R, Chasteen TG (2002) Microbial methylation of metalloids: arsenic, antimony, and bismuth. Microbiol Mol Biol Rev 66:250–271

Bhattacharjee H, Rosen BP (2000) Role of conserved histidine residues in metalloactivation of the ArsA ATPase. Biometals 13:281–288

Bhattacharjee H, Li J, Ksenzenko MY, Rosen BP (1995) Role of cysteinyl residues in metalloactivation of the oxyanion-translocating ArsA ATPase. J Biol Chem 270:11245–11250

Bhattacharjee H, Zhou T, Li J, Gatti DL, Walmsley AR, Rosen BP (2000) Structure–function relationships in an anion-translocating ATPase. Biochem Soc Trans 28:520–526

Bhattacharjee H, Ho YS, Rosen BP (2001) Genomic organization and chromosomal localization of the *Asna1* gene, a mouse homologue of a bacterial arsenic-translocating ATPase gene. Gene 272:291–299

Bhattacharjee H, Carbrey J, Rosen BP, Mukhopadhyay R (2004) Drug uptake and pharmacological modulation of drug sensitivity in leukemia by AQP9. Biochem Biophys Res Commun 322:836–841

Bobrowicz P, Wysocki R, Owsianik G, Goffeau A, Ulaszewski S (1997) Isolation of three contiguous genes, *ACR1*, *ACR2*, and *ACR3*, involved in resistance to arsenic compounds in the yeast *Saccharomyces cerevisiae*. Yeast 13:819–828

Boles E, Hollenberg CP (1997) The molecular genetics of hexose transport in yeasts. FEMS Microbiol Rev 21:85–111

Borgnia M, Nielsen S, Engel A, Agre P (1999) Cellular and molecular biology of the aquaporin water channels. Annu Rev Biochem 68:425–458

Bun-ya M, Shikata K, Nakade S, Yompakdee C, Harashima S, Oshima Y (1996) Two new genes, *PHO86* and *PHO87*, involved in inorganic phosphate uptake in *Saccharomyces cerevisiae*. Curr Genet 29:344–351

Busenlehner LS, Cosper NJ, Scott RA, Rosen BP, Wong MD, Giedroc DP (2001) Spectroscopic properties of the metalloregulatory Cd(II) and Pb(II) sites of *Staphylococcus aureus* pI258 CadC. Biochemistry 40:4426–4436

Busenlehner LS, Apuy JL, Giedroc DP (2002) Characterization of a metalloregulatory bismuth(III) site in *Staphylococcus aureus* pI258 CadC repressor. J Biol Inorg Chem 7:551–559

Butcher BG, Rawlings DE (2002) The divergent chromosomal *ars* operon of *Acidithiobacillus ferrooxidans* is regulated by an atypical ArsR protein. Microbiology 148:3983–3992

Butcher BG, Deane SM, Rawlings DE (2000) The chromosomal arsenic resistance genes of *Thiobacillus ferrooxidans* have an unusual arrangement and confer increased arsenic and antimony resistance to *Escherichia coli*. Appl Environ Microbiol 66:1826–1833

Carbrey JM, Gorelick-Feldman DA, Kozono D, Praetorius J, Nielsen S, Agre P (2003) Aquaglyceroporin AQP9: solute permeation and metabolic control of expression in liver. Proc Natl Acad Sci USA 100:2945–2950

Cavet JS, Meng W, Pennella MA, Appelhoff RJ, Giedroc DP, Robinson NJ (2002) A nickel–cobalt sensing ArsR–SmtB family repressor: contributions of cytosol and effector binding sites to metal selectivity. J Biol Chem 277:38441–38448

Cavet JS, Graham AI, Meng W, Robinson NJ (2003) A cadmium–lead sensing ArsR–SmtB repressor with novel sensory sites: complementary metal-discrimination by NMTR and CMTR in a common cytosol. J Biol Chem 278:44560–44566

Chen CM, Misra TK, Silver S, Rosen BP (1986) Nucleotide sequence of the structural genes for an anion pump. The plasmid-encoded arsenical resistance operon. J Biol Chem 261:15030–15038

Chen Y, Rosen BP (1997) Metalloregulatory properties of the ArsD repressor. J Biol Chem 272:14257–14262

Cole SP, Sparks KE, Fraser K, Loe DW, Grant CE, Wilson GM, Deeley RG (1994) Pharmacological characterization of multidrug-resistant MRP-transfected human tumor cells. Cancer Res 54:5902–5910

Cook WJ, Kar SR, Taylor KB, Hall LM (1998) Crystal structure of the cyanobacterial metallothionein repressor SmtB: a model for metalloregulatory proteins. J Mol Biol 275:337–346

Dembitsky VM, Levitsky DO (2004) Arsenolipids. Prog Lipid Res 43:403–448

DeMel S, Shi J, Martin P, Rosen BP, Edwards BF (2004) Arginine 60 in the ArsC arsenate reductase of *E. coli* plasmid R773 determines the chemical nature of the bound As(III) product. Protein Sci 13:2330–2340

Denu JM, Dixon JE (1995) A catalytic mechanism for the dual-specific phosphatases. Proc Natl Acad Sci USA 92:5910–5914

Dey S, Rosen BP (1995) Dual mode of energy coupling by the oxyanion-translocating ArsB protein. J Bacteriol 177:385–389

Dey S, Dou D, Rosen BP (1994) ATP-dependent arsenite transport in everted membrane vesicles of *Escherichia coli*. J Biol Chem 269:25442–25446

Drobna Z, Waters SB, Devesa V, Harmon AW, Thomas DJ, Styblo M (2005) Metabolism and toxicity of arsenic in human urothelial cells expressing rat arsenic (+3 oxidation state)-methyltransferase. Toxicol Appl Pharmacol 207:147–159

Duan G, Zhou Y, Tong YP, Mukhopadhyay R, Rosen BP, Zhu YG (2007) Two A CDC25 homologue from rice functions as an arsenate reductase. New Phytologist, in press

Ellis DR, Gumaelius L, Indriolo E, Pickering IJ, Banks JA, Salt DE (2006) A novel arsenate reductase from the arsenic hyperaccumulating fern *Pteris vittata*. Plant Physiol 141:1544–1554

Ellis PJ, Conrads T, Hille R, Kuhn P (2001) Crystal structure of the 100-kDa arsenite oxidase from *Alcaligenes faecalis* in two crystal forms at 1.64 Å and 2.03 Å. Structure 9:125–132

Endo G, Silver S (1995) CadC, the transcriptional regulatory protein of the cadmium resistance system of *Staphylococcus aureus* plasmid pI258. J Bacteriol 177:4437–4441

Fauman EB, Cogswell JP, Lovejoy B, Rocque WJ, Holmes W, Montana VG, Piwnica-Worms H, Rink MJ, Saper MA (1998) Crystal structure of the catalytic domain of the human cell cycle control phosphatase, Cdc25A. Cell 93:617–625

Ghosh M, Shen J, Rosen BP (1999) Pathways of As(III) detoxification in *Saccharomyces cerevisiae*. Proc Natl Acad Sci USA 96:5001–5006

Gihring TM, Bond PL, Peters SC, Banfield JF (2003) Arsenic resistance in the archaeon "*Ferroplasma acidarmanus*": new insights into the structure and evolution of the *ars* genes. Extremophiles 7:123–130

Gladysheva TB, Oden KL, Rosen BP (1994) Properties of the arsenate reductase of plasmid R773. Biochemistry 33:7288–7293

Gourbal B, Sonuc N, Bhattacharjee H, Legare D, Sundar S, Ouellette M, Rosen BP, Mukhopadhyay R (2004) Drug uptake and modulation of drug resistance in *Leishmania* by an aquaglyceroporin. J Biol Chem 279:31010–31017

Guo X, Li Y, Peng K, Hu Y, Li C, Xia B, Jin C (2005) Solution structures and backbone dynamics of arsenate reductase from *Bacillus subtilis*: reversible conformational switch associated with arsenate reduction. J Biol Chem 280:39601–39608

Hamson GC, Stosick AJ (1938) The molecular structure of arsenious oxide, As_4O_6, phosphorus trioxide, P_4O_6, phosphorus pentoxide, P_4O_{10}, and hexamethylenetetramine, $(CH_6)N_4$, by electron diffraction. J Am Chem Soc 60:1814–1822

Harvey CF, Swartz CH, Badruzzaman AB, Keon-Blute N, Yu W, Ali MA, Jay J, Beckie R, Niedan V, Brabander D, Oates PM, Ashfaque KN, Islam S, Hemond HF, Ahmed MF (2002) Arsenic mobility and groundwater extraction in Bangladesh. Science 298:1602–1606

Heller KB, Lin EC, Wilson TH (1980) Substrate specificity and transport properties of the glycerol facilitator of *Escherichia coli*. J Bacteriol 144:274–278

Hirano S, Kobayashi Y, Cui X, Kanno S, Hayakawa T, Shraim A (2004) The accumulation and toxicity of methylated arsenicals in endothelial cells: important roles of thiol compounds. Toxicol Appl Pharmacol 198:458–467

Huckle JW, Morby AP, Turner JS, Robinson NJ (1993) Isolation of a prokaryotic metallothionein locus and analysis of transcriptional control by trace metal ions. Mol Microbiol 7:177–187

Islam FS, Gault AG, Boothman C, Polya DA, Charnock JM, Chatterjee D, Lloyd JR (2004) Role of metal-reducing bacteria in arsenic release from Bengal delta sediments. Nature 430:68–71

Ji G, Silver S (1992) Reduction of arsenate to arsenite by the ArsC protein of the arsenic resistance operon of *Staphylococcus aureus* plasmid pI258. Proc Natl Acad Sci USA 89:9474–9478

Ji G, Garber EAE, Armes LG, Chen CM, Fuchs JA, Silver S (1994) Arsenate reductase of *Staphylococcus aureus* plasmid pI258. Biochemistry 33:7294–7299

Karkaria CE, Chen CM, Rosen BP (1990) Mutagenesis of a nucleotide-binding site of an anion-translocating ATPase. J Biol Chem 265:7832–7836

Kaur P, Rosen BP (1992) Mutagenesis of the C-terminal nucleotide-binding site of an anion-translocating ATPase. J Biol Chem 267:19272–19277

King LS, Kozono D, Agre P (2004) From structure to disease: the evolving tale of aquaporin biology. Nat Rev Mol Cell Biol 5:687–698

Krafft T, Macy JM (1998) Purification and characterization of the respiratory arsenate reductase of *Chrysiogenes arsenatis*. Eur J Biochem 255:647–653

Kuroda M, Dey S, Sanders OI, Rosen BP (1997) Alternate energy coupling of ArsB, the membrane subunit of the Ars anion-translocating ATPase. J Biol Chem 272:326–331
Lebrun E, Brugna M, Baymann F, Muller D, Lievremont D, Lett MC, Nitschke W (2003) Arsenite oxidase, an ancient bioenergetic enzyme. Mol Biol Evol 20:686–693
Leipe DD, Wolf YI, Koonin EV, Aravind L (2002) Classification and evolution of P-loop GTPases and related ATPases. J Mol Biol 317:41–72
Leslie EM, Haimeur A, Waalkes MP (2004) Arsenic transport by the human multidrug resistance protein 1 (MRP1/ABCC1). Evidence that a tri-glutathione conjugate is required. J Biol Chem 279:32700–32708
Li R, Haile JD, Kennelly PJ (2003) An arsenate reductase from *Synechocystis* sp. strain PCC 6803 exhibits a novel combination of catalytic characteristics. J Bacteriol 185:6780–6789
Lin S, Shi Q, Nix FB, Styblo M, Beck MA, Herbin-Davis KM, Hall LL, Simeonsson JB, Thomas DJ (2002) A novel *S*-adenosyl-L-methionine:arsenic(III) methyltransferase from rat liver cytosol. J Biol Chem 277:10795–10803
Lin SJ, Pufahl RA, Dancis A, O'Halloran TV, Culotta VC (1997) A role for the *Saccharomyces cerevisiae ATX1* gene in copper trafficking and iron transport. J Biol Chem 272:9215–9220
Lin YF, Walmsley AR, Rosen BP (2006) An arsenic metallochaperone for an arsenic detoxification pump. Proc Natl Acad Sci USA 103:15617–15622
Liu Z, Sanchez M, Jiang X, Boles E, Landerfear S, Rosen BP (2006) Mammalian glucose transporter GLUT1 facilitates transport of arsenic trioxide and methylarsonous acid. Biochem Biophys Res Commun 351:424–430
Liu J, Rosen BP (1997) Ligand interactions of the ArsC arsenate reductase. J Biol Chem 272:21084–21089
Liu J, Gladysheva TB, Lee L, Rosen BP (1995) Identification of an essential cysteinyl residue in the ArsC arsenate reductase of plasmid R773. Biochemistry 34:13472–13476
Liu Z, Shen J, Carbrey JM, Mukhopadhyay R, Agre P, Rosen BP (2002) Arsenite transport by mammalian aquaglyceroporins AQP7 and AQP9. Proc Natl Acad Sci USA 99:6053–6058
Liu Z, Boles E, Rosen BP (2004a) Arsenic trioxide uptake by hexose permeases in *Saccharomyces cerevisiae*. J Biol Chem 279:17312–17318
Liu Z, Carbrey JM, Agre P, Rosen BP (2004b) Arsenic trioxide uptake by human and rat aquaglyceroporins. Biochem Biophys Res Commun 316:1178–1185
Lopez-Maury L, Florencio FJ, Reyes JC (2003) Arsenic sensing and resistance system in the cyanobacterium *Synechocystis* sp. strain PCC 6803. J Bacteriol 185:5363–5371
Maciaszczyk E, Wysocki R, Golik P, Lazowska J, Ulaszewski S (2004) Arsenical resistance genes in *Saccharomyces douglasii* and other yeast species undergo rapid evolution involving genomic rearrangements and duplications. FEMS Yeast Res 4:821–832
Marger MD, Saier MH Jr (1993) A major superfamily of transmembrane facilitators that catalyse uniport, symport and antiport. Trends Biochem Sci 18:13–20
Martin P, DeMel S, Shi J, Gladysheva T, Gatti DL, Rosen BP, Edwards BF (2001) Insights into the structure, solvation, and mechanism of ArsC arsenate reductase, a novel arsenic detoxification enzyme. Structure 9:1071–1081
Meng YL, Liu Z, Rosen BP (2004) As(III) and Sb(III) uptake by GlpF and efflux by ArsB in *Escherichia coli*. J Biol Chem 279:18334–18341
Messens J, Hayburn G, Desmyter A, Laus G, Wyns L (1999) The essential catalytic redox couple in arsenate reductase from *Staphylococcus aureus*. Biochemistry 38:16857–16865
Messens J, Martins JC, Van Belle K, Brosens E, Desmyter A, De Gieter M, Wieruszeski JM, Willem R, Wyns L, Zegers I (2002) All intermediates of the arsenate reductase mech-

anism, including an intramolecular dynamic disulfide cascade. Proc Natl Acad Sci USA 99:8506–8511
Metz J, Wachter A, Schmidt B, Bujnicki JM, Schwappach B (2006) The yeast Arr4p ATPase binds the chloride transporter Gef1p when copper is available in the cytosol. J Biol Chem 281:410–417
Morby AP, Turner JS, Huckle JW, Robinson NJ (1993) SmtB is a metal-dependent repressor of the cyanobacterial metallothionein gene *smtA*: identification of a Zn inhibited DNA–protein complex. Nucleic Acids Res 21:921–925
Mukhopadhyay R, Rosen BP (1998) *Saccharomyces cerevisiae ACR2* gene encodes an arsenate reductase. FEMS Microbiol Lett 168:127–136
Mukhopadhyay R, Rosen BP (2001) The phosphatase $C(X)_5R$ motif is required for catalytic activity of the *Saccharomyces cerevisiae* Acr2p arsenate reductase. J Biol Chem 276:34738–34742
Mukhopadhyay R, Rosen BP (2002) Arsenate reductases in prokaryotes and eukaryotes. Environ Health Perspect 110(Suppl 5):745–748
Mukhopadhyay R, Shi J, Rosen BP (2000) Purification and characterization of Acr2p, the *Saccharomyces cerevisiae* arsenate reductase. J Biol Chem 275:21149–21157
Mukhopadhyay R, Rosen BP, Phung LT, Silver S (2002) Microbial arsenic: from geocycles to genes and enzymes. FEMS Microbiol Rev 26:311–325
Mukhopadhyay R, Zhou Y, Rosen BP (2003) Directed evolution of a yeast arsenate reductase into a protein–tyrosine phosphatase. J Biol Chem 278:24476–24480
Mukhopadhyay R, Ho YS, Swiatek PJ, Rosen BP, Bhattacharjee H (2006) Targeted disruption of the mouse *Asna1* gene results in embryonic lethality. FEBS Lett 580:3889–3894
Muller D, Lievremont D, Simeonova DD, Hubert JC, Lett MC (2003) Arsenite oxidase *aox* genes from a metal-resistant β-proteobacterium. J Bacteriol 185:135–141
Neyt C, Iriarte M, Thi VH, Cornelis GR (1997) Virulence and arsenic resistance in Yersiniae. J Bacteriol 179:612–619
Ordonez E, Letek M, Valbuena N, Gil JA, Mateos LM (2005) Analysis of genes involved in arsenic resistance in *Corynebacterium glutamicum* ATCC 13032. Appl Environ Microbiol 71:6206–6215
Oremland RS, Stolz JF (2003) The ecology of arsenic. Science 300:939–944
Oremland RS, Stolz JF (2005) Arsenic, microbes and contaminated aquifers. Trends Microbiol 13:45–49
Petrick JS, Ayala-Fierro F, Cullen WR, Carter DE, Vasken Aposhian H (2000) Monomethylarsonous acid (MMA(III)) is more toxic than arsenite in Chang human hepatocytes. Toxicol Appl Pharmacol 163:203–207
Qin J, Rosen BP, Zhang Y, Wang G, Franke S, Rensing C (2006) Arsenic detoxification and evolution of trimethylarsine gas by a microbial arsenite *S*-adenosylmethionine methyltransferase. Proc Natl Acad Sci USA 103:2075–2080
Ramirez-Solis A, Mukopadhyay R, Rosen BP, Stemmler TL (2004) Experimental and theoretical characterization of arsenite in water: insights into the coordination environment of As–O. Inorg Chem 43:2954–2959
Roos G, Loverix S, Brosens E, Van Belle K, Wyns L, Geerlings P, Messens J (2006) The activation of electrophile, nucleophile and leaving group during the reaction catalysed by pI258 arsenate reductase. Chembiochem 7:981–989
Rosen BP (2002) Biochemistry of arsenic detoxification. FEBS Lett 529:86–92
Rosen BP, Weigel U, Karkaria C, Gangola P (1988) Molecular characterization of an anion pump. The *arsA* gene product is an arsenite(antimonate)-stimulated ATPase. J Biol Chem 263:3067–3070

Rosenberg H, Gerdes RG, Chegwidden K (1977) Two systems for the uptake of phosphate in *Escherichia coli*. J Bacteriol 131:505–511
Rosenzweig AC (2002) Metallochaperones: bind and deliver. Chem Biol 9:673–677
Ruan X, Bhattacharjee H, Rosen BP (2006) Cys-113 and Cys-422 form a high-affinity metalloid binding site in the ArsA ATPase. J Biol Chem 281:9925–9934
Ryan D, Colleran E (2002) Arsenical resistance in the IncHI2 plasmids. Plasmid 47:234–240
Saltikov CW, Newman DK (2003) Genetic identification of a respiratory arsenate reductase. Proc Natl Acad Sci USA 100:10983–10988
Saltikov CW, Cifuentes A, Venkateswaran K, Newman DK (2003) The *ars* detoxification system is advantageous but not required for As(V) respiration by the genetically tractable *Shewanella* species strain ANA-3. Appl Environ Microbiol 69:2800–2809
Sanders OI, Rensing C, Kuroda M, Mitra B, Rosen BP (1997) Antimonite is accumulated by the glycerol facilitator GlpF in *Escherichia coli*. J Bacteriol 179:3365–3367
Santini JM, van den Hoven RN (2004) Molybdenum-containing arsenite oxidase of the chemolithoautotrophic arsenite oxidizer NT-26. J Bacteriol 186:1614–1619
Santini JM, Sly LI, Schnagl RD, Macy JM (2000) A new chemolithoautotrophic arsenite-oxidizing bacterium isolated from a gold mine: phylogenetic, physiological, and preliminary biochemical studies. Appl Environ Microbiol 66:92–97
Sato T, Kobayashi Y (1998) The *ars* operon in the skin element of *Bacillus subtilis* confers resistance to arsenate and arsenite. J Bacteriol 180:1655–1661
Schuldiner M, Collins SR, Thompson NJ, Denic V, Bhamidipati A, Punna T, Ihmels J, Andrews B, Boone C, Greenblatt JF, Weissman JS, Krogan NJ (2005) Exploration of the function and organization of the yeast early secretory pathway through an epistatic miniarray profile. Cell 123:507–519
Shen J, Hsu CM, Kang BK, Rosen BP, Bhattacharjee H (2003) The *Saccharomyces cerevisiae* Arr4p is involved in metal and heat tolerance. Biometals 16:369–378
Shi J, Vlamis-Gardikas A, Åslund F, Holmgren A, Rosen BP (1999) Reactivity of glutaredoxins 1, 2, and 3 from *Escherichia coli* shows that glutaredoxin 2 is the primary hydrogen donor to ArsC-catalyzed arsenate reduction. J Biol Chem 274:36039–36042
Shi J, Mukhopadhyay R, Rosen BP (2003) Identification of a triad of arginine residues in the active site of the ArsC arsenate reductase of plasmid R773. FEMS Microbiol Lett 227:295–301
Shi W, Wu J, Rosen BP (1994) Identification of a putative metal binding site in a new family of metalloregulatory proteins. J Biol Chem 269:19826–19829
Shi W, Dong J, Scott RA, Ksenzenko MY, Rosen BP (1996) The role of arsenic–thiol interactions in metalloregulation of the *ars* operon. J Biol Chem 271:9291–9297
Soignet SL (2001) Clinical experience of arsenic trioxide in relapsed acute promyelocytic leukemia. Oncologist 6 Suppl 2:11–16
Soignet SL, Maslak P, Wang ZG, Jhanwar S, Calleja E, Dardashti LJ, Corso D, DeBlasio A, Gabrilove J, Scheinberg DA, Pandolfi PP, Warrell RPJ Jr (1998) Complete remission after treatment of acute promyelocytic leukemia with arsenic trioxide. N Engl J Med 339:1341–1348
Stolz JF, Basu P, Santini JM, Oremland RS (2006) Arsenic and selenium in microbial metabolism. Annu Rev Microbiol 60:107–130
Sun Y, Wong MD, Rosen BP (2001) Role of cysteinyl residues in sensing Pb(II), Cd(II), and Zn(II) by the plasmid pI258 CadC repressor. J Biol Chem 276:14955–14960
Tauriainen S, Karp M, Chang W, Virta M (1998) Luminescent bacterial sensor for cadmium and lead. Biosens Bioelectron 13:931–938

Tisa LS, Rosen BP (1990) Molecular characterization of an anion pump. The ArsB protein is the membrane anchor for the ArsA protein. J Biol Chem 265:190–194
van den Hoven RN, Santini JM (2004) Arsenite oxidation by the heterotroph *Hydrogenophaga* sp. str. NT-14: the arsenite oxidase and its physiological electron acceptor. Biochim Biophys Acta 1656:148–155
VanZile ML, Cosper NJ, Scott RA, Giedroc DP (2000) The zinc metalloregulatory protein *Synechococcus* PCC7942 SmtB binds a single zinc ion per monomer with high affinity in a tetrahedral coordination geometry. Biochemistry 39:11818–11829
VanZile ML, Chen X, Giedroc DP (2002a) Allosteric negative regulation of *smt* O/P binding of the zinc sensor, SmtB, by metal ions: a coupled equilibrium analysis. Biochemistry 41:9776–9786
VanZile ML, Chen X, Giedroc DP (2002b) Structural characterization of distinct α3N and α5 metal sites in the cyanobacterial zinc sensor SmtB. Biochemistry 41:9765–9775
Walmsley AR, Zhou T, Borges-Walmsley MI, Rosen BP (1999) The ATPase mechanism of ArsA, the catalytic subunit of the arsenite pump. J Biol Chem 274:16153–16161
Walmsley AR, Zhou T, Borges-Walmsley MI, Rosen BP (2001a) Antimonite regulation of the ATPase activity of ArsA, the catalytic subunit of the arsenical pump. Biochem J 360:589–597
Walmsley AR, Zhou T, Borges-Walmsley MI, Rosen BP (2001b) A kinetic model for the action of a resistance efflux pump. J Biol Chem 276:6378–6391
Willsky GR, Malamy MH (1980a) Characterization of two genetically separable inorganic phosphate transport systems in *Escherichia coli*. J Bacteriol 144:356–365
Willsky GR, Malamy MH (1980b) Effect of arsenate on inorganic phosphate transport in *Escherichia coli*. J Bacteriol 144:366–374
Wong MD, Lin YF, Rosen BP (2002) The soft metal ion binding sites in the *Staphylococcus aureus* pI258 CadC Cd(II)/Pb(II)/Zn(II)-responsive repressor are formed between subunits of the homodimer. J Biol Chem 277:40930–40936
Wu J, Rosen BP (1991) The ArsR protein is a trans-acting regulatory protein. Mol Microbiol 5:1331–1336
Wu J, Rosen BP (1993) Metalloregulated expression of the *ars* operon. J Biol Chem 268:52–58
Wu J, Tisa LS, Rosen BP (1992) Membrane topology of the ArsB protein, the membrane subunit of an anion-translocating ATPase. J Biol Chem 267:12570–12576
Wysocki R, Chery CC, Wawrzycka D, Van Hulle M, Cornelis R, Thevelein JM, Tamás MJ (2001) The glycerol channel Fps1p mediates the uptake of arsenite and antimonite in *Saccharomyces cerevisiae*. Mol Microbiol 40:1391–1401
Xu C, Rosen BP (1999) Metalloregulation of soft metal resistance pumps. In: Sarkar B (ed) Metals and genetics. Plenum, New York, pp 5–19
Xu C, Shi W, Rosen BP (1996) The chromosomal *arsR* gene of *Escherichia coli* encodes a trans-acting metalloregulatory protein. J Biol Chem 271:2427–2432
Yang HC, Cheng J, Finan TM, Rosen BP, Bhattacharjee H (2005) Novel pathway for arsenic detoxification in the legume symbiont *Sinorhizobium meliloti*. J Bacteriol 187:6991–6997
Ye J, Kandegedara A, Martin P, Rosen BP (2005) Crystal structure of the *Staphylococcus aureus* pI258 CadC Cd(II)/Pb(II)/Zn(II)-responsive repressor. J Bacteriol 187:4214–4221
Zegers I, Martins JC, Willem R, Wyns L, Messens J (2001) Arsenate reductase from *S. aureus* plasmid pI258 is a phosphatase drafted for redox duty. Nat Struct Biol 8:843–847
Zhou T, Rosen BP (1997) Tryptophan fluorescence reports nucleotide-induced conformational changes in a domain of the ArsA ATPase. J Biol Chem 272:19731–19737

Zhou T, Liu S, Rosen BP (1995) Interaction of substrate and effector binding sites in the ArsA ATPase. Biochemistry 34:13622–13626

Zhou T, Radaev S, Rosen BP, Gatti DL (2000) Structure of the ArsA ATPase: the catalytic subunit of a heavy metal resistance pump. Embo J 19:4838–4845

Zhou T, Radaev S, Rosen BP, Gatti DL (2001) Conformational changes in four regions of the *Escherichia coli* ArsA ATPase link ATP hydrolysis to ion translocation. J Biol Chem 276:30414–30422

Zhou T, Shen J, Liu Y, Rosen BP (2002) Unisite and multisite catalysis in the ArsA ATPase. J Biol Chem 277:23815–23820

Zhou Y, Messier N, Ouellette M, Rosen BP, Mukhopadhyay R (2004) *Leishmania major* LmACR2 is a pentavalent antimony reductase that confers sensitivity to the drug pentostam. J Biol Chem 279:37445–37451

Zhou Y, Bhattacharjee H, Mukhopadhyay R (2006) Bifunctional role of the leishmanial antimonate reductase LmACR2 as a protein tyrosine phosphatase. Mol Biochem Parasitol 148:161–168

Zuniga S, Boskovic J, Jimenez A, Ballesta JP, Remacha M (1999) Disruption of six *Saccharomyces cerevisiae* novel genes and phenotypic analysis of the deletants. Yeast 15:945–953

Microbiol Monogr (6)
D. H. Nies, S. Silver: Molecular Microbiology of Heavy Metals
DOI 10.1007/7171_2006_087/Published online: 3 February 2007

Reduction and Efflux of Chromate by Bacteria

Carlos Cervantes (✉) · Jesús Campos-García

Laboratorio de Microbiologia,
Instituto de Investigaciones Quimico-Biologicas,
Universidad Michoacana de San Nicolás de Hidalgo,
Edificio B-3, Ciudad Universitaria, 58030 Morelia, Mexico
cvega1999@yahoo.com

Abstract The widespread industrial use of chromium has caused this heavy metal to be considered as a serious environmental pollutant. The most common forms of chromium in nature are the relatively innocuous trivalent form, Cr(III), and the more toxic hexavalent species, Cr(VI). Cr(VI) is usually present as the oxyanion chromate. Toxic effects of chromate for bacteria are associated with the inhibition of sulfate transport and with oxidative damage to biomolecules. The best studied bacterial mechanisms of resistance to chromate include reduction of Cr(VI) to the Cr(III) species and efflux of chromate from cell cytoplasm. Several chromate reductases have been identified in diverse bacterial species. Most characterized enzymes belong to the NAD(P)H-dependent flavoprotein family of reductases. Efflux of chromate by the ChrA membrane transporter, a plasmid-encoded protein, has been demonstrated in *Pseudomonas* and *Cupriavidus* species. Chromate efflux by ChrA consists of an energy-dependent process driven by the membrane potential. The CHR protein family, which includes putative ChrA homologs, currently contains about 135 sequences from all three domains of life. Other mechanisms of bacterial resistance to chromate involve the expression of components of the machinery for repair of DNA damage as well as free-radical scavenging enzymes.

1 Introduction: Chromium in the Environment

Chromium (Cr) is a transition metal in group VI-B of the periodic table. Cr naturally occurs in rocks, soils, plants, animals, and volcanic emissions; it is the seventh most abundant element on earth. However, most Cr is in the earth's mantle which results in Cr being only the twenty-first most abundant element in the crust (McGrath and Smith 1990). The most stable and abundant forms are the trivalent Cr(III) and the hexavalent Cr(VI) species. Cr(VI) is commonly present in solution as the water-soluble chromate (CrO_4^{2-}) or dichromate ($Cr_2O_7^{2-}$) oxyanions. Cr(VI), a strong oxidizing agent, is rapidly reduced by organic matter and other reducing agents to yield Cr(III) in soil and aquatic settings (McGrath and Smith 1990). Cr(III) derivatives are much less mobile and exist in the environment mostly in the form of stable complexes with both organic and inorganic ligands. The widespread use of Cr in diverse industrial processes has converted it into a serious contaminant of air, soil, and water.

2 Chromium Transport and Accumulation

Active transport of chromate across membranes by means of the sulfate uptake pathway has been demonstrated in a variety of bacterial species (Cervantes et al. 2001). The chemical analogy of these oxyanions is emphasized by the fact that chromate is a competitive inhibitor of sulfate transport in all bacterial species where it has been tested. In contrast, most cells are impermeable to Cr(III), which forms insoluble compounds in non-acidic aqueous solutions (Cary 1982). Once inside the cells, Cr(VI) is readily reduced, nonenzymatically or by the action of various enzymes, to Cr(III), which then may exert diverse toxic effects in the cytoplasm (Cervantes et al. 2001).

3 Chromium Toxicity

The biological effects of Cr depend on its oxidation state. Cr(VI) is considered the most toxic form of chromium (Cervantes et al. 2001). At the extracellular level, Cr(VI) is highly toxic to most bacteria, whereas Cr(III) is relatively innocuous because of its insolubility and subsequent inability to traverse cell membranes (Katz and Salem 1993). In the cytoplasm, Cr toxicity is mainly related to the process of reduction of Cr(VI) to lower oxidation states [i.e., Cr(V) and Cr(III)] in which free radicals may be formed (Kadiiska et al.

1994). Oxidative damage to DNA is probably responsible for the genotoxic effects caused by chromate (Itoh et al. 1995, Luo et al. 1996). Cr(III) may exert additional toxic effects by its ability to bind to phosphates in DNA (Bridgewater et al. 1994; Plaper et al. 2002), to carboxyl and sulfhydryl groups in proteins (Levis and Bianchi 1982), and by competing with the transport of iron (Moshtaghie et al. 1992).

4 Mechanisms of Resistance to Chromate

Chromate resistance mechanisms may be encoded either by chromosomal genes or by plasmids (Cervantes et al. 2001). Resistance systems located in bacterial chromosomes are usually related to strategies such as Cr(VI) reduction by specific or unspecific enzymes, repairing of DNA damages, detoxifying activities, and processes associated to sulfur or iron metabolism. On the other hand, plasmid genes are mostly responsible for membrane transporters, which directly mediate efflux of chromate ions from the cell's cytoplasm.

4.1 Reduction of Hexavalent Chromium

Under certain environmental conditions Cr can be interconverted as Cr(III) and Cr(VI) by oxidation-reduction reactions of biotic and abiotic nature. Microbial reduction of Cr(VI) to Cr(III) is not a plasmid-associated trait and can be considered as an additional chromate resistance mechanism (Cervantes et al. 2001). Three Cr(VI) reduction mechanisms have been described:

1. Chromate reduction under aerobic conditions is commonly associated with soluble chromate reductases that use NADH or NADPH as cofactors
2. Under anaerobic conditions, some bacteria can use Cr(VI) as an electron acceptor in the electron transport chain
3. Cr(VI) may also be reduced by unspecific reactions associated with organic compounds such as amino acids, nucleotides, sugars, vitamins, organic acids, or glutathione

4.1.1 Enzymatic Cr(VI) Reduction

Chromate reduction is carried out by diverse bacterial species although only a few enzymes have been characterized (Ohtake and Silver 1994; Cervantes et al. 2001). Cr(VI) reduction may occur under aerobic or anaerobic conditions and may be associated with the cell membrane or with the soluble fraction. Examples of the most studied chromate reductases are shown in Table 1. The broad range in size, cellular location, reduction conditions, as well as in the kinetic properties for these enzymes reflects the diversity of bacterial activities able to reduce Cr(VI).

The first analyzed chromate reductase was a membrane-associated enzyme from *Enterobacter cloacae* HO1 that transfers electrons to Cr(VI) by NADH-dependent cytochromes (Ohtake et al. 1990; Wang et al. 1990). Ishibashi et al. (1990) suggested that Cr(VI) reductases may have a primary role other than Cr(VI) reduction; this secondary function for Cr(VI) reductases may be related to the recent introduction of Cr(VI) to the environment by anthropogenic activities.

Cr(VI) reductases have been primarily characterized in the context of alternative substrates; these enzymes commonly show a NADH:flavin oxidoreductase activity and can also act as chromate reductases. This is the case for reductases NfsA/NfsB from *Vibrio harveyi*, with a nitrofurazone nitroreductase property as a primary activity and chromate reductase as a secondary function (Kwak et al. 2003), and ferric reductase FerB from *Paracoccus denitrificans* that uses Fe(III)-nitrilotriacetate and chromate as substrates (Mazoch et al. 2004).

The currently best studied chromate reductase is ChrR from *Pseudomonas putida*, a soluble flavin mononucleotide-binding enzyme able to catalyze the reduction of Cr(VI) to Cr(III) (Park et al. 2000). The ChrR enzyme functions as a 50-kDa dimer and shows a NADH-dependent reductase activity (Table 1). Studies with purified *P. putida* ChrR revealed a quinone reductase activity that generates a flavin semiquinone during chromate reduction; this reaction transfers $> 25\%$ of the NADH electrons to superoxide anion and probably produces the Cr(V) species transiently. ChrR seems to provide an antioxidant defense mechanism to *P. putida* by protecting cells against H_2O_2 stress (Gonzalez et al. 2005). ChrR in one pathway reduces Cr(VI) to Cr(III), generating intermediary Cr(V) and superoxide anion, and by an additional mechanism reduces quinones, which provide shielding from reactive oxygen species (ROS).

The *Escherichia coli* YieF protein shares sequence homology with the *P. putida* ChrR enzyme (Ackerley et al. 2004). Both soluble enzymes are members of a widespread family of proteins and show similar properties (Table 1). YieF, however, has a different reaction mechanism that involves an obligatory four-electron reduction of chromate in which the enzyme simultaneously transfers three electrons to chromate to produce Cr(III) and one electron to molecular oxygen generating ROS; no flavin semiquinone form of the enzyme is generated in this process (Ackerley et al. 2004). By forming a lower amount of ROS, YieF may protect *E. coli* cells against chromate toxicity.

Table 1 Properties of bacterial chromate reductases

Bacteria	Reduction conditions	Enzyme localization	Electron donor	Optimal temp (°C)	Optimal pH	Molecular mass [a] (kDa)	K_m [b]	V_{Max} [c]	Refs.
P. fluorescens LB300	Aer/Ana	M	NADH	nd	nd	nd	nd	nd	Bopp and Ehrlich, 1988
P. putida PRS2000	Aer	S	NAD(P)H	50	6.5–7.5	nd	40	6	Ishibashi et al. 1990
E. cloacae HO1	Ana	M	NADH	30	7.0	nd	1.1	650	Ohtake et al. 1990
P. ambigua G-1	Aer	S	NAD(P)H	50	8.6	65(25)	13	27	Suzuki et al. 1992
D. vulgaris	Ana	S	H_2	30	6.8	nd	nd	nd	Lovley and Phillips 1994
Bacillus sp. QC1-2	Aer/Ana	S	NADH	37	7.0	44(24)	350	50	Campos et al. 1995
V. harveyi	Aer	S	NADH	30	7.0	50(50)	5.4	10.7	Kwak et al. 2003
P. putida KT2440	Aer	S	NADH	70	5.0	50(22)	260	8.8	Ackerley et al. 2004
E. coli AMS6	Aer	S	NAD(P)H	35	5.0	50(22)	200	5	Ackerley et al. 2004
E. coli ATCC33456	Aer/Ana	S	NAD(P)H	37	6.5	84(42)	nd	322	Bae et al. 2005

[a] Molecular mass in native conditions, values in parentheses are from SDS-PAGE
[b] μM chromate
[c] nmol chromate $min^{-1} mg^{-1}$
Aer aerobiosis, *Ana* anaerobiosis, *M* membrane fraction, *S* soluble fraction, *nd* not determined

4.2 Efflux of Chromate Ions

Plasmids pUM505 of *Pseudomonas aeruginosa* and pMOL28 from *Cupriavidus metallidurans* (previously *Alcaligenes eutrophus* and *Ralstonia metallidurans*) (Cervantes et al. 1990; Nies et al. 1990) carry genes for the hydrophobic protein ChrA, which confers resistance to chromate. The two ChrA homologs are 29% identical at the amino acid level. They are reported to display a different topology: ten transmembrane segments (TMS) for the *C. metallidurans* protein (401 amino acids, aa; Nies et al. 1998) and 13 TMS for that of *P. aeruginosa* (416 aa; Jiménez-Mejía et al. 2006). The two ChrAs cause a reduced accumulation of chromate ions (Cervantes et al. 1990; Nies et al. 1990). In *C. metallidurans*, the chromosome contains the $chrA_2$ gene, encoding a protein 84% identical to the product of its plasmid-encoded $chrA_1$ homolog. Expression of $chrA_2$ also confers chromate resistance (Juhnke et al. 2002).

Using everted membrane vesicles and intact cells, the *P. aeruginosa* ChrA protein was shown to function as a chemiosmotic pump that effluxes chromate from the cytoplasm using the proton motive force (Alvarez et al. 1999; Pimentel et al. 2002). Random mutagenesis of the *P. aeruginosa chrA* gene showed that most essential amino acid residues are located in the amino terminal end of ChrA (Aguilera et al. 2004). In agreement with this finding, phylogenetic analysis of ChrA homologs revealed that the amino terminal halves are more conserved than the carboxyl terminal halves (Díaz-Pérez, Cervantes and Riveros-Rosas, submitted). These data suggest that the two halves of ChrA carry out different functions in their transporting duties.

The *chrA* gene from *C. metallidurans* plasmid pMOL28 forms part of the chr_1 operon that includes the *chrI*, *chrB*, and *chrF* genes, proposed to play regulatory roles for the expression of ChrA (Juhnke et al. 2002). A function of the ChrB protein (324 aa) in the inducibility of the *chrA* gene by chromate had been previously demonstrated (Nies et al. 1990).

The complete sequencing of plasmid pB4 from *Pseudomonas* sp. B13 strain revealed the presence of the *chrBAC* gene cluster, which showed sequence similarity with the chromate resistance determinants from plasmids pUM505 and pMOL28 (Tauch et al. 2003). Interestingly, the *chr* genes from pB4 are arranged as a transposable unit.

4.2.1 The CHR Family of Transporters

The CHR family of transporters was first described by Nies et al. (1998) as a small group of prokaryotic proteins involved in chromate or sulfate transport. This family has been classified as TC # 2.A.51 (Saier 2003) and includes proteins encoded in chromosomes and in plasmids. Currently the protein databases contain several dozen homologs, including proteins from eukaryotes (Díaz-Pérez et al. submitted) (Fig. 1). Apart from the *P. aeruginosa* and *C. metallidurans* ChrA proteins, the function of other CHR homologs has not yet been analyzed in detail.

Phylogenetic analysis suggests that the CHR proteins may have derived from a gene duplication event, as occurred with members of other families of transporters. CHR homologs exist in two sizes: small proteins (SCHR, about 200 aa in size) with only one domain, and large proteins (LCHR, about 400 aa, except most eukaryotic proteins with more than 500 aa; Fig. 1) with two homologous domains (Nies et al. 1998; Díaz-Pérez et al. submitted).

Of a total of 135 CHR homologous sequences analyzed, 77 consisted of LCHR proteins arranged in seven subgroups, containing mostly proteins from bacteria (70 sequences) (Fig. 1). Subgroups CHR2 and CHR5 include mainly proteins from proteobacteria, among them the ChrAs from *C. metallidurans* and *P. aeruginosa* with a demonstrated function in chromate efflux. These two subgroups, and probably the closer CHR3 subgroup, may contain functional chromate transporters. Subgroup CHR4 includes the only protein from an Archaea (*Methanococcus jannaschii*). The fungal CHR subgroup contains six proteins from fungal species most closely related to the numerous bacterial CHR1 subgroup containing all the Gram positive homologs (Fig. 1).

CHR gene clusters from different sources differ in their possession of associated potential regulatory genes. For example, most gene clusters from the CHR2 subgroup contain regulatory ChrB homologs (Fig. 2). In contrast, members of the CHR5 subgroup usually lack ChrB-encoding genes. The presence of *chrC*, a third gene encoding a probable superoxide dismutase (SOD) ChrC enzyme (see Sect. 4.3), is another variable feature of the CHR gene clusters. ChrC homologs are present in three out of nine members of the CHR2 group (Fig. 2) and in six out of 15 from those of CHR5. Fungal and CHR1 subgroups do not contain any associated genes with a related function.

4.3 Other Mechanisms of Chromate Tolerance

Besides Cr(VI) reduction and chromate efflux, bacteria display other mechanisms to deal with Cr. Since oxidative stress is responsible for most toxic effects of Cr, protection and detoxification systems against this process should be expected to be an important part of the defensive barrier. Additional protective strategies are related to sulfur or iron metabolism.

The chr_1 operon from *C. metallidurans* plasmid pMOL28, which expresses the chromate efflux pump ChrA (Fig. 2, entry CUPME), also encodes the ChrC and ChrE proteins that seem to be involved in the chromate resistance phenotype (Juhnke et al. 2002). ChrC (197 aa) has sequence similarity to iron-containing SOD enzymes able to detoxify superoxide radicals. The ChrE protein (113 aa) showed sequence similarity to members of the rho-

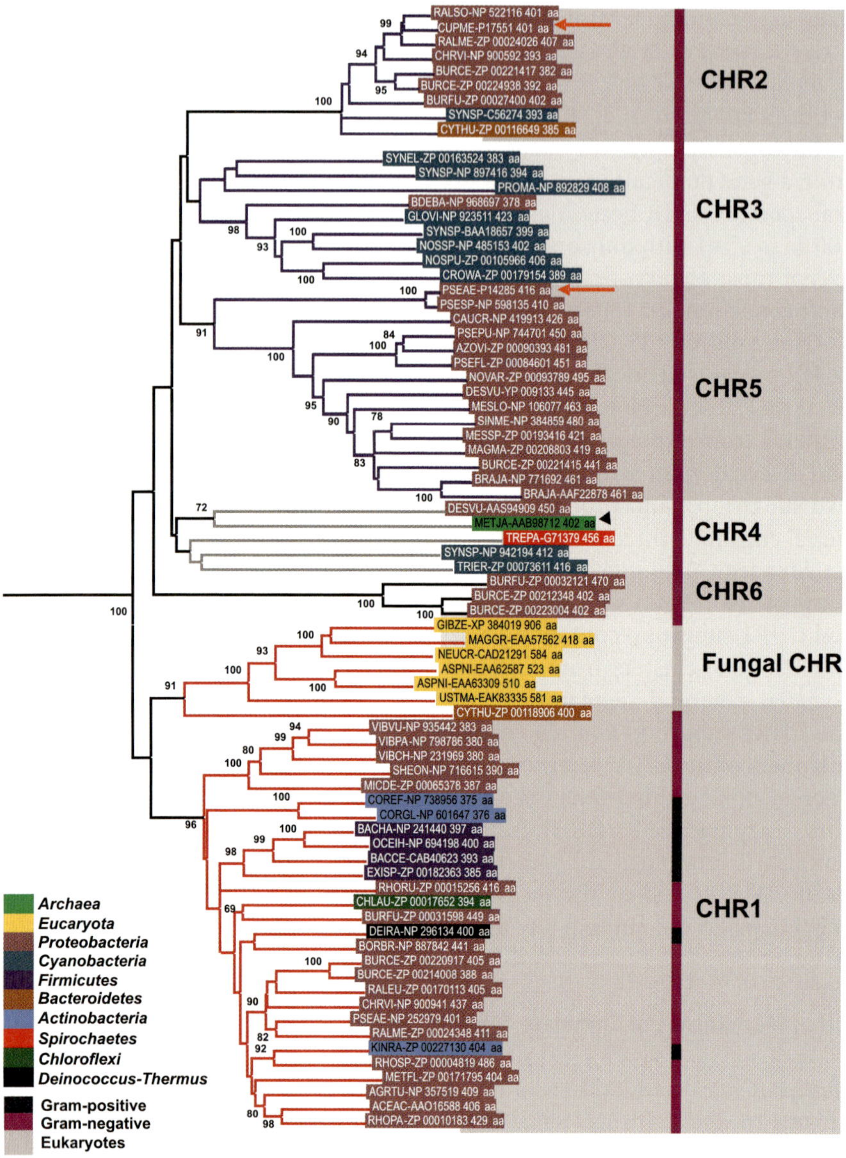

Fig. 1 Phylogenetic tree of the CHR family of proteins. The distribution of the 77 homologous sequences from LCHR proteins is shown (entries include abbreviated species name, sequence access number, and protein size). The *leftmost legends* show the taxonomic classification (color codes). The Gram types are shown by the *vertical bar to the right*. The seven main subgroups described in the text (CHR1-5 nomenclature according to Nies, 2003) are indicated. The *red arrows* indicate the characterized proteins from *C. metallidurans* (CUPME-P17551) and *P. aeruginosa* (PSEAE-P14285). The *arrowhead* marks the *M. jannaschii* archaeal sequence (METJA-AAB98712). Bootstrap values are shown for the tree branches. Data from Díaz-Pérez et al. (submitted for publication)

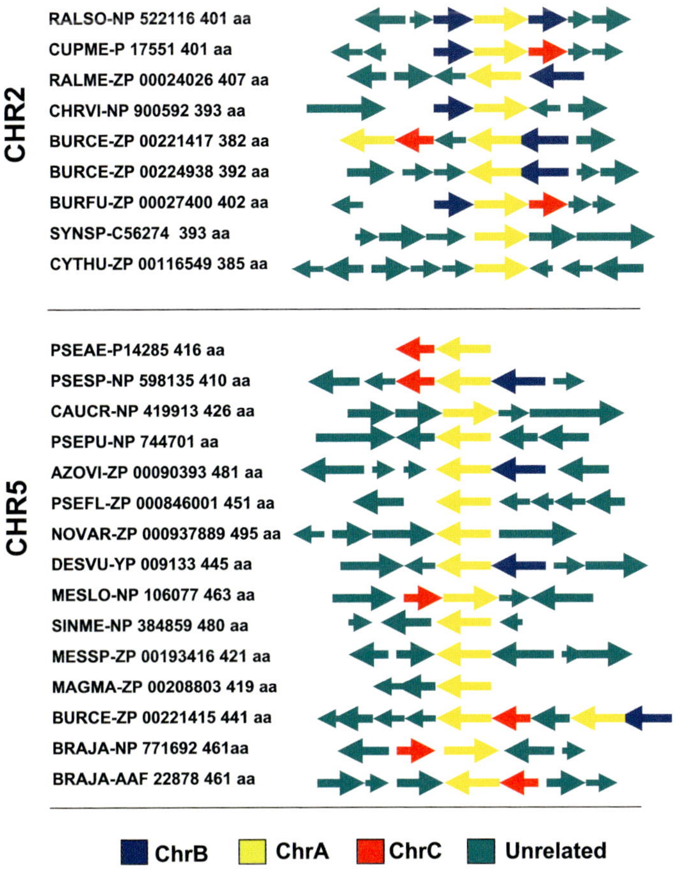

Fig. 2 Gene arrangement of the CHR2 and CHR5 subgroup clusters of the CHR family. The name, abbreviation, sequence access number, and protein size for each species are given on the *left*. *Colored arrows* indicate the direction of transcription of genes, with gene assignments given below. *Green arrows* represent genes of unknown functions that are probably not related to chromate resistance. Data from Díaz-Pérez et al. (unpublished). Abbreviations: RALSO *Ralstonia solanacearum*; CUPME *Cupriavidus metallidurans*; RALME *Ralstonia metallidurans*; CHRVI *Chromobacterium violaceum*; BURCE *Burkholderia cepacia*; BURFU *Burkholderia fungorum*; SYNSP *Synechococcus* sp.; CYTHU *Cytophaga hutchinsonii*; PSEAE *Pseudomonas aeruginosa*; PSESP *Pseudomonas* sp.; CAUCR *Caulobacter crescentus*; PSEPU *Pseudomonas putida*; AZOVI *Azotobacter vinelandii*; PSEFL *Pseudomonas fluorescens*; NOVAR *Novosphingobium aromaticivorans*; DESVU *Desulfovibrio vulgaris*; MESLO *Mesorhizobium loti*; SINME *Sinorhizobium meliloti*; MESSP *Mesorhizobium* sp.; MAGMA *Magnetospirillum magnetotacticum*; BRAJA *Bradyrhizobium japonicum*

danese superfamily, and has been proposed to participate in the cleavage of chromium–glutathione complexes, a possible protection mechanism (Juhnke et al. 2002).

Plasmid pANL from *Synechococcus* sp. encodes the sulfur-regulated SrpC protein (393 aa) that shows similarity with *C. metallidurans* and *P. aeruginosa* ChrA proteins (Nicholson and Laudenbach 1995). SrpC belongs to the CHR2 subgroup (Fig. 2, entry SYNSP) but does not seem to extrude chromate ions and may be involved in sulfate uptake.

Protection of bacterial cells from DNA damage caused by Cr is another defensive shield. Cr(VI) has long been known to induce the *E. coli* SOS repair system that protects DNA from oxidative damage (Llangostera et al. 1986). Similarly, DNA helicases RecG and RuvB, components of the recombinational DNA repair system, were shown to participate in the response to DNA damage caused by chromate in *P. aeruginosa* (Miranda et al. 2005).

Ackerley et al. (2006) found that chromate exposure of *E. coli* cells led to the depletion of the pools of glutathione and other thiols, suggesting an important detoxifying role of these compounds against Cr. These authors found that *E. coli* displays additional protective systems, including the induction of the SOS response and the activation of superoxide dismutase and catalase.

Genome-wide analysis of cellular responses to chromate stress in *Caulobacter crescentus*, a bacterium able to survive in polluted habitats, revealed the upregulation of genes encoding products involved in protection against oxidative stress, in repair of DNA damage, as well as genes for outer membrane proteins (Hu et al. 2005). *C. crescentus* seems not to have a chromate efflux system but Cr stress downregulates a sulfate transport system, probably reducing chromate uptake (Hu et al. 2005).

A more complex response was found in the metal-reducing bacterium *Shewanella oneidensis* when exposed to chromate (Brown et al. 2006). Besides the anticipated induction of genes devoted to DNA repair and cellular detoxification, *S. oneidensis* showed the upregulation of genes involved in sulfur metabolism, iron transport, and transcriptional regulation (Brown et al. 2006).

5
Concluding Remarks

Bacterial responses to Cr exposure involve direct strategies aimed at transforming Cr(VI) to innocuous Cr(III) outside the cell or at extruding toxic chromate ions from the cytoplasm. Many bacterial species are reported to reduce Cr(VI), but the properties of the Cr reductases responsible have been elucidated for only a few enzymes. The widely diverse properties of these enzymes support the hypothesis that reduction of chromate is a secondary role of Cr reductases.

A large family of probable Cr transporters has been identified in the three life domains, although only a few bacterial homologs have been well characterized.

Indirect Cr tolerance mechanisms include systems devoted to protect biomolecules from oxidative stress or to repair the damages caused by Cr derivatives. After chromate exposure, diverse bacterial species seem to deploy a varied, but usually global, regulatory network that involves the expression of genes for several different metabolic processes as a Cr stress defensive strategy. The environmental importance of Cr as a pollutant and its involvement in animal and plant toxicity makes the study of the interactions of this metal with microbial cells a fertile field for both basic and applied research.

Acknowledgements The research in our laboratories was supported by grants from CIC (Universidad Michoacana) and CONACYT (no. 41712-Q and C01-46547). We thank H. Riveros-Rosas and C. Díaz-Pérez for advice in the preparation of the manuscript.

References

Ackerley DF, Gonzalez CF, Park CH, Blake R, Keyhan M, Matin A (2004) Chromate-reducing properties of soluble flavoproteins from *Pseudomonas putida* and *Escherichia coli*. Appl Environ Microbiol 70:873–882

Ackerley DF, Barak Y, Lynch SV, Curtin J, Matin A (2006) Effect of chromate stress on *Escherichia coli* K-12. J Bacteriol 188:3371–3381

Aguilera S, Aguilar ME, Chávez MP, López-Meza JE, Pedraza-Reyes M, Campos-García J, Cervantes C (2004) Essential residues in the chromate transporter ChrA of *Pseudomonas aeruginosa*. FEMS Microbiol Lett 232:107–112

Alvarez AH, Moreno-Sánchez R, Cervantes C (1999) Chromate efflux by means of the ChrA chromate resistance protein from *Pseudomonas aeruginosa*. J Bacteriol 181:7398–7400

Bae WC, Lee HK, Choe YC, Jahng DJ, Lee SH, Kim SJ, Lee JH, Jeong BC (2005) Purification and characterization of NADPH-dependent Cr(VI) reductase from *Escherichia coli* ATCC 33456. J Microbiol 43:21–27

Bopp LH, Erlich HL (1988) Chromate resistance and reduction in *Pseudomonas fluorescens* strain LB300. Arch Microbiol 150:426–431

Bridgewater LC, Manning FC, Woo ES, Patierno SR (1994) DNA polymerase arrest by adducted trivalent chromium. Mol Carcinog 9:122–133

Brown SD, Thompson MR, VerBerkmoes NC, Chourey K, Shah M, Zhou J, Hettich RL, Thompson DK (2006) Molecular dynamics of the *Shewanella oneidensis* response to chromate stress. Mol Cell Proteomics 5:1054–1071

Campos J, Martínez-Pacheco M, Cervantes C (1995) Hexavalent chromium reduction by a chromate-resistant *Bacillus* sp strain. Antonie van Leeuwenhoek 68:203–208

Cary EE (1982) Chromium in air, soil and natural waters. In: Langard S (ed) Biological and environmental aspects of chromium. Elsevier, Amsterdam, pp 48–64

Cervantes C, Ohtake H, Chu L, Misra T, Silver S (1990) Cloning, nucleotide sequence, and expression of the chromate resistance determinant of *Pseudomonas aeruginosa* plasmid pUM505. J Bacteriol 172:287–291

Cervantes C, Campos-García J, Devars S, Gutiérrez-Corona F, Loza-Tavera H, Torres-Guzmán JC, Moreno-Sánchez R (2001) Interactions of chromium with microorganisms and plants. FEMS Microbiol Rev 25:335–347

Gonzalez CF, Ackerley DF, Lynch SV, Matin A (2005) ChrR, a soluble quinone reductase of *Pseudomonas putida* that defends against H_2O_2. J Biol Chem 280:22590–22595

Hu P, Brodie EL, Suzuki Y, McAdams HH, Andersen GL (2005) Whole-genome transcriptional analysis of heavy metal stresses in *Caulobacter crescentus*. J Bacteriol 187:8437–8449

Ishibashi Y, Cervantes C, Silver S (1990) Chromium reduction in *Pseudomonas putida*. Appl Environ Microbiol 56:2268–2270

Itoh M, Nakamura M, Suzuki T, Kawai K, Horitsu H, Takamizawa K (1995) Mechanism of chromium(VI) toxicity in *Escherichia coli*: is hydrogen peroxide essential in Cr(VI) toxicity. J Biochem 117:780–786

Jiménez-Mejía R, Campos-García J, Cervantes C (2006) Membrane topology of the chromate transporter ChrA of *Pseudomonas aeruginosa*. FEMS Microbiol Lett 262:178–184

Juhnke S, Peitzsch N, Hubener N, Groβe C, Nies DH (2002) New genes involved in chromate resistance in *Ralstonia metallidurans* strain CH34. Arch Microbiol 179:15–25

Kadiiska MB, Xiang QH, Mason RP (1994) In vivo free radical generation by chromium (VI): An electron resonance spin-trapping investigation. Chem Res Toxicol 7:800–805

Katz SA, Salem H (1993) The toxicology of chromium with respect to its chemical speciation: a review. J Appl Toxicol 13:217–224

Kwak YH, Lee DS, Kim HB (2003) *Vibrio harveyi* nitroreductase is also a chromate reductase. Appl Environ Microbiol 69:4390–4395

Levis AG, Bianchi V (1982) Mutagenic and cytogenetic effects of chromium compounds. In: Langard S (ed) Biological and environmental aspects of chromium. Elsevier, Amsterdam, pp 171–208

Llangostera M, Garrido S, Guerrero R, Barbé J (1986) Induction of SOS genes of *Escherichia coli* by chromium compounds. Environ Mutagen 8:571–577

Lovley DR, Phillips EJP (1994) Reduction of chromate by *Desulfovibrio vulgaris* and its c3 cytochrome. Appl Environ Microbiol 60:726–728

Luo H, Lu Y, Shi X, Mao Y, Dalal NS (1996) Chromium (IV)-mediated Fenton-like reaction causes DNA damage: Implication to genotoxicity of chromate. Ann Clin Lab Sci 26:185–191

Mazoch J, Tesarik R, Sedlacek V, Kucera I, Turanek J (2004) Isolation and biochemical characterization of two soluble iron (III) reductases from *Paracoccus denitrificans*. Eur J Biochem 271:553–562

McGrath SP, Smith S (1990) Chromium and nickel. In: Alloway BJ (ed) Heavy metals in soils. Wiley, New York, pp 125–150

Miranda AT, González MV, González G, Vargas E, Campos-García J, Cervantes C (2005) Involvement of DNA helicases in chromate resistance by *Pseudomonas aeruginosa* PAO1. Mutat Res 578:202–209

Moshtaghie AA, Ani M, Bazrafshan MR (1992) Comparative binding study of aluminum and chromium to human transferrin. Effect of iron. Biol Trace Elem Res 32:39–46

Nicholson ML, Laudenbach DE (1995) Genes encoded on a cyanobacterial plasmid are transcriptionally regulated by sulfur availability and CysR. J Bacteriol 177:2143–2150

Nies DH (2003) Efflux-mediated heavy metal resistance in prokaryotes. FEMS Microbiol Rev 27:313–339

Nies A, Nies DH, Silver S (1990) Nucleotide sequence and expression of a plasmid-encoded chromate resistance determinant from *Alcaligenes eutrophus*. J Biol Chem 265:5648–5653

Nies DH, Koch S, Wachi S, Peitzsch N, Saier MH (1998) CHR, a novel family of prokaryotic proton motive force-driven transporters probably containing chromate/sulfate antiporters. J Bacteriol 180:5799–5802

Ohtake H, Fujii E, Toda K (1990) Bacterial reduction of hexavalent chromium: Kinetic aspects of chromate reduction by *Enterobacter cloacae* HO1. Biocatalysis 4:227–235

Ohtake H, Silver S (1994) Bacterial detoxification of toxic chromate. In: Chaudhry GR (ed) Biological degradation and bioremediation of toxic chemicals. Dioscorides, Portland, OR, pp 403–415

Park CH, Keyhan M, Wielinga B, Fendorf S, Matin A (2000) Purification to homogeneity and characterization of a novel *Pseudomonas putida* chromate reductase. Appl Environ Microbiol 66:1788–1795

Pimentel BE, Moreno-Sánchez R, Cervantes C (2002) Efflux of chromate by cells of *Pseudomonas aeruginosa* expressing the ChrA protein. FEMS Microbiol Lett 212:249–254

Plaper A, Jenko-Brinovec S, Premzl A, Kos J, Raspor P (2002) Genotoxicity of trivalent chromium in bacterial cells. Possible effects on DNA topology. Chem Res Toxicol 15:943–949

Saier MH Jr (2003) Tracing pathways of transport protein evolution. Mol Microbiol 48:1145–1156

Suzuki T, Miyata N, Horitsu H, Kawai K, Takamizawa K, Tai Y, Okazaki M (1992) NAD(P)H-dependent chromium (VI) reductase of *Pseudomonas ambigua* G-1: a Cr(V) intermediate is formed during the reduction of Cr(VI) to Cr(III). J Bacteriol 174:5340–5345

Tauch A, Schluter A, Bischoff N, Goesmann A, Meyer F, Puhler A (2003) The 79 370-bp conjugative plasmid pB4 consists of an IncP-1β backbone loaded with a chromate resistance transposon, the *strA-strB* streptomycin resistance gene pair, the oxacillinase gene bla_{NPS-1}, and a tripartite antibiotic efflux system of the resistance-nodulation-division family. Mol Gen Genomics 268:570–584

Wang P, Mori T, Toda K, Ohtake H (1990) Membrane-associated chromate reductase activity from *Enterobacter cloacae*. J Bacteriol 172:1670–1672

Microbiol Monogr (6)
D. H. Nies, S. Silver: Molecular Microbiology of Heavy Metals
DOI 10.1007/7171_2006_088/Published online: 1 February 2007

Molybdate and Tungstate: Uptake, Homeostasis, Cofactors, and Enzymes

Guenter Schwarz[1] (✉) · Peter-Leon Hagedoorn[2] · Katrin Fischer[1]

[1]Institute of Biochemistry, University of Cologne,
Otto-Fischer-Str. 12–14, 50674 Köln, Germany
gschwarz@uni-koeln.de

[2]Department of Biotechnology, Delft University of Technology, Julianalaan 67,
2628BC Delft, The Netherlands

Abstract Molybdenum (Mo) and tungsten (W) are trace elements that catalyze, upon binding to the appropriate cofactors, diverse and important redox reactions in the global carbon, nitrogen, and sulfur cycles. Mo is found in two forms of oxygen-labile metal cofactors, a pterin-based and a Fe – S-cluster-based scaffold, while W naturally only occurs in association with pterin cofactors and FeW-nitrogenases have been generated artificially. Both oxyanions enter the cell via an ABC-type high affinity uptake system and are subsequently processed by a multistep biosynthetic machinery forming either Mo- and W-pterin cofactors (Moco or Wco) in a large variety of Mo- and W-containing enzymes or the FeMo cofactor (FeMo-co) in nitrogenase-catalyzed nitrogen fixation. The functional diversity of pterin-based Mo and W cofactors is reflected by a large number of enzymes such as nitrate reductase, dimethyl sulfoxide reductase, formate dehydrogenase, aldehyde oxidoreductase and CO dehydrogenase. In these enzymes Mo and W are bound via thiolates to one or two unique tricyclic pterin moieties, commonly referred to as molybdopterin but the term "metal binding pterin" (MPT) is more appropriate due to its association with both, Mo and W. It is commonly believed, but still not demonstrated, that Moco and Wco are synthesized by a similar and highly conserved pathway. Synthesis of the Moco can be divided into four major steps, according to the biosynthetic intermediates cyclic pyranopterin monophosphate, MPT, and adenylated MPT. Differences in the final metal insertion step(s) between Moco and Wco synthesis will be discussed. In contrast, FeMo-co biosynthesis is less understood in terms of reaction intermediate and mechanisms of different reactions catalyzed by the involved proteins. It starts with the formation of Fe – S cluster core structures that are assembled and arranged to a topology similar to mature FeMo-co. In the next steps, Mo and homocitrate are transferred before the mature cofactor is inserted into nitrogenase. Finally, a brief overview about Mo- and W-pterin enzymes as well as FeMo- and FeW-nitrogenases is given.

Abbreviations

AOR	Aldehyde oxidoreductase
cPMP	Cyclic pyranopterin monophosphate
DMSO	Dimethyl sulfoxide
DMSOR	Dimethyl sulfoxide reductase
FDH	Formate dehydrogenase
FeMo-co	FeMo cofactor
HVOR	(2*R*)-Hydroxy-carboxylate:viologen oxidoreductase
Mo	Molybdenum
Moco	Molybdenum cofactor
Mop	Molybdate-binding protein
MCD	MPT cytosine dinucleotide
MGD	MPT guanine dinucleotide
MPT	Molybdopterin/metal-binding pterin
MPT-AMP	Adenylated MPT
N_2	Dinitrogen
NR	Nitrate reductase
SAM	*S*-Adenosylmethionine
SO	Sulfite oxidase
TMAO	Trimethylamine-*N*-oxide
TMAOR	Trimethylamine-*N*-oxide reductase
V	Vanadium
W	Tungsten
Wco	Tungsten cofactor
XOR	Xanthine oxidoreductase

1 Introduction

The oxyanions molybdate and tungstate show similar geometry and present the metalliferous precursor molecules for cofactors that use molybdenum (Mo) and tungsten (W) as catalytically active metals for a diverse array of redox reactions (Stiefel 2002). While Mo is found in all kingdoms associated with different types of Mo-dependent enzymes (Mo-enzymes), W is restricted to prokaryotes and mainly found in archaea. Genome data base analyses indicate that approximately 60% of the archaea but only 3% of bacteria encode for putative W-dependent enzymes (W-enzymes). From the evolutionary point of view it is commonly accepted that W presents the ancestor of Mo-enzymes because W-enzymes occur almost exclusively in anaerobic prokaryotes and are only known to catalyze low potential redox chemistry of carbon substrates. Mo-enzymes, however, occur in all forms of life and can catalyze redox reactions with higher potentials, such as sulfate oxidation and nitrate reduction.

It is generally supposed that the primitive earth was reducing due to the presence of H_2, CH_4, and H_2S in the atmosphere. H_2S will reduce Mo(VI) to Mo(IV), but it will not reduce W(VI). Mo(IV) forms an insoluble sulfide, MoS_2, while W remains as WO_4^{2-} or forms the soluble WS_4^{2-}. Mineral deposits of Mo are often MoS_2 (scheelite), while W is $CaWO_4^{2-}$ (wolframite). Mo likely became available when oxygen was introduced into the atmosphere and sulfide was oxidized to sulfate (3.5 billion years ago). Later, Mo took over the role of W and catalyzed redox reactions at higher potentials with substrates such as sulfate and nitrate (Williams and Frausto da Silva 2002). Mo and W are very similar in terms of atomic and ionic radii, valence states, and coordination numbers. The redox properties of isostructural Mo and W compounds are different. In general the redox potential of a W compound is lower than of its Mo analog.

Mo and W gain their biological activity upon incorporation into cofactors that serve as an anchor to provide the appropriate environment enabling efficient control of the redox properties of the metal (Rees 2002). These cofactors have in common that in most cases sulfur atoms form three ligands of the respective metal together with two oxygen atoms (Fig. 1). Both ligands are either terminal or part of covalent bonds to the apo-enzyme or other molecules, while at least two sulfurs are always derived from the cofactor. Nature has evolved two systems for holding these metals, one is the Fe – S-cluster-based Fe-metal-cofactor exclusively found in nitrogenases (Dos Santos et al. 2004), while the other is a pterin-based cofactor with a unique dithiolene function (Schwarz 2005). Both cofactors are oxygen-sensitive and very unstable outside of the corresponding apo-enzymes. Besides positioning and redox-control it is also believed that both cofactors participate in electron transfer to or from the catalytic metal.

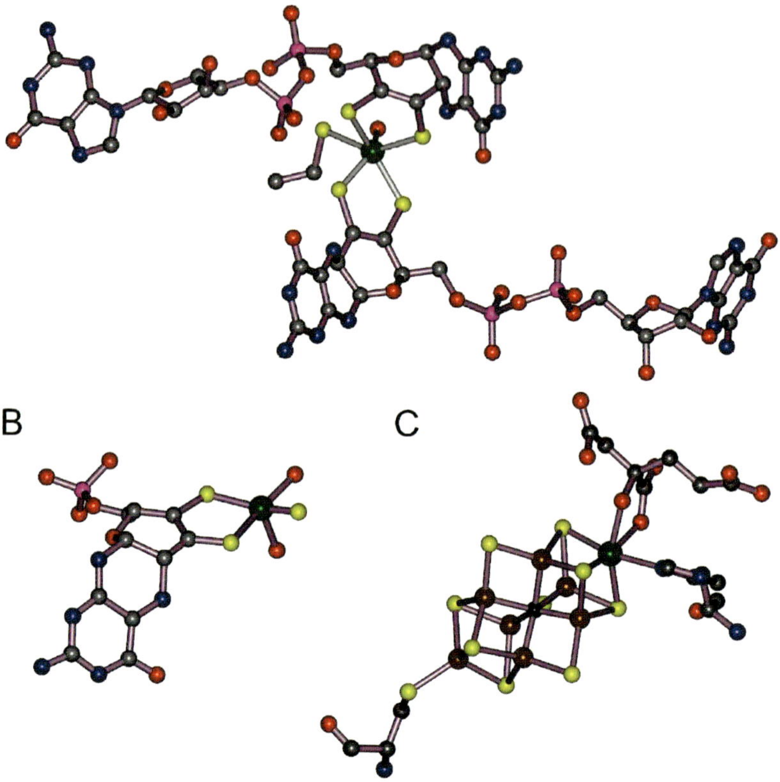

Fig. 1 Three-dimensional structures of pterin- and Fe – S-cluster-based Mo cofactors. *A* Bis-molybdopterin guanine dinucleotide Mo cofactor of periplasmic nitrate reductase (2nap). *B* Molybdopterin-type cofactor from xanthine dehydrogenase (1jro, modified according to 1v97). *B* Fe – S-cluster-based iron Mo cofactor from nitrogenase (1m1n). Structures are shown in ball-and-stick mode, generated with MOLSCRIPT (Esnouf 1997) and rendered with POVRAY (http://www.povray.org)

2
Uptake of Molybate and Tungstate

Molybdenum is the most abundant transition metal in oceans. Due to the oxygen-rich and slightly alkaline environment Mo is almost exclusively present as the dianionic molybdate ion. With 10 μg/L it is more than ten times more concentrated than zinc and 100 times more than copper (Cox 1995). In contrast, concentrations of W in seawater are 0.1 μg/L but are dramatically increased in hydrothermal vents, where many organisms expressing W-dependent enzymes are found. In any case the oxyanions molybdate and tungstate are the main source of Mo and W that enter the cell.

Mo and W oxyanions are taken up by high-affinity transporters of the ABC-type (Pau and Lawson 2002). They are characterized by a periplasmic oxyanion-binding protein, a transmembrane protein involved in translocation of anions, and a cytoplasmic membrane-associated protein that drives the active transport machinery by ATP hydrolysis. Many transporters of that type are known in bacteria and they usually function with highest affinity for their specific oxyanions, but they also act on other substrates with lower affinity. For example, molybdate and tungstate are co-transported at higher concentrations by sulfate and phosphate transporters (Grunden and Shanmugam 1997). From the evolutionary point of view such high affinity and specificity uptake systems allow bacteria to scavenge the less abundant molybdate or tungstate in the presence of competing anions, such as sulfate, whose concentration in sea water is about 10^5 times higher than that of molybdate.

2.1 Molybdate Uptake

The first mutants with a defect in Mo metabolism were identified with a molybdate-repairable phenotype. These mutants could be complemented by high concentrations of molybdate in the growth medium and consequently they were believed to be impaired in molybdate uptake or Mo processing during cofactor biosynthesis. Mutants with complete suppression of their phenotype had defects in the *mod* locus encoding a high-affinity uptake system (Glaser and DeMoss 1971). Depending on the organism studied, the *mod* locus encodes for four to six proteins from diverging operons. In *E. coli* the ABC-type molybdate transporter is composed of ModA (24–25 kDa), a periplasmic monomeric oxyanion-binding protein, ModB (25 kDa), a dimeric integral transmembrane protein forming the anion channel, and ModC (38–39 kDa), a dimeric molybdate-binding peripheral membrane-binding protein hydrolyzing ATP. The function of the fourth protein in the *modABCD* operon, ModD, is unknown. It shows homology to nicotinate-nucleotide pyrophosphorylases, and deletion of ModD in *Rhodobacter capsulatus* resulted in no significant deficiency of Mo-dependent enzymes (Wang et al. 1993).

The best characterized component of the molybdate transporter is the periplasmic molybdate-binding protein ModA. It binds molybdate with high affinity in stoichiometric amounts. Using different methods K_D values between 3 μM and 20 nM have been obtained. The latter fits well with the k_M of molybdate for the entire cell (50 nM) (Imperial et al. 1998). The ModA structure consists of two globular domains connected by strands at the bottom of a deep cleft between the domains (Hu et al. 1997). Molybdate and also tungstate bind between both domains in a completely dehydrated anion-binding pocket via seven hydrogen bonds. Although the binding mode of sulfate in homologous sulfate-binding proteins is similar, the residues involved are different because they are not strictly conserved within the family of periplasmic anion-binding proteins. Substrate specificity is based on the volume of the anion-binding pocket that reflects the actual size of the ligand. In ModA and homologous proteins an average volume of 72 $Å^3$ perfectly matches the sphere of the tetragonal molybdate molecule, while in sulfate-binding proteins this pocket is reduced to 59 $Å^3$. Therefore, ModA is also able to bind tungstate, which has almost the same size as molybdate (Stiefel 2002).

In analogy to the integral membrane protein of other ABC transporters, ModB is believed to form a dimer. Sequence hydropathy analysis supports the formation of five transmembrane helices within ModB. A conserved motif within the cytoplasmic loop may interact with the peripheral membrane protein ModC. Following the topology of ModB, a dimeric arrangement and interaction with ModB is also proposed for ModC. The first 200 residues form the highly conserved ATP-binding domain followed by a C-terminal domain, which is homologous to other cytoplasmic molybdate-binding proteins due to their common Mop domains (see below).

2.2 Cytoplasmic Molybdate-Binding Proteins

Additional proteins encoded by different *mod* operons have functions in intracellular molybdate binding and regulation of gene expression. *E. coli* ModE controls gene expression of the *mod* (Anderson et al. 1997) as well as the *moa* operon (Anderson et al. 2000) by sensing molybdate concentrations. Dependent on the molybdate status, the expression of operons in molybdate uptake or molybdopterin (MPT) synthesis are regulated. In the presence of high concentrations of molybdate the *mod* operon is repressed while the *moa* operon is induced in order to produce sufficient MPT scaffolds to bind molybdate. ModE comprises two functional domains, an N-terminal DNA-binding domain and a C-terminal molybdate-binding domain. Upon high-affinity binding of molybdate to ModE ($K_D = 0.8$ μM) the affinity of ModE for DNA (*modA* operator) increases about 25 times (Anderson et al. 1997) due to extensive conformational changes within the molecule, causing a large quench in fluorescence.

The C-terminal domain of ModE comprises 140 residues and can be divided into two subdomains homologous to cytoplasmic molybdate-binding proteins called Mop. Mops were first discovered in *Clostridium pasteurianum* (Hinton und Mortenson 1985) and initially mis-annotated as MPT-binding proteins due to the observation of fluorescence. Later it was clarified that Mops do not bind MPT. In *C. pasteurianum* in total three Mops (7 kDa each) with more than 90% sequence identity were found. They are believed to be important for molybdate homeostasis during nitrogenase biogenesis. Proteins that contain such Mop domains are called molbindins. ModE and ModG contain two Mop domains while ModC exhibits only one. The crystal structure of another molybdate-binding protein (MopII) from *C. pasteurianum* shows a hexameric arrangement of trimers of dimers with eight oxyanion binding sites in total (Schuttelkopf et al. 2002). In contrast to ModE, molybdate binding does not alter the conformation of MopII dramatically. All molybdate binding sites are located at inter-subunit interfaces with two additional sites at the molecular threefold axis within the hexamer.

2.3 Tungstate Uptake

The antagonistic effect of tungstate on Mo-enzymes was known long before the first naturally occurring W-enzymes were discovered. With the discovery of the bacterial molybdate transporter ModABC, it became clear that the active transport of tungstate was as efficient as for molybdate.

Only recently the first, highly specific, tungstate transporter TupABC (tungsten uptake protein) was discovered in the bacterium *Eubacterium acidaminophilum* (Makdessi et al. 2001). The periplasmic tungstate binding protein, TupA, has an apparent K_d of 0.5 μM for tungstate, and shows only weak binding of molybdate and sulfate when present in more than 1000-fold molar excess.

Interestingly, this tungstate transporter is not present in all organisms that produce W-enzymes. The hyperthermophilic archaea seem to contain neither TupABC nor ModABC on their genome. This opens the possibility for a novel transporter system different from TupA and ModA. Recently a candidate for this novel tungstate transporter has been identified in the genome of *Pyrococcus furiosus* that was designated as WtpABC (W transport protein). Furthermore, this WtpABC is present in organisms that express W-enzymes and do not contain TupABC. Genomic information suggests that this WtpABC transporter is almost exclusively present in archaea, while TupA is predominantly present in bacteria. Recently, for the soluble component WtpA, high-affinity binding was demonstrated using isothermal titration calorimetry with a k_D for tungstate of 17 pM, while the affinity for molybdate was at least one order of magnitude lower (Bevers et al. 2006).

3 Biosynthesis of Mo- and W-Pterin Cofactors

Elucidation of the chemical nature of pterin-based metal cofactors goes back to the pioneering work of Rajagopalan, Johnson and coworkers (Rajagopalan and Johnson 1992). By demonstrating that this cofactor contains a unique C6-substituted four-carbon side chain a novel biosynthetic route in pteridine metabolism was identified. In Moco and Wco, this four-carbon side chain forms a pyrano ring between the *C3′* hydroxy group and the pterin C7 atom (Fig. 2). In addition, they contain a phosphate at the *C4′* carbon and an ene-

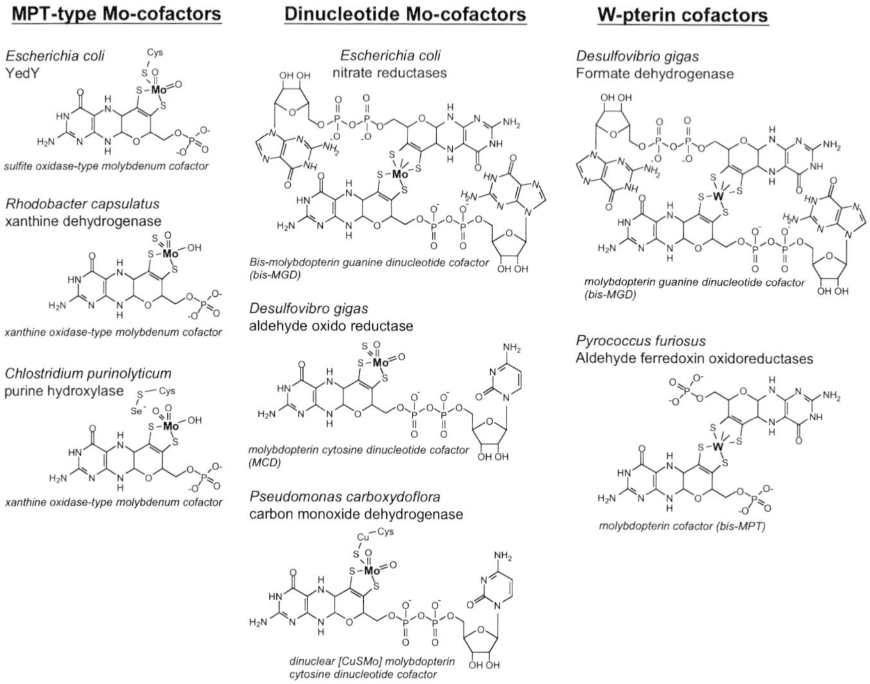

Fig. 2 Different pterin-based Mo- and W-cofactors. Mo-cofactors are classified into MPT- and dinucleotide-type, while the latter can be divided into bis- and mono-MPT-based forms. W-pterin cofactors belong to the dinucleotide-type forms. For each cofactor type an example of an enzyme is given

dithiolate function, which links the redox-active metal to the reduced pterin. As it was first isolated from Mo-dependent enzymes, this pterin-backbone was named molybdopterin (MPT) but as it also coordinates W a more correct name of the molecule would be metal-binding pterin. Depending on the type of cofactor, one (mono-MPT) or two pterin (bis-MPT) equivalents coordinate Mo, while for W-dependent enzymes only bis-MPT cofactors have been identified so far (Fig. 2). Additional modifications occur with cofactors of the dimethyl sulfoxide reductase (DMSOR) family as well as carbon monoxide dehydrogenases (CODH) where nucleotides, GMP or CMP, are bound to the phosphate of the pterin. Due to structural similarities between MPT and the attached nucleotides, these cofactors were named dinucleotide cofactors (MPT guanine dinucleotide: MGD).

3.1 Mo-Pterin Cofactors in *E. coli*

Biosynthesis of Mo-pterin cofactors was intensively studied in *E. coli*. However, due to the overall conservation of the cofactor itself and its biosynthetic machinery knowledge from Mo-pterin cofactor biosynthesis in higher eukaryotes also significantly contributed to our understanding of Mo-pterin cofactor biogenesis (Rajagopalan and Johnson 1992; Schwarz 2005). In addition to the *mod* locus (molybdate uptake) there are four additional operons (*moaABCDE*, *mobAB*, *moeAB*, *mogA*) that express ten gene products, of which eight have been shown to be required for Mo-enzyme activity in *E. coli*. Bacterial Mo-pterin cofactor biosynthesis (Figs. 3 and 4) starts with GTP and is divided into four steps according to the biosynthetic intermediates cyclic pyranopterin monophosphate (cPMP), MPT, and adenylated MPT (MPT-AMP). The biosynthesis of bacterial and eukaryotic Mo-pterin cofactors has been reviewed in detail recently (Schwarz 2005),

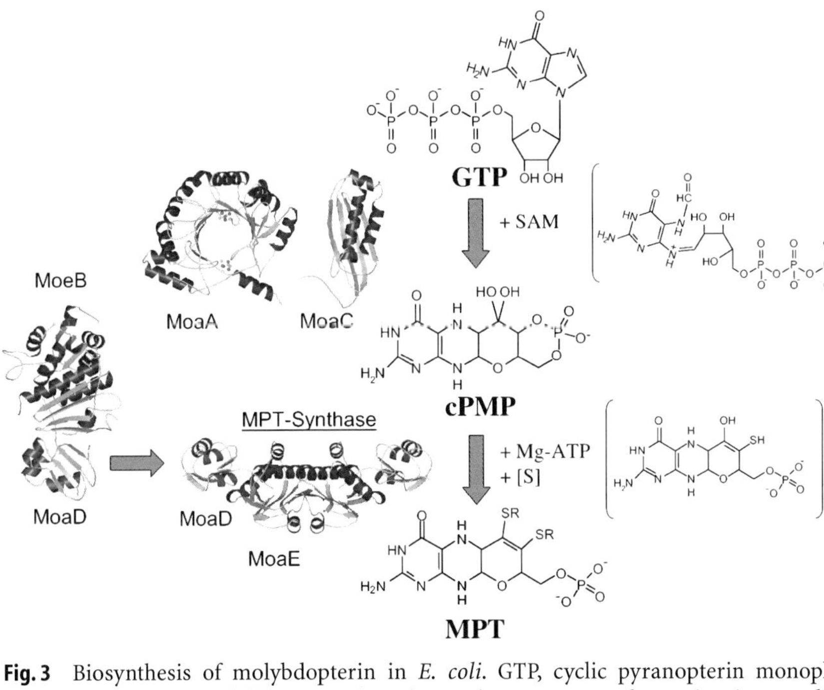

Fig. 3 Biosynthesis of molybdopterin in *E. coli*. GTP, cyclic pyranopterin monophosphate (cPMP), and molybdopterin (MPT) are shown. Proposed reaction intermediates of cPMP (Hanzelmann and Schindelin 2006) and MPT synthesis (Wuebbens and Rajagopalan 2003) are shown in *brackets*. Crystal structures of proteins involved in MPT synthesis (1tv8), MoaC (1ekr), MoeB–MoaD complex (1jw9), and MPT synthase (1fm0) are shown in ribbon presentation and prepared as described in the legend to Fig. 1

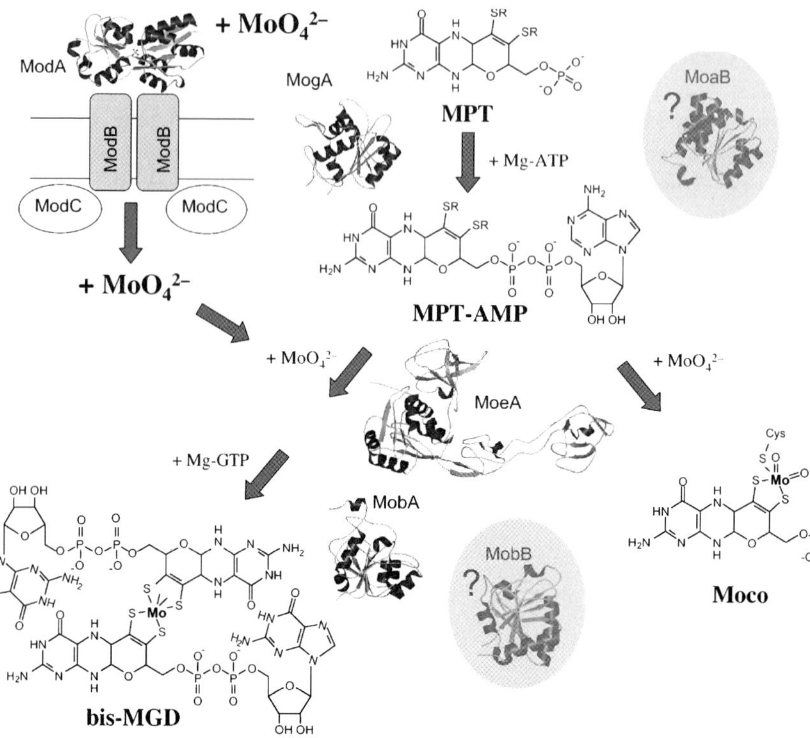

Fig. 4 Mo insertion and maturation of different Mo-cofactors in *E. coli*. Based on functional reconstitution of *mogA* mutants by Cnx1G it is believed that MogA also functions as adenylyl transferase producing adenylated MPT (MPT-AMP) that is subsequently converted into bis-MGD- or MPT-type cofactors (one example is shown). Due to the lack of experimental evidence no intermediate of bis-MGD synthesis is shown. Crystal structures of proteins involved in MPT-AMP synthesis (MogA, 1di6), Mo insertion (MoeA, 1fc5), and dinucleotide formation (MobA, 1ek5) are shown. For the ABC-transporter of molybdate only the ModA structure is available (1amf). Proteins with unknown function are shown in *gray* (MoaB, 1mkz, MobB, 1np6). Structures are shown in ribbon presentation and prepared as described in the legend to Fig. 1

therefore we present some special aspects of the bacterial pathway only in brief.

3.1.1 Synthesis of Cyclic Pyranopterin Monophosphate cPMP

In the first step, GTP is converted by a complex reaction sequence into cPMP. The main difference to other pteridine pathways (producing three-carbon side chains) is that the C8 atom of the purine is inserted between the 2′ and 3′ ribose carbon atoms. Each carbon of the ribose and the ring carbons of the guanine are incorporated into cPMP (Wuebbens and Rajagopalan 1995).

Already the first intermediate cPMP is a pyranopterin similar to the mature cofactor, and carries a geminal diol in the *C1′* position of the side chain as well as a cyclic phosphate (Santamaria-Araujo et al. 2004). Proteins MoaA and MoaC (Fig. 3) synthesize cPMP. MoaA is an Fe – S protein and belongs to the family of *S*-adenosylmethionine (SAM)-dependent radical enzymes that catalyze the formation of radicals by reductive cleavage of SAM (Sofia et al. 2001). The structure of MoaA is characterized by an incomplete triosephosphate isomerase barrel, which binds an N-terminal [4Fe – 4S] cluster typical for proteins that belong to SAM-dependent enzymes while the C-terminal domain contains a second, MoaA-specific [4Fe – 4S] cluster (Hänzelmann and Schindelin 2004). Both [4Fe – 4S] clusters are oxygen-sensitive and coordinated by only three cysteines. In each case, the fourth ligand is provided by the substrate (C-terminal cluster: GTP) or co-substrate (N-terminal cluster: SAM), respectively. The function of the other protein, MoaC, is unknown but it is believed that the hexameric molecule (Wuebbens et al. 2000) participates in pyrophosphate release upon the rearrangement reaction of MoaA.

3.1.2 Sulfur Transfer by MPT Synthase

To form the MPT dithiolate, two sulfur atoms have to be incorporated into cPMP by MPT synthase, a heterotetrameric complex of two small (MoaD) and two large subunits (MoaE) (Fig. 3). Both small subunits carry a single sulfur atom as thiocarboxylate at the highly conserved C-terminal double-glycine motif (Gutzke et al. 2001), which is deeply buried into the large subunit to form the active site (Rudolph et al. 2001). As one sulfur atom is bound per small subunit, a two-step mechanism for the formation of the MPT dithiolate has been proposed involving the formation of a mono-sulfurated intermediate (Gutzke et al. 2001; Wuebbens and Rajagopalan 2003). In a separate reaction, sulfur is transferred to the small subunit of MPT synthase (Fig. 3). MoeB catalyzes the adenylation of the C-terminal glycine residue in MoaD (Lake et al. 2001; Leimkuehler et al. 2001). Free sulfide or persulfide-loaded cysteine desulfurases are needed to cleave the acyl-adenylate, thus forming the MoaD thiocarboxylate (Leimkuehler and Rajagopalan 2001).

3.1.3 Metal Insertion

In the following steps, Mo is attached to MPT by the action of two proteins, MogA and MoeA, that are fused into multidomain proteins in higher eukaryotes, such as Cnx1 from plants (Schwarz et al. 2000). MogA forms a trimer (Liu et al. 2000), while MoeA is an elongated dimer with the active site located at the interface between two monomers (Xiang et al. 2001). Detailed biochemical characterization is still lacking for MogA but it is known that *mogA* mutants accumulate MPT and that they are complemented by high concentration of molybdate in the growth medium (Joshi et al. 1996). For *E. coli* MoeA, Moco synthesis was demonstrated using an in vitro assay that contained MPT and molybdate (Nichols and Rajagopalan 2005). In this assay MogA showed inhibiting activity, which was due to a competitive binding of MPT, as had been previously demonstrated for the MogA-homologous G-domain of plant Cnx1 (Cnx1G) (Schwarz et al. 1997).

Functional conservation between MogA and its plant homolog Cnx1 (Schwarz et al. 2000) served as the basis for study of Cnx1 as a model system for Mo transfer to MPT, which has been summarized recently (Schwarz 2005). In brief, besides its MPT-binding ability Cnx1G also exhibits a catalytic function (Kuper et al. 2000), which remained enigmatic for many years. A crystal structure of Cnx1G in complex with co-purified pterins revealed a novel intermediate in Moco synthesis, MPT-AMP (Kuper et al. 2004), which was subsequently proven to be the reaction product of Cnx1G (Llamas et al. 2004). MPT-AMP is transferred cooperatively with molybdate from Cnx1G to Cnx1E and rapidly hydrolyzed in the presence of magnesium or zinc. Simultaneously, Mo derived from bound molybdate is inserted and Moco as well as AMP are released (Llamas et al. 2006). Finally, in different Cnx1G-pterin complexes copper was identified as MPT dithiolate ligand (Kuper et al. 2004), but to date its way and nature of binding and its functional role remained unknown.

In *E. coli*, the majority of Mo-enzymes belong to the DMSOR family where one Mo atom is chelated by two MPT equivalents. The fact that Mo insertion into mono-MPT of plants already requires a multistep reaction mechanism might suggest that attaching two MPT molecules is even more complicated and needs a different enzymatic reaction than in eukaryotes.

3.1.4
Dinucleotide Formation

In the absence of molybdate no MPT dinucleotide cofactors are synthesized (Hänzelmann and Meyer 1998; Joshi et al. 1996) because MPT-AMP accumulates and all subsequent modifications are blocked, which confirms that Mo is inserted before the final modification takes place. Except for the intermediate MPT-AMP, two other nucleotide modifications are known that result in MPT guanine and cytosine dinucleotide cofactors (MGD and MCD).

The *mob* locus is essential for the dinucleotide formation in *E. coli* (Palmer et al. 1994). MobA catalyzes the conversion of MPT and GTP to MGD (Palmer et al. 1998) while MobB is a GTP-binding protein with weak GTPase activity (Eaves et al. 1997) but dispensable, at least in vitro (Temple and Rajagopalan 2000). In vivo though, other factors such as chaperones (Blasco et al. 1998) may play an important role for this step. The crystal structure of *E. coli* MobA (Lake et al. 2000; Stevenson et al. 2000) shows an a/b architecture with a typical nucleotide-binding Rossmann fold. MobA was found to be monomeric (Stevenson et al. 2000) and the active site was defined by conserved residues, structure-guided mutagenesis (Guse et al. 2003), as well as co-crystallization of MobA with GTP, which is bound in the N-terminal part (Lake et al. 2000).

3.1.5 Additional Modifications and Cofactor Transfer to Apo-Enzymes

For enzymes of the xanthine oxidoreductase (XOR) family, attachment of a terminal sulfur to the Mo is crucial for their activity. In eukaryotes, this reaction is catalyzed by the Moco sulfurase, which is best studied in plants (Bittner et al. 2001). In bacteria, only few years back it became evident that enzymes of the XOR family contain the eukaryotic type of sulfurated mono-pterin cofactor (Leimkühler et al. 1998). However, the bacterial protein responsible for Moco sulfuration is not known yet, but it is believed that a cysteine desulfurase of the NifS family provides the sulfur for Moco sulfuration, similar to the situation in eukaryotes (Bittner et al. 2001).

After synthesis and maturation, the cofactor has to be incorporated into the appropriate apo-enzyme either directly from the Moco-synthetic machinery or by a Moco carrier and/or storage protein. Studies in *E. coli* demonstrated a complex network of interactions between proteins catalyzing the final steps of Moco biosynthesis and downstream components such as apo-enzymes and enzyme-specific chaperones (Magalon et al. 2002; Vergnes et al. 2004). In the case of the bacterial nitrate reductase NarGHI, the chaperone NarJ was shown to be essential for cofactor insertion and correct targeting to the plasma membrane (Blasco et al. 1998). Recently, a Moco carrier protein from a lower eukaryote (*Chlamydomonas reinhardtii*) has been characterized biochemically and structurally (Fischer et al. 2006). As it shares significant homologies to bacterial proteins with unknown function one can expect that these homologs exhibit similar properties, thereby controlling cofactor homeostasis.

3.2 Bis-MPT Cofactors with Tungsten

In archaea, two types of W-pterin cofactors are known that form the so-called bis-MPT type of cofactors (Fig. 2). Enzymes of the W formate dehydrogenase family have the same cofactor as their eubacterial Mo homologs (DMSOR family) (Moura et al. 2004). A new class of W-pterin enzymes is formed by aldehyde ferredoxin oxidoreductases (AOR), which contain the only known bis-MPT cofactors (Chan et al. 1995).

Little is known about the enzymes that synthesize both types of W-cofactors. Genomic information from organisms that produce W-enzymes suggests that the biosynthetic machinery is basically similar to that of the Mo-cofactor (Table 1). However, there are interesting features in each of the biosynthetic steps that need to be highlighted and might indicate slight differences in the corresponding reactions. Despite the presence of two [4Fe – 4S] clusters in MoaA it has also been shown that the C-terminal highly conserved double-glycine motif is functionally important (Hänzelmann et al. 2002). This motif is missing, for example, in MoaA proteins from all *Pyrococcus* species. For MoaD in total three different homologs were found in *Pyrococcus horikoshi* and other archaea with all of them carrying the highly conserved and functionally crucial C-terminal double-glycine motif. These three proteins might all either function as MPT synthase small subunits or they participate in other transfer reactions that are based on glycine-adenylate intermediates. For MoeB, the MoaD adenylyl transferase, again two copies were identified in *P. horikoshi* with one containing an N-terminal extension of 100 residues with unknown function. While in *E. coli* different NifS-like proteins where shown to act as sulfur donors of MoaD thiocarboxylation (Leimkuehler and Rajagopalan 2001), whereas only one cysteine desulfurase (SufS) is known in *Pyrococcus furiosus*, which is located in close proximity to the *moeB* locus (Robb et al. 2001).

A striking difference to *E. coli* is seen for proteins involved in metal transfer. The genomes of *P. horikoshi* and many related thermophiles lack MogA but instead they all contain MoaB, a protein of unknown function with high

Table 1 The presence of Moco biosynthesis gene products on the genome of *Pyrococcus furiosus*, an organism known to produce Wco-containing enzymes

Step	*E. coli* [a]	*P. furiosus* [a]	Identity/similarity (%)
cPMP synthesis	MoaA (329)	PF0090 (307)	28/47
	MoaC (160)	PF1854 (156)	40/55
MPT synthesis	MoaD (81)	PF0543 (90)	33/51
		PF0345 (82)	29/43
		PF1482 (73)	18/22 [b]
	MoaE (150)	PF0100 (145)	38/54
	MoeB (249)	PF0009 (230)	35/54
		PF0003 (366)	19/33
Adenylation MPT	MogA (195)	–	
	MoaB (170)	PF0372 (169)	31/50
Metal insertion	MoeA (411)	PF0542 (401)	32/48
		PF1783 (395)	32/51
Nucleotide attachment	MobA (194)	PF0618 (196)	29/51
	MobB (175)	PF1954 (244)	11/21 [c]

[a] The number of residues is given in parentheses
[b] Homology only in the 30 C-terminal amino acids (48% identity/59% similarity)
[c] Homology only in the 120 N-terminal amino acids (25% identity/46% similarity)

homologies on primary and structural levels to *E. coli* MogA (Sanishvili et al. 2004). The structure of *E. coli* MoaB shows a hexameric arrangement and, as it is mainly found in archaea, it might indicate that formation of higher ordered oligomers increases thermal stability, as known for other metabolic enzymes. In *E. coli*, MoaB is dispensable for cofactor synthesis because its deletion does not cause Moco deficiency (Schwarz et al. unpublished results). However, if MoaB is the only enzyme of this MogA/MoaB family, it might take over the function of MogA in further processing MPT for subsequent W insertion in archaea. It remains to be elucidated if MoaB also catalyzes adenylyl transfer or if other nucleotides are used. Sanishvili et al. (2004) have assumed that due to its genomic localization *E. coli* MoaB participates in cPMP synthesis (step 1) and consequently they showed a weak GTP binding and hydrolysis. However, it is known that cPMP synthesis is only dependent on MoaA and MoaC (Hänzelmann and Schindelin 2004). Therefore, guanyl transfer to MPT might be considered as alternative nucleotide transfer during the metal insertion process.

In addition, MoeA, the protein catalyzing metal insertion, is found in duplicate in a number of hyperthermophilic archaea, such as *P. horikoshii*, for which both unpublished crystal structures are available (MoeA-1, 1uz5 and MoeA-2, 1wu2). It remains to be elucidated if this duplication points toward metal selectivity or if they are related to both forms of bis-MPT-based W-cofactors known so far. Besides duplication, another variation is found in *Archeoglobus fulgidus* where one of the two MoeA proteins shows a C-terminal extension of 150 residues with significant homologies to tungstate and molybdate-binding proteins of the Mop family. This finding clearly supports the function of MoeA in metal insertion and might point to an efficient channeling of the anion substrates to the site of metal insertion.

3.3
The FeMo-Cofactor

As mentioned above, the other type of Mo-containing cofactor is the FeMo-co of nitrogenase. Here, Mo is anchored again by sulfur atoms but they are held in position via the most complex metal cluster known in nature, comprising seven iron, nine sulfur, and one atom of unidentified nature (Dos Santos et al. 2004). In addition to the three scaffold sulfurs, homocitrate provides two additional oxygens as non-proteinogenic Mo ligands. The cofactor is bound to the MoFe protein (NifDK) via a cysteine that binds an iron at one end of the cluster while on the other end a histidine forms the sixth ligand of Mo (Fig. 1).

Similar to the synthesis of pterin-based metal cofactors, the scaffold holding the Mo is synthesized first. Although it is commonly accepted that FeMo-co is synthesized by a multistep pathway before it is inserted into apo-FeMo protein, the exact sequence of events, the nature and structure of intermediates, as well as the catalytic mechanism of individual steps are still not well understood. Recent reviews cover many aspects of FeMo-co biosynthesis in depth (Dos Santos et al. 2004; Frazzon and Dean 2002). Therefore, we only give a very brief summary, show some of the latest findings on FeMo-co maturation and draw lines of similarities to Mo-pterin cofactor synthesis.

FeMo-co biosynthesis is best studied in *Azotobacter vinelandii* where 21 *nif* (nitrogen fixation) (Rubio and Ludden 2005) genes have been identified as encoding for proteins involved in or being associated with FeMo-co/-protein maturation. First, Fe – S fragments are synthesized by the NifU/NifS protein complex (Yuvaniyama et al. 2000). In this reaction the pyridoxal phosphate-dependent cysteine-desulfurase (NifS) generates a reactive protein-bound persulfide (Zheng et al. 1993) serving as sulfur source that is subsequently transferred to NifU, an iron-binding protein, which is believed to form a scaffold for the formation of labile [2Fe – 2S] clusters (Yuvaniyama et al. 2000). These building blocks are then used by the NifB protein to produce the core Fe – S fragment, for which very recent data support a similar topology to mature FeMo-co except for the presence of Mo and homocitrate (Corbett et al. 2006). It was also shown only recently that NifB, which carries typical features of radical SAM-proteins (Sofia et al. 2001), depends on SAM, indicating that radical chemistry is required during FeMo-co synthesis (Curatti et al. 2006). For example, the nature of the newly identified central atom bridging irons of the [4Fe – 3S] to the [3Fe – 3S] fragment is still unknown (Einsle et al. 2002) and it is not clear either by which mechanism both clusters are linked via this central atom during the formation of the NifB-associated precursor of FeMo-co (Corbett et al. 2006). It is important to note that radical SAM enzymes bind oxygen-sensitive [4Fe – 4S] clusters with one iron being coordinated by SAM. In parallel, SAM-dependent radical chemistry is also involved in Mo-pterin biosynthesis, where the pterin backbone of MPT is synthesized (Hänzelmann and Schindelin 2004).

The NifB cofactor, a Fe – S core fragment, is transferred to the NifEN complex, which is structurally similar to the MoFe protein (NifDK) and forms a scaffold for the final maturation of FeMo-co (Goodwin et al. 1998). This process is associated with NifX, an escort protein transferring a FeMo-co precursor (Rangaraj et al. 2001). Although molybdate uptake is similar between Mo-pterin and FeMo-co biosynthesis and well understood, nothing is known about how Mo is inserted into the Fe – S core fragment and which protein catalyzes this reaction. As Mo was detected with FeMo-cos derived from mutants defective in homocitrate synthesis (*nifV* mutants) it is believed that metal insertion takes place first, and subsequently homocitrate is transferred. The latter was found to be dependent on NifX, which was shown to play multiple roles in escorting precursors of FeMo-co or even the mature cofactor (Rangaraj and Ludden 2002). Determination of ^{99}Mo labeling indicated that NifEN, NifX, and NifH bind Mo-containing FeMo-co precursors (Rangaraj and Ludden 2002). In contrast to Mo-pterin biosynthesis, escort proteins seem to play a prominent role in FeMo-co biosynthesis as there are at least three proteins (NifX, NifH, and NifY) participating in different and/or overlapping transfer steps. Depending on the starting protein/compound used, minimal requirements were defined for FeMo-co synthesis including ATP (Hu et al. 2005). Knowing the ATP-dependent activation of MPT (Kuper et al. 2004) and the subsequent molybdate insertion reaction (Llamas et al. 2006), one can question if ATP might play a similar role in Mo insertion in FeMo-co with the involvement of any adenylated molybdate intermediate (Pau et al. personal communication).

4 Mo- and W-Pterin Cofactor-Dependent Enzymes

Mo-pterin enzymes catalyze key reactions in the global nitrogen, carbon, and sulfur cycles (Hille 2002). More than 50 different enzymes have been described throughout all organisms with most of them being found in prokaryotes. The research on the biochemistry of W was pioneered by the discovery of a W-containing formate dehydrogenase from *Clostridium thermoaceticum* (recently renamed *Moorella thermoacetica*) by Andreesen and Ljungdahl in 1973 (Andreesen and Ljungdahl 1973). Since then well over 40 research articles and five review articles on W-containing enzymes have appeared.

4.1 Mo-Enzymes

Bacterial Mo-enzymes show a great variety and based on their cofactor type, coordination, and domain arrangement they were classified into three families (Hille 1996). The largest and most diverse family is found in bacteria and formed by dimethyl sulfoxide reductase (DMSOR)-like enzymes. Enzymes of the xanthine oxidoreductase (XOR) and the sulfite oxidase (SO) families are less abundant in bacteria and are also found in eukaryotes. The only enzyme not characterized by a mononuclear metal center is carbon monoxide dehydrogenase, which can be structurally classified as XOR enzyme, but contains an additional copper atom forming a unique dinuclear CuSMo center that is attached to a mono-MPT cytosine dinucleotide cofactor (Dobbek et al. 2002).

4.1.1 Dimethyl Sulfoxide Reductase Family

The active site of this family is characterized by a bis-MGD-type of cofactor in addition to either two oxygen ligands or one oxygen ligand and one proteinogenic ligand such as cysteine, Se-cysteine, serine or aspartate (Hille

2002). Proteins such as DMSOR, nitrate reductase (NR), formate dehydrogenase (FDH), and trimethylamine *N*-oxide reductase (TMAOR) are the main representatives in this family.

DMSORs constitute a structurally diverse group of enzymes catalyzing the reduction of dimethyl sulfoxide (DMSO) as final reaction in anaerobic respiration. DMSORs range from monomeric forms with bis-MGD as sole redox active center in *Rhodobacter capsulatus* (Schindelin et al. 1996) to more complex enzymes with additional subunits as well as additional cofactors, such as the membrane-bound heterotrimeric DMSOR from *E. coli* with various Fe – S clusters (Weiner et al. 1992). TMAOR represents a close member of the DMSORs. Like DMSOR, it is localized in the periplasm and assists in the terminal electron transfer reaction during anaerobic respiration with trimethylamine *N*-oxide as electron acceptor.

Three forms of bacterial nitrate reductase (NRs) with distinct appearance and functions are known. The membrane-associated NR on the cytoplasmic side (Nar) participates in anaerobic respiratory processes and the generation of a proton gradient via the plasma membrane. Membrane-associated NRs facing the periplasm (Nap) usually serve to acquire nitrogen under nitrate-limiting conditions as well as redox-balancing in the presence of reduced carbon substrates or anaerobic respiration without the generation of proton motif force (Stolz and Basu 2002). Both Nap and Nar obtain the required reducing equivalents for nitrate reduction from the membrane-intrinsic quinol pool via enzyme-specific membrane-associated subunits. The third group of NRs comprises cytoplasmic heterodimeric enzymes catalyzing nitrate assimilation when reduced nitrogen sources are scarce (Richardson et al. 2001). Crystal structures revealed that Nap contains a hydroxyl group as well as a cysteine-derived sulfur as Mo ligand (Arnoux et al. 2003; Dias et al. 1999), whereas in *E. coli* NarG the metal is coordinated by one or two oxygen atoms derived from a conserved aspartate (Bertero et al. 2003; Jormakka et al. 2004). FDHs oxidize formate to CO_2 and can occur as monomers (Boyington et al. 1997) or as membrane-bound heterotrimers (Jormakka et al. 2002) in the same organism and usually contain a Se-cysteine ligating the Mo-bis-MGD cofactor.

4.1.2
Xanthine Oxidoreductase Family

Members of the XOR family catalyze the hydroxylation of carbon centers from aromatic, heterocyclic compounds or aldehydes. While the aldehyde oxidoreductase from *Desulfovibrio gigas* is monomeric with one MCD and two [2Fe – 2S] clusters (Rebelo et al. 2001), the xanthine dehydrogenase from *Rhodobacter capsulatus* occurs as a heterotetramer with two Moco-containing XdhB subunits and two XdhA subunits, each of which harbors two [2Fe – 2S] clusters and FAD as prosthetic groups (Leimkühler et al. 1998). For XdhB, a specific chaperone (XdhC) was described that assists in the insertion of the XOR-specific sulfurated Mo-pterin cofactor (Fig. 2) (Neumann et al. 2006).

4.1.3 Sulfite Oxidase Family

SOs perform the final step in dissimilatory oxidation of reduced sulfur compounds. In contrast to eukaryotic SOs, which are obligate homodimers, the sulfite dehydrogenase from *Starkeya novella* is a periplasmic heterodimeric enzyme consisting of a Mo-containing subunit SorA and a cytochrome *c*-containing subunit SorB (Kappler et al. 2000). SorA contains a conserved domain similar to the eukaryotic one responsible for dimerization, nevertheless, an intrinsic disulfide bond sterically hinders the formation of dimers without limiting the activity (Kappler and Bailey 2005). Based on genome analyses, another sulfite oxidoreductase was identified, the heterodimeric, membrane-associated YedYZ enzyme from *E. coli* (Loschi et al. 2004). Although the Mo-containing catalytic domain of YedY shows overall structural similarity to the catalytic domain of chicken SO (Kisker et al. 1997), significant differences exist in the substrate-binding site. The residues in the catalytic center of YedY resemble more the active sites of DMSOR and TMAOR, and subsequently reductase activity of YedYZ was observed with a variety of substrates such as DMSO or TMAO (Loschi et al. 2004).

4.2 W-Enzymes

All W-containing enzymes that have been isolated can be classified into only two enzyme families based on their amino acid sequence: the aldehyde:ferredoxin oxidoreductases (AOR) family and the formate dehydrogenase (FDH) family. Table 2 gives an overview of the molecular properties of the W-pterin enzymes that have been isolated to date.

4.2.1 AOR Family

From Table 2 several general properties of the AOR family of enzymes can be deduced:

(i) The catalytic subunit is 62–80 kDa.
(ii) The enzymes are mono-, di-, tri- or tetramers of the catalytic subunit.
(iii) The enzymes contain a single bis-MPT cofactor and a single [4Fe – 4S] cluster per catalytic subunit.
(iv) The enzymes use ferredoxin as a redox partner.
(v) The enzymes are oxygen sensitive.

Table 2 Properties of W-containing enzymes that have been isolated to date

Enzyme	Representative organism	SU[d]	SU mass (kDa)	Redox partner	Metals per SU	CF	Substrates	Refs.
Aldehyde oxidoreductase family								
Aldehyde oxidoreductase	*Pyrococcus furiosus*	α_2	67	Ferredoxin	1W, [4Fe – 4S]	MPT	Aromatic and aliphatic aldehydes, carboxylic acids	Mukund and Adams 1991
Formaldehyde oxidoreductase	*Pyrococcus furiosus*	α_4	69	Ferredoxin	1W, [4Fe – 4S], Ca	MPT	C1-C3 aldehydes, semi- and di-aldehydes	Roy et al. 1999
Glyceraldehyde-3-phosphate oxidoreductase	*Pyrococcus furiosus*	α_1	74	Ferredoxin	1W, [4Fe – 4S], 2Zn	MPT	Glyceraldehyde-3-phosphate	Mukund and Adams 1995
WOR4	*Pyrococcus furiosus*	α_2	70	n.r.	0.7 W, [4Fe – 4S], 0.8 Ca, 0.3 Zn	n.r.	–[e]	Roy and Adams 2002
WOR5	*Pyrococcus furiosus*	α_2	65	Ferredoxin	0.1 W, [4Fe – 4S], 1.4 Ca	n.r.	Aromatic and aliphatic aldehydes	Bevers et al. 2005
HVOR[a]	*Proteus vulgaris*	α_n	80	(Viologen)	0.9 Mo, 4Fe, 4S	MPT	2*R*-Hydroxy-carboxylates[f]	Trautwein et al. 1994
Formate dehydrogenase family								
Formate dehydrogenase	*Moorella thermoacetica*[c]	α_2, β_2	96, 76	$NADP^+$	1 W, [4Fe – 4S] [2Fe – 2S] 1 Se	MGD	Formate	Leonhardt and Andreesen 1977

Table 2 (continued)

Enzyme	Representative organism	SU[d]	SU mass (kDa)	Redox partner	Metals per SU	CF	Substrates	Refs.
Formylmethano-furan dehydrogenase	*Methano-bacterium wolfei*	α, β, γ	64, 51, 35	(Viologen)	0.4 W, 5 Fe	MGD	CO_2, methanofuran	Schmitz et al. 1992
Acetylene hydratase	*Pelobacter acetylenicus*	α	83	–	0.5 W, [4Fe – 4S]	MGD	Acetylene	Rosner and Schink 1995
Dimethyl-sulfoxide reductase[b]	*Rhodobacter capsulatus*	α	85	Cyto-chrome c	1 W	MGD	DMSO	Stewart et al. 2000
Trimethylamine-*N*-oxide reductase[b]	*Escherichia coli*	α_2	90	Cyto-chrome c	2.4 W	MGD	TMAO, DMSO, other *N*-oxides and *S*-oxides[g]	Buc et al. 1999

The molecular properties of an enzyme from only one representative organism are given
[a] Molybdenum enzyme with significant sequence homology with the aldehyde oxidoreductase family
[b] W-substituted Mo-enzymes
[c] Previously named *Clostridium thermoaceticum* (Collins et al. 1994)
[d] Subunit [e] WOR4 did not exhibit aldehyde oxidation, carboxylic acid reduction, 2-hydroxycarboxylic acid oxidation or aldehyde reduction activities for the substrates tested
[f] Pyruvate, phenylpyruvate, (*R*)-lactate, D-gluconate, D-glucose 6-phosphate and *N*-acetylneuraminate
[g] Additional substrates were: hydroxylamine *N*-oxide, 4-methylphorpholine *N*-oxide, 2-picoline *N*-oxide, 4-picoline *N*-oxide, pyridine *N*-oxide, diphenyl sulphoxide, tetramethylene sulphoxide (Buc et al. 1999)
MPT bis-pterin cofactor, *MGD* guanine dinucleotide bis-pterin cofactor, *n.r.* not reported, *WOR* tungsten-containing (aldehyde) oxidoreductase, *HVOR* (2*R*)-hydroxycarboxylate-viologen-oxidoreductase

(vi) The enzymes can catalyze aldehyde oxidation to carboxylic acids, and in some cases carboxylic acid reduction to aldehydes. The latter reaction is of particular interest for potential biotechnological applications of these enzymes, since aldehydes are highly valued compounds for the fine-chemical industry.

Attempts to incorporate Mo or V into *Pyrococcus furiosus* AOR, formaldehyde:ferredoxin oxidoreductase and glyceraldehyde-3-phosphate oxidoreductase have been unsuccessful. V was found to inhibit the incorporation of contaminating traces of W into the W-enzymes. Mo did not inhibit the incorporation of W, not even when Mo was present in a 6500-fold molar excess (Mukund and Adams 1996).

Although Mo cannot replace W in AOR family enzymes, the family does contain a Mo-enzyme member. Of this enzyme, (2*R*)-hydroxycarboxylate:viologen oxidoreductase (HVOR), the N-terminal amino acid sequence (20 amino acids) was determined and was found to share significant sequence similarity to the N-termini of other AOR family enzymes. The presence of Mo was determined using a colorimetric assay and confirmed by measuring the incorporation of radioactive ^{99}Mo into the enzyme (Trautwein et al. 1994). HVOR is an unusual member of the AOR family, since it is the only membrane-bound enzyme.

4.2.2
FDH Family

Members of the FDH family of enzymes share the following properties:

(i) The enzymes contain a single bis-MGD and additional [4Fe – 4S] or [2Fe – 2S] clusters (with the exception of DMSOR and TMAOR, which do not contain Fe – S clusters)

(ii) The enzymes exhibit a great variety of redox partners, and in most cases the physiological redox partner has not been identified

(iii) Often Se is found, and in the case of *Desulfovibrio gigas* FDH the crystal structure showed that Se is a ligand to the W center (Raaijmakers et al. 2002)

FDH also occurs as a Mo-containing enzyme in nature. In fact, the amino acid sequences of the Mo- and W-containing FDHs show that they are from the same enzyme family. The W-enzyme FDH family can therefore not be seen as separate from the Mo-enzyme DMSOR family.

Table 2 shows two Mo-enzymes, DMSOR and TMAOR, which retain their catalytic activity after substitution of Mo with W. In both cases the substitution was achieved by growing the organism on tungstate. The activity profiles, however, were changed for the W-substituted enzymes. DMSOR was found to be more active in reducing DMSO, but was inactive in DMS oxidation. TMAOR was found to have a broader substrate specificity, since it was able to reduce sulfoxides (DMSO, diphenyl sulfoxide, and tetramethylene sulfoxide), unlike its Mo-containing counterpart (Buc et al. 1999).

Acetylene hydratase is a special W-enzyme since it catalyzes the non-redox reaction: the hydratation of acetylene to acetaldehyde. A strong reductant like sodium dithionite is necessary to activate the enzyme. Therefore it has been thought that the reaction mechanism may involve several redox steps, even though the net reaction is non-redox. This would explain the presence of the redox-active W center and Fe – S cluster. Recently it was found that Mo could be incorporated into the *P. acetylenicus* acetylene hydratase (Boll et al. 2005). The activity of the Mo-enzyme was 60% reduced compared to the W-enzyme.

The fact that W and Mo are apparently interchangeable in enzymes of the FDH/DMSOR family enzymes marks a clear difference to enzymes of the AOR family.

5 Nitrogenases

Nitrogenases catalyze the energy-consuming conversion of atmospheric N_2 to ammonia, a central reaction in the global biogeochemical nitrogen cycle. Besides N_2 fixation, nitrogenases carry out the reduction of protons to H_2 as well as non-physiological reactions such as the reduction of acetylene to ethylene. In contrast to the V- or Fe-containing nitrogenases, Mo-nitrogenases are the best studied group (Lawson and Smith 2002).

5.1 FeMo Nitrogenase

The Mo nitrogenase complex consists of two proteins, the MoFe-protein or component 1, and the Fe-protein or component 2 (Burgess and Lowe 1996). Component 1 consists of an a_2b_2 heterotetramer harboring one $MoFe_7S_9X$-homocitrate cluster (FeMoco) within each α subunit as well as one Fe_8S_7 cluster (P-cluster) localized in each of the two ab interfaces. The homodimeric component 2 contains a single [4Fe – 4S] cluster in the dimer interface as well as one Mg-ATP binding site per subunit. Starting with the crystallization of both components (Chan et al. 1993; Georgiadis et al. 1992) as well as the enzyme complex (Schindelin et al. 1997), structural data has contributed in great part to elucidate the nature and arrangement of the cofactors in nitrogenase. The high-resolution structure of the *Azotobacter vinelandii* FeMo-protein (Einsle et al. 2002) revealed that an additional small atom is situated in the center of the inorganic metal cluster of FeMoco (Fig. 1), which is labeled X, because its nature has not yet been unambiguously determined, with nitrogen as the most likely candidate (Barney et al. 2006).

N_2 reduction is concomitant with the production of H_2 and requires at least the hydrolysis of 16 Mg-ATP molecules and the transfer of eight electrons and eight protons (Rees and Howard 2000). Two Mg-ATPs are hydrolyzed in the Fe-protein in order to transfer one electron via the [4Fe – 4S] cluster to the associated MoFe-protein (Lawson and Smith 2002). Afterwards, the two proteins dissociate, and re-reduction of the Fe-protein takes place by ferredoxin or flavodoxin, as well as the exchange of the co-substrates. Thus, at least eight cycles of association and dissociation of the two components are necessary for the production of one mole of ammonia. In the MoFe-protein, the electrons are transferred via the P-cluster to FeMoco, where N_2 reduction is carried out. In contrast to the binding of alternative substrates, N_2 binding is constrained to a more negative redox state of the FeMo-protein (Igarashi and Seefeldt 2003). It remains to be elucidated where substrate binding exactly occurs or how the reaction mechanism takes place in detail.

5.2
FeW Nitrogenase

In 1973, the same year as the discovery of the first naturally occurring W-enzyme, the incorporation of W into nitrogenase was established (Benemann et al. 1973). W does not naturally occur in nitrogenases, but attempts to incorporate W in nitrogenase were undertaken in order to study the role of Mo and V in the nitrogenase catalytic mechanism. W-containing nitrogenases have been produced in *Azotobacter vinelandii* and *Rhodobacter capsulatus* grown under Mo-deficient and high W concentrations (1 mM) (Hales and Case 1987; Siemann et al. 2003). The enzyme from *A. vinelandii* contained Mo and W in a 1:1 ratio. The enzyme appeared to exhibit regular nitrogenase activities despite the fact that 50% of its nitrogenase cofactor contained W instead of Mo. The W-containing nitrogenase from *R. capsulatus* is of particular interest since it did not contain Mo. The *R. capsulatus* W-nitrogenase was inactive in dinitrogen reduction, marginally active in acetylene reduction, but exhibited a significant proton reduction activity.

The changed catalytic properties of W-nitrogenase are consistent with a change in redox potential of the FeWco. It is known from inorganic chemistry that W compounds are more difficult to reduce than isostructural Mo compounds. This has also been shown to be true for a W-substituted Mo-pterin enzyme, DMSOR (Hagedoorn et al. 2003). Nitrogenase is not an exception; the redox potential of the FeWco is at least 150 mV lower than of the FeMoco. Due to the lower redox potential it becomes difficult to reduce the FeWco with more than two electrons, which is sufficient for the reduction of protons to H_2, but not for N_2 reduction. Furthermore, it is generally believed that Mo-nitrogenase has to be reduced with three electrons and three protons before it can bind nitrogen upon the release of H_2 (Igarashi and Seefeldt 2003).

References

Anderson LA, McNairn E, Lubke T, Pau RN, Boxer DH (2000) ModE-dependent molybdate regulation of the molybdenum cofactor operon moa in *Escherichia coli.* J Bacteriol 182:7035–7043

Anderson LA, Palmer T, Price NC, Bornemann S, Boxer DH, Pau RN (1997) Characterisation of the molybdenum-responsive ModE regulatory protein and its binding to the promoter region of the modABCD (molybdenum transport) operon of *Escherichia coli.* Eur J Biochem 246:119–126

Andreesen JR, Ljungdahl LG (1973) Formate dehydrogenase of Clostridium thermoaceticum: incorporation of selenium-75, and the effects of selenite, molybdate, and tungstate on the enzyme. J Bacteriol 116:867–873

Arnoux P, Sabaty M, Alric J, Frangioni B, Guigliarelli B, Adriano JM, Pignol D (2003) Structural and redox plasticity in the heterodimeric periplasmic nitrate reductase. Nat Struct Biol 10:928–934

Barney BM, Lee HI, Dos Santos PC, Hoffman BM, Dean DR, Seefeldt LC (2006) Breaking the N_2 triple bond: insights into the nitrogenase mechanism. Dalton Trans pp 2277–2284

Benemann JR, Smith GM, Kostel PJ, McKenna CE (1973) Tungsten incorporation into *Azotobacter vinelandii* nitrogenase. FEBS Lett 29:219–221

Bertero MG, Rothery RA, Palak M, Hou C, Lim D, Blasco F, Weiner JH, Strynadka NC (2003) Insights into the respiratory electron transfer pathway from the structure of nitrate reductase A. Nat Struct Biol 10:681–687

Bevers LE, Bol E, Hagedoorn PL, Hagen WR (2005) WOR5: A novel tungsten containing aldehyde oxidoreductase from *Pyrococcus furiosus* with a broad substrate specificity. J Bacteriol 187:7056–7061

Bevers LE, Hagedoorn PL, Krijger GC, Hagen WR (2006) Tungsten transport protein A (WtpA) in Pyrococcus furiosus: the first member of a new class of tungstate and molybdate transporters. J Bacteriol 188:6498–6505

Bittner F, Oreb M, Mendel RR (2001) ABA3 is a molybdenum cofactor sulfurase required for activation of aldehyde oxidase and xanthine dehydrogenase in Arabidopsis thaliana. J Biol Chem 276:40381–40384

Blasco F, Dos Santos JP, Magalon A, Frixon C, Guigliarelli B, Santini CL, Giordano G (1998) NarJ is a specific chaperone required for molybdenum cofactor assembly in nitrate reductase A of *Escherichia coli.* Mol Microbiol 28:435–447

Boll M, Schink B, Messerschmidt A, Kroneck PMH (2005) Novel bacterial molybdenum and tungsten enzymes: three-dimensional structure, spectroscopy, and reaction mechanism. Biol Chem 386:999–1006

Boyington JC, Gladyshev VN, Khangulov SV, Stadtman TC, Sun PD (1997) Crystal structure of formate dehydrogenase H: catalysis involving Mo, molybdopterin selenocysteine, and an Fe4S4 cluster. Science 275:1305–1308

Buc J, Santini C-L, Giordani R, Czjzek M, Wu L-F, Giordano G (1999) Enzymatic and physiological properties of the tungsten-substituted molybdenum TMAO reductase from *Escherichia coli.* Mol Microbiol 32:159–168

Burgess BK, Lowe DJ (1996) Mechanism of molybdenum nitrogenase. Chem Rev 96:2983–3012

Chan MK, Kim J, Rees DC (1993) The nitrogenase FeMo-cofactor and P-cluster pair: 2.2 Å resolution structures. Science 260:792–794

Chan MK, Mukund S, Kletzin A, Adams MWW, Rees DC (1995) Structure of a hyperthermophilic tungstopterin, aldehyde ferredoxin oxidoreductase. Science 267:1463–1469

Collins MD, Lawson PA, Willems A, Cordoba JJ, Fernandez-Garayzabal J, Garcia P, Cai J, Hippe H, Farrow JA (1994) The phylogeny of the genus Clostridium: proposal of five new genera and eleven new species combinations. Int J Syst Bacteriol 44:812–826

Corbett MC, Hu Y, Fay AW, Ribbe MW, Hedman B, Hodgson KO (2006) Structural insights into a protein-bound iron-molybdenum cofactor precursor. Proc Natl Acad Sci USA 103:1238–1243

Cox PA (1995) The elements on earth. Oxford University Press, Oxford, UK

Curatti L, Ludden PW, Rubio LM (2006) NifB-dependent in vitro synthesis of the iron-molybdenum cofactor of nitrogenase. Proc Natl Acad Sci USA 103:5297–5301

Dias JM, Than ME, Humm AE, Huber R, Bourenkov GP, Bartunik HD, Bursakov S, Calvete J, Caldeira J, Carneiro C, Moura JJG, Moura I, Romao MJ (1999) Crystal structure of the first dissimilatory nitrate reductase at 1.9 Å solved by MAD methods. Structure 7:65–79

Dobbek H, Gremer L, Kiefersauer R, Huber R, Meyer O (2002) Catalysis at a dinuclear [CuSMo(=O)OH] cluster in a CO dehydrogenase resolved at 1.9 Å resolution. Proc Natl Acad Sci USA 99:15971–15976

Dos Santos PC, Dean DR, Hu Y, Ribbe MW (2004) Formation and insertion of the nitrogenase iron-molybdenum cofactor. Chem Rev 104:1159–1173

Eaves DJ, Palmer T, Boxer DH (1997) The product of the molybdenum cofactor gene mobB of *Escherichia coli* is a GTP-binding protein. Eur J Biochem 246:690–697

Einsle O, Tezcan FA, Andrade SL, Schmid B, Yoshida M, Howard JB, Rees DC (2002) Nitrogenase MoFe-protein at 1.16 Å resolution: a central ligand in the FeMo-cofactor. Science 297:1696–1700

Esnouf RM (1997) An extensively modified version of MolScript that includes greatly enhanced coloring capabilities. J Mol Graph Model 15:132–134

Fischer K, Llamas A, Tejada-Jimenez M, Schader N, Kuper J, Ataya FS, Galvan A, Mendel RR, Fernandez E, Schwarz G (2006) Function and structure of the molybdenum cofactor carrier protein MCP from chlamydomonas reinhardtii. J Biol Chem 281:30186–30194

Frazzon J, Dean DR (2002) Biosynthesis of the nitrogenase iron-molybdenum-cofactor from *Azotobacter vinelandii*. Met Ions Biol Syst 39:163–186

Georgiadis MM, Komiya H, Chakrabarti P, Woo D, Kornuc JJ, Rees DC (1992) Crystallographic structure of the nitrogenase iron protein from *Azotobacter vinelandii*. Science 257:1653–1659

Glaser JH, DeMoss JA (1971) Phenotypic restoration by molybdate of nitrate reductase activity in chlD mutants of *Escherichia coli*. J Bacteriol 108:854–860

Goodwin PJ, Agar JN, Roll JT, Roberts GP, Johnson MK, Dean DR (1998) The *Azotobacter vinelandii* NifEN complex contains two identical [4Fe – 4S] clusters. Biochemistry 37:10420–10428

Grunden AM, Shanmugam KT (1997) Molybdate transport and regulation in bacteria. Arch Microbiol 168:345–354

Guse A, Stevenson CE, Kuper J, Buchanan G, Schwarz G, Giordano G, Magalon A, Mendel RR, Lawson DM, Palmer T (2003) Biochemical and structural analysis of the molybdenum cofactor biosynthesis protein MobA. J Biol Chem 278:25302–25307

Gutzke G, Fischer B, Mendel RR, Schwarz G (2001) Thiocarboxylation of molybdopterin synthase provides evidence for the mechanism of dithiolene formation in metal-binding pterins. J Biol Chem 276:36268–36274

Hagedoorn PL, Hagen WR, Stewart LJ, Docrat A, Bailey S, Garner CD (2003) Redox characteristics of the tungsten DMSO reductase of *Rhodobacter capsulatus*. FEBS Lett 555:606–610

Hales BJ, Case EE (1987) Nitrogen fixation by *Azotobacter vinelandii* in tungsten-containing medium. J Biol Chem 262:16205–16211

Hänzelmann P, Meyer O (1998) Effect of molybdate and tungstate on the biosynthesis of CO dehydrogenase and the molybdopterin cytosine-dinucleotide-type of molybdenum cofactor in Hydrogenophaga pseudoflava. Eur J Biochem 255:755–765

Hanzelmann P, Schindelin H (2006) Binding of 5′-GTP to the C-terminal Fe – S cluster of the radical *S*-adenosylmethionine enzyme MoaA provides insights into its mechanism. Proc Natl Acad Sci USA 103:6829–6834

Hänzelmann P, Schindelin H (2004) Crystal structure of the *S*-adenosylmethionine-dependent enzyme MoaA and its implications for molybdenum cofactor deficiency in humans. Proc Natl Acad Sci USA 101:12870–12875

Hänzelmann P, Schwarz G, Mendel RR (2002) Functionality of alternative splice forms of the first enzymes involved in human molybdenum cofactor biosynthesis. J Biol Chem 277:18303–18312

Hille R (1996) The mononuclear molybdenum enzymes. Chem Rev 96:2757–2816

Hille R (2002) Molybdenum enzymes containing the pyranopterin cofactor: an overview. Met Ions Biol Syst 39:187–226

Hu Y, Fay AW, Ribbe MW (2005) Identification of a nitrogenase FeMo cofactor precursor on NifEN complex. Proc Natl Acad Sci USA 102:3236–3241

Hu Y, Rech S, Gunsalus RP, Rees DC (1997) Crystal structure of the molybdate binding protein ModA. Nat Struct Biol 4:703–707

Igarashi RY, Seefeldt LC (2003) Nitrogen fixation: the mechanism of the Mo-dependent nitrogenas. Crit Rev Biochem Mol Biol 38:351–384

Imperial J, Hadi M, Amy NK (1998) Molybdate binding by ModA, the periplasmic component of the *Escherichia coli* mod molybdate transport system. Biochim Biophys Acta 1370:337–346

Jormakka M, Richardson D, Byrne B, Iwata S (2004) Architecture of NarGH reveals a structural classification of Mo-bisMGD enzymes. Structure 12:95–104

Jormakka M, Tornroth S, Byrne B, Iwata S (2002) Molecular basis of proton motive force generation: structure of formate dehydrogenase-N. Science 295:1863–1868

Joshi MS, Johnson JL, Rajagopalan KV (1996) Molybdenum cofactor biosynthesis in *Escherichia coli* mod and mog mutants. J Bacteriol 178:4310–4312

Kappler U, Bailey S (2005) Molecular basis of intramolecular electron transfer in sulfite-oxidizing enzymes is revealed by high resolution structure of a heterodimeric complex of the catalytic molybdopterin subunit and a c-type cytochrome subunit. J Biol Chem 280:24999–25007

Kappler U, Bennett B, Rethmeier J, Schwarz G, Deutzmann R, McEwan AG, Dahl C (2000) Sulfite: cytochrome c oxidoreductase from *Thiobacillus novellus* - purification, characterization, and molecular biology of a heterodimeric member of the sulfite oxidase family. J Biol Chem 275:13202–13212

Kisker C, Schindelin H, Pacheco A, Wehbi WA, Garrett RM, Rajagopalan KV, Enemark JH, Rees DC (1997) Molecular basis of sulfite oxidase deficiency from the structure of sulfite oxidase. Cell 91:973–983

Kuper J, Llamas A, Hecht HJ, Mendel RR, Schwarz G (2004) Structure of molybdopterin-bound Cnx1G domain links molybdenum and copper metabolism. Nature 430:803–806

Kuper J, Palmer T, Mendel RR, Schwarz G (2000) Mutations in the molybdenum cofactor biosynthetic protein Cnx1G from Arabidopsis thaliana define functions for molybdopterin bind, Mo-insertion and molybdenum cofactor stabilization. Proc Natl Acad Sci USA 97:6475–6480

Lake MW, Temple CA, Rajagopalan KV, Schindelin H (2000) The crystal structure of the *Escherichia coli* MobA protein provides insight into molybdopterin guanine dinucleotide biosynthesis. J Biol Chem 275:40211–40217

Lake MW, Wuebbens MM, Rajagopalan KV, Schindelin H (2001) Mechanism of ubiquitin activation revealed by the structure of a bacterial MoeB–MoaD complex. Nature 414:325–329

Lawson DM, Smith BE (2002) Molybdenum nitrogenases: a crystallographic and mechanistic view. Met Ions Biol Syst 39:75–119

Leimkuehler S, Rajagopalan KV (2001) A sulfurtransferase is required in the transfer of cysteine sulfur in the in vitro synthesis of molybdopterin from precursor Z in *Escherichia coli.* J Biol Chem 276:22024–22031

Leimkuehler S, Wuebbens MM, Rajagopalan KV (2001) Characterization of *Escherichia coli* MoeB and its involvement in the activation of molybdopterin synthase for the biosynthesis of the molybdenum cofactor. J Biol Chem 276:34695–34701

Leimkühler S, Kern M, Solomon PS, McEwan AG, Schwarz G, Mendel RR, Klipp W (1998) Xanthine dehydrogenase from the phototrophic purple bacterium Rhodobacter capsulatus is more similar to its eukaryotic counterparts than to prokaryotic molybdenum enzymes. Mol Microbiol 27:853–869

Leonhardt U, Andreesen JR (1977) Some properties of formate dehydrogenase accumulation and incorporation of ^{185}W-tungsten into proteins of *Clostridium formicoaceticum.* Arch Microbiol 115:277–284

Liu MT, Wuebbens MM, Rajagopalan KV, Schindelin H (2000) Crystal structure of the gephyrin-related molybdenum cofactor biosynthesis protein MogA from *Escherichia coli.* J Biol Chem 275:1814–1822

Llamas A, Mendel RR, Schwarz G (2004) Synthesis of adenylated molybdopterin: an essential step for molybdenum insertion. J Biol Chem 279:55241–55246

Llamas A, Otte T, Multhaup G, Mendel RR, Schwarz G (2006) The mechanism of nucleotide-assisted molybdenum insertion into molybdopterin. A novel route toward metal cofactor assembly. J Biol Chem 281:18343–18350

Loschi L, Brokx SJ, Hills TL, Zhang G, Bertero MG, Lovering AL, Weiner JH, Strynadka NC (2004) Structural and biochemical identification of a novel bacterial oxidoreductase. J Biol Chem 279:50391–50400

Magalon A, Frixon C, Pommier J, Giordano G, Blasco F (2002) In vivo interactions between gene products involved in the final stages of molybdenum cofactor biosynthesis in *Escherichia coli.* J Biol Chem 7:7

Makdessi K, Andreesen JR, Pich A (2001) Tungstate uptake by a highly specific ABC transporter in Eubacterium acidaminophilum. J Biol Chem 276:24557–24564

Moura JJ, Brondino CD, Trincao J, Romao MJ (2004) Mo and W bis-MGD enzymes: nitrate reductases and formate dehydrogenases. J Biol Inorg Chem 9:791–799

Mukund S, Adams MWW (1995) Glyceraldehyde-3-phosphate ferredoxin oxidoreductase, a novel tungsten-containing enzyme with a potential glycolytic role in the hyperthermophilic Archaeon *Pyrococcus furiosus.* J Biol Chem 270:8389–8392

Mukund S, Adams MWW (1996) Molybdenum and vanadium do not replace tungsten in the catalytically active forms of the three tungstoenzymes in the hyperthermophilic Archaeon *Pyrococcus furiosus.* J Bacteriol 178:163–167

Neumann M, Schulte M, Junemann N, Stocklein W, Leimkuhler S (2006) Rhodobacter capsulatus XdhC is involved in molybdenum cofactor binding and insertion into xanthine dehydrogenase. J Biol Chem 281:15701–15708

Nichols JD, Rajagopalan KV (2005) In vitro molybdenum ligation to molybdopterin using purified components. J Biol Chem 280:7817–7822

Palmer T, Goodfellow IP, Sockett RE, McEwan AG, Boxer DH (1998) Characterisation of the mob locus from Rhodobacter sphaeroides required for molybdenum cofactor biosynthesis. Biochim Biophys Acta 1395:135–140

Palmer T, Vasishta A, Whitty PW, Boxer DH (1994) Isolation of protein FA, a product of the mob locus required for molybdenum cofactor biosynthesis in *Escherichia coli.* Eur J Biochem 222:687–692

Pau RN, Lawson DM (2002) Transport, homeostasis, regulation, and binding of molybdate and tungstate to proteins. Met Ions Biol Syst 39:31–74

Raaijmakers H, Macieira S, Dias JM, Teixeira S, Bursakov S, Huber R, Moura JJG, Moura I, Romão MJ (2002) Gene sequence and the 1.8 Å crystal structure of the tungsten-containing formate dehydrogenase from *Desulfovibrio gigas.* Structure 10:1261–1272

Rajagopalan KV, Johnson JL (1992) The pterin molybdenum cofactors. J Biol Chem 267:10199–10202

Rangaraj P, Ludden PW (2002) Accumulation of 99Mo-containing iron-molybdenum cofactor precursors of nitrogenase on NifNE, NifH, and NifX of *Azotobacter vinelandii.* J Biol Chem 277:40106–40111

Rangaraj P, Ruttimann-Johnson C, Shah VK, Ludden PW (2001) Accumulation of 55Fe-labeled precursors of the iron-molybdenum cofactor of nitrogenase on NifH, NifX of *Azotobacter vinelandii.* J Biol Chem 276:15968–15974

Rebelo JM, Dias JM, Huber R, Moura JJ, Romao MJ (2001) Structure refinement of the aldehyde oxidoreductase from Desulfovibrio gigas (MOP) at 1.28 Å. J Biol Inorg Chem 6:791–800

Rees DC (2002) Great metalloclusters in enzymology. Annu Rev Biochem 71:221–246

Rees DC, Howard JB (2000) Nitrogenase: standing at the crossroads. Curr Opin Chem Biol 4:559–566

Richardson DJ, Berks BC, Russell DA, Spiro S, Taylor CJ (2001) Functional, biochemical and genetic diversity of prokaryotic nitrate reductases. Cell Mol Life Sci 58:165–178

Robb FT, Maeder DL, Brown JR, DiRuggiero J, Stump MD, Yeh RK, Weiss RB, Dunn DM (2001) Genomic sequence of hyperthermophile, Pyrococcus furiosus: implications for physiology and enzymology. Methods Enzymol 330:134–157

Rosner BM, Schink B (1995) Purification and characterization of acetylene hydratase of *Pelobacter acetylenicus*, a tungsten iron-sulfur protein. J Bacteriol 177:5767–5772

Roy R, Adams MWW (2002) Characterization of a fourth tungsten-containing enzyme from the hyperthermophilic Archaeon *Pyrococcus furiosus.* J Bacteriol 184:6952–6956

Roy R, Mukund S, Shut GJ, Dunn DM, Weiss R, Adams MWW (1999) Purification and Molecular characterization of the tungsten-containing formaldehyde ferredoxin oxidoreductase from the hyperthermophilic Archaeon *Pyrococcus furiosus*: the third of a putative five-member tungstoenzyme family. J Bacteriol 181:1171–1180

Rubio LM, Ludden PW (2005) Maturation of nitrogenase: a biochemical puzzle. J Bacteriol 187:405–414

Rudolph MJ, Wuebbens MM, Rajagopalan KV, Schindelin H (2001) Crystal structure of molybdopterin synthase and its evolutionary relationship to ubiquitin activation. Nat Struct Biol 8:42–46

Sanishvili R, Beasley S, Skarina T, Glesne D, Joachimiak A, Edwards A, Savchenko A (2004) The crystal structure of *Escherichia coli* MoaB suggests a probable role in molybdenum cofactor synthesis. J Biol Chem 279:42139–42146

Santamaria-Araujo JA, Fischer B, Otte T, Nimtz M, Mendel RR, Wray V, Schwarz G (2004) The tetrahydropyranopterin structure of the sulfur-free and metal-free molybdenum cofactor precursor. J Biol Chem 279:15994–15999

Schindelin H, Kisker C, Hilton J, Rajagopalan KV, Rees DC (1996) Crystal structure of DMSO reductase: redox-linked changes in molybdopterin coordination. Science 272:1615–1621

Schindelin H, Kisker C, Schlessman JL, Howard JB, Rees DC (1997) Structure of ADP × AIF4(-)-stabilized nitrogenase complex and its implications for signal transduction. Nature 387:370–376

Schmitz RA, Richter M, Linder D, Thauer RK (1992) A tungsten-containing active formylmethanofuran dehydrogenase in the thermophilic Archaeon *Methanobacterium wolfei*. Eur J Biochem 207:559–565

Schuttelkopf AW, Harrison JA, Boxer DH, Hunter WN (2002) Passive acquisition of ligand by the MopII molbindin from Clostridium pasteurianum: structures of apo and oxyanion-bound forms. J Biol Chem 277:15013–15020

Schwarz G (2005) Molybdenum cofactor biosynthesis and deficiency. Cell Mol Life Sci 62:2792–2810

Schwarz G, Boxer DH, Mendel RR (1997) Molybdenum cofactor biosynthesis. The plant protein Cnx1 binds molybdopterin with high affinity. J Biol Chem 272:26811–26814

Schwarz G, Schulze J, Bittner F, Eilers T, Kuper J, Bollmann G, Nerlich A, Brinkmann H, Mendel RR (2000) The molybdenum cofactor biosynthetic protein Cnx1 complements molybdate- repairable mutants, transfers molybdenum to the metal binding pterin, and is associated with the cytoskeleton. Plant Cell 12:2455–2472

Siemann S, Schneider K, Oley M, Müller A (2003) Characterization of a tungsten-substituted nitrogenase isolated from Rhodobacter capsulatus. Biochemistry 42:3846–3857

Sofia HJ, Chen G, Hetzler BG, Reyes-Spindola JF, Miller NE (2001) Radical SAM, a novel protein superfamily linking unresolved steps in familiar biosynthetic pathways with radical mechanisms: functional characterization using new analysis and information visualization methods. Nucleic Acids Res 29:1097–1106

Stevenson CEM, Sargent F, Buchanan G, Palmer T, Lawson DM (2000) Crystal structure of the molybdenum cofactor biosynthesis protein MobA from *Escherichia coli* at near-atom resolution. Structure 8:1115–1125

Stewart LJ, Bailey S, Bennet B, Charnock JM, Garner CD, McAlpine AS (2000) Dimethylsulfoxide reductase: an enzyme capable of catalysis with either molybdenum or tungsten at the active site. J Mol Biol 299:593–600

Stiefel EI (2002) The biogeochemistry of molybdenum and tungsten. Met Ions Biol Syst 39:1–29

Stolz JF, Basu P (2002) Evolution of nitrate reductase: molecular and structural variations on a common function. Chem Biochem 3:198–206

Temple CA, Rajagopalan KV (2000) Mechanism of assembly of the bis(molybdopterin guanine dinucleotide)molybdenum cofactor in Rhodobacter sphaeroides dimethyl sulfoxide reductase. J Biol Chem 275:40202–40210

Trautwein T, Krauss F, Lottspeich F, Simon H (1994) The (2*R*)-hydroxycarboxylate-viologen-oxidoreductase from *Proteus vulgaris* is a molybdenum-containing iron-sulphur protein. Eur J Biochem 222:1025–1032

Vergnes A, Gouffi-Belhabich K, Blasco F, Giordano G, Magalon A (2004) Involvement of the molybdenum cofactor biosynthetic machinery in the maturation of the *Escherichia coli* nitrate reductase A. J Biol Chem 279:41398–41403

Wang G, Angermuller S, Klipp W (1993) Characterization of Rhodobacter capsulatus genes encoding a molybdenum transport system and putative molybdenum-pterin-binding proteins. J Bacteriol 175:3031–3042

Weiner JH, Rothery RA, Sambasivarao D, Trieber CA (1992) Molecular analysis of dimethylsulfoxide reductase: a complex iron-sulfur molybdoenzyme of *Escherichia coli*. Biochim Biophys Acta 1102:1–18

Williams RJ, Frausto da Silva JJ (2002) The involvement of molybdenum in life. Biochem Biophys Res Commun 292:293–299

Wuebbens MM, Liu MTW, Rajagopalan KV, Schindelin H (2000) Insights into molybdenum cofactor deficiency provided by the crystal structure of the molybdenum cofactor biosynthesis protein MoaC. Structure 8:709–718

Wuebbens MM, Rajagopalan KV (1995) Investigation of the early steps of molybdopterin biosynthesis in *Escherichia coli* through the use of in vivo labeling studies. J Biol Chem 270:1082–1087

Wuebbens MM, Rajagopalan KV (2003) Mechanistic and mutational studies of *Escherichia coli* molybdopterin synthase clarify the final step of molybdopterin biosynthesis. J Biol Chem 278:14523–14532

Xiang S, Nichols J, Rajagopalan KV, Schindelin H (2001) The crystal structure of *Escherichia coli* MoeA and its relationship to the multifunctional protein gephyrin. Structure 9:299–310

Yuvaniyama P, Agar JN, Cash VL, Johnson MK, Dean DR (2000) NifS-directed assembly of a transient [2Fe – 2S] cluster within the NifU protein. Proc Natl Acad Sci USA 97:599–604

Zheng L, White RH, Cash VL, Jack RF, Dean DR (1993) Cysteine desulfurase activity indicates a role for NIFS in metallocluster biosynthesis. Proc Natl Acad Sci USA 90:2754–2758

Subject Index

Printing: Krips bv, Meppel
Binding: Stürtz, Würzburg